Sustainable Water Systems

Sustainable Water Systems

Miklas Scholz
University of Johannesburg, Johannesburg, South Africa
District of Herzogtum Lauenburg, Ratzeburg, Germany
Kunststoff-Technik Adams, Elsfleth, Germany
Nexus by Sweden, Västerås, Sweden

WILEY

This edition first published 2026

Registered Office(s)
John Wiley & Sons, Inc., 111 River Street, Hoboken, NJ 07030, USA
John Wiley & Sons Ltd, New Era House, 8 Oldlands Way, Bognor Regis, West Sussex, PO22 9NQ, UK

For details of our global editorial offices, customer services, and more information about Wiley products visit us at www.wiley.com.

The manufacturer's authorized representative according to the EU General Product Safety Regulation is Wiley-VCH GmbH, Boschstr. 12, 69469 Weinheim, Germany, e-mail: Product_Safety@wiley.com.

Wiley also publishes its books in a variety of electronic formats and by print-on-demand. Some content that appears in standard print versions of this book may not be available in other formats.

Library of Congress Cataloging-in-Publication Data Applied for:

Hardback ISBN: 9781394294121

Cover Design: Wiley
Cover Images: © Miklas Scholz, © Lasse Johansson/Shutterstock

Set in 9.5/12.5pt STIXTwoText by Straive, Pondicherry, India
Printed and bound by CPI Group (UK) Ltd, Croydon, CR0 4YY
C9781394294121_280825

Dedication

I dedicate this book to my wider family and friends, who supported me during my studies and career.
Special thanks go to my partner Ramilya Galimova; children Philippa Scholz, Jolena Scholz, Felix Hedmark, and Jamie Hedmark; twin sister Ricarda Lorey; and mother Gudrun Spieshöfer.

Contents

Foreword *xv*
Preface *xvii*
About the Author *xxiii*
Acknowledgements *xxv*

1 Natural Wetland Systems *1*
1.1 Hydraulics, Water Quality and Vegetation Characteristics of Ditches *1*
1.1.1 Introduction *1*
1.1.2 Experimental *2*
1.1.3 Results *5*
1.1.3.1 Characteristics of Open Ditches *5*
1.1.3.2 Water Quality and Vegetation *7*
1.1.4 Discussion *11*
1.1.4.1 Watercourse Classification *11*
1.1.4.2 Total Roughness and Summer Flooding *12*
1.1.4.3 Water Quality Influenced by Vegetation *12*
1.1.5 Conclusions *13*
1.2 Planted Soil Infiltration Systems for Treatment of Log Yard Runoff *13*
1.2.1 Introduction *13*
1.2.2 Experimental Setup and Methodology *14*
1.2.3 Results and Discussion *15*
1.2.3.1 Log yard Runoff Compared with Drainage Water *15*
1.2.3.2 Comparison of Drainage Waters *17*
1.2.3.3 Comparison of Treatment Efficiencies *17*
1.2.4 Conclusions and Recommendations *18*
1.3 Anthropogenic Land Use Change Impacts on Nutrient Concentrations in Waterbodies *18*
1.3.1 Introduction *18*
1.3.2 Land Use Changes and Surface Water Quality *19*
1.3.2.1 Investigations *19*
1.3.2.2 Key Variables *24*
1.3.2.3 Modelling Outcomes *28*
1.3.3 Proposals for Water Quality Conservation *29*
1.3.4 Conclusions and Outlook *31*
1.4 Peatland Response to Climate Change and Water Level Management *31*

1.4.1 Introduction *31*
1.4.2 Materials and Methods *33*
1.4.2.1 Mesocosm Experiments *33*
1.4.2.2 Climate Scenario Simulations *33*
1.4.2.3 Water-level Management *36*
1.4.2.4 Flux Measurements *36*
1.4.2.5 Plant Analysis *37*
1.4.2.6 Statistical Analysis *37*
1.4.3 Results *38*
1.4.3.1 Climate Variables Under Different Scenarios *38*
1.4.3.2 Climate, Water Level and Respiration *39*
1.4.3.3 Climate Change, Water Level and Plant Distribution *43*
1.4.4 Discussion *45*
1.4.4.1 Interactions *45*
1.4.4.2 Effect of Climate on the Carbon Sink Function of Unmanaged Mesocosms *45*
1.4.4.3 Impact of Drought on the Carbon Sink Function of Unmanaged Mesocosms *46*
1.4.4.4 Interactive Effects on the Carbon Sink Function of Managed Mesocosms *48*
1.4.5 Conclusions and Recommendations *50*
References *51*

2 Urban Water and Sustainable Drainage Systems *63*
2.1 Full Silt Traps Discharging into Watercourses *63*
2.1.1 Introduction *63*
2.1.2 Study Site, Materials and Methods *64*
2.1.3 Results and Discussion *65*
2.1.3.1 Water and Sediment Quality of the Silt Trap *65*
2.1.3.2 Water Quality of the Full Silt Trap During Dry- and Wet-Weather Flow *68*
2.1.3.3 Water Quality of the Receiving Watercourse *68*
2.1.4 Conclusions *69*
2.2 Filter Media, Plant Communities and Microbiology within Constructed Wetlands *69*
2.2.1 Introduction *69*
2.2.2 Materials and Methods *70*
2.2.2.1 Environmental Conditions and Operation *70*
2.2.2.2 Filter Media Composition *72*
2.2.2.3 Analytical Procedures *72*
2.2.2.4 Micro-biological and Plant Examinations *73*
2.2.3 Results and Discussion *74*
2.2.3.1 Filter Media Costs *74*
2.2.3.2 Comparison of Treatment Efficiencies *74*
2.2.3.3 Metal Analysis *78*
2.2.3.4 Water Quality and Micro-biology *80*
2.2.4 Conclusions *81*
2.3 Vertical Subsurface Flow-Constructed Wetlands with Different Substrates *82*
2.3.1 Introduction *82*
2.3.2 Materials and Methods *83*
2.3.3 Results and Discussion *86*
2.3.3.1 Adsorption Isotherm Experiments *86*

2.3.3.2 Column Experiments *88*
2.3.4 Conclusions *93*
2.4 Treatment of Gully Pot Effluent with Constructed Wetlands *93*
2.4.1 Introduction *93*
2.4.2 Materials and Methods *94*
2.4.2.1 Study Site and Filter Media Composition *94*
2.4.2.2 Environmental Conditions and Operation *97*
2.4.2.3 Metal and Other Variable Determinations *98*
2.4.3 Results *100*
2.4.4 Discussion *105*
2.4.4.1 Performance Comparison *105*
2.4.4.2 Constructed Wetlands, Sustainability and Filterability *106*
2.4.4.3 Alternative Treatment *107*
2.4.5 Conclusions and Further Research *107*
2.5 Wetland and Dry Pond System *108*
2.5.1 Introduction *108*
2.5.2 Case Study and Methodology *108*
2.5.3 Results and Discussion *112*
2.5.3.1 Water Quality *112*
2.5.3.2 Capacity of the System *114*
2.5.3.3 Economics *115*
2.5.3.4 Risk of Flooding and Infiltration *116*
2.5.3.5 Sustainable Water Industry Asset Resource Decisions Analysis *116*
2.5.4 Conclusions *117*
2.6 Permeable Pavement and Ground Source Heating Pump Systems *117*
2.6.1 Introduction *117*
2.6.2 Methodology *119*
2.6.2.1 Environmental Conditions *119*
2.6.2.2 Components of Rigs *119*
2.6.2.3 Operational Conditions *120*
2.6.2.4 Water Quality *122*
2.6.3 Results and Discussion *123*
2.6.3.1 Water Quality *123*
2.6.3.2 Prevention of Water-related Diseases *128*
2.6.3.3 Influence of Temperature *130*
2.6.4 Conclusions and Recommendations *130*
2.7 Permeable Pavement and Photocatalytic Titanium Dioxide Oxidation System *131*
2.7.1 Introduction *131*
2.7.2 Materials and Methods *132*
2.7.2.1 Permeable Pavement Engineering *132*
2.7.2.2 Photocatalytic Processes with Titanium Dioxide *132*
2.7.2.3 Experimental Permeable Pavement Set-up *134*
2.7.2.4 Experimental Set-up of the Photochemical Experiments *135*
2.7.2.5 Analytical and Microbiological Procedures *136*
2.7.2.6 Photocatalytic Disinfection Model *136*
2.7.3 Results and Discussion *137*
2.7.3.1 Permeable Pavement Contaminant Removal Efficiency *137*

2.7.3.2 Microbial Photochemical Deactivation *138*
2.7.4 Conclusions *141*
2.8 Refurbishment and Improvement of Screen Systems for Flood Control and Water Protection *142*
2.8.1 Introduction and Background *142*
2.8.1.1 Local Flooding Due to Heavy Rain *142*
2.8.1.2 Existing Screen Systems *143*
2.8.1.3 Screen Maintenance *143*
2.8.1.4 Screen Design *143*
2.8.1.5 Automated Screens *144*
2.8.1.6 Supporting Structures *144*
2.8.1.7 Objectives *145*
2.8.2 Methodology *145*
2.8.2.1 Locality *145*
2.8.2.2 Digital Recording and Geo-systems *146*
2.8.2.3 Inspection of the Locality *146*
2.8.2.4 Follow-up and Analysis *146*
2.8.2.5 Measures of Prioritisation *146*
2.8.3 Results and Discussions *148*
2.8.3.1 Characteristics of Screen Systems *148*
2.8.3.2 Screen System Maintenance *150*
2.8.3.3 Recommendations for Action *153*
2.8.4 Conclusions and Outlook *154*
References *154*

3 Sustainable Flood Retention Basins including Integrated Constructed Wetlands *163*
3.1 Sustainable Flood Retention Basin Management *163*
3.1.1 Introduction *163*
3.1.2 Methodology *164*
3.1.2.1 Data *164*
3.1.2.2 Variograms *166*
3.1.2.3 Kriging *166*
3.1.3 Results and Discussion *168*
3.1.3.1 Findings Based on Ordinary Kriging *168*
3.1.3.2 Findings Based on Disjunctive Kriging *171*
3.1.3.3 Consequences for Flood Risk Management *175*
3.1.4 Conclusions *177*
3.2 Nutrient Release from Integrated Constructed Wetland Sediment *177*
3.2.1 Introduction *177*
3.2.2 Materials and Methods *178*
3.2.2.1 Site and Experimental Set-up *178*
3.2.2.2 Sampling and Analytical Methods *184*
3.2.3 Results and Discussion *184*
3.2.3.1 Comparison of Treatment Performances *184*
3.2.3.2 Vegetation *194*
3.2.3.3 Groundwater Contamination *194*
3.2.3.4 Sediment Management *194*

3.2.4 Conclusions and Recommendations *195*
3.3 Groundwater Quality Impacts from an Integrated Constructed Wetland *196*
3.3.1 Introduction *196*
3.3.2 Materials and Methods *197*
3.3.2.1 Study Site Description *197*
3.3.2.2 Monitoring Wells and Groundwater Sampling *199*
3.3.2.3 Monitoring and Analysis *200*
3.3.2.4 Statistical Analyses *201*
3.3.3 Results and Discussion *202*
3.3.3.1 Piezometer Hydrographs and Seasonal Fluctuations *202*
3.3.3.2 Contaminant Concentrations in Groundwater *204*
3.3.3.3 Factors Influencing the Variability in Groundwater Quality *210*
3.3.4 Conclusions *213*
References *213*

4 Water and Wastewater Treatment Technology and Modelling *221*
4.1 Biological Activated Carbon Beds *221*
4.1.1 Introduction *221*
4.1.2 Materials and Methods *222*
4.1.3 Results and Discussion *224*
4.1.3.1 Biomass Growth Monitoring and Control *224*
4.1.3.2 Biological Indicator Significance *226*
4.1.3.3 Spreadsheet Modelling of Filter Effluents *229*
4.1.4 Conclusions *230*
4.2 Constructed Wetlands Treating Sewage *231*
4.2.1 Introduction *231*
4.2.2 Case Studies and Methods *231*
4.2.2.1 Description of Wetland Systems *231*
4.2.2.2 Water Quality Monitoring and Limitations *235*
4.2.3 Results and Discussion *236*
4.2.3.1 Water Quality and Performance Efficiency *236*
4.2.3.2 Seasonal Variations and Wetland Aging *240*
4.2.4 Conclusions and Recommendations *240*
4.3 Neural Network Simulation of the Chemical Oxygen Demand Reduction *242*
4.3.1 Introduction *242*
4.3.2 Materials and Methods *244*
4.3.2.1 Source of Data *244*
4.3.2.2 Neural Network Model *245*
4.3.3 Results and Discussion *247*
4.3.4 Conclusions and Recommendations *253*
References *253*

5 Industrial Wastewater Treatment and Modelling *257*
5.1 Membrane Bioreactors and Constructed Wetlands Treating Rendering Wastewater *257*
5.1.1 Introduction *257*
5.1.2 Materials and Methods *258*
5.1.2.1 Industrial Rendering Plant *258*

5.1.2.2 Constructed Wetland *259*
5.1.2.3 Membrane Bioreactor *260*
5.1.2.4 Water Quality Analysis *260*
5.1.3 Results and Discussion *261*
5.1.3.1 Water Quality of the Dissolved Air Flotation Plant *261*
5.1.3.2 Water Quality of the Membrane Bioreactor *262*
5.1.3.3 Water Quality of the Constructed Wetland *264*
5.1.3.4 Comparison of Treatment Performances *265*
5.1.3.5 Water Quality Variability of the Membrane Bioreactor *266*
5.1.3.6 Sampling Optimisation *266*
5.1.4 Conclusions *267*
5.2 Benzene Removal with Constructed Treatment Wetlands *269*
5.2.1 Introduction *269*
5.2.1.1 Constructed Treatment Wetlands *269*
5.2.1.2 Benzene Removal *270*
5.2.1.3 Aim and Objectives *271*
5.2.2 Materials and Methods *272*
5.2.2.1 System Design and Operation *272*
5.2.2.2 Biodegradation and Volatilisation *273*
5.2.3 Results and Discussion *274*
5.2.3.1 Treatment Performance *274*
5.2.3.2 Impact of Volatilisation *280*
5.2.4 Conclusions *280*
5.3 Diesel Oil Spillage Removal Using Agricultural Waste Products *281*
5.3.1 Introduction *281*
5.3.2 Materials and Methods *282*
5.3.2.1 Materials *282*
5.3.2.2 Methods *282*
5.3.3 Results and Discussion *284*
5.3.3.1 Sorption Capacity of the Agricultural Wastes *284*
5.3.3.2 Floating Performance of the Wastes *285*
5.3.3.3 Orthogonal Design *286*
5.3.4 Conclusions and Recommendations *290*
5.4 Kohonen Self-Organising Map to Predict Biochemical Oxygen Demand *290*
5.4.1 Introduction *290*
5.4.2 Literature Review *291*
5.4.3 Methodology *292*
5.4.4 Case Study *294*
5.4.5 Numerical Analysis and Modelling *296*
5.4.6 Results *297*
5.4.7 Discussion *302*
5.4.8 Conclusions and Recommendations *304*
References *304*

6 Sludge Dewatering Tests *311*
6.1 Dewaterability Assessment Including the Capillary Suction Time Test *311*
6.1.1 Introduction *311*
6.1.2 Materials and Methodology *312*

6.1.2.1 Capillary Suction Time Devices and Measurements *312*
6.1.2.2 Testing of Different Papers *313*
6.1.2.3 Testing of Different Sludges *314*
6.1.2.4 Stirred Capillary Suction Time Test *315*
6.1.3 Results and Discussion *316*
6.1.3.1 Single Radii Test and Paper Application *316*
6.1.3.2 Multi-radii Test and Paper Application *319*
6.1.3.3 Rectangular Testing Device *319*
6.1.3.4 Stirring of Sludge to Avoid Sedimentation *319*
6.1.3.5 Primary Sludge *320*
6.1.3.6 Surplus Activated Sludge *320*
6.1.3.7 Synthetic Sludge *320*
6.1.3.8 Selection of an Alternative Test Apparatus, Procedure and Paper *321*
6.1.4 Conclusions and Further Research *321*
6.2 Improved Design and Precision of the Capillary Suction Time Testing Device *322*
6.2.1 Introduction *322*
6.2.2 Methodology *322*
6.2.2.1 Preparation of Synthetic Sludge *322*
6.2.2.2 Physical and Chemical Properties of Synthetic Sludge *323*
6.2.2.3 Funnel Sealant *324*
6.2.2.4 Experimental Design and Testing *325*
6.2.3 Results and Discussion *326*
6.2.3.1 Synthetic Sludge Properties *326*
6.2.3.2 Effect of Sealant on Test Accuracy *327*
6.2.3.3 Quantifying the Leakage *332*
6.2.3.4 Specific Resistance to Filtration Accuracy *333*
6.2.4 Conclusions and Recommendations *334*
6.3 Sludge Floc Size and Water Composition Impact on Dewaterability *334*
6.3.1 Introduction *334*
6.3.2 Materials and Methods *336*
6.3.2.1 Dewaterability Tests *336*
6.3.2.2 Coagulants and Mixers *336*
6.3.2.3 Sample Preparation *337*
6.3.2.4 Coagulation Experiments *337*
6.3.3 Results and Discussions *338*
6.3.3.1 Impact of Synthetic Raw Water Flocs on Dewaterability *338*
6.3.3.2 Impact of Synthetic Domestic Wastewater Flocs on Dewaterability *341*
6.3.3.3 Synthetic Raw Water Floc Size Distribution *342*
6.3.3.4 Synthetic Domestic Wastewater Floc Size Distribution *342*
6.3.3.5 Natural Water Floc Size Distribution *343*
6.3.4 Conclusions and Recommendations *343*
References *344*

7 Water Availability and Public Health *349*
7.1 Introduction *349*
7.2 Methodology *351*
7.3 Results and Discussion *353*
7.3.1 Water Resources and Usage *353*

7.3.2 Water Availability and Public Health *354*
7.3.3 Potential for Community Involvement *355*
7.4 Conclusions *355*
Appendix The Questionnaire *356*
Section A: Personal Data *356*
Section B: Household Water Collection and Use *356*
Section C: Water and Water-Related Diseases *357*
Section D: Community Water *358*
Section E: Community Participation in Water Supply Development *359*
References *360*

Index *363*

Foreword

Our first encounter: I first met Miklas Scholz during the inaugural meeting of WETPOL (International Symposium on Wetland Pollutant Dynamics and Control) held at Ghent University, Belgium, in 2005. A few years later, we started collaborating, engaging in various fields of applied wetland science, including those aspects exploring how wetlands function and their supporting sciences. The importance of 'landscape-fit' and aesthetics with regard to accommodating wetland infrastructure was integral. With a career based in applied biology and the evolutionary sciences, I demonstrated to him the relevance and associated benefits to be gained from the use of 'integrating' constructed wetlands into associated lands. When particular attention and focus were given to sustainable ecological considerations, particular additional benefits accrued, including improved efficacy of treatment and many additional elements not usually sought in water treatment generally. These had special relevance in Ireland with its relatively high precipitation and where I work and live. This publication is now Miklas's 11th book on water management and demonstrates his diverse understanding of the subject. I am most pleased to have been an inspiring mentor!

My career as a wetland scientist: When I first met Miklas, I was working for the Irish Government's Department of Environment, Culture and Local Government. Together with colleagues, my son Caolan, daughter Aila and various graduates, Miklas and I undertook a range of applied research focused on 'blue green' polluted water treatment solutions. After official retirement, I worked as Senior Resident Engineer for Waterford City and County Council and currently as a Senior Scientist for VESI Environmental Limited. We were partners, along with others, in the WATER JPI project 'RainSolutions', which Miklas headed. We have continued to engage, exchanging ideas and insights into how water from a diversity of sources can be optimally managed, including those located in rural, industrial and urban environments.

Today, sustainable water systems and their associated management needs are ever more important: This new textbook by Miklas supports, inspires and provides the evidence needed through applied research and associated case studies. It includes cutting-edge insights across various geographical regions of the world and water types/sources and includes innovative associated technologies. It delivers a wide-ranging overview and associated insights into current water treatment in a way that is most readable. Through its particular focus on wetland systems, it advances the central importance and relevance of all wetland ecosystems, both natural and constructed. His approach promotes the role of wetlands and their relevance towards achieving a more deserving and central worldwide role through his demonstration of a suite of applications, innovations and benefits. Importantly, this book reflects their increasing inter-dependable approach and importance in the pursuit of reliable and sustainable 'blue green' infrastructures across the planet, including those in agriculture, urban and industrial environments.

Special focus is given to how the integration of multi-functional values might be achieved through land-based water treatments with respect to the enhancement of landscapes: obtaining many social, economic and other wide-ranging environmental benefits – beyond those of water treatment alone: The reader can expect to gain enhanced and necessary confidence from the proposed sustainable water systems presented, such as the constructed wetlands covered in this book. For this purpose, the well-presented integrated constructed wetland (ICW) concept that emerged and has continued to evolve in Ireland and elsewhere, over the past nearly 40 years, that has increasingly become 'best practice' in many countries across the world, is presented (examples in Chapters 1, 2 and 4). Furthermore, this integrated approach is increasingly becoming an integral part of the overall 'sustainable flood retention basin concept' addressing the needs of multi-stakeholders for complex water resources challenges (Chapter 3).

Dr Rory Harrington
Dunhill, 8 September 2024

Preface

What is this book about? This new textbook has a broad focus and is therefore entitled 'Sustainable Water Systems' to attract a wide audience of academics and practitioners. The book covers broad water and environmental engineering system aspects relevant to water resources management as well as the treatment of storm water and wastewater, providing a descriptive overview of complex 'black box' systems and general design issues involved. Fundamental science and engineering principles are explained to address the student and professional markets. Standard and novel design recommendations account for the interests of professional engineers and environmental scientists. The latest research findings in water treatment systems are discussed to attract academics and senior consultants who could recommend the textbook to final year and postgraduate students as well as graduate engineers, respectively.

The book deals comprehensively with the design, operation, maintenance as well as water quality monitoring and modelling of traditional and novel wetland systems. It also provides an analysis of asset performance, the modelling of treatment processes and the performance of existing infrastructure in both developed and developing countries, as well as the sustainability and economic issues involved.

The explained underlying scientific principles will also be of interest to all concerned with the built environment, including town planners, developers, engineering technicians, agricultural engineers and public health workers. The book has been written for a wide readership, but sufficient hot research topics have been addressed to guarantee a long shelf life of the book. Therefore, case study topics are diverse and research projects are multidisciplinary, holistic, experimental and modelling-oriented.

What is the target audience? The textbook is essential for undergraduate and postgraduate students, lecturers, and researchers in the civil and environmental engineering, environmental science, agriculture and ecological fields of sustainable water management. It is a standard reference for the design, operation and management of wetlands by engineers and scientists working for the water industry, local authorities, non-governmental organizations and governmental bodies. Moreover, consulting engineers will be able to apply practical design recommendations and refer to a large variety of practical international case studies, including large-scale field studies.

What are the key selling features? This new textbook has a broad focus on all applied aspects of both sustainability and system research. The sustainable treatment and management of all water and wastewater types is covered. Applied research case studies independent from countries and climatic regions illustrate the application on innovative technologies and methodologies. Both urban and rural case studies with applied and academic aspects are addressed.

How is this book structured and what questions does it answer? The textbook is split into seven inter-related chapters and corresponding sub-chapters to increase its readability. Each chapter answers topical ***'How can ...?'*** overarching questions as outlined below:

How can natural wetland systems be effectively used to treat different types of waters?
Chapter 1.1 outlines a peatland management strategy to utilise the high nutrient retention potential of degenerated peatlands. The effect of raised water levels and extensive land use management on hydraulic properties, water quality and vegetation characteristics of heavily vegetated and groundwater-fed open ditches was investigated at a river valley. Within-ditch vegetation and other hydraulic obstructions increase the hydraulic residence time and lead to an improvement in the water quality along the open ditch.

Chapter 1.2 is concerned with the treatment of log yard runoff, which is required to avoid contamination of receiving watercourses. The research aim was to assess if infiltration of log yard runoff through planted soil systems is successful and if different plant species affect the treatment performance. The infiltration treatment was effective in reducing total organic carbon and total phosphorus concentrations in the runoff.

Chapter 1.3 aims to review studies about impacts of anthropogenic land use changes on levels of nutrient concentrations in surface waterbodies. Anthropogenic land use such as agricultural and urban areas usually enhances nutrient concentrations much more than natural lands such as forest and barren.

Chapter 1.4 assesses the effect of climate change on carbon dioxide exchange of temperate peatlands. Climate chamber simulations were conducted using experimental peatland mesocosms exposed to current and future representative concentration pathway (RCP) climate scenarios (RCP 2.6, 4.5 and 8.5). Water level management is necessary for RCP 8.5, beneficial for RCP 4.5 and unimportant for RCP 2.6 and the current climate.

How can sustainable drainage systems and techniques be effectively used in urban water management and treatment?
Chapter 2.1 aims to assess the influence of a full silt trap at the end of a stormwater drainage pipe on the water quality of stormwater discharged into an urban watercourse. Suspended solids for treated stormwater were often too high, but pollutants accumulated in the silt trap.

Chapter 2.2 investigates the treatment efficiency of passive vertical-flow wetland filters containing *Phragmites australis* and/or *Typha latifolia* and granular media of different adsorption capacities. There appears to be no additional benefit in using macrophytes and expensive adsorption media in constructed wetlands to enhance metal reduction during the set-up period of five months.

Phosphorus retention by experimental unplanted vertical-flow constructed wetlands depends on substrates, influent quality and hydraulic residence time according to Chapter 2.3. Phosphorus adsorption capacities of shale, ironstone and hornblende were best explained by Langmuir adsorption isotherms. In comparison, Freundlich adsorption isotherms fitted gravel well.

Chapter 2.4 aims to assess the treatment efficiencies for gully pot effluent of experimental vertical-flow constructed wetland filters containing macrophytes and granular media of different adsorption capacities in cold climate. For those filters receiving metals, an obvious breakthrough of nickel was recorded after road gritting (containing salt).

Pond structures as cost-effective source-control drainage techniques can be applied to reduce the downstream risk of flooding according to Chapter 2.5. A case study was assessed based upon a combined attenuation wetland and dry pond construction for roof-water runoff.

Chapter 2.6 assesses the concept of a combined traditional permeable pavement with a ground source heat pump. The great system stability of the innovation and minor water quality data variability between individual experimental pavement systems provide good evidence for the controlled engineered application.

Chapter 2.7 aims to assess the efficiency of a batch flow combined titanium dioxide and ultraviolet light photocatalytic reactor. The purpose was to remove waterborne microbial contaminants from the effluent of permeable pavement systems.

Finally, Chapter 2.8 aims to protect below-ground channels and people, prevent flooding, improve water quality and save personnel costs through screen automation and monitoring optimization. Results show that repairing or enlarging screens optimizes their functionality and reduces the risk of flooding. A particular focus is on increasing the screen dimension from one- and two-dimensional to three-dimensional screens.

How can the sustainable flood retention basin and integrated constructed wetland concepts be successfully applied to support the build environment?

Chapter 3.1 analyses sustainable flood retention basins (SFRB) as adaptive structures contributing to water resources management and flood risk control. A dataset of 371 potential SFRB (including many operating reservoirs) characterized by 40 variables have been surveyed across central Scotland. Geostatistical techniques are applied on the dataset. Spatial analysis showed that ordinary kriging, which is a spatial interpolation method, could be successfully applied to estimate numerical values for all key flood control variables everywhere in the study area. Moreover, the probability that certain threshold values relevant to flood control managers were exceeded can also be calculated by using disjunctive kriging. The findings provide an effective screening tool in assessing flood control using SFRB.

Furthermore, Chapter 3.2 highlights that constructed wetland sediments are frequently contaminated with nitrogen and phosphorus, and that there is a potential that these accumulated constituents could be remobilized to reach surface or ground waters. Five configured identical mesocosms, each filled with sub-soil collected from the most contaminated first cells of two 10-year-old fully operational integrated constructed wetland (ICW), were used to examine nutrient removal within sediment layers. The results indicated that accumulated nutrients remobilised into the inflow (ambient) surface water and that the sediment capacity for nutrient retention decreased as the wetlands aged.

Finally, Chapter 3.3 assesses the extent of groundwater quality deterioration due to the establishment of a full-scale ICW system treating domestic wastewater in Ireland. The ICW consists of two sedimentation ponds and a sequence of five shallow vegetated wetland cells. The ICW cells were lined with 500 mm^{-1} thick local subsoil material, which comprised a mixture of alluvium, organic soils, tills and gravel. Overall, the quality of groundwater underlying the ICW system was slightly contaminated with bulk organic matter and some inorganic nutrients. Significantly higher contaminant concentrations were recorded in monitoring wells up-gradient and near to the distal wetland cells than down-gradient ones, which were near to the proximal cells. For the down-gradient piezometers, concentrations seldomly exceeded the natural background levels. Findings suggest that ICW systems pose a minimal risk to the groundwater quality.

How can various modelling approaches be successfully applied to solve timely water and wastewater treatment technology challenges?

The optimisation of water purification with biological activated carbon (BAC) is described in Chapter 4.1. Procedures are suggested to control biofilm growth and to use bio-indicators to predict chemical oxygen demand (COD) removal efficiencies. There was a strong positive correlation between the abundance of some protozoa in the liquid phase of the BAC bed and COD concentration in the effluent. Mathematical spreadsheet models were constructed to estimate COD removal efficiency of BAC filters with different loading rates, dissolved oxygen, pH, nutrient requirements and populations of micro-organisms.

For Chapter 4.2, data from five Wildfowl and Wetlands Trust constructed wetland systems offering a range of styles and inflow water types between 2005 and 2009 were examined and compared to identify long-term trends in nutrient removal. Ammonia-nitrogen concentrations were reduced between 31.9% and 96.8%. In contrast, the concentrations of nitrate-nitrogen and total oxidised nitrogen in the effluent exceeded the influent in many of the systems. Reduction in ortho-phosphate-phosphorus and total phosphorous were between −10.5% and 87.6% and between 6.9% and 92.5%, respectively. Removal efficiencies of biochemical oxygen demand were between 0.0% and 87.3%, and reductions of total suspended solids ranged from −249.3% to 57.6%. Ammonia-nitrogen reduction was effective during summer. Long-term nitrogen removal has been efficient and consistent. However, phosphorous reduction was only sufficient during the early stages of operation and generally declined as the wetland aged.

Finally, Chapter 4.3 simulates the performance of BAC filters based on microbiological investigations as well as measurements of pH and dissolved oxygen during the bio-regeneration mode with untreated river water. The performance parameters measured include COD. The neural network model could estimate the COD reduction in a BAC filter based on the pH, dissolved oxygen and count of micro-organisms in water samples taken from the liquid phase surrounding the carbon granules. The average absolute deviation in the case of the training set was 1.1% and in the case of the testing set it was 14.6%.

How can simple sludge dewatering tests help to design and manage large dewatering equipment units and processes?

Chapter 6.1 critically assess the standard capillary suction time (CST) test and proposes a modified device (prototype) and a revised CST procedure. The empirical CST test (using a circular funnel) is well established as the leading method for the determination of sludge dewaterability despite of its current shortcomings such as restricted modelling possibilities, and therefore the ability to predict physical processes such as the amount of water bounded by the paper. Nevertheless, the CST apparatus is portable, and the method is easy to conduct, quick, cost-effective and accurate, if the product of solid concentration and average-specific resistance to filtration is of interest. A novel prototype with a rectangular instead of a circular funnel incorporating a stirrer (optional) and using a cheaper paper with similar or improved characteristics is proposed to reduce consumable costs and improve dewaterability interpretation.

Chapter 6.2 aims to improve the CST product design by testing the effectiveness of using a funnel sealant to reduce variability. The use of a funnel sealant resulted in increased test precision. There was a reduction of up to 63% in the coefficient of variation and a substantial improvement in the predictability of the specific resistance to filtration test.

Finally, Chapter 6.3 assesses the impact of different water compositions on sludge dewaterability, assessments of floc sizes using a particle size analyzer and sludge dewaterability based on CST test. Synthetic raw water had small floc sizes, and synthetic domestic wastewater had both larger median floc sizes and a better correlation between sludge dewaterability and median floc sizes. Floc size distribution results showed that synthetic raw water is associated with a narrow particle size distribution. In comparison, synthetic domestic wastewater produced a wider distribution.

However, CST values were similar for both waters. Compared to synthetic wastewater, natural wastewater had the largest distribution with generally larger particle sizes.

How can water availability influence public health? Chapter 7 presents a case study describing water availability challenges and corresponding public health implications for a rural area in Nigeria. A water availability assessment was carried out in eight villages controlled by the Jalingo Local Government of Taraba State. Questionnaires were applied to obtain data from 60 households (approximately 650 participants) concerning issues like water supply sources, water quantity, water quality and community participation in water resources projects. It was found that community well water is the major source of water: 35% of the inhabitants depend on well water during the wet season, while 69% depend on it during the dry season. About 80% of the households have access to less than 30 L of water per person per day. Due to low water availability, there is a prevalence of water-related diseases. This study can be useful to agencies from the developed world involved in rural water supply projects.

About the Author

DProf. Miklas Scholz, Cand Ing, BEng (equiv), PgC, MSc, PhD, DSc, CWEM, CEnv, CSci, CEng, FHEA, FIEMA, FCIWEM, FICE, Fellow of IWA, Fellow of IETI, is a Distinguished Professor at Johannesburg University, South Africa. Miklas is the Head of the Department of Water Management at the District of Herzogtum Lauenburg, Germany. He is also a Technical Specialist for Nexus by Sweden and a Hydraulic Engineer at Kunststoff-Technik Adams, Germany.

He has published 10 books and 346 journal articles in 136 different target journals. Prof. Scholz has a total of about 18,300 (11,500 citations since 2020) citations, resulting in an h-index of 66 and an i10-index of 252. He ranks among the top 2% of academics globally, based on his i10-index in the past five years. Miklas is also ranked among the world's top 2% scientists, according to Stanford University's ranking. His net worth score is about 79% for 2024. The corresponding relative social fame wealth (not income) as a brand name is estimated to be $928 million USD.

A bibliometric analysis of all constructed wetland-related publications and corresponding authors with a minimum number of 20 publications and 100 citations indicates that Miklas ranks fifth among 70 authors (including those who have sadly passed away).

In 2019, Prof. Scholz was awarded EURO 7M for the EU H2020 REA project, Water Retention and Nutrient Recycling in Soils and Streams for Improved Agricultural Production (WATERAGRI). He received EURO 1.52M for the JPI Water 2018 project, Research-based Assessment of Integrated approaches to Nature-based SOLUTIONS (RAINSOLUTIONS).

Acknowledgements

This book is predominantly based on 31 previously published papers. Special credits go to all authors associated with these articles. All support received has been acknowledged in the individual original articles.

1

Natural Wetland Systems

1.1 Hydraulics, Water Quality and Vegetation Characteristics of Ditches

1.1.1 Introduction

According to the European Water Framework Directive (European Commission 2000) a 'good surface water status' of all water bodies had to be achieved by 2015. The Convention for the Protection of the Marine Environment of the North-East Atlantic (OSPAR Convention) and the Helsinki Convention on the Protection of the Marine Environment of the Baltic Sea Area (HELCOM Convention) were entered into force in 1998 and 2000, respectively. However, this ambitious goal of nutrient load reduction was often not met (Grimvall et al. 2000; Umweltbundesamt 2000).

Assessing hydraulic flow patterns in natural and constructed wetlands, and constructed treatment reed beds is an important scientific and environmental task when enhancing water quality (Mitsch and Gosselink 2000; Scholz 2003b; Scholz and Xu 2002; Vymazal et al., 1998). In general, the abiotic and biotic conditions in rivers and wetlands support several physical and biogeochemical processes resulting in reduced nutrient concentrations in the out-flowing water compared to the inflowing water.

Wetland and river restoration are frequently suggested as an effective measure to combat eutrophication of aquatic ecosystems and to reduce nutrient loads (particularly nitrate-nitrogen) to the sea. This ecohydrological water management approach supported by the European Water Framework Directive may result in the reduction of drains and open ditches or at least in a change of their management to restore lateral flow pattern from the upland areas through the wetland to the river. For example, ceasing macrophyte mowing is usually a cost-effective (in fact doing nothing results in savings) method to restore wetlands by creating localised flooding during summer and autumn.

The hydraulic and biochemical functioning of ditch ecosystems and their contribution to the overall catchment dynamic of a riparian wetland located in a valley will be studied. The data gathered should help society to understand the challenges of diffuse pollution and holistic catchment management.

This case study aims to assess the hydraulic properties, water qualities and vegetation characteristics for selected open ditches. The selected open ditches should be representative for open-ditch groups differing in characteristics including hydraulic resistance time caused by macrophyte growth. It was anticipated that the federal state governments' current water management plan that aims to cease the regular mowing of in-stream and in-ditch macrophyte vegetation to increase the hydraulic residence time and subsequently the water level would influence findings.

Sustainable Water Systems, First Edition. Miklas Scholz.

The study should also help to assess the effectiveness and consequences of the new management strategy and highlight areas of further work. The objectives are (i) to characterise all watercourses, including open ditches at a selected stretch of the River Eider; (ii) to carry out experiments with temporary weirs and a tracer such as sodium chloride at different hydraulic loading rates for different flow obstruction covers in open ditches; (iii) to assess the effect of macrophytes and channel morphology on flow conditions; (iv) to describe river, floodplain and ditch interaction during summer floods; (v) to assess the water quality within the open ditches and the River Eider; and (vi) to characterise the vegetation at the case study site particular at those areas close to the open ditches. Chapter 1.1 is based on an original article by Scholz (2005).

1.1.2 Experimental

The River Eider Valley is a riparian freshwater wetland of 257 ha in Northern Germany (10°3′ East and 54°11′ to 54°15′ North). The valley is located 10 km south of Kiel (capital of the federal state of Schleswig-Holstein).

The River Eider Valley peatland is hydrologically characterised by groundwater inflow from the surrounding hills and river water inflow from the 120 km^2 upstream catchment basin (Trepel and Kluge 2002). Figure 1.1a gives an overview of the wider study area showing the River Eider, the main open ditches with measurable discharge since June 2003 and the selected representative Ditches 14, 22–24 (Figure 1.1b). Photographs of Ditch 22 are shown in Figure 1.1c and Figure 1.1d. The River Eider has usual discharge values between 0.3 and 4.0 m^3 s^{-1}. Since the 18th century, reed cutters have mowed the vegetation in the River Eider and the adjacent open ditches to keep the riverbed open and to reduce the risk of late summer flooding.

A new water management plan aimed to cease the regular mowing of in-stream and in-ditch macrophyte vegetation to increase the hydraulic residence time (Trepel et al. 2003). Consequently, the river water level increased in the riparian wetland valley. The overall purpose of the current water management plan is the reduction of peat mineralisation by raising water levels, which may lead in the future to the establishment of new peat-forming vegetation types (Schrautzer et al. 1996; Trepel and Kluge 2002).

A total of 160 small watercourses including open ditches in the River Eider Valley were hierarchically classified (six levels) according to their geometry (width and depth), vegetation cover and type, hydrogeology, hydrochemistry and hydraulic flow properties (Table 1.1). The classification was based on expert water engineering understanding rather than on a statistical evaluation: An open ditch has been defined as (i) a small individual watercourse (not part of a wider network) with a reasonable and measurable amount of water discharging into the River Eider throughout the year, (ii) length >50 m, (iii) width ≥10 cm, (iv) water depth ≥3 cm throughout the open ditch and (v) without excess flow obstruction cover (≥95%). It follows that ≤25% of all watercourses were classified as open ditches according to this restrictive definition.

The following four representative open ditches out of 39 individual open ditches have been selected for further studies: Ditches 14, 22 (Figure 1.1c, d), 23 and 24 (overview; Figure 1.1b). The open-ditch numbers relate to the corresponding geographical location. All selected watercourses show common characteristics of groundwater-fed, rainwater-fed and surface water-fed open ditches.

Ditch 22 represented main open ditches with a width ≥150 cm and a depth ≥100 cm. Ditches 14, 23 and 24 were examples for small open ditches. Moreover, Ditches 14 and 23 were examples for small open ditches with low vegetation cover (<50%). In comparison with other selected open

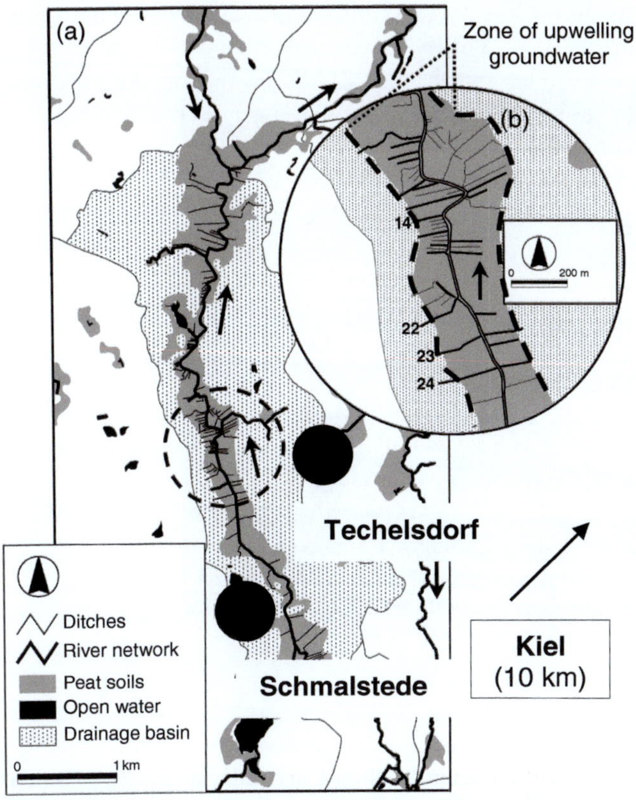

(c)

Figure 1.1 (a) River eider valley; (b) ditches; (c) Ditch 22; and (d) Ditch 22 (flooded). Photographs: Comparison of the water level near the outlet arrangement of Ditch 22 (c) before (11 August 2003) and (d) during flooding after heavy rainfall two days before and on 1 September 2003. A river with many specified and unspecified old drainage ditches is shown supported by an example ditch that becomes flooded. *Source:* Scholz (2005)/with permission of John Wiley & Sons.

(d)

Figure 1.1 (Continued)

Table 1.1 Classification of 160 small watercourses in the River Eider Valley on six classification levels (two sub-groups each)[a].

1a. Ditch width < 150 cm (**145**)							
2a. Ditch depth < 100 cm (**140**)							
3a. Flow obstruction cover < 50% (**17**)			3b. Flow obstruction cover ≥ 50% (**123**)				
4a. Groundwater-fed (**16**)			4a. Groundwater-fed (**54**)		4b. Not groundwater-fed (**69**)		
5a. Conductivity < 500 µS (**5**)	5b. Conductivity ≥ 500 µS (**11**)		5b. Conductivity ≥ 500 µS (**46**)		5a. Conductivity < 500 µS (**25**)		5b. Conductivity ≥ 500 µS (**44**)
6a. Ditch (**1**) = **Ditch 14**	6a. Ditch (**1**) = **Ditch 23**	6b. Not a ditch (**10**)	6a. Ditch (**9**) includes **Ditch 24**	6b. Not a ditch (**37**)	6a. Ditch (**6**)	6b. Not a ditch (**19**)	6b. Not a ditch (**37**)
1b Ditch width ≥ 150 cm (***15***)							
2a. Ditch depth < 100 cm (**10**)					2b. Ditch depth ≥ 100 cm (**5**)		
					3a. Flow obstruction cover < 50% (**3**)		
					4a. Groundwater-fed (**3**)		
					5a. Conductivity < 500 µS (**2**)		
					6a. Ditch (**1**) = **Ditch 22**		

[a] Only individual groups with ≥10 watercourses and groups containing sub-groups that include a selected representative ditch (Ditches 14, 22–24) are shown. Highlighted values (bold and italic) in brackets indicate the within-group number of watercourses.

ditches, Ditch 24 had a high vegetation cover of >50%. The vegetation distribution was estimated in May 2003. The conductivity of Ditch 14 in comparison to Ditch 23 was low (<500 μS).

The following Gauckler–Manning–Strickler open channel Eq. (1.1), according to Chaudhry (1996), was used to determine the theoretical velocity of the selected Ditches 14, 22–24:

$$v = k_M \times R^{2/3} \times S^{1/2} \tag{1.1}$$

where v = average velocity of flow (m/s), k_M = total roughness (or total resistance) coefficient ($m^{1/3} \times s$), R = hydraulic radius of the cross-section (m) and S = channel gradient (no unit).

Temporary triangle-shaped galvanised metal weirs were used to estimate the discharge of the selected open ditches at different cross-sections. The time taken for a small bucket of 12 L or a large bucket of 20 L with selection depending on ditch geometry to fill up was timed. In addition to visual k_M estimations, both the measured discharges and Eq. (1.1) were used to calculate the real ones.

General analytical procedures applied by the authors agreed with the American Standard Methods (American Public Health Association 1995). The water analysis for all nutrients was undertaken using a four-channel continuous flow auto-analyser with auto-sampler and automatic pre-dilution (Skalar Manufacturer, San plus System, Model 4000, Type 1121). This work was undertaken according to German Standard Methods: nitrate-nitrogen, nitrite-nitrogen, ortho-phosphate and ammonia-nitrogen in water samples were determined directly after sampling according to the German Standard Methods DIN EN ISO 13395, DIN EN 1189 and DIN EN ISO 11732 (Deutsches Institute für Normung 1996a,b, 2003), respectively.

The following further water quality parameters (instruments used are indicated in brackets) were measured in the field at 5-m intervals along the open ditches: conductivity (WTW TetraCon 325), pH (WTW SenTix 41), redox potential (WTW SenTix ORP), dissolved oxygen (WTW OxiCal SL) and temperature (WTW TetraCon 325).

Concerning the tracer experiments for Ditches 22–24, the Driesen and Kern μS-Log540 data logger for conductivity and temperature was used at open-ditch cross-sections at 15, 50 and 80 m intervals. The tracer (approximately 1 kg pure sodium chloride [NaCl]) was introduced at the open-ditch cross-sections 110 m, where sufficient flow was visually obvious. In contrast to Ditch 23, the discharge of Ditch 22 upstream of cross-section 110 m was low during the experiment. The tracer calibration curve (NaCl versus conductivity) was associated with a coefficient of determination R^2 of 1.00.

Vegetation relevees (species lists ordered by sampling quadrants) were carried out near the outlet into the channel of the River Eider and at 50-m intervals along each representative ditch. Plants located not more than 1 m away from each wetted ditch embankment were included in the survey. Classes were indicated as either rare = 1 (>0 and <1% cover), present = 2 (1–30% cover) or dominant = 3 (>30% cover). The nomenclature of the plants coincides with recent documentation (Wisskirchen and Häupler 1998).

The data analysis was performed using Microsoft Excel and Minitab. Statistical methods outlined elsewhere (Fowler and Cohen 1998) were applied.

1.1.3 Results

1.1.3.1 Characteristics of Open Ditches

Figure 1.1a gives an overview of the study site showing the River Eider, the representative small open ditches and the selected representative Ditches 14, 22 (Figure 1.1c, d), 23 and 24 (overview; Figure 1.1b). Table 1.1 gives a further overview of the characterisation of small watercourses in the

River Eider Valley. Most open ditches were small and overgrown. The estimated k_M values of both the vegetated and excavated Ditch 23 are about 1 and 10, respectively.

An analysis of variance for the Ditches 14, 22–24 with respect to standard geometrical (depth, width, etc.) and water quality (conductivity, dissolved oxygen, etc.) variables were performed. The findings confirm that the ditches were statistically significantly different in terms of their characteristics ($P < 0.01$).

The discharges of the groundwater-fed open ditches at the outlet near the River Eider were relatively constant. Ditches 14, 22 and 24 had discharges of approximately 0.4 (standard deviation (*SD*): 0.06), 1.2 (*SD*: 0.22) and 0.7 (*SD*: 0.27) $L\,s^{-1}$, respectively. In comparison, Ditch 23 had a discharge of 2.5 (*SD*: 0.67) $L\,s^{-1}$.

Figure 1.2 shows the profile of Ditch 23 at different cross-sections and gives an indication of the River Eider water level variations (Figure 1.2a). The excavation depths at the corresponding cross-sections are shown in Figure 1.2b. It is apparent that the River Eider discharged into the lower stretch of the open ditch.

Figure 1.3 shows a comparison of the measured and calculated discharges for Ditch 23 preceding (28 July 2003; Figure 1.3a) and following (11 August 2003; Figure 1.3b) the removal of aquatic vegetation and associated silt. The total roughness k_M can either be chosen to optimise the fit or be forced to be within an acceptable range of logical values based on expert judgement. For Ditch 23 with vegetation, this logical range was set to be between 1 and 4, and for Ditch 23 after vegetation

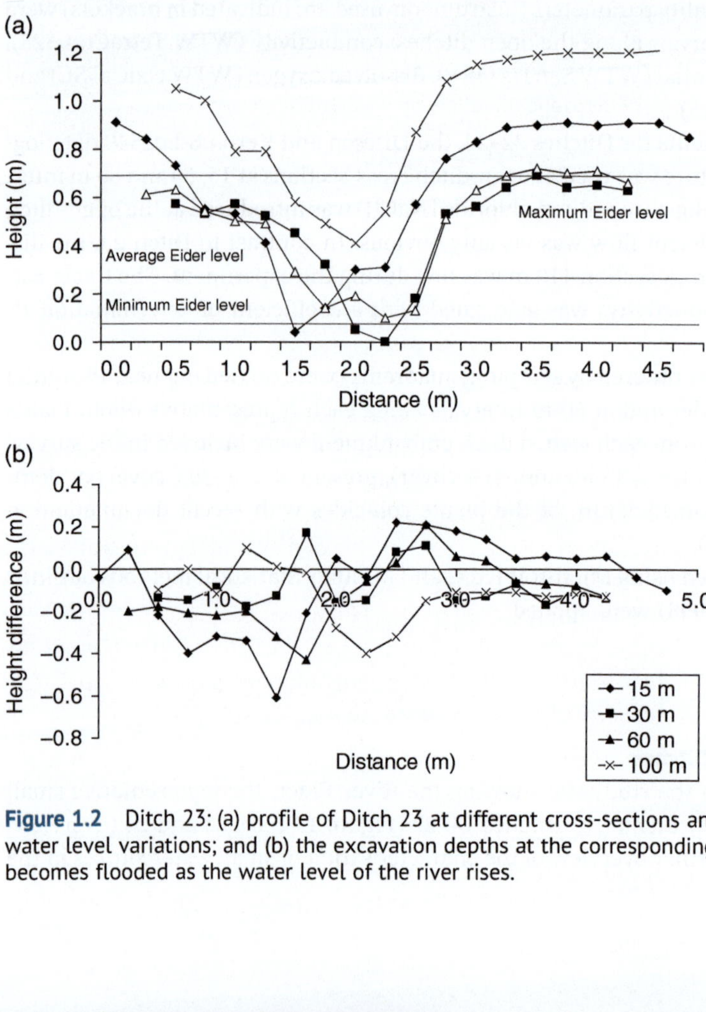

Figure 1.2 Ditch 23: (a) profile of Ditch 23 at different cross-sections and an indication of the River Eider water level variations; and (b) the excavation depths at the corresponding cross-sections. The ditch becomes flooded as the water level of the river rises.

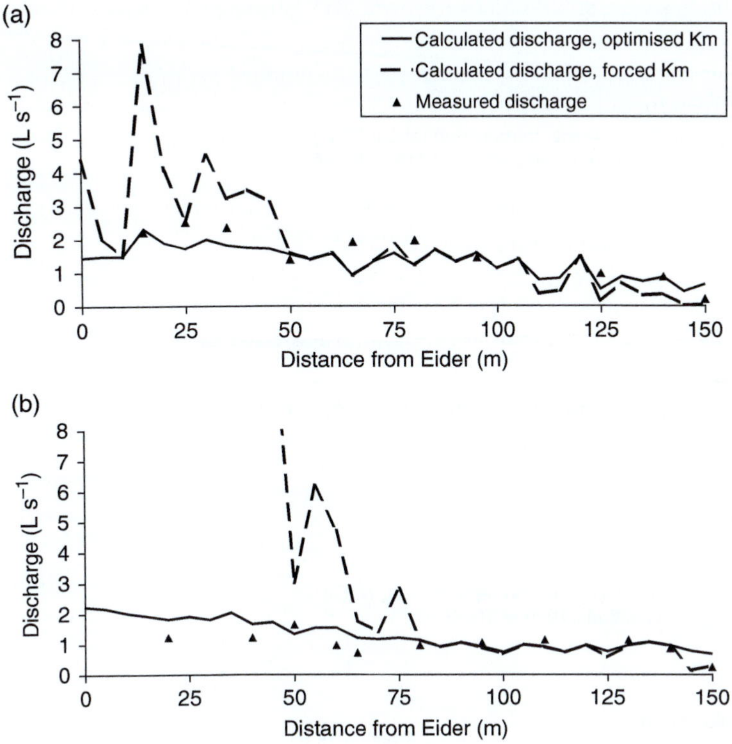

Figure 1.3 Discharges for Ditch 23: (a) preceding; and (b) following vegetation removal. Vegetation removal leads to water flowing faster.

removal, this range was set to be between 4 and 10. The value for k_M is actually rather a crude factor depending predominantly on the wetted perimeter rather than on the total roughness. This contrasts with the theory presented in the literature (Chaudhry 1996).

Figure 1.4 shows findings of the tracer experiment for the non-excavated Ditch 22 (14 August 2003; Figure 1.4a), the flooded partly excavated Ditch 24 (16 September 2003; Figure 1.4b) and the flooded full excavated Ditch 23 (1 September 2003; Figure 1.4c). The conductivity data shown are raw, not temperature-adjusted measurements. The calculated discharges based on the tracer experiments for Ditch 22 at the cross-sections 80, 50 and 15 m were 0.43, 0.41 and 0.42 L s^{-1}, respectively. In comparison, the calculated discharge based on the tracer experiments for Ditch 23 at the cross-sections 80, 50 and 15 m were 3.79, 1.81 and 0.63 L s^{-1}, respectively. If individual conductivity baselines are used for each of these three discharge calculations, the corresponding values are 3.72, 3.12 and 3.21 L s^{-1}. These discharges agree with the ones measured with experimental weirs.

The dispersions downstream of the open ditches are elevated (Figure 1.4). However, the dispersion at cross-section 15 m is influenced by the high-water level of the River Eider (Ditch 22; Figure 1.4c). For example, flooding of a 10 and 40 m wide embankment strip close to Ditch 14 and the group of Ditches 22–24, respectively, was recorded on 4 September 2003.

1.1.3.2 Water Quality and Vegetation

Table 1.2 summarises the water quality of the small watercourses and the river in the River Eider Valley. Total nitrogen (particularly nitrate) concentrations in these waters were high (Mitsch and

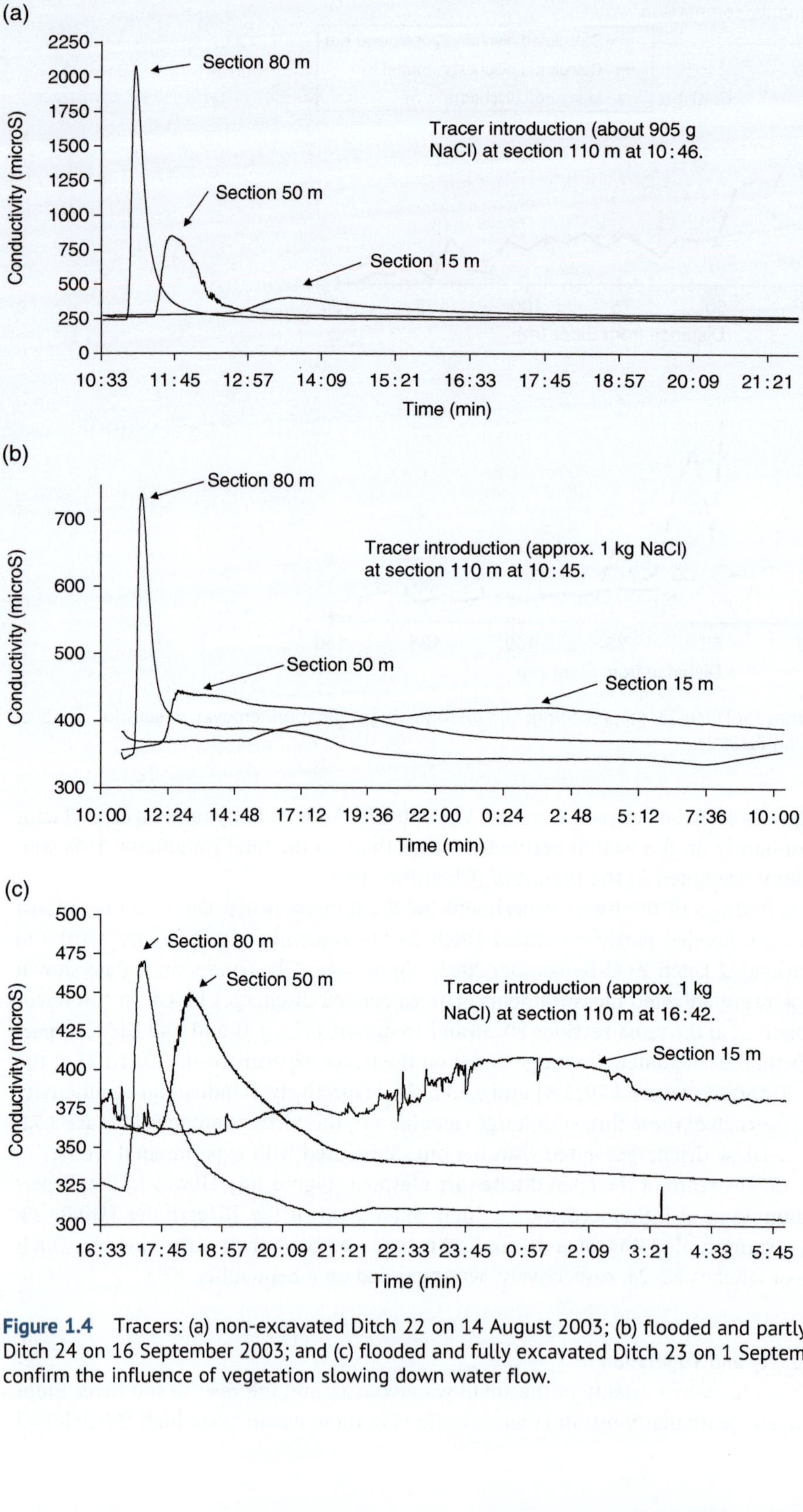

Figure 1.4 Tracers: (a) non-excavated Ditch 22 on 14 August 2003; (b) flooded and partly excavated Ditch 24 on 16 September 2003; and (c) flooded and fully excavated Ditch 23 on 1 September 2003. Tracers confirm the influence of vegetation slowing down water flow.

Table 1.2 Water quality comparison.

Variable	Statistics	Schmalstede[a] (Upstream)	Watercourses[a] (In the valley)	Flintbek[a] (Downstream)
Water temperature	Mean	13.8	12.0	13.6
(°C)	Standard deviation	1.68	2.39	1.86
	Count	17	72	17
pH	Mean	7.9	7.5	7.7
(–)	Standard deviation	0.10	0.26	0.12
	Count	17	56	17
Ammonia-N	Mean	0.1	0.2	0.1
(mg/L)	Standard deviation	0.03	0.56	0.05
	Count	17	101	17
Nitrite-N	Mean	0.0	0.0	0.0
($mg\,L^{-1}$)	Standard deviation	0.01	0.02	0.01
	Count	12	29	12
Nitrate-N[b]	Mean	3.1	1.5	2.2
($mg\,L^{-1}$)	Standard deviation	0.53	2.39	0.28
	Count	17	101	17
Ortho-phosphate-P	Mean	0.0	0.0	0.0
($mg\,L^{-1}$)	Standard deviation	0.02	0.02	0.01
	Count	17	32	17

[a] Water quality comparison between small watercourses located within the River Eider Valley and the Eider near the villages of Schmalstede and Flintbek located north-east of Techelsdorf (Figure 1.1a); data were collected in May and June between 1999 and 2003, and May and June between 1999 and 2002, respectively.
[b] Considering screening data for groundwater-fed open ditches collected in summer 2003 only, the nitrate-N concentration for these open ditches is higher than the corresponding one for the River Eider.

Gosselink 2000). The open ditches and other small watercourses were actually diluting the River Eider water in terms of nitrate-nitrogen leading to an overall water quality improvement. However, there are exceptions particularly for groundwater-fed open ditches in areas influenced by agricultural use in 2003.

Figure 1.5 shows a comparison between standard water quality variables (easy and cost-efficient to measure) and nitrate-N (difficult and cost-intensive to measure). The correlation coefficients between nitrate-nitrogen, and the redox potential and temperature were 0.62 and −0.55, respectively.

Table 1.3 gives an overview of the vegetation characteristics of the selected open Ditches 14, 22–24. The plant names are a combination of genus (e.g., *Carex*), species (e.g., *acutiformis*) and usually the name of the person (e.g., EHRH. as an abbreviation for Ehrenberg) who first described the species. The ditch vegetation is dominated by tall and productive plant species such as *Carex acutiformis* EHRH., *Phalaris arundinacea* L., *Glyceria maxima* (HARTM.) HOLMB. or *Urtica dioica* L. s. str. These species indicate nutrient-rich conditions.

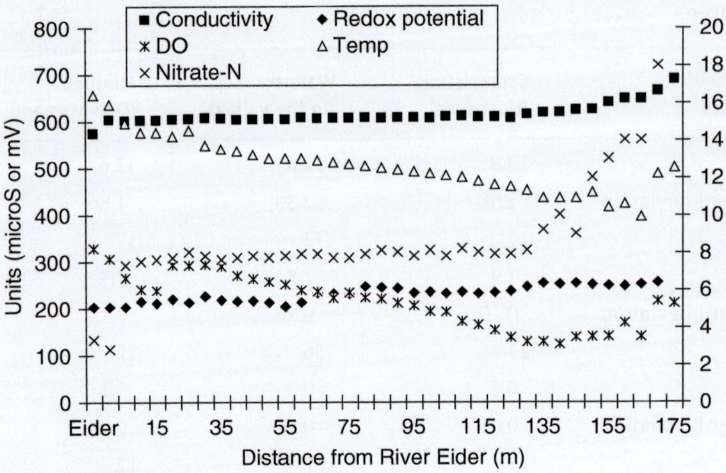

Figure 1.5 Ditch 24: Comparison of standard water quality variables and nitrate-nitrogen (N). DO, dissolved oxygen; Temp, temperature. Conductivity and redox potential slightly increase, DO and temperature decrease and nitrate-N increases considerably.

Table 1.3 Vegetation characteristics[a].

Ditch number/distance to Eider (m)		**14**	**14**	**14**	**22**	**22**	**22**	**23**	**23**	**23**	**24**	**24**	**24**
Species	Group	0	50	100	0	50	100	0	50	100	0	50	100
Poa trivialis L. s. l.	MA[b]					2	3				1		
Ranunculus repens L.	MA[b]	1	1										
Caltha palustris L.	MO[c]							1	1				
Carex cespitosa L.	MO[c]	2	2					1					
Epilobium hirsutum L.	MO[c]			2									
Filipendula ulmaria (L.) MAXIM.	MO[c]		2	1		2	2		2	1			
Juncus effusus L.	MO[c]		1				2						
Lysimachia nummularia L.	MO[c]									1		1	
Myosotis scorpioides agg.	MO[c]				2			1	1		2		
Scirpus sylvaticus L.	MO[c]						2						
Berula erecta (HUDS.) COVILLE	PH[d]		1		1								
Calamagrostis canescens agg.	PH[d]		2										
Carex acutiformis EHRH.	PH[d]	2	3	2					2			2	
Carex paniculata L.	PH[d]			2		3			2				
Equisetum fluviatile L.	PH[d]					1	2						

Table 1.3 (Continued)

Ditch number/distance to Eider (m)		14	14	14	22	22	22	23	23	23	24	24	24
Galium palustre L. s. l.	PH[d]		1						1			1	
Glyceria maxima (HARTM.) HOLMB.	PH[d]			3	1					1			
Nasturtium officinale R. Br.	PH[d]			1	1			2	3		2		3
Persicaria hydropiper (L.) DELARBRE	PH[d]	1			2			2	1		2	1	
Phalaris arundinacea L.	PH[d]	3	3	2	3	2	2	3	3	3	3	3	1
Rorippa palustris (L.) BESSER	PH[d]				1						1		
Stellaria alsine GRIMM	PH[d]	1			1						1		
Valeriana officinalis L. s. str.	PH[d]		1						1				
Galeopsis tetrahit L.	AR[e]				1	1					1		
Galium aparine L.	AR[e]		1	1	1	1	1			2			
Urtica dioica L. s. str.	AR[e]	1		2		2	2			3			
Alopecurus geniculatus L.	LP[f]				1			1					
Glyceria fluitans (L.) R. BR.	LP[f]		1		1								
Lemna minor L.	LE[g]				2			2					

[a] Vegetation characteristics of selected representative ditches (Ditches 14, 22–24). Classes were indicated as rare = 1 (<1% cover). present = 2 (1–30% cover) and dominant = 3 (>30% cover).
[b] *Molinio-Arrhenatheretea.*
[c] *Molinietalia.*
[d] *Phragmitetea australis.*
[e] *Artemisietea vulgaris.*
[f] *Lolio-Potentillion.*
[g] *Lemnetea.*

1.1.4 Discussion

1.1.4.1 Watercourse Classification

Due to the absence of an overall open-ditch maintenance plan, ≥85% of all watercourses in the River Eider Valley were small in terms of their width (<150 cm) and depth (<100 cm). Greater than 75% of all watercourses (121 out of 160) were heavily vegetated (≥50% flow obstruction cover) predominantly with macrophytes in May 2003 (Table 1.1). Nevertheless, approximately 25% of all watercourses were classified as individual open ditches with a major drainage purpose according to expert water engineering interpretation.

The depths of the less managed Ditches 14 and 22 in comparison to the depths of the more managed Ditches 23 and 24 increased relatively uniform from the springs towards the River Eider.

A V-profile was still apparent for all open ditches even though more than a decade has passed since the last open-ditch dredging measures took place (top graph of Figure 1.2). Manual vegetation removal undertaken by the author of Ditches 23 and 24 resulted in deeper V-profiles (lower graph of Figure 1.2).

1.1.4.2 Total Roughness and Summer Flooding

The values for total roughness k_M could be optimised by backwards calculations if the discharge was directly measured with a temporary weir or estimated with a tracer experiment. Optimised k_M ranges for Ditch 23 preceding and following excavation were between 0 and 128 (mean: 9) $m^{1/3} \times s$, and between 0 and 21 (mean: 5) $m^{1/3} \times s$, respectively. However, the k_M value should have increased after excavation (and not decreased as observed). In comparison, the corresponding k_M ranges for similar open ditches described elsewhere (Chaudhry 1996) before and after excavation were between 7 and 13 (mean: 10) $m^{1/3} \times s$, and 33 and 43 (mean: 40) $m^{1/3} \times s$, respectively. It follows that the experimental and literature-based theoretical k_M values for the excavated open ditches did not match. A possible explanation could be that Eq. (1.1) accounts only indirectly for size and shape of the channel, stage (water level) and bed load (sediment). These variables should be integrated into Eq. (1.1), but this proposed work is beyond the scope of Chapter 1.1.

In comparison to findings reported by Trepel et al. (2003) that the macrophyte growth within the River Eider is responsible for high water levels (flooding), the absence of vegetation (after manual excavation) within shallow and small open ditches results in virtually no increase of the k_M value. This indicates the difficulties of applying the open channel Eq. (1.1) for the small silted-up and particularly flooded open ditches in the River Eider Valley.

The impact of a flooded floodplain on the variations of the conductivity baseline, dispersion within the ditch and subsequently on the discharge and velocity within the corresponding ditch can be assessed with the help of Figure 1.4. The upper graph indicates the dispersion for a heavily vegetated open ditch. Fluctuations of conductivity due to high vegetation cover and other obstructions or due to diffuse pollution (Figure 1.4a) would therefore not be unusual. Nevertheless, the conductivity curves for Ditch 22 are smooth, which is ideal for dispersion, discharge and velocity estimations.

In comparison, the bottom graph of Figure 1.4 shows the corresponding results for an excavated open ditch with an average width of approximately 35 m due to flooding. Considering the recent removal of vegetation, the individual curves should be smooth and theoretically relatively smoother than the curves presented in the upper graph of Figure 1.4. However, the curve representing the cross-section at 15 m is influenced by considerably summer flooding (open ditch with: approximately 185 m) and therefore erratic. This can be explained with the variable local hydraulic conditions and the different water chemistry of the flooded floodplain.

1.1.4.3 Water Quality Influenced by Vegetation

The presence of the open ditches reduced nutrient retention and biochemical transformation (e.g., denitrification) within the peatland as indicated previously for similar regions in Northern Germany (Bockholt et al. 1992). Nevertheless, heavily vegetated open ditches had also a limited potential to retain nutrients via adjusted plant uptake.

A comparison between the overall water quality of the River Eider and the water quality of most small watercourses indicated that the discharge from the open ditches usually does not pollute the River Eider in terms of nutrient input (Table 1.2). This was partially the case due to the good water quality (except for high nitrate contamination of the groundwater) of the spring water and the presence of heavy aquatic vegetation that takes up nutrients.

Nitrate-nitrogen concentrations above 1 mg L^{-1} (see also Table 1.2) are usually an indication for man-made pollution; e.g., peatland use for agricultural purposes requires drainage and fertilisation. Nitrate concentrations were higher for grazed (Ditches 23 and 24) than ungrazed open ditches (Ditches 14 and 22). A discussion on the effective monitoring of nitrate change is beyond the scope of Chapter 1.1, but this issue has been discussed elsewhere (Bockholt et al. 1992; Burt 2003).

Table 1.3 indicates the cover and dominance of the Phragmitetea australis group (macrophytes) close to the open ditches that still have an obvious drainage purpose. The coverage of *C. acutiformis* increased when the occurrence of *Urtica diocia* L. s. str. was limited due to wet conditions. *U. diocia* and *P. arundinacea* (Table 1.3) stands were typical for drained peat soil. *Urtica dioica* increased with nutrient availability due to higher mineralisation rates in drained peat soils.

1.1.5 Conclusions

The classification of all watercourses indicated the dominance of small heavily vegetated open ditches partly supplied with groundwater (high in conductivity and nitrate-nitrogen). In contrast to the theory, the total roughness k_M was actually a crude 'fudge factor' depending predominantly on the hydraulic radius rather than on the vegetation cover. The traditional estimation of total bed roughness by the naked eye may lead to great errors in determining the discharge.

The absence of a river-mowing scheme resulted in heavy vegetation growth within open ditches and river stretches leading to a rise of the corresponding water levels and subsequent flooding in late summer. Tracer experiments showed high dispersion for the lower part of the open ditch (up to cross-section 30 m) due to the presence of vegetation, a flooded floodplain and a mixing zone between open ditch and River Eider water during late summer.

Elevated nitrate concentrations for upstream open-ditch cross-sections were apparent. A comparison between the overall water quality of the River Eider and the water quality of most groundwater-fed open ditches indicated that the discharge from the ditches usually did not pollute the river.

Six socio-ecological plant groups in total were identified for the selected representative open ditches. The dominant Phragmitetea australis-group comprised 13 species associated with tall and productive macrophytes.

1.2 Planted Soil Infiltration Systems for Treatment of Log Yard Runoff

1.2.1 Introduction

Contaminated runoff from log yards is generated when water encounters wood, woody debris, equipment and vehicles at outdoor wood sorting, processing and storage facilities (Woodhouse and Duff 2004). It also is common to sprinkle logs with water to protect them from cracking and biological attack (Hedmark and Scholz 2008). The contamination of the runoff depends on the specific conditions at the individual log yard; for example, the species of tree stored on the log yard, proportion of the runoff that has met stored wood and dilution or pretreatment of the runoff before sampling.

Typical runoff has the following characteristics (Hedmark and Scholz 2008; Hedmark et al. 2009): high concentration of organic substances, some of which can be toxic to aquatic plants and animals (100–6000 mg L^{-1}); relatively high amount of total suspended solids (TSS) (100–5000 mg L^{-1}); medium-to-high concentrations of total phosphorus (0.5–5 mg L^{-1}); relatively low concentration of total nitrogen (0.5–2.5 mg L^{-1}); low concentrations of (heavy) metals (0.00–0.05 mg L^{-1} per metal) and low-to-neutral pH (4–8).

The objectives of this study are to assess if infiltration of log yard runoff through soil systems reduces the concentrations of pollutants, such as organic carbon and phosphorus, in the runoff, and if different species of plants affect the treatment performance when log yard runoff is infiltrated through vegetated soil systems. Chapter 1.2 is based on the original article by Hedmark et al. (2010).

1.2.2 Experimental Setup and Methodology

The study site is in a field near Heby Sawmill in central Sweden (59°56′N, 16°50′E). The area had previously been used for agricultural production. Soil analysis from another experimental site in close vicinity of this study site showed that the organic matter in the topsoil was 3.8% (by mass) and that the pH increased from 5.5 in the topsoil to 7.0 at a soil depth of 2.5 m. The soil texture was loam (18% clay, 38% silt and 44% sand), which is known to have a relatively high soil water retention capability and absorptivity, and therefore can easily adsorb inorganic chemicals such as phosphate (Scholz 2006). The soil texture was relatively uniform throughout the profile. No surface runoff or ponding was observed, and drainage was sufficient to allow for good infiltration. Furthermore, no cracking or fracturing of the soil surface was observed.

The site consisted of eight square (20 m × 20 m) and flat planted soil infiltration systems or compartments (called plots from here onwards), covering a total area of 40 m × 80 m. The plots were arranged in a matrix of two rows, with four columns each. The two plots of the first column were planted with *A. glutinosa* and *S. schwerinii × viminalis*, respectively. The following two plots of the second column were planted with *S. schwerinii × viminalis* and *A. glutinosa*, in that order. The plots of the third column were planted with *P. arundinacea* and *L. perenne*, respectively. Finally, the remaining plots of the last column were planted with *L. perenne* and *P. arundinacea*, correspondingly. No buffer strips between the plots and no replicate plots were constructed.

During May and June 2003, five standard drainage pipes (outer and inner diameters of 74 and 65 mm) per plot were placed 0.9 m below ground level. The slots in the pipes were covered with a 10-μm polyester mesh to prevent clogging. Gravel was placed around the pipes, and most of the previously removed soil was put back again. The drainage system was designed to drain each flat plot separately. Water could not drain easily from one plot into the next, because of the relatively good soil drainage characteristics. The distance between the drainage pipes was 4 m, and the distance from the drainage pipes to the edge of each plot was at least 2 m. The strips of land between the plots that were not irrigated functioned as indirect buffer strips. However, there was no control, with respect to root growth, beyond the border of each plot.

The eight plots were planted with four different species in 2003; two plots were planted with bare-rooted 50-to-80-cm-high *A. glutinosa*; two plots were planted with approximately 50-cm-high *S. schwerinii×viminalis*; two were sown with *L. perenne* seeds; and two were sown with *P. arundinacea* seeds. *Salix schwerinii×viminalis* planting was performed in a double-row system, with an approximate spacing of 0.75 and 1.5 m, respectively, between the rows and approximately 0.6 m within the row, resulting in a plant density of approximately 1.6 plants m^{-2}. The plant density for *A. glutinosa* was roughly 1 plant m^{-2}. The grasses were sown so that complete ground cover was achieved rapidly. Two replicates for each planted strip were considered appropriate, because the case study site was sufficiently homogenous throughout the soil profile.

The study site was left to mature until early spring 2005. By then, *A. glutinosa* and *S. schweriniix-viminalis* had reached a height between 1.5 and 1.8 m. *Lolium perenne* had a height of approximately 0.4 m, and *P. arundinacea* measured 0.5 m. A plastic hose was placed on the surface of the plots for irrigation purposes. The hose was laid down in two arms on every plot, parallel to and between the drainage pipes. The hose was perforated to allow the irrigation water to trickle out as uniformly over each plot as practically possible. The drainage water was collected by the below-ground pipe system, which introduced preferential flow conditions, because of low flow resistance within the pipes compared with the soil.

A previous study by Jonsson et al. (2006), discussing the runoff quality at this site, showed that the runoff was too low in nitrogen (approximately 1.3 mg L^{-1}) compared with carbon and phosphorus to allow for sufficient biodegradation to take place (Scholz 2006). To address this potential problem, the plots were fertilised with 80 kg N ha^{-1} in early July 2005. Each plot received 20 kg of fertiliser (15.5% nitrogen and 18% potassium) comprising nitrate of lime (nitrocalcite) and other nutrients. No fertiliser was applied thereafter. The fertiliser contained no phosphorus and no organic matter.

During all sprinkling seasons, the plots were irrigated with log yard runoff, which originated from the adjacent sawmill, where logs of *Picea abies* (L.) Karst (Norway spruce) were stored on an asphalted log yard comprising 8 ha (80 000 m^2). During the sprinkling seasons in 2005, 2006 and 2007, the plots were irrigated with 8, 12 and 12 mm d^{-1}, respectively. The irrigation started at 8 : 00 a.m. and continued until the preset amount of water had been released, that is, until approximately 11 : 20 a.m. in 2005 and 1 : 00 p.m. in 2006 and 2007.

A weather-monitoring station was placed directly at the study site and connected to a data logger. The estimated mean annual precipitation for the area is 600 mm (based on a 30-year mean of a nearby weather station), and the approximate mean temperature is 5°C (SMHI 2009). During the periods when log yard runoff was infiltrated at the site, data obtained from the station showed that the mean daily rainfalls and temperatures were 1.2, 2.4 and 1.6 mm, and 12.6, 14.3 and 13.3°C in 2005, 2006 and 2007, respectively. Log yard runoff was infiltrated during the log-sprinkling season, which lasts from approximately the end of April to the beginning of October. Analytical procedures applied were based on American Standard Methods (APHA et al. 1998).

1.2.3 Results and Discussion

1.2.3.1 Log yard Runoff Compared with Drainage Water

Contaminant means and standard deviations for the log yard runoff and the drainage water from the plots have been summarised in Table 1.4. A comparison based on a one-way analysis of variance (ANOVA) between irrigation and drainage water for mean concentrations total inorganic carbon of total carbon, total inorganic carbon (TIC), total organic carbon (TOC), total phosphorus and TSS for the four different species (two plots per species) was performed. The ANOVA showed that total carbon, TOC and total phosphorus concentrations in the captured drainage water were lower and significantly different ($p < 0.05$) from the log yard runoff during the entire study period. This confirms that infiltration can be used successfully as a treatment method to reduce organic matter and phosphorus in log yard runoff, but findings indicate that the treatment efficiency over time could be limited for phosphorus.

For TIC, there was no significant treatment effect. Furthermore, the infiltration of the runoff also did not result in a statistically significant decrease in TSS concentrations. Despite the clear differences in mean concentrations of TSS in the log yard runoff and the drainage water from the planted

Table 1.4 Water quality parameters for the inflow and outflow of experimental soil infiltration systems planted with different species.

		TC		TIC		TOC			TP			TSS	
Sampling water type	**Statistics**	**2006**	**2007**	**2006**	**2007**	**2005**	**2006**	**2007**	**2005**	**2006**	**2007**	**2006**	**2007**
Irrigation water (inflow)	Mean	218.4	246.0	28.5	21.5	120.3	171.6	224.5	1.2	1.5	2.3	108.0	117.1
	Stdev	79.84	64.16	24.80	4.59	44.88	73.33	65.29	0.19	0.58	1.11	39.23	63.32
	n	3	7	3	7	5	6	7	5	6	7	4	7
Outflow for Alnus glutinosa	Mean	109.4	131.6	20.9	23.3	52.0	88.5	108.4	0.6	0.8	1.2	68.4	156.0
	Stdev	49.37	42.18	5.09	3.62	13.39	45.89	41.26	0.29	0.27	0.76	38.66	209.63
	n	16	14	16	14	7	16	14	7	16	14	16	14
Outflow for Salix schweriniixviminalis	Mean	114.4	134.0	23.0	24.8	46.7	91.3	109.1	0.6	0.9	1.4	54.8	144.4
	Stdev	47.06	48.37	4.43	4.88	14.29	47.79	48.68	0.14	0.42	0.75	36.64	171.37
	n	16	14	16	14	7	15	14	7	15	14	15	14
Outflow for Phalaris arundinacea	Mean	108.8	144.4	21.3	22.9	43.0	87.5	121.5	0.5	0.9	1.5	70.9	125.1
	Stdev	41.54	40.62	5.37	3.46	7.07	40.28	38.46	0.17	0.49	0.52	68.92	106.59
	n	16	14	16	14	8	16	14	8	16	14	16	14
Outflow for Lolium perenne	Mean	100.1	121.7	18.1	21.4	43.1	82.0	100.3	0.5	0.7	1.2	51.3	87.7
	Stdev	44.46	43.06	6.71	3.78	12.23	41.96	41.33	0.17	0.37	0.57	22.23	51.02
	n	16	14	16	14	7	16	14	7	16	14	15	14

Alnus glutinosa (L.) Gärtner, common alder; *Lolium perenne*; *n*, sample number; *Phalaris arundinacea*, reed canary grass; rye grass; *Salix schweriniixviminalis* 'Gudrun', willow; stdev, standard deviation; TC, total carbon (mg L^{-1}); TIC, total inorganic carbon (mg L^{-1}); TOC, total organic carbon (mg L^{-1}); TP, total phosphorus (mg L^{-1}); TSS, total suspended solids (mg L^{-1}).

plots (Table 1.4), the difference was not statistically significant for 2006. The low number of runoff samples ($n = 4$) and the generally high variability of TSS in the drainage water may explain the lack of statistical significance at $p < 0.05$.

In 2007, the infiltration did not lead to a decrease in TSS concentrations. On the contrary, the concentrations in the drainage water were higher than in the runoff for four of the eight plots, possibly because of a breakthrough of partly degraded organic matter, such as detritus and litter (Scholz 2006).

The pH values were slightly higher for the drainage water (annual means = 6.8–7.5) compared with the log yard runoff (annual means = 6.3–7.4). However, both the runoff and the drainage water were in the range of what is commonly interpreted as an expected pH range (6–8) for receiving watercourses (Scholz 2006). The drainage water generally had lower concentrations of total nitrogen (annual means = 0.8–3.4) than the log yard runoff (annual means = 1.3–4.1). Previous studies have shown that the concentration of total nitrogen in log yard runoff generally is low and that no treatment to reduce the concentration is necessary, except when the receiving watercourse is classified as 'very nitrogen vulnerable' (Hedmark and Scholz 2008). On the contrary, in this study, the soil was fertilised with nitrate in 2005. Modest fertilisation was undertaken to prevent a shortage of bioavailable nitrogen, which could have led to reduced plant growth.

1.2.3.2 Comparison of Drainage Waters

The results from the infiltration experiment were analysed to assess if there were any statistically significant differences in treatment efficiency for the different plant species. However, the concentrations of TOC, total phosphorus and TSS in the drainage water from the plots with different plant species did not differ significantly at $p < 0.05$. The fact that the use of different plants did not significantly affect the removal efficiency (particularly regarding total phosphorus and TOC) strengthens the assumption that the microbial biomass contributed to the purification performance (Scholz 2006).

There also were no differences in concentrations of total carbon for the species in the infiltration experiment. However, a two-sample *t*-test showed that the drainage water from the plots vegetated with *S. schwerinii* × *viminalis* had significantly higher concentrations of TIC than the drainage water from the plots planted with *L. perenne*. This was unexpected, because the infiltration treatment had no removal effect on TIC. This difference between the plant species could not be explained in this case study.

A statistical comparison with the two-sample *t*-test also was made for the water quality parameters of the drainage water from the plots vegetated with tree species (*A. glutinosa* and *S. schwerinii* × *viminalis*) compared with grasses (*L. perenne* and *P. arundinacea*). None of the assessed parameters (total carbon, TIC, TOC, total phosphorus and TSS) were statistically significantly different ($p < 0.05$).

1.2.3.3 Comparison of Treatment Efficiencies

The findings confirmed that the infiltration treatment efficiency for log yard runoff was high regarding TOC and that the removal of total phosphorus was high and relatively continuous during the experiment (Table 1.4). The removal of total carbon was relatively high, but the infiltration treatment was not successful for TIC and TSS. Furthermore, the treatment efficiencies decreased slightly, but not significantly, for TOC and total phosphorus during the study. However, the loads increased considerably from year to year, and the amount of TOC and total phosphorus that was removed by the system also increased. The TOC loads on the system were

1630, 3461 and 4124 kg ha^{-1} in 2005, 2006 and 2007, respectively. The loads of total phosphorus for these years were 15, 30 and 43 kg ha^{-1}, respectively.

The treatment efficiencies of TSS decreased in 2007, which cannot be explained by an increased load of suspended solids in the irrigation water. The load of suspended solids only increased by 1.3% from 2006 to 2007. The change from sink to source of TSS cannot be explained by the small increase in load. However, it is known that some infiltration systems are sources for TSS, which is common for vertical downflow filtration systems (Al-Senafy and Al-Otaibi 2002; Scholz 2003a).

1.2.4 Conclusions and Recommendations

The treatment efficiency for total phosphorus was relatively good, but further field-scale experiments are required to determine the long-term treatment effect at different phosphorus loads. Findings indicated no significant difference in treatment efficiencies when comparing four species of plants with each other. There also were no significant differences between the tree and grass species. The research indicates that, under similar boundary conditions, an area of approximately 0.5 ha of infiltration system would be required for every 10 000 m^3 of yearly runoff from a log yard, to achieve the same retention of contaminants as in this study. Further research is required to address the inherent limitations of the study (i.e., no buffer strips between the plots, limited number of replicate plots, no unplanted control and no nutrient analysis within each soil column), which was not designed as a rigorous experiment, but rather as an applied engineering evaluation.

1.3 Anthropogenic Land Use Change Impacts on Nutrient Concentrations in Waterbodies

1.3.1 Introduction

Water quality challenges have become an increasing global concern due to the adverse impacts of rapid population growth and corresponding anthropogenic activities such as unsustainable irrigation practices, mining and urbanisation (Buck et al. 2004). Man-made disturbances produce different types of pollutants such as nutrients (e.g., nitrogen and phosphorus), which play an important role in plant growth, and are often discharged either untreated or partially treated into surface waterbodies (Donner et al. 2004). The excess of nutrient levels in waterbodies accelerates their eutrophication process that has harmful effects on drinking water supplies, recreation, fisheries and wildlife (Scholz 2010).

The increased pressure on water sources to meet current and anticipated demands is critically dependent on pollution resulting from various human-induced activities. However, finding a promising relation between land use (LU) and surface water quality (SWQ) is challenging (Goldstein et al. 2007). Search results using the keywords 'land use' and 'surface water quality' are shown in Figure 1.6. Annual publication rates are ascending, and a review of recent findings is therefore useful.

Chapter 1.3 aims to assess the impacts of anthropogenic LU changes on nutrient concentrations in surface waterbodies. This review highlights key points regarding the input dataset parameters such as LU, climatic conditions and landscape features for LU and SWQ interaction studies. Reported correlations between nutrient concentrations and anthropogenic LU changes are assessed to explain the logical relationships. Finally, some practices that have been recommended to

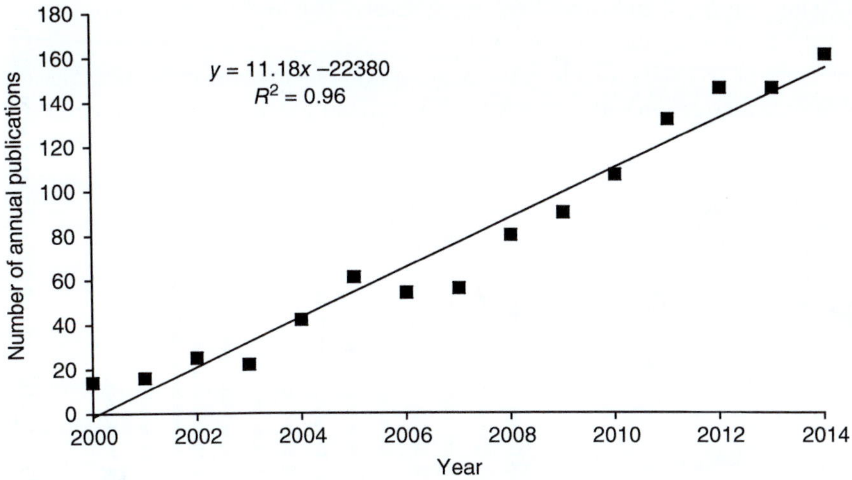

Figure 1.6 Annual publication rates for the keywords 'land use change' and 'surface water quality' using scopus.com. A positive linear regression line with a R^2 of 0.96 is displayed.

preserve water quality despite negative impacts of LU changes are introduced as well as key conclusions are drawn from this review. Delkash et al. (2018) provided the scientific basis for Chapter 1.3.

1.3.2 Land Use Changes and Surface Water Quality

1.3.2.1 Investigations

Alteration of land to provide shelter for humans and agricultural products are the primary anthropogenic LU changes (DeFries and Eshleman 2004). The LU changes associated with an increase in population growth form complex interrelationships, which require comprehensive spatial and temporal studies (Mouri et al. 2011; Wang et al. 2013). Classification of LU changes, for example, in Portugal between 1958 and 2007 is shown in Figure 1.7.

Anthropogenic LU changes release significant nutrient loads into waterbodies, which worsen SWQ conditions (Duncan 2014). Agricultural LU usually releases nutrients into waterbodies through non-point sources (NPS), leading to diffuse pollution, which has become a major environmental concern (Duncan 2014). Numerous diffuse nutrient pollution studies focused on the effects of runoff flowing over agricultural land (Scholz 2010). These studies reported that agricultural land coverage strongly influences nitrogen (Ahearn et al. 2005; Duncan 2014), phosphorus (Liu et al. 2014) and total suspended solid concentrations (Ahearn et al. 2005) as well as sediment loads in watercourses (Ma et al. 2011).

Urban LU affects SWQ through two different ways: point sources (PS) and NPS of nutrients discharging into waterbodies. PS contribute to surface water pollution by directly discharging domestic and industrial wastewater into the corresponding waterbodies. Nitrate concentration usually has a statistically significantly positive correlation with urban density due to wastewater PS discharges (Krause et al. 2008).

Urban LU can contribute to nutrient loads discharged into waterbodies through rainfall conversion to runoff and/or drained water as well as washing-off non-point nutrient pollution sources (Lee et al. 2009). An increase in impermeable area due to urbanisation may lead to higher runoff

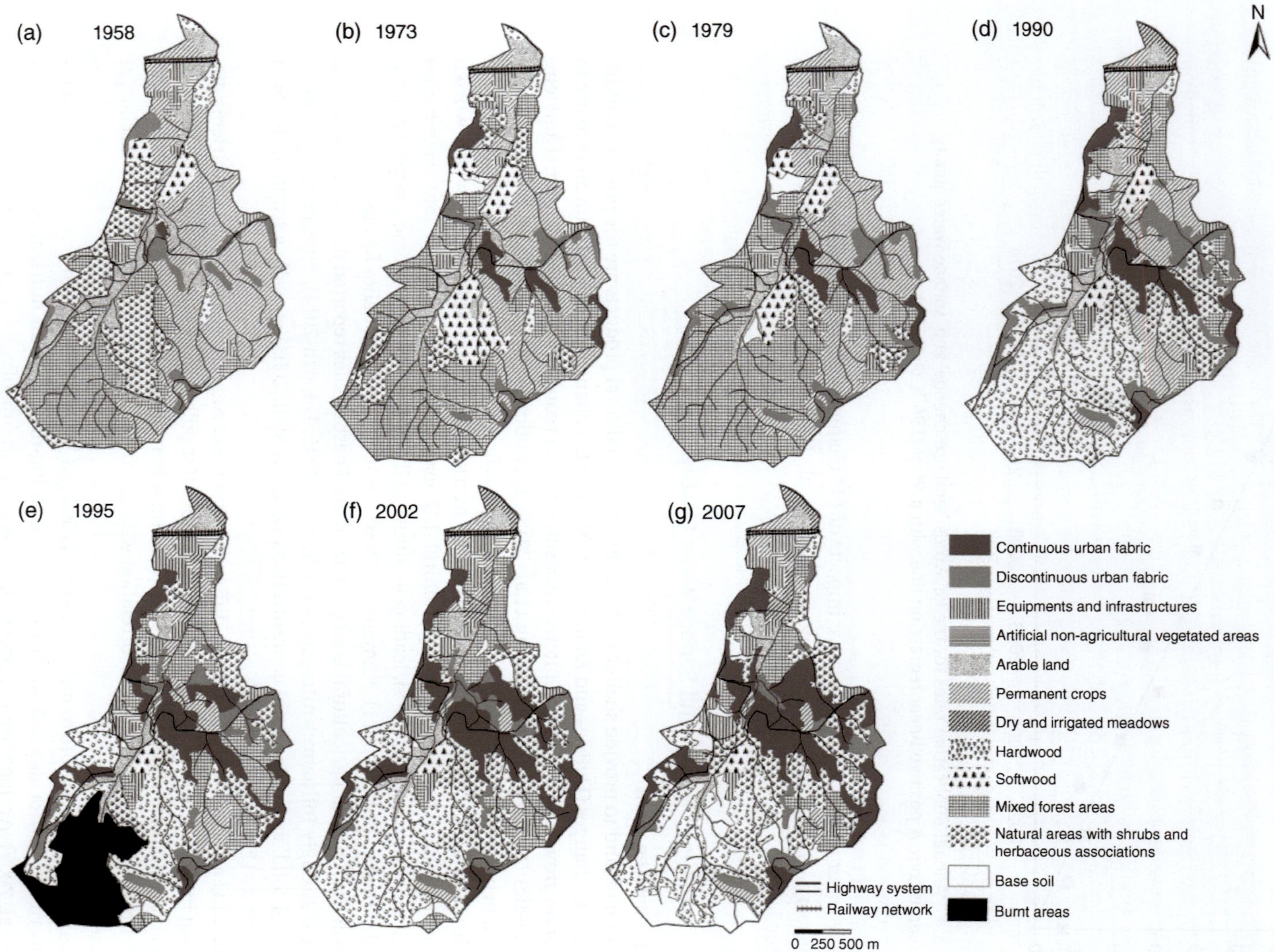

Figure 1.7 Temporal land use map changes in central Portugal in (a) 1958; (b) 1973; (c) 1979; (d) 1990; (e) 1995; (f) 2002 and (g) 2007 (Tavares et al. 2012). A progressive increase in urban areas becomes apparent.

volume and lower percolation (Nie et al. 2011). Increased runoff washes out nutrients from surfaces, eventually entering a stream. The absence of vegetation often plays a central role in soil erosion and wash-out of nutrients (Li et al. 2008). Ghaffari et al. (2010) have highlighted the significant contribution of increases in urban areas to the water quality deterioration and the alteration of flow regimes. They asserted that LU changes from 1967 to 2007 led to a 33% increase in urban runoff volume and a 22% decrease in groundwater recharge. Runoff from urbanised areas carries pollutants such as increased loads of nutrients and solids (Biggs et al. 2004).

Researchers reported that between 66% and 75% of deforestation leads to noticeable signals in nutrient concentrations in surface waterbodies. Washing-out the leaves originating from urban trees by runoff was identified as the main source of total nutrients (Meyer et al. 2005). They found a statistically significant correlation between urban density and leaf breakdown rate ($R = 0.91$). Increases in urban density from 10% to 30% would increase the leaf breakdown rate by threefold. Walsh et al. (2005) pointed out the ecological impacts on SWQ. They underlined that urbanity enhances surface water nutrient and toxicant concentrations as well as changes surface water hydrology. Larger peak flow, shorter lag time and a disturbing frequency of overland flow are the influences of urbanity on SWQ. Figure 1.8 exhibits the difference between natural and urban land water balance (British Precast Concrete Federation 2003). Runoff makes the greatest contribution in water balance in urban LU (55%). However, just 10% of water is attributed to surface runoff in rural LU.

Many soils comprise clay and organic matter. Clay having a negative charge can immobilise compounds with positive charge such as potentially toxic metals (Martins et al. 2015). The soil organic fraction adsorbs organic compounds into runoff passing over soil (Ahmed et al. 2015). Thus, natural treatment of diffuse pollutants plays an important role in water quality management of natural lands. Meyer et al. (2005) calculated the soluble reactive phosphorus and ammonium cation uptake rates as a function of urban LU density. They found statistically significant negative correlations between both of them. Their results revealed that a 10% increase in urban density may half soluble reactive phosphorus and ammonium cation uptake rates.

Although SWQ field measurements are the most reliable data, they are sometimes costly and hard to perform (Jajarmizadeh et al. 2012). Computer modelling aids field measurements to generate a field-validated model that can predict the fate of nutrients under different conditions after calibration and validation (Panagopoulos et al. 2011). Developing a computer model needs a

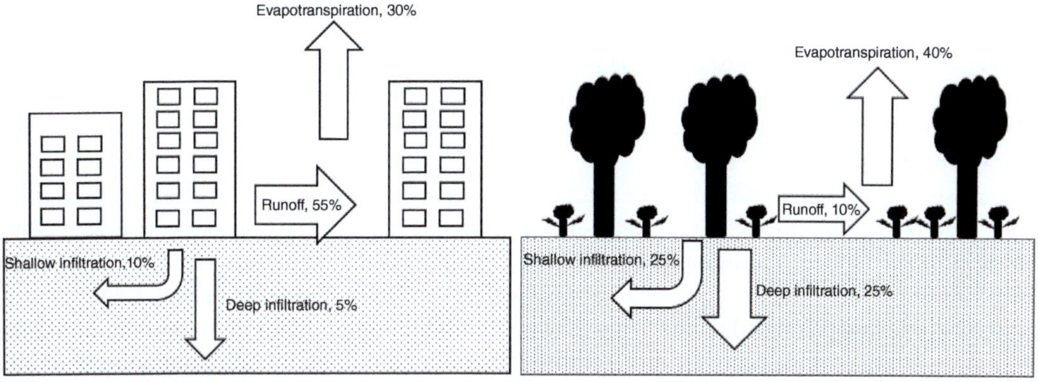

Figure 1.8 Impacts of land use on the natural drainage cycle (British Precast Concrete Federation 2003). An urban and a natural landscape as well as corresponding water flows are shown to indicate the influence of soil sealing. *Source:* John Wiley & Sons/CC BY 4.0.

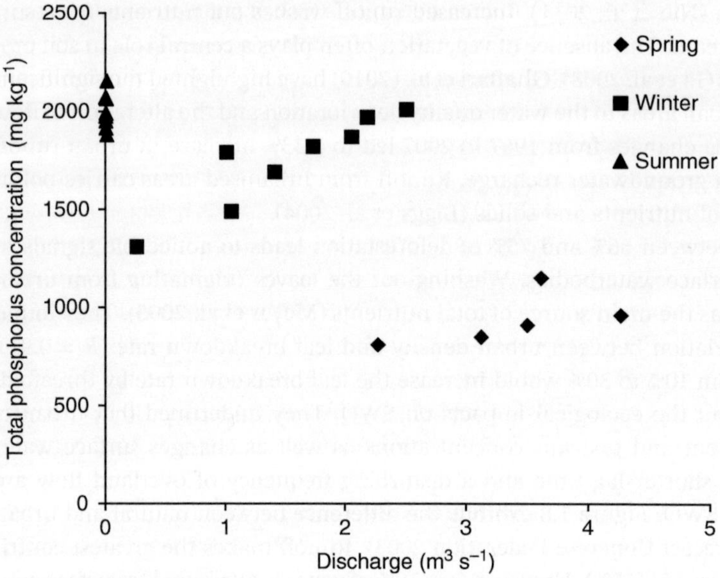

Figure 1.9 Total phosphorus content in soil around (less than 20 m) a river in Iran (Delkash et al. 2014). Three distinctive seasonal clusters (spring, summer and winter) are shown. *Source:* Delkash et al. 2014/Springer Nature.

comprehensive understanding about on-going processes affecting the nutrients' fate, which might be difficult and complex (Abbaspour et al. 2007). First, the researcher needs to identify the active processes and sources of nutrient transport in a watershed as shown in Figure 1.9 (Delkash et al. 2014). Second, someone can predict the impacts of change in the watershed. Third, authorities and watershed managers can assess water quality promotion strategies for watersheds. Finally, using field-validated computer modelling tools reduce the need for field measurements under difficult conditions and at high expense (Zhang et al. 2013).

The target of LU-SWQ models is to make connections between LU maps and SWQ. These models are considered a strong tool for managers to study the impacts of LU on SWQ and find the best practice to reduce nutrient loads. Several models have been suggested and refined to achieve this goal (Jajarmizadeh et al. 2012). Mathematical models utilise simulated processes (representing natural processes) to model the hydrology of watersheds. Higher speed and larger calculation volumes, clearer chronological sets of relations as well as logical steps between input and output make the mathematical models promising for simulating complicated processes in the future (Frevert and Singh 2006). Mathematical models are divided into two different groups: physical-based and empirical approaches (Kamphuis 1982).

Because of the limited current understanding regarding hydrological processes governing the field, it is hard to develop a fully physical-based mathematical model. Common procedure using physical-based models is initiation by a sensitivity analysis to recognise sensitive parameters and understand the dominant processes among several probable processes during model calibration using one independent data set. Then another independent data set is used to validate the calibrated model. A validated model can be utilised to predict the different LU planning impacts on SWQ (Frevert and Singh 2006).

Tu (2009) developed a geographical information system-based watershed model simulator to predict the condition of nitrogen loads and water quantity for eastern Massachusetts, USA. The better assessment science integrating (BASIN) function is included in this type of simulator and has been tested to predict nutrient loads in Ohio State (Tong and Chen 2002). Kling et al. (2014) developed a model called LUMINATE to assess impacts of agricultural LU on water quality with decision-making features. Zhang et al. (2013) coupled the soil and water assessment tool (SWAT) and the conversion of LU and its effect at small regional extent (CLUE-S) models to predict the future LU impacts on nitrogen and phosphorus loads in China. They reported that agricultural activities increase nitrogen and phosphorus loads by 14 and 10%, respectively. SWAT is a semi-continuous model in space (some parameters fed to the model as distributed parameters while some others have one value for the whole of the watershed) and continuous in time simulating nutrient, pesticide, pathogen and hydrological responses to LU changes according to water mass balance equations (Arnold et al. 1994).

Despite being physical-based models, empirical models usually require fewer input data to establish empirical relationships between LU and SWQ. The export coefficient (E_k) approach is a popular approach in this model category (Worral and Burt 1999), and E_k are derived by monitoring station measurements of discharge and nutrient loads to represent the effluent mass under specific climatic and LU conditions. In addition to LU, meteorological parameters, soil type and topography may affect the results in this approach. Steady and uniform distributions of parameters are among major assumptions. Generally, this method tries to find the load of pollutants based on Eq. (1.1).

$$L = \int_0^T QC \, dt \tag{1.2}$$

where L is the pollution load, Q is the river discharge, C is the concentration of pollutant and T is the time of interval.

But in general, the discharge and concentration data are not continuous functions in time. Therefore, the integral in Eq. (1.3) should be converted to summation. Equation (1.3) can be used to determine the role of each LU item in the pollution of a watershed (Hodge and Armstrong 1993) and is usually solved by applying a multiple linear regression method.

$$L = PS + \sum_{i=1}^{n} \sum_{k=1}^{m} A_i E_k \tag{1.3}$$

where L is the expected output load, A_i is the area of each LU category, E_k is the export coefficient, m is the total number of area and n is the number of LU classes.

This method was utilised to study the nutrient loads discharged into Fuji River in Japan. It was reported that forest LU with 62% and agricultural LU with 20% of the TP loads have the most significant contributions in phosphorus pollution of the watershed (Shrestha et al. 2008). In another investigation, the phosphorus pollution in the south plain of England was studied by E_k. It was noted that domestic wastewater has the greatest share in this area and 80% of the treatment would lead to a 52% reduction in TP load (Bowes et al. 2005). Some results of phosphorus E_k reported in the literature are given in Table 1.5 (Ierodiaconou et al. 2005; Johnes 1996; May et al. 2001).

Multiple (Maillard and Santos 2008) and bivariate (Eshleman et al. 2009) linear regression models can be used to achieve good correlations between water quality parameters and LU with high level of accuracies. Stepwise multiple linear regression models, which select the most effective LU parameters, were also developed to achieve this goal (Ren et al. 2003). Pearson correlation and analysis of variance were applied to assess the variation of water quality with different LU (Li et al. 2008, 2009). The exploratory spatial data analysis and the geographically weighted regression methods were used to express changes of pollutants in any LU spatially (Tu 2011).

Table 1.5 Phosphorus export coefficients values for several land usages.

Land cover classification	Phosphorus export coefficient	Reference
Agriculture	0.42	Abbaspour et al. (2007)
Irrigated pasture	5.8	Ierodiaconou et al. (2005)
Land with coniferous cover	0.02	Johnes (1996)
Market garden	7	Ierodiaconou et al. (2005)
Native vegetation	0.015	Ierodiaconou et al. (2005)
Orchard	0.02	Johnes (1996)
Urban	1.4	Ierodiaconou et al. (2005)
Urban	0.83	May et al. (2001)
Urban	1.73	Shrestha et al. (2008)

Generally, physical-based models have better accuracies than empirical ones. However, they need larger input datasets (Frevert and Singh 2006). A comparison of SWAT (physical-based model) and E_k approaches revealed that the application of SWAT for hydrological models seems more promising due to lower errors. It has been reported that SWAT and E_k methods predict phosphorus loads with about 7% and between 9% and 33% error, respectively, in a watershed in Iran (Delkash et al. 2014).

1.3.2.2 Key Variables

LU, meteorological parameters, landscape features and pollutant sources play a central role in determining nutrient concentration levels in waterbodies. Changes in LU affect water balances, paths and flow rates in watersheds (Somura et al. 2012). Because evapotranspiration and infiltration rates vary with LU changes (Neris et al. 2012), LU maps help in understanding flow paths and infiltration rates. Neglecting a nutrient source may lead to lack of knowledge of the full cycle of the nutrients in the examined watershed (Bowes et al. 2005). Heavy rainfall may transfer the nutrients over a long distance. Furthermore, landscape features such as slope, soil texture as well as shape and size of a watershed influence the impacts of LU on SWQ.

LU maps are necessary to study the impacts of LU changes on water quality (Singh et al. 2013). These maps may offer detailed information (layers) for each type of land cover and each water quality parameter measured in appropriate time steps. Updating of LU maps showing the intra- and inter-LU temporal and spatial variations is essential for promising predictions of water quality (Tu 2011). There are several methods to incorporate remote images with water quality results such as satellite images (Moderate-resolution Imaging Spectroradiometer, Landsat and Satellite for Observation of Earth) and aerial photographs (Singh et al. 2013).

Many LU-SWQ relationships have been successfully modelled using satellite imagery based on LU and land cover data on various spatial scales (Laurent and Ruelland 2011). There are two main approaches: The first one is referred to as 'one image' (i.e., one time step), which is the most used method to assign the impacts of different LU to water quality (Stanley and Maxted 2008). The second approach is based on successfully identifying a relationship between differences of two images and the changes in concentrations to assess the effect of LU variations on water quality (Schilling and Spooner 2006). Obtaining raster images from satellites is the first step of this type of study, followed by characterising LU, which can be done using several parameters such as the normalised difference vegetation index (Ma et al. 2011), forest disturbance (Buma 2012) and arranging LU (Lopez et al. 2008).

Heavy rainfall can lead to the washing-out of nutrients from the ground. Therefore, the time during which water samples are collected may significantly affect the correlation between LU and water quality parameters. The effect of LU on water quality changes according to season (Jha et al. 2013). Some researchers suggest that measurements should be undertaken during storm periods when different runoff paths join (Brown and Clasen 2012). Runoff scrubbing nitrate originating from manure and fertilisers is dominant during the wet season, while PS pollution from urban LU is dominant during the dry season (Smith et al. 2013). Measuring the phosphorus content in the soil near Kan River, Iran, seasonal variations were observed as shown in Figure 1.10 (Delkash et al. 2014). Positive correlations between soil phosphorus content and river discharge are the highest in autumn and winter, when the corresponding river discharge is relatively high. The author associated this positive correlation with soil erosion occurring by runoff and accumulation of phosphorus around the river.

The amount of nitrogen in waterbodies is directly related to the volume of urban stream flow (Tu 2009). Rainfall may lead to runoff that washes out nutrients from soil that subsequently enters waterbodies (Wu et al. 2012). In a large Belgian catchment (114 km^2), nitrate concentrations had two annual peaks: one during winter storms (51% total nitrate load) and the other when groundwater flow joins surface waters, which represents 28% of the total load (Van Herpe and Troch 2000). Comparing urban and forest LU revealed that nutrient measurements have a peak in urban LU that

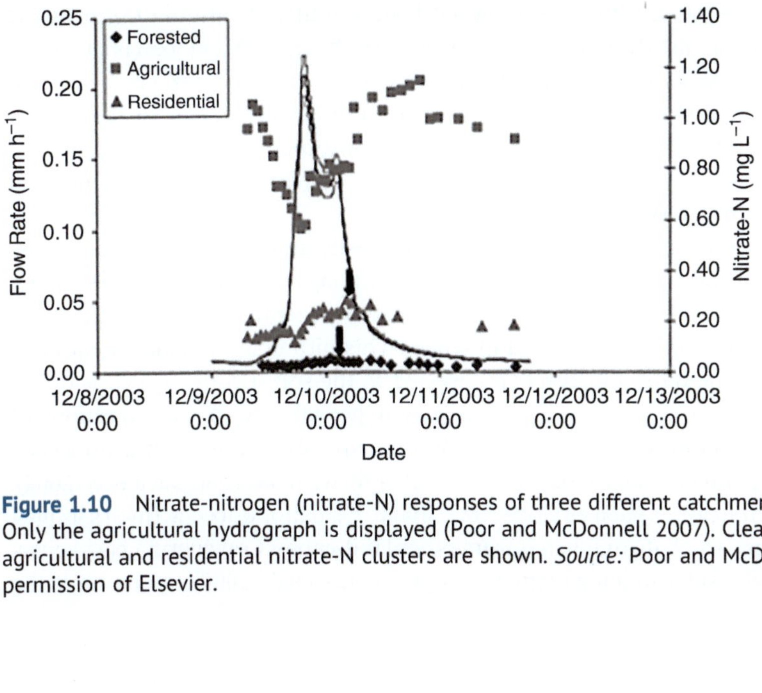

Figure 1.10 Nitrate-nitrogen (nitrate-N) responses of three different catchments to a storm in 2003. Only the agricultural hydrograph is displayed (Poor and McDonnell 2007). Clearly separated forested, agricultural and residential nitrate-N clusters are shown. *Source:* Poor and McDonnell 2007/with permission of Elsevier.

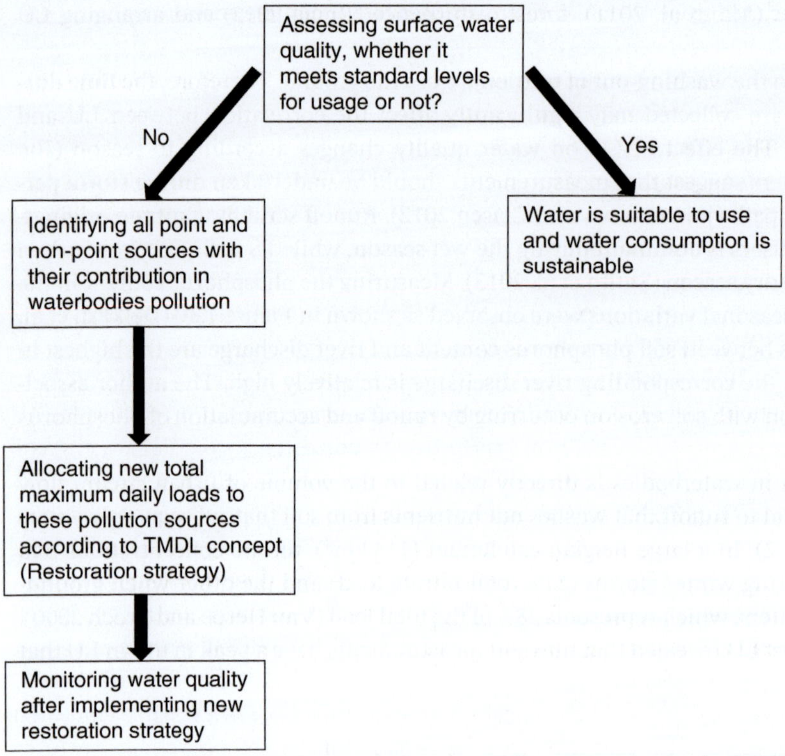

Figure 1.11 Schematic procedure for the total maximum daily load. A simple yes–no flow chart is shown.

is related to high water flows (Delkash et al. 2014). However, nutrient loads can follow an almost constant trend in both urban and forest LU (Poor and McDonnell 2007); e.g., Figure 1.11 shows the nitrate responses of three different catchments to a storm in 2003.

Most of the nutrient loads are usually discharged to waterbodies from agricultural LU during the initial part of a storm event (Poor and McDonnell 2007). This initial but significant load is called the first (foul) flush (Scholz 2010). As shown in Figure 1.10, Poor and McDonnell (2007) revealed that forest LU has the lowest nitrate export during storm periods. In contrast, urban LU has the highest export during all storms; it can be referred to as a PS load due to anthropogenic activities. Agricultural LU has higher exports during storms after long dry periods (first flushes) and the load decreases during wet periods progressively.

Details such as type and size of agricultural activity, crop intensity and pattern, soil type, meteorological data such as rainfall and temperature, irrigation and drainage system are required to predict water quality accurately (Nash et al. 2015). Akhavan et al. (2010) reported that higher nitrate leachate occurred during potato rotation compared to wheat. This nitrate leachate for potato cultivation was about 30–42% of the total nitrogen (TN) applied to the soil. Upstream LU, width of catchment, fertiliser application and number of stocks play a pivotal role in concentrations of higher-order streams, while lower-order streams (i.e., local ones) are affected by other factors such as location of stock crossing, point of access to stream and level of riparian damage (Buck et al. 2004).

Each landscape has its own features such as slope, size, shape, human-induced activities as well as number and order of rivers, which distinguish it from other landscapes. Considering basin characteristics can improve model results, making them more reliable. Lee et al. (2009) reported that the

Table 1.6 Non-point sources (NPS) contribution at different slopes for various references.

Zhang et al. (2000)		Zhang et al. (2003)		Huang et al. (2003)	
Slope (%)	**NPS load (kg)**	**Slope (%)**	**NPS load (kg)**	**Slope (%)**	**NPS load (kg)**
5	0.43	6	0.51	5	0.47
10	0.84	12	0.99	10	0.78
13.67	1	13.67	1	13.67	1
20	1.09	18	1.02	—	—
25	1.6	24	1.06	—	—

responses of TN and total phosphorus (TP) are in accord with the topographical properties of LU. Impacts of LU on water quantity (surface runoff) have been studied for a range of watersheds between 1 and 73 km^2 in size located in the Loess Plateau of China (Zhang et al. 2010). They found that LU changes have similar impacts in conversion of precipitation to runoff in large watersheds, regardless of precipitation variations. While LU changes in smaller watersheds have fewer impacts on runoff generations during higher precipitations. Impacts of landscape slope in NPS discharging loads into waterbodies have been given in Table 1.6 (Huang et al. 2003; Zhang et al. 2000, 2003), which indicates that the slope of a watershed plays an important role in nutrient release rates in a watershed.

The LU impacts on SWQ vary spatially and chronically with landscape scale in a watershed located in Japan (Buck et al. 2004). It was noted that two first-order stream zones with some similar features such as vegetation, topography and size had significant differences in nutrient levels; nitrate, ammonia and TP were 20, 10 and 2 times larger in one of them, respectively. The author related this difference to the number of stocks that were grazing during the measurements. Lee et al. (2009) pointed out that spatial pattern metrics support better the prediction of water quality as a function of surrounding LU. They figured out that edge and patch densities and the nearest neighbour are the most effective metrics in TP and TN predictions in a watershed located in South Korea.

A mass balance equation is concerned with the conservation of mass considering accumulation, influent, effluent, sink and source of a pollutant in a controlled volume (Chapra 2008). Any LU leaves its own impacts on a watershed, which comprises inflow, outflow, several characteristic reaches and potentially several tributaries. Measured values of any contaminant at a certain site are not necessarily indicators of the role of the associated LU. However, these measurements show the total nutrient concentrations of target pollutants at this site. Ignoring potential sources of contamination might lead to a significant deviation from reality (Chapra 2008; Huang et al. 2003; Zhang et al. 2000, 2003, 2010).

A nutrient load may originate from the upstream watershed of a basin, entering the basin as part of the inflow. It was reported in Indiana, USA, that 19% of the non-urban LU conversion to urban area caused 17% and 13% reductions in nitrogen and phosphorus loads, respectively (Bhaduri et al. 2000). This result was mainly related to a sparse presence of agricultural LU (just 2.7%) in urban LU neighbourhoods. Studying LU-SWQ in the USA with a weighted regression technique, it was found that recreation LU in low-density urban areas has a positive relationship with dissolved nutrient runoff loads, while this LU somehow dilutes nutrients when it is adjacent to dense urban LU (Tu 2011). These studies highlighted the importance of adjacent LU and mass balance. Different pathways such as atmospheric deposition (Huston et al. 2009) and groundwater interactions with surface water have been assessed (Panagopoulos et al. 2007).

Neglecting wastewater treatment plant effluents often decrease the goodness of correlation between the LU category urban area and water quality parameters (Garnier et al. 2013). For instance, 36% of TP load is associated with domestic wastewater released into a river in Iran (Shrestha et al. 2008). Therefore, SWQ investigations should always consider pollutant mass balances in watersheds or control volumes (Qaiser et al. 2011).

1.3.2.3 Modelling Outcomes

Natural LU such as a forest functions often as a pollution sink. Zhang et al. (2013) reported that LU conversion from forest to orchards caused 5% and 4% reductions in nitrogen and phosphorus loads, respectively. Statistically significant negative correlations found in Georgia (USA) between forest LU and TN (−0.43; $p < 0.007$), organic nitrogen (−0.45; $p < 0.005$), nitrate (−0.4; $p < 0.018$) and phosphorus (−0.47; $p < 0.002$) support this observation (Tu 2013). Nutrients in natural LU such as forest are usually part of suspended solids (Goonetilleke et al. 2005).

Agricultural activities may lead to the discharge of nutrients into surface waterbodies and subsequently stimulate eutrophication of waterbodies such as lakes and rivers (Poor and McDonnell 2007). These nutrient sources include but are not limited to the following sources: applying fertilisers, insecticides and herbicides on farmlands and residential areas; sediment from improperly managed construction sites, crop and forest lands and eroding stream banks; salt from irrigation practices; deposition of pollutants in atmosphere and hydro-modifications. Nitrate is the major element in agricultural runoff (Somura et al. 2012). The application of fertiliser to agricultural land has a statistically significantly positive correlation with the corresponding nitrogen export load (Akhavan et al. 2010). In the Han River basin, China, weak Pearson correlation coefficients were found between agricultural LU and ammonia (0.21) and nitrate (0.22) (Li et al. 2008). The authors reported a moderately positive correlation between agricultural LU and dissolved phosphorus (0.45). Phosphorus is usually related to NPS pollution due to agricultural activities.

During a study in Australia, a principal component analysis showed that TN and TP had weak correlations with suspended solids, which renders that nitrogen and phosphorus are present in dissolved forms in urban and rural areas (Goonetilleke et al. 2005). However, an increase in the proportion of urban area improves this correlation. A significant correlation ($p < 0.05$) was found between dissolved nutrients and urbanised area (Tu 2013). In a study in China, low Pearson correlations figures were found between both nitrate and ammonia with urban LU, which were 0.25 and 0.14, respectively, inferring that there is no correlation between urban LU and these nutrients (Li et al. 2008). On the other hand, the authors reported a statistically significantly correlation between dissolved phosphorus and urban LU (0.80). Total phosphorus has a linear relationship with the level of urbanisation (Yin et al. 2011). It was noted that urban LU in 1973 contributed about 9% and 18% in TN and TP loads, respectively, while these contributions in 1991 enhanced to 20% and 36% for TN and TP in this order. The results of urban LU correlation with different nutrient forms reveal that urban LU increases nutrient loads (Tu 2013). It was reported that urban LU has a positive correlation with TN (0.49), organic nitrogen (0.52), nitrate (0.53) and phosphorus (0.56).

Some of the results of LU-SWQ are shown in Table 1.7 (Yin et al. 2005), which displays some empirical equations derived between LU and nutrients as well as correlations between different LU changes and nutrient loads. Table 1.7 indicates the debilitating impacts of anthropogenic LU changes on SWQ.

Despite some common relationships between LU and SWQ, there are some contradictory findings in terms of correlations between LU and nutrient loads released into surface waterbodies in the literature. These controversial results given in Table 1.8 might be due to neglecting the key variables discussed in Chapter 1.3.2.2.

Table 1.7 Summary of investigations about impacts of land use (LU) changes on nutrient concentrations in soil water.

LU	Study period	Area (km^2)	Pollutant	Result	Reference
VEG[a], AGR[b] and URB[c]	2005–2006	95 200	Dissolved phosphorus	Higher dissolved phosphorus during the dry season $= 0.001 + 0.047 \times$ URB[c] (km^2); $R^2 = 0.6$	Li et al. (2008)
			Nitrate	Higher nitrate loading during the rainy season $= 1 + 0.15 \times$ Bareland (km^2); $R^2 = 0.75$	
FRS[d], AGR[b]	1995–2004	63.4–104.9	Nitrogen	Significant correlation between nitrogen load and stream flow for seven watersheds ($R^2 > 0.85$)	Tu (2009)
URB[c], FRS[d] and AGR[b]	1990–1998	5840	Total nitrogen	Spearman correlations for URB[c] $= 0.23$ and AGR[b] $= 0.19$	Tong and Chen (2002)
			Total phosphorus	Spearman correlations for URB[c] $= 0.34$ and AGR[b] $= 0.16$	
URB[c] and AGR[b]	1997–2007	156 141	Total nitrogen	Monthly loads based on monthly runoff depth $= 0.205 \times e^{0.0054 * MRD}$; $R^2 = 0.95$	Wu et al. (2012)
			Total phosphorus	Monthly loads based on monthly runoff depth $= 0.185 \times e^{0.0067 * MRD}$; $R^2 = 0.96$	
URB[c]	1995	6400	Total nitrogen	Concentration based on buffer size ($B = 2000$ m) $= 2.321 \times e^{2.173 * B}$	Yin et al. (2005)
			Total phosphorus	Concentration based on buffer size ($B = 2000$ m) and river rank (R) $= -2.69 + 1.43 \times B + 0.211 \times R$	
FRS[d], URB[c] and AGR[b]	2001–2002	1010–1309	Nitrate	Correlation with URB[c] for the Shibetsu area in Japan = 0.51, and correlation with URB[c] for the Akkeshi area, Japan = 0.50	Woli et al. (2004)
URB[c], GRS[e], WTL[f], FRS[d] and AGR[b]	2002	~27	Total nitrogen	Pearson correlations in autumn: URB[c] $= 0.19$ and AGR[b] $= -0.03$	Lee et al. (2009)
			Total phosphorus	Pearson correlations in autumn: AGR[b] $= 0.28$ and AGR[b] $= 0.10$	

[a] Vegetation.
[b] Agriculture.
[c] Urban.
[d] Forest.
[e] Grassland.
[f] Wetland.

1.3.3 Proposals for Water Quality Conservation

Water quality treatment and conservation measures including wastewater treatment (Wu et al. 2012), development of urban instead of rural and barren areas (Tu 2011), applying pollutant adsorbents such as zeolite to field (Delkash et al. 2015), fertiliser management (Panagopoulos et al. 2011), crop rotation in agricultural LU (Panagopoulos et al. 2007), optimisation of the fertilising time (Woli et al. 2004) and construction of artificial wetlands (Goonetilleke et al. 2005) have been proposed.

Table 1.8 Summary of contradictory results about land use (LU) and surface water quality (SWQ) relations.

Water contamination/LU	Model	Statement	Reference
Nutrients/agricultural LU	BASIN	Positive correlation	Tong and Chen (2002)
	Statistical analysis	No correlation	Williams et al. (2005)
Phosphorus/agricultural LU	Geographical weighted Regression	No correlation	Tu (2011)
	SWAT, export coefficient Model	Positive correlation	Delkash et al. (2014)
Nitrogen/agricultural LU	Pearson correlation	No correlation	Lee et al. (2009)
	SWAT, CLUE-S	Positive correlation	Zhang et al. (2013)
Nitrate/urban LU	Geographical weighted Regression	Negative correlation	Tu (2011)
	Regression analysis	Positive correlation	Woli et al. (2004)
TP/residential LU	Linear regression	No correlation	Dauer et al. (2000)
	Bivariate regression	Positive correlation	Yin et al. (2011)

LU changes close to a stream can have major effects on water quality (Dauer et al. 2000; Delkash et al. 2015; Williams et al. 2005; Zampella et al. 2007). Riparian vegetation and buffer zones reduce the nutrient loads, sorption and denitrification as well as control sediment transport (Zhang et al. 2010). Riparian vegetation is important for stream water quality (Smart et al. 2001). However, there is no agreement about the optimum wetland dimensions and distances to the stream (Sliva and Williams 2001). A buffer zone to control diffuse pollution is defined as a permanent green zone close to a waterbody and maintained separately from other land. Buffer zones are promising for keeping nutrients such as phosphorus and nitrogen in the soil and can capture pollutants travelling through air (Reichenberger et al. 2007). Vegetation buffer zones can reduce nitrate loads by 30% (Sahu and Gu 2009).

Low-impact development practices are utilised to mitigate reverse impacts of developing urbanisation and its consequences on water quality. They include constructed wetlands (Kovacic et al. 2006), permeable pavements (Tota-Maharaj and Scholz 2010) and grassed swales (Deletic and Fletcher 2006).

The total maximum daily load (TMDL) allocates the acceptable value of discharge for all sources to meet the essential requirement of water quality (Zhang and Yu 2004). Generally, 5–10% of TMDL is assigned to the margin of safety (MOS) covering some uncertainties in this method. The TMDL procedure is outlined in Figure 1.8 (British Precast Concrete Federation 2003). Firstly, someone compares the current nutrient concentrations to standards to assess SWQ. In case of pollution (higher concentrations than standard criteria), new loads are allocated to both PS and NPS, and the new nutrient concentrations are compared to standard criteria. If they meet the criteria, this maximum daily load is allocated to each source as maximum allowable load.

$$\mathrm{TMDL} = \sum\nolimits_{i} \mathrm{NPS_i} \sum\nolimits_{j} \mathrm{PS_j} + \mathrm{MOS} \tag{1.4}$$

where TMDL is the total maximum daily load and MOS represents the margin of safety.

1.3.4 Conclusions and Outlook

LU changes alter surface runoff, evapotranspiration and infiltration, affecting the fate of nutrients. The author understands that beside anthropogenic nutrient loads, which are released into the waterbodies, developing these types of lands leads to decreasing natural self-purification, which occurs for natural LU. For instance, phosphorus compounds can be attached to clay soils. This natural self-purification process is restricted as the LU changes.

There is general agreement that anthropogenic LU change demotes SWQ by discharging nutrient loads. For instance, agricultural LU releases considerable nutrient loads into waterbodies. Impermeable surfaces are increased during urban LU development which causes increases in surface runoff volume and washing out of nutrients. However, assessments revealed that there is some disagreement among reports regarding impacts of LU on SWQ. These contradictions might be due to different parameters and processes that can affect the results of LU impacts on SWQ. Considering mass balances, meteorological parameters such as precipitation and landscape features including size, shape and slope of watersheds usually play central roles in identifying LU impacts.

Reported statistically significant correlations between nutrient concentrations and anthropogenic LU changes were assessed. There is considerable evidence for positive correlations between urban and agricultural LU on one hand and elevated nitrogen and phosphorus concentrations in corresponding receiving watercourses on the other hand.

Practices that have been recommended to preserve water quality despite negative impacts of LU changes have been reviewed. Water quality treatment and conservation measures including promotion of riparian vegetation, grassed swales and construction of artificial wetlands as buffer zones are most promising.

Sustainable wastewater treatment with integrated constructed wetlands and the sustainable use of fertilisers are the most promising low-cost methods to conserve the water quality of receiving watercourses that are subject to runoff from intensively used agricultural areas. The TMDL program can lead to better watershed management in terms of controlling nutrient loads that are released to waterbodies. In addition, vegetated wet buffer zones around the most sensitive waterbodies are promising passive measures.

1.4 Peatland Response to Climate Change and Water Level Management

1.4.1 Introduction

Peatlands are one of the largest carbon storage areas in the biosphere (Yu 2012), holding a carbon pool equivalent to around half of the current atmospheric carbon pool (Gorham 1991). A high water table and the consequent anoxia along with low soil temperature and low pH are key causes of low decomposition of organic matter and accumulation of carbon in peatlands (Laiho 2006). However, there is a growing concern that peatlands in many regions may change from carbon sinks to carbon sources in the face of global change (Fenner and Freeman 2011; Ise et al. 2008; Loisel et al. 2021). This is especially a concern for temperate regions, which contain a significant share of the world's degrading peatlands (that lose carbon), and in addition stand out as being intensively utilised by humans (Leifeld and Menichetti 2018). Thus, whereas most of the global peat carbon stock is in boreal and polar wetlands, the fate of temperate peatland carbon is of relevance in the face of climate change.

Higher temperatures influence the carbon dioxide release from fungal and microbial activity (Asemaninejad et al. 2018) as well as carbon dioxide uptake by plant photosynthesis, which are both mainly thermophilic processes (Frolking et al. 2001; Poorter et al. 2012; Weltzin et al. 2000). Increased temperatures elevate the primary production of plants in peatlands until the point where temperature becomes a stress factor (Weltzin et al. 2000); e.g., up to 27°C in the case of *Sphagnum spp.* (Johansson and Linder 1980). Besides direct effects of temperature, warming can indirectly affect peatlands through alterations in water level and associated oxygen (Buttler et al. 2015; Munir et al. 2015; Laine et al. 2019a). Due to the combined positive effects of warming and oxygenation, a substantial increase in heterotrophic respiration has been observed in peatlands when elevated temperature has been accompanied by water-level drawdown (Laine et al. 2019b; Samson et al. 2018; Zhong et al. 2020).

Elevated temperature along with a more frequent occurrence of extremes such as floods and droughts as well as changes in nutrient regime may shift plant composition (Dieleman et al. 2015; Walker et al. 2006). Change in cover and composition of dominant plant functional groups (ericoid [*Erica*-like] dwarf-shrubs, graminoids [herbaceous s with a grass-like morphology]) and bryophytes (moisture-loving non-vascular land plants) in peatlands has consequences for the carbon dioxide flux (Kuiper et al. 2014; Ward et al. 2009; Zhao et al. 2017). Lower *Sphagnum* productivity can be expected because of climate-change-induced water stress (Bragazza et al. 2016; Kuiper et al. 2014). Furthermore, global warming will increase vascular plant productivity (graminoids and ericoid shrubs) (Ward et al. 2009), thereby maintaining the peatland carbon dioxide sink role. However, it is not clear whether the expansion of vascular plants producing more degradable litter can preserve the peatland sink role in the long term (Del Giudice and Lindo 2017; Robroek et al. 2016).

Water-level management has been regarded as an important climate change mitigation strategy as it can regulate net carbon dioxide (CO_2) exchange and sustain peatland natural net CO_2 sink role (Beyer et al. 2021; Leifeld et al. 2011; Salimi et al. 2021a). However, estimates of water-level management efficiency for climate change mitigation do not take into account how future climate change will change the sink function of peatlands (Beyer et al. 2021). In addition, there is a need to understand the effectiveness of water-level management in the face of more frequent and severe climatic extreme events. There is a risk that drought severity can hinder climate change mitigation goals for managed peatlands (Harris et al. 2006). Furthermore, little is known about how restored and intact peatlands respond to the altered temperature and hydrology under future climate change scenarios (Günther et al. 2020; Loisel et al. 2021).

Ultimately, these altered factors will drive a shift in plant functional types, and it is not yet clear how the new plant composition will respond to climate perturbations (Kuiper et al. 2014; Munir et al. 2015; Sulman et al. 2010). Therefore, studying the response of plant communities to climate change along with extreme events in both managed and unmanaged peatlands is critical to understand the trend of net ecosystem CO_2 exchange (Salimi et al. 2021a).

Since 2017, the author and his team have performed continuous experiments with mesocosms; peatland samples obtained from an ombrotrophic (rain-fed) bog in southern Sweden located within climate chambers. In this study, researchers simulated different representative concentration pathway (RCP) future climate scenarios for the experimental bog mesocosms for the first time. The author also simulates the current climate (CC) scenario based on the hydrological years (starting from the first of October) 2017, 2018 and 2019. These climate data were simulated within the experiment in 2018, 2019 and 2020, respectively. It should be noted that 2018 was recorded in Scania (Skåne in Swedish), Sweden, as the warmest and driest summer since 1950. The year 2019 had the second warmest summer, but rather typical precipitation. The author also examined the effect of water-level management in all climate scenarios.

The aim of Chapter 1.4 is to understand the impact of different levels of climate change (current and future RCP climate scenarios including RCP 2.6, RCP 4.5 and RCP 8.5) and water-level management on the carbon dioxide sink function of peatlands. The researchers hypothesise that (i) the sink function of unmanaged mesocosms decreases from the coldest (current) to the warmest (RCP 8.5) climate scenario and that (ii) water-level management enhances the sink function of the system compared to the unmanaged systems under all climate scenarios. An original article by Salimi et al. (2021b) provided the material for this chapter.

1.4.2 Materials and Methods

1.4.2.1 Mesocosm Experiments

The peatland samples were collected from an ombrotrophic bog called Fäjemyr, which is in the province of Scania (latitude of 56°15′N, longitude of 13°33′E and altitude of 140 m). The samples were extracted using spades and shovels from the top layer of the peatland and were placed directly in glass tanks (30 cm in length; 22 cm in width; 24 cm in height) in the field. The peatland mesocosms consist of 20 cm of the top bog vegetation community with some young peat at the bottom.

All mesocosms were representative of the wider sampling site and comprised an approximately equal proportion of dominant plant types in Fäjemyr: dwarf shrubs (*Calluna vulgaris* and *Erica tetralix*), sedges (*Eriophorum vaginatum*) and *Sphagnum* spp. (*S. magellanicum* and *S. rubellum*) (Lund et al. 2012).

The researchers randomly distributed 16 peatland mesocosms into four climate chambers (four mesocosms per chamber). Moreover, the four mesocosms in each climate chamber were split into two managed and two unmanaged systems (two replicates for each group). The author managed the water level for the managed mesocosms and skipped the water-level adjustment for the unmanaged ones (see water-level management section) to understand how water-level adjustment can change the response of peatlands in terms of respiration, net ecosystem exchange (NEE), gross primary production (GPP) and climate scenario (Salimi et al. 2021a).

Four climate chambers were used to simulate four different climate scenarios for peatland mesocosms: one CC scenario as a control and three future potential climate scenarios. The climate chambers allowed for an accurate and dynamic climate scenario simulation for the mesocosms by applying four different climate variables of temperature, precipitation, relative humidity and radiation simultaneously (Salimi et al. 2021a).

1.4.2.2 Climate Scenario Simulations

To derive better and more realistic estimates of CO_2 fluxes (Salimi et al. 2021a), simulations of radiative forcing climate scenarios have been conducted in this study. To create future climate scenarios for the climate chambers, hourly data of 10 different regional climate models (RCM) were collected from the Rossby Centre of the Swedish Meteorological and Hydrological Institute (SMHI) (Table 1.9). The data for the RCM were collocated for different potential future climate scenarios, which are based on different radiative forcing target levels predicted for the future (Anno 2100) and are released by the Intergovernmental Panel on Climate Change (IPCC) (AR5; IPCC 2007).

According to the IPCC's Fifth Assessment Report (AR5; IPCC 2007), there are four different scenarios, which are called RCP2.6, RCP4.5, RCP6 and RCP 8.5 (Van Vuuren et al. 2011). In this project, the data of the RCMs were collected for RCP 2.6, RCP 4.5 and RCP 8.5 (Table 1.9) to simulate low, moderate, and extreme future climate change scenarios in the climate chambers. The data were collected for the domain Scania County located in southern Sweden. The four climate variables temperature, humidity, radiation and precipitation were used to simulate the different scenarios within the climate chambers dynamically (Figure 1.12). Hourly data of these variables for

Table 1.9 List of the regional climate models (RCM), which have been used for the simulation of future scenarios for RCP 2.6, 4.5 and 8.5.

Model	Full name of the RCM model	RCP	Reference
CCCma-CanESM2	Canadian Centre for Climate Modelling and Analysis- Canadian earth system model	4.5 and 8.5	Arora et al. (2011)
CNRM-CERFACS-CNRM-CM5	Centre National de Recherches Météorologiques, France – Centre Europe´en De Recherche et de Formation Avance´e-Component Model 5	4.5 and 8.5	Voldoire et al. (2013)
CSIRO-QCCCE-CSIRO-Mk3-6-0	Commonwealth Scientific and Industrial Research Organisation, Mark 3.6.0	4.5 and 8.5	Jeffrey et al. (2013)
ICHEC-EC-EARTH	Irish Centre for High-end Computing *k* – European Community-Earth-system Model	2.6, 4.5 and 8.5	Hazeleger et al. (2012)
IPSL-IPSL-CM5A-MR	Institut Pierre Simon Laplace Climate Model 5A-Medium Resolution	4.5 and 8.5	Hourdin et al. 2013)
MIROC-MIROC5	Model for Interdisciplinary Research on Climate-	2.6, 4.5 and 8.5	Watanabe et al. (2010)
MOHC-HadGEM2-ES	Met Office Hadley Centre-Hadley Centre Global Environment Model version 2	2.6, 4.5 and 8.5	Collins et al. (2011)
MPI-M-MPI-ESM-LR	Max Planck Institute for Meteorology – Earth System Model – low resolution	2.6, 4.5 and 8.5	Giorgetta et al. (2013)
NCC-NorESM1-M	Norwegian Climate Centre – Norwegian Earth System Model, Intermediate Resolution	2.6, 4.5 and 8.5	Bentsen et al. (2013)
NOAA-GFDL-GFDL-ESM2M	National Oceanic Atmospheric Administration – Geophysical Fluid Dynamics Laboratory – Earth System Model	4.5 and 8.5	Dunne et al. (2012)

the present climate scenario (2016–2019 data from Malmö A station) were downloaded from SMHI (http://opendata-download-metobs.smhi.se/explore/) for climate scenario simulations.

Since there is uncertainty regarding all RCM outputs, no specific RCM was used for creating the climate scenarios; instead, the delta change approach was employed to include all existing RCM to reduce uncertainty. In this method, the difference between the output of each RCM for the last 30 years of the century (2069–2098) and the historical data of the same models for the same number of years (1976–2005) has been estimated. The results were then averaged across all RCM models for all RCP, and, subsequently, the estimated differences (for temperature) and ratios (for precipitation, relative humidity and radiation) were calculated for each month resulting in monthly delta change coefficients. The estimated monthly delta change coefficients were applied to the hourly observation data of Malmö station to obtain the future climate scenario values. The three-hourly values for temperature, relative humidity and radiation were calculated and used for the climate change scenario simulation in the chambers.

To simulate current and future climates, the team used the four climate chambers KK 750 (Pol-Eko-Aparatura Wodzisław Śląski, Poland; https://www.pol-eko.com.pl/model/climatic-chambers-kk/climatic-chamber-kk-750/). These chambers were equipped with a phytotronic system to regulate temperature and humidity. Each chamber was also supplied with a fluorescent

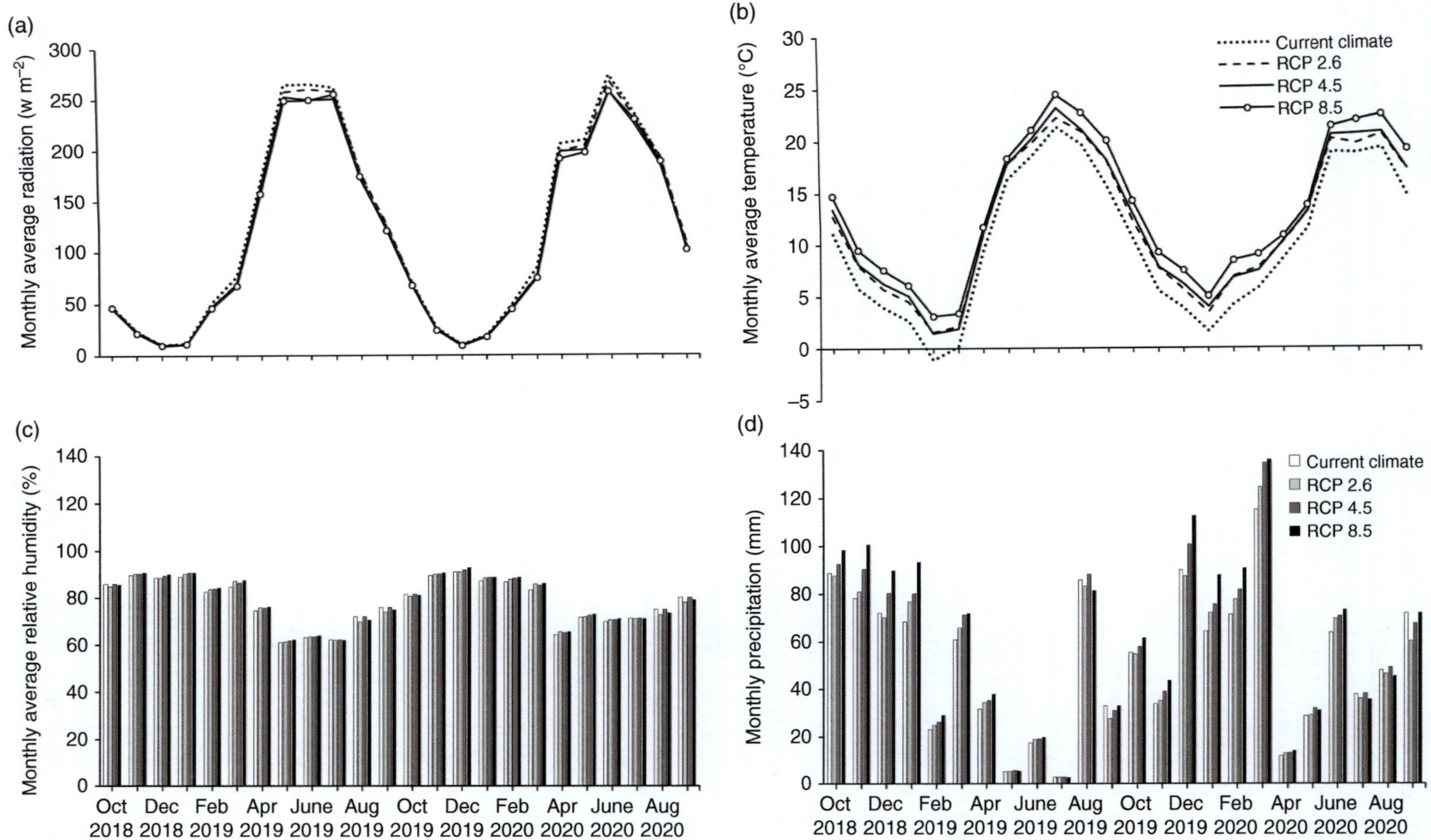

Figure 1.12 Monthly average of simulated climate variables: (a) radiation; (b) temperature; (c) relative humidity and (d) precipitation for different climate scenarios (current climate and representative concentration pathways (RCP) 2.6, RCP 4.5 and RCP 8.5). Trends are visible for temperature and precipitation: A higher RCP is usually linked to higher temperature and rainfall.

lamp 840 (daylight) for day and night simulation. Both illumination intensity and duration were controlled by the chamber as well. The climate chamber regulates the temperature from −10°C to +60°C with the light switched-off, and from 0°C to +50°C with the light on. The chambers are equipped with an ultrasonic humidifier, which must be supplied with deionised water to provide the required humidity in the chambers. There is also an air flap and a ventilator in each chamber to allow for a stable condition in terms of air extraction and circulation. The climate chambers are remotely programmable allowing the user to regulate the temperature and relative humidity, as well as radiation at a three-hour resolution. Precipitation simulation was conducted manually on a weekly basis.

1.4.2.3 Water-level Management

Precipitation was simulated for all mesocosms in all chambers on a weekly basis based on precipitation data from the created climate scenarios (explained above). Rainwater was gathered from nearby greenhouse glass roofs and manually added to the top surface of the mesocosms, where it infiltrated into the peatland soil. The water level in the mesocosms was regulated to prevent adverse effects of mesocosm floods or drought stress on the carbon sink function of the mesocosm. The water levels between 10 (±0.5) and 18 (±0.5) cm (from the bottom of the tank) were considered as acceptable thresholds for mesocosms to provide all types of plants with water for photosynthesis and avoid nutrient loss due to flooding as they can boost the rate of photosynthesis.

Water-level management was carried out either by adding water from the storm water pond called Lake Lake (Sjön Sjön in Swedish), which is located on the campus of the Faculty of Engineering at Lund University, to the mesocosms when the water level dropped to less than 10 cm due to evapotranspiration or by removing excess water (runoff or management outflow) from the mesocosms, when the water level exceeded 18 cm. Water was added for management purposes to the top surface of the mesocosms, while excess water was removed from the vertical standing pipe located in the centre of each mesocosm where in a practical application the outflow of the mesocosm would be collected. For the unmanaged mesocosms, the author and his team skipped the water adjustment procedures. It follows that rainwater simulated based on climate scenario predictions was the only inflow to the unmanaged mesocosms. Therefore, some mesocosms encountered extreme events such as drought and flooding (maximum 4 cm above the topsoil).

1.4.2.4 Flux Measurements

The carbon dioxide fluxes of the peatland mesocosms were measured at a monthly time step inside all climate chambers. The EGM-5 Portable CO_2 Gas Analyzer (PP Systems, Amesbury, Massachusetts, USA) was used to measure the concentration of carbon dioxide in the closed chamber. The analyser was connected to a 36-L transparent Plexiglas static measurement box.

The measurement box was equipped with a circulation fan to homogenise the air inside the box. For each measurement, the measurement box created a closed ecosystem over each mesocosm located in the climate chamber. During all measurements, the following parameters were recorded: mesocosm water level, soil temperature and approximate volume of vegetation in the chamber. To measure the volume of the total air that contributed to the carbon dioxide flux in the chamber, the air-filled porosity of the mesocosm had to be measured.

Carbon dioxide concentrations were measured over time for both day and night to estimate the NEE of the mesocosm (day measurement) and then the ecosystem respiration (night measurement), respectively. The GPP, which is an indication of photosynthetic uptake, can be calculated by subtracting respiration from the NEE. A negative value represents carbon dioxide uptake (carbon dioxide sink) into the ecosystem and a positive value indicates carbon dioxide release to the atmosphere (carbon dioxide source).

The carbon dioxide concentration (ppm) was measured inside the measurement chamber for 180 seconds. The flux of carbon dioxide was calculated using Eq. 1.1 assuming that one kg mol^{-1} of gas (equals 44.01 kg of carbon dioxide) occupies a volume of 22.41 m^3 at standard conditions for a temperature of 0°C and at a pressure of 1013.25 mbar.

$$\text{Flux} = b * \frac{p}{1013.25} * \frac{273}{273 + T_{\text{air}}} * \frac{44.01}{22.41} * \frac{V}{A} * 10^{-2} \tag{1.5}$$

where the Flux is presented in mg $CO_2\,m^{-2}\,s^{-1}$; b indicates the slope (ppm s^{-1}) determined by linear regression of carbon dioxide concentration (ppm) change inside the measurement chamber over time (s); $\frac{p}{1013.25}$ is the correction of barometric pressure with p measured in mbar; $\frac{273}{273 + T_{\text{air}}}$ is the correction for air temperature in the chamber with T_{air} measured in °C; $\frac{44.01\ \text{kg}}{22.41\ \text{m}^3}$ is the molar volume and ideal gas constant at standard temperature and pressure (0°C and 100 kpa); V is the volume (cm^3) of the measurement chamber and A is the inner surface area (cm^2) of the chamber and the remaining term (10^{-2}) is a unit conversion factor.

1.4.2.5 Plant Analysis

The mesocosms were analysed for different plant functional type (*Sphagnum* spp., dwarf shrubs and sedges) coverage at the start of the experiment in July 2017 (July), and the following years including 2018 (August), 2019 (July) and 2020 (September). The unbiased estimates of the areal covers of each plant functional type were analysed. A 15×11 grid with a cell size of 2 cm times 2 cm was superimposed on each mesocosm for subsequent plant species analysis. The presence or absence of each species at each grid intersection was noted. The coverage of each plant functional type was calculated relative to the tank area as a percentage.

1.4.2.6 Statistical Analysis

A one-way ANOVA has been used to find the significant differences between different climate variables under different climate scenarios. The Tukey's test was then used to assess pairwise differences. The Sen's slope estimator (Sen 1968) was applied to the CC data to examine whether there is a significant trend for the climate variables under different climate scenarios over the experimental period.

The author used a mixed-effects model to test the significant effect of different climate scenarios, water-level management as well as their interactions on respiration, NEE and GPP responses over time. Climate scenarios and water-level management were considered as the main effects in this model. The strength of the mixed-effects model is that the term random effect is incorporated into the model. As samples (mesocosms) were randomly collected from the peatland (Fäjemyr) and assigned randomly to four climate scenarios, the replicates (mesocosms) were defined as random effects in the model. Moreover, time expressed as season was involved as a random effect as well (Bolker et al. 2009).

The results of the model were explored to assess if it met the assumptions; i.e., that residuals have a normal distribution and constant variance. For multiple comparison analysis, the Bonferroni adjustment test (Sokal and Rohlf 1995) was conducted as a post hoc analysis to determine the statistically significant differences between the managed and unmanaged systems under different climate scenarios. A two-way ANOVA was used to analyse the effect of different climate scenarios, water-level management and their interactions on the distribution different plant functional types (*Sphagnum* spp., dwarf shrubs and sedges). The Bonferroni adjustment test was used for the multiple comparisons. For all tests in this study, the significance level was set to $p < 0.05$.

1.4.3 Results

1.4.3.1 Climate Variables Under Different Scenarios

Sen's slope estimator results revealed no significant ($p > 0.05$) trend for the annual and monthly time series of all climate variables except radiation, which showed a significant ($p < 0.05$) increasing trend in April over the experimental period. All Sen's slope values for the time series have been shown in Table 1.10. These results are valid for all climate scenarios as the future climate scenarios have been generated from the CC scenario applying the delta change coefficients. Therefore, the trends of all future climate scenarios follow the CC scenario.

A multiple comparison (Tukey's test) between the climate scenarios showed that all climate scenarios were significantly different ($p < 0.05$) from each other in terms of temperature (Figure 1.13b). There was an increasing trend from the CC towards RCP 8.5. The differences between the annual average temperature of the future climate scenarios RCP 2.6, RCP 4.5 and RCP 8.5 relative to the CC (control scenario) were 1.8 ± 0.52°C, 2 ± 0.43°C and 3.2 ± 0.75°C, respectively (Figure 1.12b).

There was neither a significant difference ($p > 0.05$) in relative humidity between the CC scenario and RCP 2.6, nor between RCP 4.5 and RCP 8.5. Nevertheless, both the CC scenario and RCP 2.6 had a significantly ($p < 0.05$) lower relative humidity than RCP 4.5 and RCP 8.5 (Figure 1.13). Although precipitation showed an increasing trend from the colder scenario (the CC) to the warmest climate scenario (RCP 8.5), no significant difference ($p > 0.05$) was identified between the four climate scenarios (Figure 1.13).

Table 1.10 Results of the Sen's slope estimator representing the trend of different climate variables over the period of the experiment. These results are applicable to all climate scenarios.

Time series	Temperature (°C)	Relative humidity (%)	Radiation (w/m^2)	Precipitation (mm)
January	4	0	0	2
February	0	−2	2	2
March	2	−2	−2	4
April	2	−4	6*	−4
May	−2	0	−2	0
June	4	−4	4	−2
July	2	−4	2	−4
August	4	−4	2	−2
September	−2	0	−2	2
October	0	0	2	0
November	4	4	−4	−2
December	0	2	−2	2
Annual	4	−2	2	0

Positive values indicate an increasing trend, whereas negative ones a decreasing trend. Zero indicates no trend. The asterisk (*) indicates a significant trend.

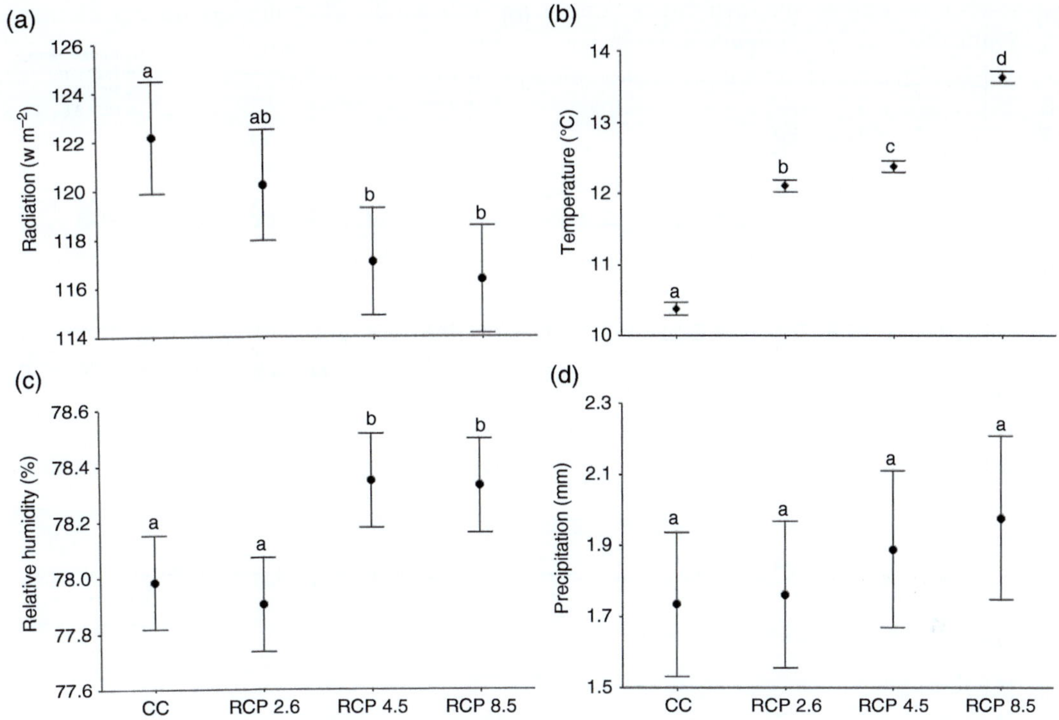

Figure 1.13 Simulated climate variable means: (a) radiation; (b) temperature; (c) relative humidity and (d) precipitation for different climate scenarios (current climate (CC) and the representative concentration pathways (RCP) 2.6, RCP 4.5 and RCP 8.5). Means that do not share a superscripted letter are significantly different at $\alpha = 0.05$ (Tukey's significance test). A higher RCP is linked to both higher temperature and precipitation but lower radiation.

Radiation showed a decreasing trend from the coldest scenario (the CC) to the warmest climate scenario (RCP 8.5) (Figure 1.13a). The CC had significantly ($p < 0.05$) higher radiation than RCP 4.5 and RCP 8.5, but not RCP 2.6. Moreover, there were no significant ($p > 0.05$) differences in radiation between all future climate scenarios (Figure 1.13a).

1.4.3.2 Climate, Water Level and Respiration

The results of the two-year climate change simulation showed no statistically significant effect of climate scenario on NEE or its components (GPP and respiration). The effect of water-level management was significant ($p = 0.013$) on NEE, but not significant for respiration and GPP ($p > 0.05$).

The respiration ranged from 4×10^{-7} to 0.3 mg CO_2 m^{-2} s^{-1}. The lowest seasonal average of respiration (0.0023 mg CO_2 m^{-2} s^{-1}) was noted for winter 2019 for the managed mesocosms under the CC scenario and the highest seasonal average of respiration (0.16 mg CO_2 m^{-2} s^{-1}) was recorded for summer 2019 in the managed mesocosms under RCP 2.6 (Figure 1.14a). The Bonferroni test did not show significant differences between the respiration of managed and unmanaged mesocosms for different climate scenarios (Figure 1.15a). However, the respiration in the managed mesocosms was 1.8 times more than in the unmanaged ones for RCP 8.5 (Figures 1.15a and 1.16a). There was a small difference between the respiration of managed and unmanaged mesocosms under RCP 4.5 (Figures 1.15a and 1.16a). The ratio of managed to unmanaged mesocosm respiration was 0.6 and 0.7 for RCP 2.6 and the CC scenario, respectively (Figures 1.15a and 1.16a).

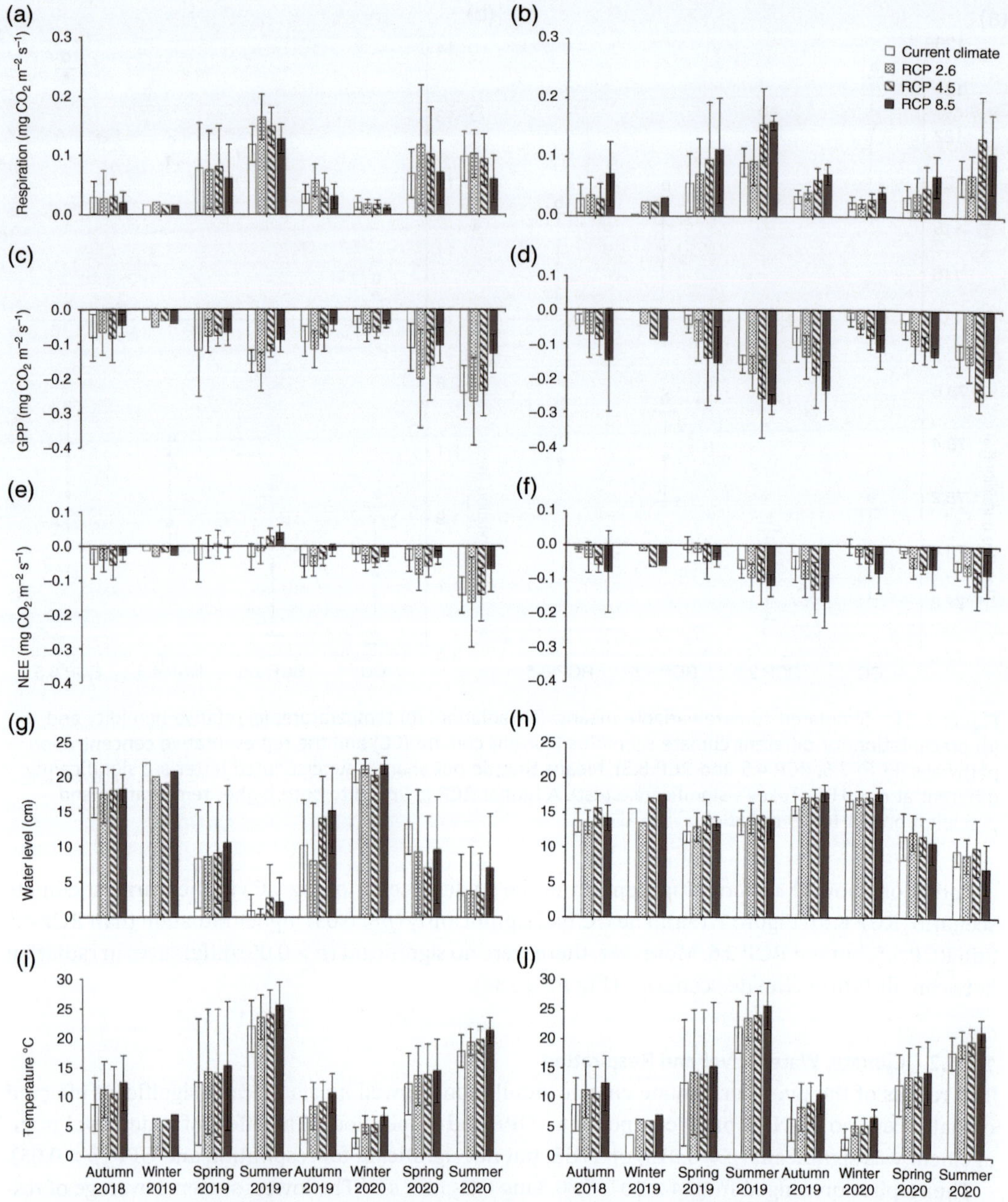

Figure 1.14 Seasonal averages for (a) managed respiration; (b) unmanaged respiration; (c) managed gross primary production (GPP); (d) unmanaged GPP; (e) managed net ecosystem exchange (NEE); (f) unmanaged NEE; (g) managed water level; (h) unmanaged water level; (i) managed temperature and (j) unmanaged temperature of mesocosms under different climate scenarios (current climate (CC) and representative concentration pathways (RCP) 2.6, RCP 4.5 and RCP 8.5) are illustrated over period of experiment from 2019 to 2020. Except for GPP, clear repeating seasonal trends are shown.

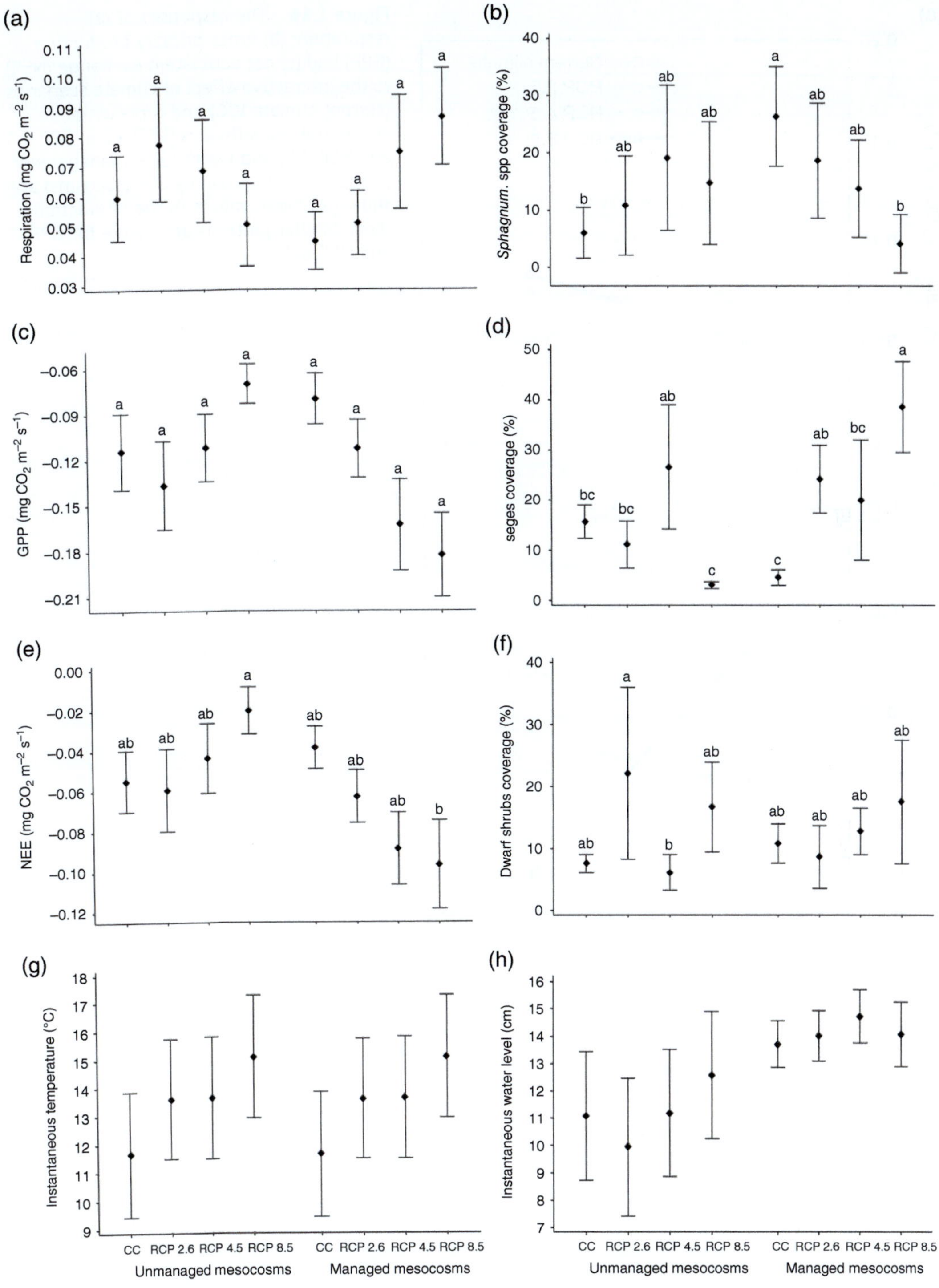

Figure 1.15 (a) Respiration means; (b) Sphagnum spp. coverage means; (c) gross primary production (GPP) means; (d) sedges coverage; (e) net ecosystem exchange (NEE) means; (f) dwarf shrubs coverage; (g) instantaneous temperature and (h) instantaneous water level for managed and unmanaged mesocosms under different climate scenarios (current climate (CC) and representative concentration pathways (RCP) 2.6, RCP 4.5 and RCP 8.5). Means that do not share a superscripted letter are significantly different at $\alpha = 0.05$ (Bonferroni adjusted significance test). Note: The significant difference test was only performed for the responses (a–f) and not for the environmental parameters (g and h). Significant differences can be seen for Sphagnum spp. coverage means, sedges coverage, NEE means and dwarf shrubs coverage.

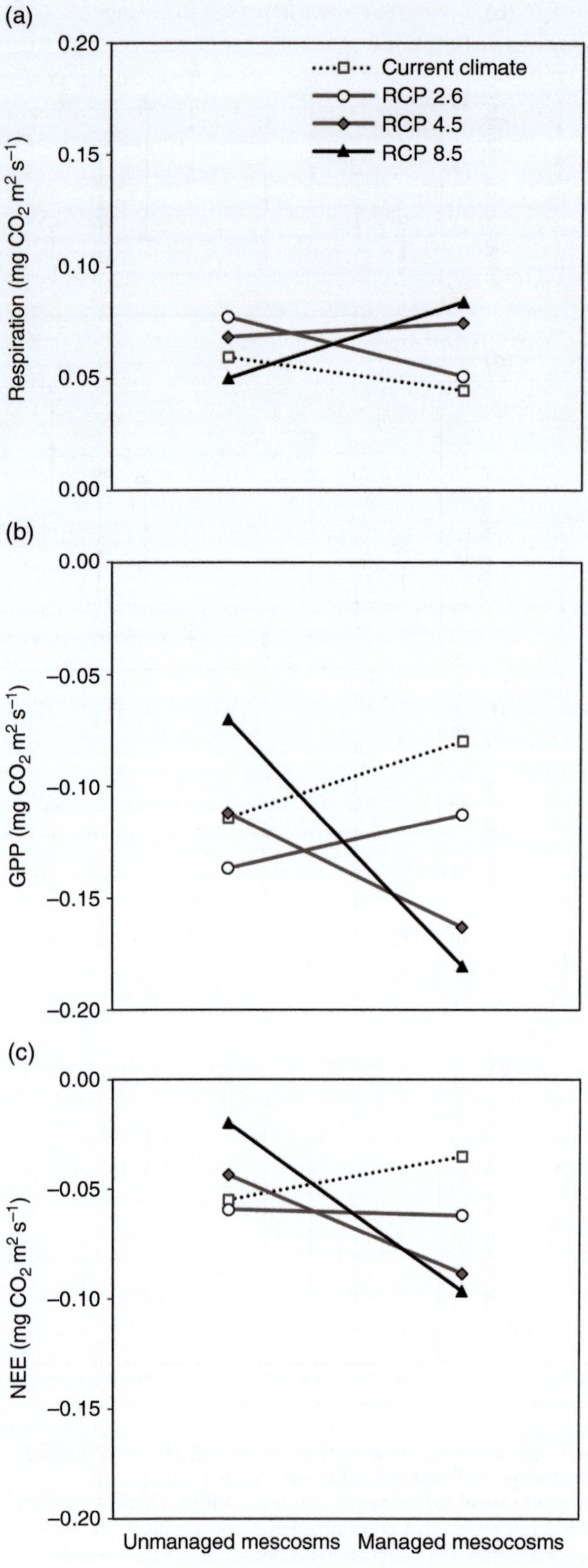

Figure 1.16 The responses of (a) respiration; (b) gross primary production (GPP) and (c) net ecosystem exchange (NEE) to the interactive effect of climate scenarios (current climate [CC] and representative concentration pathways [RCPs] 2.6, RCP 4.5 and RCP 8.5) and water-level management (managed and unmanaged mesocosms) are shown as mean values in the interaction plots. Similar patterns are visible for GPP and NEE only.

The GPP ranged from −0.48 to 0.023 mg CO_2 m^{-2} s^{-1}. The lowest seasonal average of GPP (−0.280 mg CO_2 m^{-2} s^{-1}) was noted for summer 2019 for the managed mesocosms under RCP 8.5, whereas the highest seasonal average of GPP (−0.003 mg CO_2 m^{-2} s^{-1}) was recorded for winter 2019 for the managed mesocosms under the CC scenario (Figure 1.14c, d). No significant (Bonferroni test; $p > 0.05$) differences were found between the GPP of the managed and unmanaged mesocosms for different climate scenarios (Figure 1.15c). The highest average of GPP (lowest average of photosynthesis) was noted for the unmanaged mesocosms subjected to RCP 8.5 followed by the managed mesocosms for the CC scenario (Figures 1.15c and 1.16b). The ratio of GPP for the managed to the unmanaged mesocosm showed that the effect of water-level management is more critical for the warmest climate scenario: RCP 8.5 had a ratio of 2.8 compared to the ratio of, for example, RCP 4.5, which was only 1.5 (Figure 1.16b). Surprisingly, the effect of water-level management did not improve the rate of GPP in the managed mesocosms for RCP 2.6 and the CC scenario, as the ratio of GPP for the managed to unmanaged mesocosms regarding RCP 2.6 and the CC were 0.8 and 0.7, respectively (Figure 1.16b).

The NEE ranged from −0.420 up to 0.084 mg CO_2 m^{-2} s^{-1} over two years. The lowest seasonal average of NEE (−0.170 mg CO_2 m^{-2} s^{-1}) was noted for autumn 2019 regarding the managed mesocosms under RCP 8.5, while the highest seasonal average of NEE (0.040 mg CO_2 m^{-2} s^{-1}) was recorded for summer 2019 concerning the unmanaged mesocosms under RCP 8.5 (Figure 1.14e, f). The highest average of NEE (lowest average of CO_2 uptake) was recorded for the unmanaged mesocosms under RCP 8.5 followed by the managed mesocosms under the CC scenario. Moreover, the results of the Bonferroni test showed a significant difference ($p = 0.041$) between the NEE average of the managed and unmanaged mesocosms under the climate scenario RCP 8.5 (Figure 1.15e). The ratio of managed mesocosm NEE to the unmanaged mesocosm NEE was the highest for RCP 8.5 (5.8) followed by RCP 4.5 (2.1), RCP 2.6 (1.1) and CC (0.6) (Figure 1.16c).

The interactive effect of water-level management and climate scenario was significant for all measured variables: NEE ($p = 0.013$), GPP ($p = 0.014$) and respiration ($p = 0.018$). Comparison of the managed and unmanaged mesocosms revealed that the interactive effect of water-level management and climate change resulted in greater variability of NEE, GPP and respiration in the managed mesocosms (Figure 1.16). The author found an inverse interactive effect of water-level management and climate change on GPP and respiration; i.e., GPP and respiration showed a decreasing trend from the unmanaged mesocosms to the managed mesocosms under RCP 4.5 and RCP 8.5, whereas this trend was increasing for RCP 2.6 and the CC scenario.

Regarding NEE, the interactive effect of water-level management and climate change caused the managed mesocosms to have a lower NEE rate than the unmanaged ones under RCP 4.5 and 8.5. However NEE between the managed and unmanaged mesocosms had no substantial change under RCP 2.6, and the CC showed an increasing trend from the unmanaged to managed system in contrast to other climate scenarios.

1.4.3.3 Climate Change, Water Level and Plant Distribution

The results of a two-way ANOVA showed that the effect of climate scenario (F = 19.27, $p < 0.001$), water-level management (F = 18.96, $p < 0.001$) and their interaction (F = 52.59, $p < 0.001$) was statistically significant for *Sphagnum* spp. The coverage of *Sphagnum* spp. decreased gradually for both the current and future climate scenarios, but the decline was more drastic for RCP 8.5 (Figure 1.17a). The significant effect of water-level management caused a significantly higher ($p < 0.05$) coverage of *Sphagnum* spp. in the managed compared to the unmanaged mesocosms under the CC scenario (Figure 1.15b). The alleviation effect of water-level management was the greatest for the CC (Figures 1.15b and 1.17a).

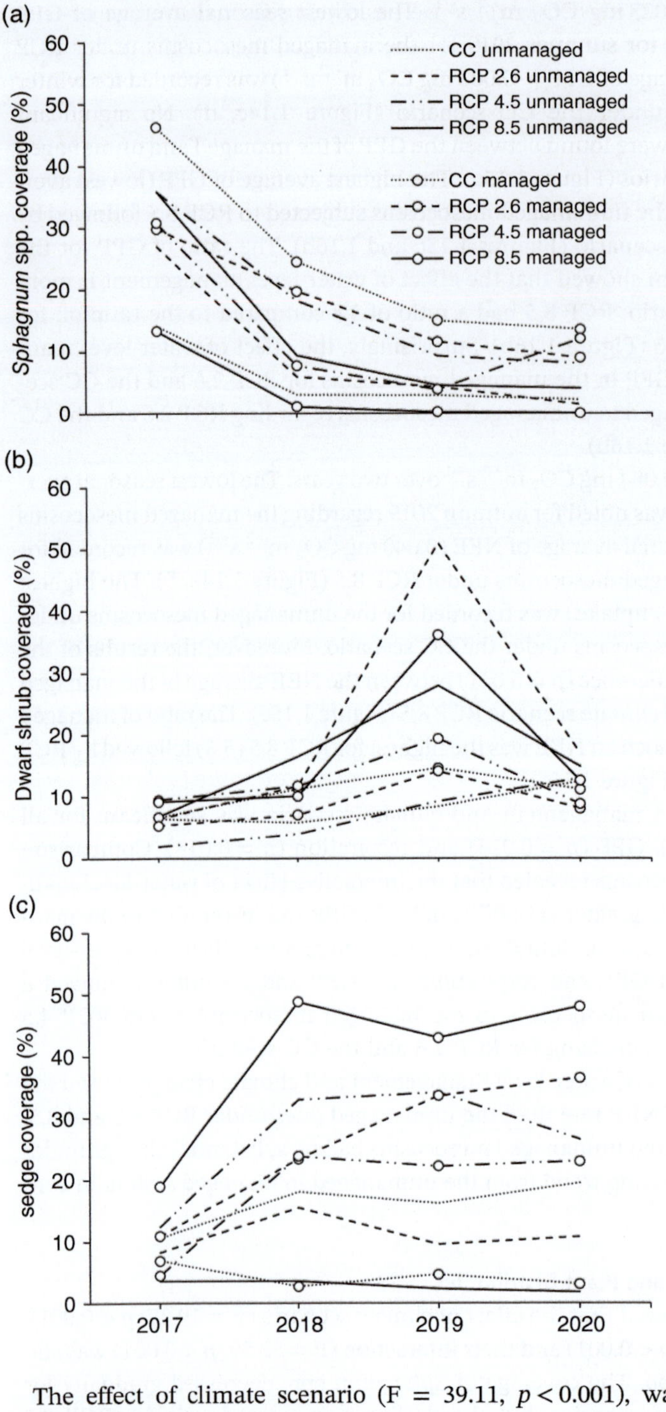

Figure 1.17 Changes in coverage of plant functional types including (a) *Sphagnum* spp.; (b) dwarf shrubs and (c) sedges over the entire period of the experiment. Clear trends except of that *Sphagnum* spp. declines over time are not apparent.

The effect of climate scenario (F = 39.11, $p < 0.001$), water-level management (F = 74.32, $p < 0.001$) and their interaction (F = 123.81, $p < 0.001$) was statistically significant for *Eriophorum vaginatum* (hare's-tail cotton grass). *Eriophorum vaginatum* increased over time in all managed mesocosms subjected to future climate scenarios, but not for the CC scenario (Figure 1.17c). The increase was drastic for the managed mesocosms concerning the scenario RCP 8.5 (Figures 1.15d and 1.17c). All unmanaged mesocosms have undergone a gradual increase in *E. vaginatum* coverage, except for those for RCP 8.5, which have experienced a decline (Figures 1.15d and 1.17c).

The effect of climate scenario (F = 25.60, $p < 0.001$) as well as climate scenario and water-level management interaction (F = 27.02, $p < 0.001$) were found to be significant for dwarf shrubs. Dwarf shrubs increased its coverage significantly in the unmanaged systems for the RCP 2.6 ($p > 0.05$) compared to other managed and unmanaged systems for the other climate scenarios and considerably for the managed and unmanaged system under the RCP 8.5 scenario (Figures 1.15f and 1.16b). However, the author observed a decline in these systems for 2020 (Figure 1.17b). Gradual increases in both managed and unmanaged systems for the other scenarios were not substantial.

1.4.4 Discussion

1.4.4.1 Interactions

The author and his team demonstrated that warmer future climate scenarios alone cannot trigger a significant shift in peatland ecosystem sink function (Figures 1.15e and 1.18). However, water availability (water-level management), particularly in interaction with temperature (climate scenario), may play an important role in changing the CO_2 sink function of peatlands (Figure 1.16c). These findings are novel, yet in line with previous studies (Lohila et al. 2011; Sullivan et al. 2008), which emphasise the critical interactive impact of temperature and wetness on GPP and respiration response as either factor alone. Furthermore, the author realised that this interactive effect is essential in shifting plant composition, which is a key factor in determining the response of net CO_2 ecosystem exchange of peatlands subject to global warming and the resulting decrease in water level (Moore et al. 2002; Weltzin et al. 2003).

1.4.4.2 Effect of Climate on the Carbon Sink Function of Unmanaged Mesocosms

The author expected that the unmanaged system would be the largest CO_2 sink under the CC scenario, with the lowest temperature, sufficient water level and less water stress during drought. However, this was not the case (Figures 1.14e and 1.15e). Lower temperatures in this climate scenario may reduce plant productivity and vascular plant expansion, and thus CO_2 uptake (Dieleman et al. 2015; Weltzin et al. 2000). Unmanaged mesocosms under RCP 2.6 had the best CO_2 sink

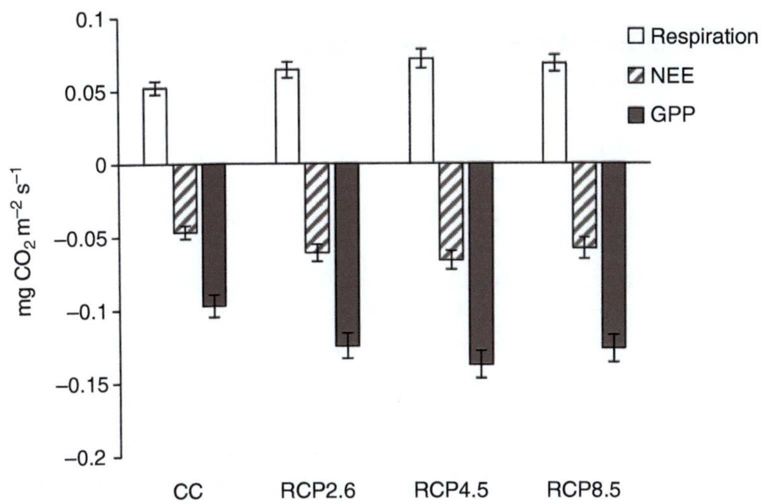

Figure 1.18 Average of respiration, net ecosystem exchange (NEE) and gross primary production (GPP) for all mesocosms (both managed and unmanaged) under different climate scenarios (current climate (CC) and representative concentration pathways (RCP) 2.6, RCP 4.5 and RCP 8.5). For the first three RCP scenarios, respiration increases with RCP and both NEE and GPP decrease with an increase of RCP.

function (Figures 1.14e and 1.15e), despite having the lowest water level during the growing season during summer and autumn 2019 (Figures 1.14g and 1.15h). This finding implies that a slightly higher temperature for RCP 2.6 could induce CO_2 uptake through increased photosynthetic capacity of plants in the unmanaged mesocosms, especially dwarf shrubs, which had the greatest coverage (Figures 1.15f and 1.17b). This increased photosynthetic capacity could offset the highest rate of respiration, resulting in the largest CO_2 sink for this scenario. Other studies (Breeuwer et al. 2010; Weltzin et al. 2003) also reported an increased NEE of CO_2 because of higher temperature and lower water level. They indicated that a lower water level could improve oxygen supply to plant roots and nutrient availability, which are the factors that can lead to a higher vascular plant productivity (Ratcliffe et al. 2019; Walker et al. 2015) and higher respiration rates (Ise et al. 2008).

A lower CO_2 sink function under the warmer climate scenarios RCP 4.5 and RCP 8.5 was observed as predicted (Figures 1.14e and 1.15e). Possibly, a lower CO_2 sink would be in part due to a higher rate of decomposition and respiration during drought (Figure 1.14a) because of higher temperature (Bubier et al. 2003; Davidson and Janssens 2006). In addition, a lower CO_2 sink function seems to be associated with the gradual decrease in plant GPP, which declined due to the drought of 2019 and high temperatures in 2019 and 2020 during the growing seasons, over the course of the experiment (Figures 1.14c and 1.17).

In general, findings demonstrated that under future climate scenarios, all unmanaged systems will continue to serve as CO_2 sinks throughout the experiment period (Figures 1.14e and 1.15e). However, under the stress of drought, the unmanaged systems under RCP 4.5 and RCP 8.5 lost their photosynthetic efficiency and switched to becoming CO_2 sources (Figures 1.14e and 1.19). The unmanaged system under RCP 2.6 could have a stronger net CO_2 sink function compared to the ones under other future climate scenarios and be able to act as CO_2 sinks even during droughts (Figures 1.14e and 1.19), owing to a boosted dwarf shrub primary production. These results support the findings by Wu and Roulet (2014), who modelled both fens and bogs, and found that bogs are less sensitive to the future IPCC climate scenarios (A1B, A2, B1 and Commit) until 2100. They indicated that bogs might still be a carbon sink up to 2100 for almost all future climate scenarios through an enhanced GPP, but that their sink role will be reduced relative to baseline fluxes under the present climate conditions.

1.4.4.3 Impact of Drought on the Carbon Sink Function of Unmanaged Mesocosms

The unmanaged mesocosms in both CC and RCP 2.6 scenarios preserved their sink function during the 2019 drought (Figures 1.14e and 1.19), suggesting less drought tension under these scenarios due to lower temperature (Figures 1.14i and 1.19), even at zero water level (Figure 1.19). Furthermore, Ratcliffe et al. (2019) suggested that the resistance of the dominated vegetation group to lower water levels was a critical factor in maintaining the resilience of carbon dioxide sink efficiency within peatlands.

The high rate of carbon dioxide uptake in the unmanaged system under RCP 2.6 decreased during the extreme and long drought of 2019 (end of May to July), but the systems maintained their CO_2 sink function (Figure 1.19). In the following summer of 2020, when the plants were not under drought stress, the rate of carbon dioxide uptake increased due to the recovery of plants (particularly dwarf shrubs) to the extent that they were the largest carbon dioxide sink compared to all other managed and unmanaged systems in other climate scenarios (Figures 1.14e and 1.19). Reduced sensitivity of shrubs to hydrology (Bragazza et al. 2013; Ratcliffe et al. 2019) and elevated sensitivity to temperature has been reported in other studies (Dieleman et al. 2015; Munir et al. 2015; Radu and Duval 2018) as this also appears to be happening for the unmanaged mesocosms under RCP 2.6 in this study. Furthermore, Flanagan and Syed (2011) noted that nearly equal increases in

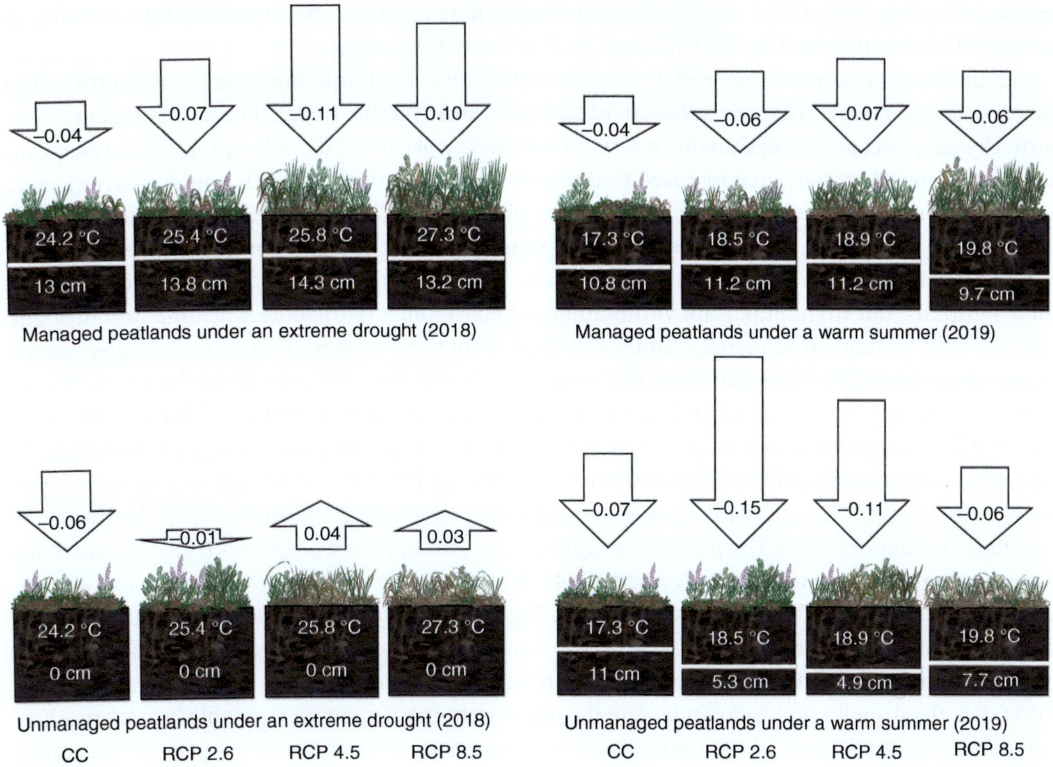

Figure 1.19 Magnitude of carbon dioxide sink function of peatland mesocosms under different climate scenarios (current climate (CC) and representative concentration pathways (RCP) 2.6, RCP 4.5 and RCP 8.5) during the warmest months (May, June and July) for two extreme summers in 2018 and 2019 (real time) simulated in 2019 and 2020 in the experiment, respectively. The values indicate net ecosystem exchange (NEE); the net CO_2 sink has been expressed as a negative value, while the net CO_2 source expressed as a positive value. Water-level averages are shown by white lines and the values underneath them. The temperature averages indicate the temperatures within the climate chambers. Unmanaged peatlands under extreme droughts become sources of carbon for RCP 4.5 and RCP 8.5.

photosynthesis and respiration of an ecosystem exposed to climate change can keep the system balanced in terms of net CO_2 ecosystem exchange. Thus, with a slightly warmer climate such as RCP 2.6, increased photosynthesis offset higher respiration, and the negative effects of drought are compensated in subsequent years with normal precipitation.

The severe drought in 2019 caused unmanaged mesocosms under the RCP 4.5 and RCP 8.5 scenarios to switch temporarily from sinks to sources (Figures 1.14e and 1.19). Vascular plants respond to drought, physiologically and structurally, to minimise water loss, and this causes a major decrease in the photosynthetic efficiency of unmanaged mesocosms under RCP 4.5 and RCP 8.5 scenarios (Cai et al. 2010). In addition, the exposure of these mesocosms to severe droughts under these two climate scenarios resulted in *Sphagnum* spp. mortality (Figure 1.17a) and a consequent shrinkage and subsidence of the top surface of the mesocosms (Dieleman et al. 2015; Leifeld and Menichetti 2018).

In line with the findings of this study, Lund et al. (2012) confirmed a shift from sink to source during a long and early drought during the growing season in Fäjemyr, where the mesocosms in this study were sampled, which could have an adverse influence on canopy development and thus reduce the system GPP. Thus, water stress may have a more serious impact on plant functionality in warmer climates indicated by RCP 4.5 and RCP 8.5 during drought.

Both unmanaged systems under RCP 4.5 and RCP 8.5 recovered from the drought and returned to the carbon sink state during the subsequent autumn rains (Figure 1.14e). These findings are supported by Kuiper et al. (2014) showing a rapid recommence of carbon dioxide uptake after rewetting the mesocosms. However, they reported that the mesocosms without ericoids could recover faster. In comparison, in this study, more or equal carbon dioxide uptake by the mesocosms with relatively more ericoid plants (e.g., RCP 2.6 scenario) was noted (Figure 1.19). Moreover, Jassey and Signarbieux (2019) documented that those boreal systems benefited from transitional warming-induced photosynthesis, but highlighted the vulnerability of ecosystems to extreme warming and drought that can have lasting consequences. In this study, during the summer of 2020, unmanaged mesocosms under the RCP 8.5 scenario have lost some of their sink function relative to the other climate scenarios (Figures 1.14e and 1.19). The unmanaged mesocosms under RCP 8.5 have undergone irreversible changes for plants such as *Sphagnum* spp. and sedges (Figure 1.17a, c), which slightly diminished the photosynthetic efficiency (Figure 1.14c), and they could not recover as efficient as the ones for the RCP 2.6 and RCP 4.5 scenarios (Figures 1.14e and 1.19). In line with the results in this study, Buttler et al. (2015) reported that *Sphagnum* moss coverage of an ombrotrophic peatland decreased in response to an interactive impact of warming and reduced soil wetness due to a change in their moss structure affecting its moisture-holding capacity (Dorrepaal et al. 2004). Therefore, for extreme scenarios such as RCP 8.5, natural recovery of peatlands from drought may be unlikely.

The recovery of the carbon sink function of unmanaged mesocosms during the growing season of 2020 (after the drought of 2019) can be due the capacity of these systems to establish more productive plant communities (vascular plants). It was surprising that the unmanaged system could recover to the extent that a year after the drought, they showed a greater sink capacity compared to the managed systems (Figures 1.14e, f and 1.17). This can be attributed to the introduction of nutrients (Shaver et al. 1992) into the unmanaged system due to the litter decomposition during the drought of 2019. These nutrients were available to plants in the following growing season and boosted their growth rate (Laiho 2006; Munir et al. 2015; Straková et al. 2012). Shaver et al. (1992) documented that higher carbon dioxide sequestration rates in plant biomass can occur in nutrient-limited northern peatlands, as higher temperatures can increase nutrient mineralisation, making it available for vascular plants. The crucial impact of vascular plants on net carbon dioxide assimilation in peatlands was also found in other studies (Kuiper et al. 2014; Laine et al. 2012; Ward et al. 2013). However, a further group of studies has stated that an increase in vascular plant coverage can stimulate soil microbes by providing labile carbon, which can mobilise additional energy from the decomposition of relatively old organic matter (Fontaine et al. 2007; Schmidt et al. 2011).

1.4.4.4 Interactive Effects on the Carbon Sink Function of Managed Mesocosms

The author and his researchers found that the response of peatland is a function of the interaction of climate change and water-level management. At different circumstances, each of them may play a prominent role. For example, he observed that NEE and its components (absolute values for NEE and GPP) for the managed system followed the sequence of the climate scenarios (Figure 1.15a, c, e, g). This implies a strong association of the system to temperature, when the water level was virtually constant for the managed systems. Moreover, when water was continuously available for the managed system, lower temperatures during the 2020 growing season caused lower rates of carbon uptake compared to the 2019 growing season with about 4–5°C higher temperature (Figure 1.19). In comparison to the unmanaged systems, the trend of NEE and its components (GPP and reversed respiration) followed the trend of water level (Figure 1.15a, c, e, h).

In accordance with the author's observation, Bubier et al. (2003) revealed that the water table position within a peatland was the strongest control on respiration during a dry summer, whereas surface peat temperature could explain most of the respiration variability during a wet summer.

The author found that the effect of water-level management differed depending on the climate scenario. He demonstrated that water-level management could not enhance the photosynthetic capacity of the mesocosms under the CC scenario and RCP 2.6 (Figure 1.16b). This finding contradicts the idea that increasing the water level will promote CO_2 uptake (Karki et al. 2016; Swenson et al. 2019). However, in both cases, the managed mesocosms had a lower respiration than the unmanaged ones (Figure 1.16a), but they still had a lower and nearly equivalent CO_2 sink power (Figure 1.16c). In general, higher water levels under the CC scenario could only result in slightly higher primary production of *Sphagnum* spp. (Chaudhary et al. 2018) compared to the unmanaged mesocosms (Figure 1.15b). Since the coverage of vascular plants, particularly dwarf shrubs, did not increase notably (Figures 1.15f and 1.17b), the team could not have observed an improvement in sink function of the managed systems (Munir et al. 2015; Ward et al. 2013) relative to the unmanaged ones under the CC scenario and RCP 2.6.

Water-level management had the greatest positive impact on the sink function of mesocosms under RCP 8.5 followed by RCP 4.5 (Figure 1.16e) by stimulating the growth of vascular plants (sedges and dwarf shrubs) during the growing season over the two-year experiment (Figure 1.17b, c) (Beyer et al. 2021). Other studies showed a lower NEE (higher CO_2 uptake) in the restored areas after rewetting compared to drained peatlands (Karki et al. 2016; Planas-Clarke et al. 2020; Schimelpfenig et al. 2014; Swenson et al. 2019). For example, Nugent et al. (2018) indicated that internal hydrological controls can strengthen the stability and strong carbon sink function of peatlands during summer drawdowns. As a result, in warmer climates, particularly during drought, water-level management can have a profound impact on maintaining the sink function of peatlands.

The author observed a marginal decline in the sink function of managed mesocosms under RCP 8.5 from 2019 to 2020 (Figure 1.14f). This is due to a gradual decrease in productive vascular plant coverage (Figure 1.17b, c), because of high temperatures over two years. These results were expected for sedges, which are an arctic-boreal plant group that does not benefit from a long-term extreme warm climate as the one simulated for RCP 8.5. In comparison, continuous growth of sedges in managed mesocosms under RCP 4.5 was observed (Figure 1.17c), indicating more pleasant conditions for this plant functional type (Olefeldt et al. 2017). In agreement with the results of Chapter 1.4, Dieleman et al. (2015) reported an abundance of vascular plants, especially graminoids (up to 15 times more), at a temperature increase of +4°C compared to ambient temperatures. The results indicate that, while water-level management can significantly improve the sink role of managed mesocosms, decreased tolerance of vascular plants (especially graminoids) to higher temperatures during longer warm seasons can reduce water-level management efficiency under the warmest climate scenario (RCP 8.5) in the long term.

Overall, the author highlights the beneficial effects of water-level management on plant GPP and sink function of the wetland systems under warmer climate scenarios. They are likely to cope with the adverse impact of climate change and widespread droughts in the future (Beyer et al. 2021; Nugent et al. 2018; Wilson et al. 2016). In agreement with results discussed in this chapter, Jarveoja et al. (2018) found in their partitioning study that the daily variations in NEE were primarily regulated by plant productivity and the dynamics in respiration were mainly determined by plants rather than by microbial respiration.

Although Chapter 1.4 emphasises the important role of plant productivity, which can be manipulated by water-level management, a gradual decrease in GPP for a warmer climate scenario like

RCP 8.5 could be expected over the long-term even with active water-level management. However, to meet the Paris Agreement target (i.e., limiting the rise in global temperature to well below 2°C), the peatlands at risk need to be rewetted to reduce the emission of too much CO_2 (Günther et al. 2020).

1.4.5 Conclusions and Recommendations

The CO_2 sink function of peatlands is largely controlled by plant functional types. The plant population proportions shift over time in response to climate change, and that this shift can cause uncertainty in the ecosystem response as it might vary considerably depending on both the magnitude of climate change and the severity of management actions.

The effect of climate scenarios on NEE and its components (GPP and respiration) depends upon whether the system is managed or unmanaged in terms of water-level regulation. Temperature was the main driver for the sink function, when the water is almost equally available to the managed wetland systems. However, for the unmanaged systems, the lowest water level was associated with the largest sink. This unexpected result is due to the differential impacts of drought on various climate scenarios, plant succession and vegetation photosynthetic capacity. For example, systems impacted on by the RCP 2.6 scenario combined with low temperature will experience less drought stress, and vascular plants will benefit from the extra nutrients provided by more oxic conditions caused by drought, boosting the CO_2 sink function. In comparison, plants in warmer climate scenarios experience higher drought stress for RCP 4.5 and 8.5 and will lose a large amount of photosynthetic capacity.

The author concludes that changes in vegetation communities on the peat top surface play a great role in the CO_2 flux of the peatland ecosystem. This was revealed by comparing GPP and respiration of both managed and unmanaged mesocosms during the growing seasons (spring and summer) between 2019 and 2020. The unmanaged mesocosms under RCP 8.5 had the lowest respiration and photosynthesis rates. These findings are surprising as one might expect that the higher rate of respiration from RCP 8.5 is due to exposure to more extreme drought (2019) and higher temperatures (2020) compared to other climate scenarios. The researcher therefore concluded that the lowest respiration for RCP 8.5 can be attributed to the mortality of the plants (mainly sedges and *Sphagnum* spp.), because of the stress factor of higher temperature and drought. This might have an adverse irreversible consequence for RCP 8.5, but not for other future climate scenarios.

The significant difference between the managed and unmanaged systems under RCP 8.5 implies the necessity of water-level management for this scenario. In addition, the positive impact of water-level management for the RCP 4.5 scenario was evident as well. Although the advantageous impact of water-level management has been demonstrated for warmer climate scenarios in this experiment, the impact of plant succession, dominance of plants with higher decomposition sensitivity as well as change in microbial community on the carbon sink function of peatlands is not clear for the long-term, and the experiments supported by large-scale field trials should be continued to assess these influences.

Water-level management, according to the findings of this and other studies, is an effective method for mitigating climate change. Referring to the observed trend of NEE in response to the interaction of climate change and water-level management, the team concludes that water-level management is necessary for RCP 8.5, beneficial for RCP 4.5 (particularly during drought), and not essential for RCP 2.6 and the CC.

In this study, a critical impact of plant functional type on peatland CO_2 sink function was noted. Therefore, the author recommends that a new experimental setup should be established to investigate the impact of different plant functional types on CO_2 emission and water availability. In such a future study, plant functional types need to be regarded as a treatment (independent variable) affecting the CO_2 emission and water availability of peatland ecosystem as the response (dependent variable).

References

Abbaspour, K., Vejdani, M., and Haghighat, S. (2007). *SWAT-CUP Calibration and Uncertainty Programs for SWAT*, 1596–1602. Duebendorf: Swiss Federal Institute of Aquatic Science and Technology.

Ahearn, D.S., Sheibley, R.W., Dahlgren, R.A. et al. (2005). Land use and land cover influence on water quality in the last free-flowing river draining the western Sierra Nevada, California. *Journal of Hydrology* 313 (3): 234–247.

Ahmed, A.A., Thiele-Bruhn, S., Aziz, S.G. et al. (2015). Interaction of polar and nonpolar organic pollutants with soil organic matter: sorption experiments and molecular dynamics simulation. *Science of the Total Environment* 508: 276–287.

Akhavan, S., Abedi-Koupai, J., Mousavi, S. et al. (2010). Application of SWAT model to investigate nitrate leaching in Hamadan–Bahar Watershed, Iran. *Agriculture, Ecosystems and Environment* 139 (4): 675–688.

Al-Senafy, M. and Al-Otaibi, M. (2002). Replenishing groundwater aquifers through recharge of treated wastewater. *Acta Universitatis Carolinae, Geologica* 46 (2–3): 674–676.

American Public Health Association (1995). *Standard Methods for the Examination of Water and Wastewater*, 19e. Washington DC, USA: American Public Health Association, American Waterworks Association and Water and Environment Federation.

American Public Health Association; American Water Works Association; Water Environment Federation (1998). *Standards Methods for the Examination of Water and Wastewater*, 20e. Washington, DC: American Public Health Association.

Arnold, J.G., Williams, J.R., and Srinivasan, R. (1994). *SWAT (Soil and Water Assessment Tool)*. Washington, DC: Grassland, Soil and Water Research Laboratory, USDA, Agricultural Research Service.

Arora, V.K., Scinocca, J.F., Boer, G.J. et al. (2011). Carbon emission limits required to satisfy future representative concentration pathways of greenhouse gases. *Geophysical Research Letters* https://doi.org/10.1029/2010GL046270.

Asemaninejad, A., Thorn, R.G., Branfireun, B.A., and Lindo, Z. (2018). Climate change favours specific fungal communities in boreal peatlands. *Soil Biology and Biochemistry* 120: 28–36. https://doi.org/10.1016/j.soilbio.2018.01.029.

Bentsen, M., Bethke, I., Debernard, J.B. et al. (2013). The Norwegian earth system model, NorESM1-M – part 1: description and basic evaluation of the physical climate. *Geoscientific Model Development* https://doi.org/10.5194/gmd-6-687-2013.

Beyer, F., Jansen, F., Jurasinski, G. et al. (2021). Drought years in peatland rewetting: rapid vegetation succession can maintain the net CO_2 sink function. *Biogeosciences* 18 (3): 917–935. https://doi.org/10.5194/bg-18-917-2021.

Bhaduri, B., Harbor, J., Engel, B., and Grove, M. (2000). Assessing watershed-scale, long-term hydrologic impacts of land-use change using a GIS-NPS model. *Environmental Management* 26 (6): 643–658.

Biggs, T., Dunne, T., and Martinelli, L. (2004). Natural controls and human impacts on stream nutrient concentrations in a deforested region of the Brazilian Amazon basin. *Biogeochemistry* 68 (2): 227–257.

Bockholt, R., Koch, G., Ebert, W., and Fedderwitz, E. (1992). Nutrient contents of drainwater and drainage ditchwater on agriculturelly used areas in the drinking water territory of the river Kösterbeck. *Zeitschrift für Kulturtechnik und Landentwicklung* 33: 178–185. (in German).

Bolker, B.M., Brooks, M.E., Clark, C.J. et al. (2009). Generalized linear mixed models: a practical guide for ecology and evolution. *Trends in Ecology & Evolution* 24 (3): 127–135. https://doi.org/10.1016/j.tree.2008.10.008.

Bowes, M.J., House, W.A., Hodgkinson, R.A., and Leach, D.V. (2005). Phosphorus–discharge hysteresis during storm events along a river catchment: the River Swale, UK. *Water Research* 39 (5): 751–762.

Bragazza, L., Parisod, J., Buttler, A., and Bardgett, R.D. (2013). Bio- geochemical plant–soil microbe feedback in response to climate warming. *Nature Climate Change* 3: 273–277. https://doi.org/10.1038/nclimate1781.

Bragazza, L., Buttler, A., Robroek, B.J.M. et al. (2016). Persistent high temperature and low precipitation reduce peat carbon accumulation. *Global Change Biology* 22 (12): 4114–4123. https://doi.org/10.1111/gcb.13319.

Breeuwer, A., Heijmans, M.M., Robroek, B.J., and Berendse, F. (2010). Field simulation of global change: transplanting northern bog mesocosms southward. *Ecosystems* 13 (5): 712–726. https://doi.org/10.1007/s10021-010-9349-y.

British Precast Concrete Federation (2003). *Guide to the Design Construction and Maintenance of Concrete Block Permeable Pavement*. Leicester: British Precast Concrete Federation.

Brown, J. and Clasen, T. (2012). High adherence is necessary to realize health gains from water quality interventions. *PLoS One* 7 (5): e36735.

Bubier, J., Crill, P., Mosedale, A. et al. (2003). Peatland responses to varying interannual moisture conditions as measured by automatic CO_2 chambers. *Global Biogeochemical Cycles* 17 (2): https://doi.org/10.1029/2002GB001946.

Buck, O., Niyogi, D.K., and Townsend, C.R. (2004). Scale-dependence of land use effects on water quality of streams in agricultural catchments. *Environmental Pollution* 130 (2): 287–299.

Buma, B. (2012). Evaluating the utility and seasonality of NDVI values for assessing post-disturbance recovery in a subalpine forest. *Environmental Monitoring and Assessment* 184 (6): 3849–3860.

Burt, T.P. (2003). Monitoring change in hydrological systems. *Science of the Total Environment* 310 (1–3): 9–16.

Buttler, A., Robroek, B.J., Laggoun-Défarge, F. et al. (2015). Experimental warming interacts with soil moisture to discriminate plant responses in an ombrotrophic peatland. *Journal of Vegetation Science* 26 (5): 964–974. https://doi.org/10.1111/jvs.12296.

Cai, T., Flanagan, L.B., and Syed, K.H. (2010). Warmer and drier conditions stimulate respiration more than photosynthesis in a boreal peatland ecosystem: analysis of automatic chambers and eddy covariance measurements. *Plant, Cell & Environment* 33 (3): 394–407. https://doi.org/10.1111/j.1365-3040.2009.02089.x.

Chapra, S.C. (2008). *Surface Water-quality Modeling*. Long Grove: Waveland Press.

Chaudhary, N., Miller, P.A., and Smith, B. (2018). Biotic and abiotic drivers of peatland growth and microtopography: a model demonstration. *Ecosystems* 21 (6): 1196–1214. https://doi.org/10.1007/s10021-017-0213-1.

Chaudhry, M.H. (1996). Principles of flow of water. In: *Resources Handbook* (ed. L.W. Mays), 2.1–2.43. New York, USA: McGraw-Hill.

Collins, W.J., Bellouin, N., Doutriaux-Boucher, M. et al. (2011). Development and evaluation of an earth-system model – HadGEM2. *Geoscientific Model Development* https://doi.org/10.5194/gmd-4-1051-2011.

Dauer, D.M., Ranasinghe, J.A., and Weisberg, S.B. (2000). Relationships between benthic community condition, water quality, sediment quality, nutrient loads, and land use patterns in Chesapeake Bay. *Estuaries* 23 (1): 80–96.

Davidson, E.A. and Janssens, I.A. (2006). Temperature sensitivity of soil carbon decomposition and feedbacks to climate change. *Nature* 440 (7081): 165–173. https://doi.org/10.1038/nature04514.

DeFries, R. and Eshleman, K.N. (2004). Land-use change and hydrologic processes: a major focus for the future. *Hydrological Processes* 18 (11): 2183–2186.

Del Giudice, R. and Lindo, Z. (2017). Short-term leaching dynamics of three peatland plant species reveals how shifts in plant communities may affect decomposition processes. *Geoderma* 285: 110–116. https://doi.org/10.1016/j.geoderma.2016.09.028.

Deletic, A. and Fletcher, T.D. (2006). Performance of grass filters used for stormwater treatment – a field and modelling study. *Journal of Hydrology* 317 (3): 261–275.

Delkash, M., Al-Faraj, F.A., and Scholz, M. (2014). Comparing the export coefficient approach with the soil and water assessment tool to predict phosphorous pollution: the Kan watershed case study. *Water, Air, and Soil Pollution* 225 (10): 1–17.

Delkash, M., Al-Faraj, F.A.M., and Scholz, M. (2018). Impacts of anthropogenic land use changes on nutrient concentrations in surface waterbodies: a review. *CLEAN - Soil, Air, Water* 46 (5): 1800051.

Delkash, M., Bakhshayesh, B.E., and Kazemian, H. (2015). Using zeolitic adsorbents to cleanup special wastewater streams: a review. *Microporous and Mesoporous Materials* 214: 224.

DEUTSCHES INSTITUTE FÜR NORMUNG (1996a). *DIN EN ISO 13395: Water Quality - Determination of Nitrite Nitrogen and Nitrate Nitrogen and the Sum of Both by Flow Analysis (CFA and FIA) and Spectrometric Detection*. Berlin, Germany: Deutsches Institute für Normung e.V., Beuth Verlag.

DEUTSCHES INSTITUTE FÜR NORMUNG (1996b). *DIN EN 1189: Wasserbeschaffenheit – Bestimmung von Phosphor – Photometrisches Verfahren mittels Ammoniummolybdat*. Berlin, Germany: Deutsches Institute für Normung e.V., Beuth Verlag (in German).

DEUTSCHES INSTITUTE FÜR NORMUNG (2003). *DIN EN ISO 11732: Wasserbeschaffenheit – Bestimmung von Ammoniumstickstoff – Verfahren mittels Fließanalyse (CFA und FIA) und spektrometrischer Detektion (ISO/DIS 11732:2003)*. Berlin, Germany: Deutsches Institute für Normung e.V., Beuth Verlag (in German).

Dieleman, C.M., Branfireun, B.A., McLaughlin, J.W., and Lindo, Z. (2015). Climate change drives a shift in peatland ecosystem plant community: implications for ecosystem function and stability. *Global Change Biology* 21 (1): 388–395. https://doi.org/10.1111/gcb.12643.

Donner, S.D., Kucharik, C.J., and Foley, J.A. (2004). Impact of changing land use practices on nitrate export by the Mississippi River. *Global Biogeochem Cycles* 18 (1): 1028.

Dorrepaal, E., Aerts, R., Cornelissen, J.H. et al. (2004). Summer warming and increased winter snow cover affect Sphagnum fuscum growth, structure and production in a sub-arctic bog. *Global Change Biology* 10 (1): 93–104. https://doi.org/10.1111/j.1365-2486.2003.00718.x.

Duncan, R. (2014). Regulating agricultural land use to manage water quality: the challenges for science and policy in enforcing limits on non-point source pollution in New Zealand. *Land Use Policy* 41: 378–387.

Dunne, J.P., John, J.G., Adcroft, A.J. et al. (2012). GFDL's ESM2 global coupled climate-carbon earth system models. Part I: physical formulation and baseline simulation characteristics. *Journal of Climate* https://doi.org/10.1175/JCLI-D-11-00560.1.

Eshleman, K.N., McNeil, B.E., and Townsend, P.A. (2009). Validation of a remote sensing based index of forest disturbance using streamwater nitrogen data. *Ecological Indicators* 9 (3): 476–484.

European Commission (2000). Directive 2000/60/EC of the european parliament and of the council of 23 October 2000 establishing a framework for community action in the field of water policy. *Official Journal of the European Union* L327: 1–72.

Fenner, N. and Freeman, C. (2011). Drought-induced carbon loss in peatlands. *Nature Geoscience* 4 (12): 895–900. https://doi.org/10.1038/ngeo1323.

Flanagan, L.B. and Syed, K.H. (2011). Stimulation of both photosynthesis and respiration in response to warmer and drier conditions in a boreal peatland ecosystem. *Global Change Biology* 17 (7): 2271–2287. https://doi.org/10.1111/j.1365-2486.2010.02378.x.

Fontaine, S., Barot, S., Barré, P. et al. (2007). Stability of organic carbon in deep soil layers controlled by fresh carbon supply. *Nature* 450 (7167): 277–280. https://doi.org/10.1038/nature06275.

Fowler, J. and Cohen, L. (1998). *Practical Statistics for Field Biology*. West Sussex, UK: Wiley LTD.

Frevert, D.K. and Singh, V. (2006). *Watershed Models*. Boca Raton: Taylor & Francis.

Frolking, S., Roulet, N.T., Moore, T.R. et al. (2001). Modeling northern peatland decomposition and peat accumulation. *Ecosystems* 4 (5): 479–498. https://doi.org/10.1007/s10021-001-0105-1.

Garnier, J., Brion, N., Callens, J. et al. (2013). Modeling historical changes in nutrient delivery and water quality of the Zenne River (1790s–2010): the role of land use, waterscape and urban wastewater management. *Journal of Marine Systems* 128: 62–76.

Ghaffari, G., Keesstra, S., Ghodousi, J., and Ahmadi, H. (2010). SWAT-simulated hydrological impact of land-use change in the Zanjanrood basin, Northwest Iran. *Hydrological Processes* 24 (7): 892–903.

Giorgetta, M.A., Jungclaus, J., Reick, C.H. et al. (2013). Climate and carbon cycle changes from 1850 to 2100 in MPI-ESM simulations for the coupled model intercomparison project phase 5. *Journal of Advances in Modeling Earth Systems* https://doi.org/10.1002/jame.20038.

Goldstein, R.M., Carlisle, D.M., Meador, M.R., and Short, T.M. (2007). Can basin land use effects on physical characteristics of streams be determined at broad geographic scales? *Environmental Monitoring and Assessment* 130 (1–3): 495–510.

Goonetilleke, A., Thomas, E., Ginn, S., and Gilbert, D. (2005). Understanding the role of land use in urban stormwater quality management. *Journal of Environmental Management* 74 (1): 31–42.

Gorham, E. (1991). Northern peatlands: role in the carbon cycle and probable responses to climatic warming. *Ecological Applications* 1 (2): 182–195.

Grimvall, A., Stålnacke, P., and Tonderski, A. (2000). Time scales of nutrient losses from land to sea – a european perspective. *Ecological Engineering* 14 (4): 363–371.

Günther, A., Barthelmes, A., Huth, V. et al. (2020). Prompt rewetting of drained peatlands reduces climate warming despite methane emissions. *Nature Communications* 11 (1): 1–5. https://doi.org/10.1038/s41467-020-15499-z.

Hazeleger, W., Wang, X., Severijns, C. et al. (2012). EC-Earth V2.2: description and validation of a new seamless earth system prediction model. *Climate Dynamics* https://doi.org/10.1007/s00382-011-1228-5.

Hedmark, A. and Scholz, M. (2008). Review of environmental effects and treatment of runoff from storage and handling of wood. *Bioresource Technology* 99 (14): 5997–6009.

Hedmark, A., Scholz, M., and Elowson, T. (2009). Treatment of log yard runoff impacted by aged logs in a free water surface constructed wetland. *Environmental Engineering Science* 26 (11): 1623–1632.

Hedmark, Å., Scholz, M., Aronsson, P., and Elowson, T. (2010). Comparison of planted soil infiltration systems for treatment of log yard runoff. *Water Environment Research* 82 (7): 666–669.

Hodge, T.A. and Armstrong, L. (1993). Use of a multiple linear regression model to estimate stormwater pollutant loading. In: *New Techniques for Modeling the Management of Stormwater Quality Impacts* (ed. W. James), 201–214. Boca Raton: Lewis Publishers.

Hourdin, F., Foujols, M.A., Codron, F. et al. (2013). Impact of the LMDZ atmospheric grid configuration on the climate and sensitivity of the IPSL-CM5A coupled model. *Climate Dynamics* https://doi.org/10.1007/s00382-012-1411-3.

Huang, M., Zhang, G., Zhang, X., and Zhou, C. (2003). Phosphorus export by surface runoff from agricultural field plots in Guanting watershed. *Ecology and Environment* 12: 139–144.

Huston, R., Chan, Y., Gardner, T. et al. (2009). Characterisation of atmospheric deposition as a source of contaminants in urban rainwater tanks. *Water Research* 43 (6): 1630–1640.

Ierodiaconou, D., Laurenson, L., Leblanc, M. et al. (2005). The consequences of land use change on nutrient exports: a regional scale assessment in south-west Victoria, Australia. *Journal of Environmental Management* 74 (4): 305–316.

Ise, T., Dunn, A.L., Wofsy, S.C., and Moorcroft, P.R. (2008). High sensitivity of peat decomposition to climate change through water-table feedback. *Nature Geoscience* 1 (11): 763–766. https://doi.org/10.1038/ngeo331.

Jajarmizadeh, M., Harun, S., and Salarpour, M. (2012). A review on theoretical consideration and types of models in hydrology. *Journal of Environmental Science and Technology* 5 (5): 249.

Jarveoja, J., Nilsson, M.B., Gazovic, M. et al. (2018). Partitioning of the net CO2 exchange using an automated chamber system reveals plant phenology as key control of production and respiration fluxes in a boreal peatland. *Global Change Biology* 24 (8): 3436–3451. https://doi.org/10.1111/gcb.14292.

Jassey, V.E.J. and Signarbieux, C. (2019). Effects of climate warming on Sphagnum photosynthesis in peatlands depend on peat moisture and species-specific anatomical traits. *Global Change Biology* 25 (11): 3859–3870. https://doi.org/10.1111/gcb.14788.

Jeffrey, S., Rotstayn, L., Collier, M. et al. (2013). Australia's CMIP5 submission using the CSIRO-Mk3.6 model. *Australian Meteorological and Oceanographic Journal* https://doi.org/10.22499/2.6301.001.

Jha, M.K., Gassman, P.W., and Panagopoulos, Y. (2013). Regional changes in nitrate loadings in the Upper Mississippi River Basin under predicted mid-century climate. *Regional Environmental Change* 15 (3): 449–460.

Johansson, L.G. and Linder, S. (1980). Photosynthesis of Sphagnum in different microhabitats on a subarctic mire. *Ecological Bulletins* 181–190. https://www.jstor.org/stable/20112772.

Johnes, P.J. (1996). Evaluation and management of the impact of land use change on the nitrogen and phosphorus load delivered to surface waters: the export coefficient modelling approach. *Journal of Hydrology* 183 (3): 323–349.

Jonsson, M., Dimitriou, I., Aronsson, P., and Elowson, T. (2006). Treatment of log yard runoff by irrigation of grass and willows. *Environmental Pollution* 139 (1): 157–166.

Kamphuis, J. (1982). Modeling hydraulic phenomena: a glossary of terms. *Journal of the Hydraulics Division* 108 (7): 845–852.

Karki, S., Elsgaard, L., Kandel, T.P., and Lærke, P.E. (2016). Carbon balance of rewetted and drained peat soils used for biomass production: a mesocosm study. *GCB Bioenergy* 8 (5): 969–980. https://doi.org/10.1111/gcbb.12334.

Kling, C.L., Panagopoulos, Y., Rabotyagov, S.S. et al. (2014). LUMINATE: linking agricultural land use, local water quality and Gulf of Mexico hypoxia. *European Review of Agricultural Economics* 41 (3): 431–459.

Kovacic, D.A., Twait, R.M., Wallace, M.P., and Bowling, J.M. (2006). Use of created wetlands to improve water quality in the Midwest – lake Bloomington case study. *Ecological Engineering* 28 (3): 258–270.

Krause, S., Jacobs, J., Voss, A. et al. (2008). Assessing the impact of changes in landuse and management practices on the diffuse pollution and retention of nitrate in a riparian floodplain. *Science of the Total Environment* 389 (1): 149–164.

Laiho, R. (2006). Decomposition in peatlands: reconciling seemingly contrasting results on the impacts of lowered water levels. *Soil Biology and Biochemistry* 38 (8): 2011 2024. https://doi.org/10.1016/j.soilbio.2006.02.017.

Laine, A.M., Bubier, J., Riutta, T. et al. (2012). Abundance and composition of plant biomass as potential controls for mire net ecosytem CO_2 exchange. *Botany* 90 (1): 63–74. https://doi.org/10.1139/b11-068.

Laine, A.M., Mäkiranta, P., Laiho, R. et al. (2019a). Warming impacts on boreal fen CO_2 exchange under wet and dry conditions. *Global Change Biology* 25 (6): 1995–2008. https://doi.org/10.1111/gcb.14617.

Laine, A.M., Mehtätalo, L., Tolvanen, A. et al. (2019b). Impacts of drainage, restoration and warming on boreal wetland greenhouse gas fluxes. *Science of the Total Environment* 647: 169–181. https://doi.org/10.1016/j.scitotenv.2018.07.390.

Laurent, F. and Ruelland, D. (2011). Assessing impacts of alternative land use and agricultural practices on nitrate pollution at the catchment scale. *Journal of Hydrology* 409 (1): 440–450.

Lee, S., Hwang, S., Lee, S. et al. (2009). Landscape ecological approach to the relationships of land use patterns in watersheds to water quality characteristics. *Landscape and Urban Planning* 92 (2): 80–89.

Leifeld, J., Müller, M., and Fuhrer, J. (2011). Peatland subsidence and carbon loss from drained temperate fens. *Soil Use and Management* 27 (2): 170–176. https://doi.org/10.1111/j.1475-2743.2011.00327.x.

Li, S., Gu, S., Liu, W. et al. (2008). Water quality in relation to land use and land cover in the upper Han River Basin, China. *Catena* 75 (2): 216–222.

Li, S., Gu, S., Tan, X., and Zhang, Q. (2009). Water quality in the upper Han River basin, China: the impacts of land use/land cover in riparian buffer zone. *Journal of Hazardous Materials* 165 (1): 317–324.

Liu, R., Wang, J., Shi, J. et al. (2014). Runoff characteristics and nutrient loss mechanism from plain farmland under simulated rainfall conditions. *Science of the Total Environment* 468: 1069–1077.

Lohila, A., Minkkinen, K., Aurela, M. et al. (2011). Greenhouse gas flux measurements in a forestry-drained peatland indicate a large carbon sink. *Biogeosciences* 8 (11): 3203–3218. https://doi.org/10.5194/bg-8-3203-2011.

Loisel, J., Gallego-Sala, A.V., Amesbury, M.J. et al. (2021). Expert assessment of future vulnerability of the global peatland carbon sink. *Nature Climate Change* 11 (1): 70–77. https://doi.org/10.1038/s41558-020-00944-0.

Lopez, R.D., Nash, M.S., Heggem, D.T., and Ebert, D.W. (2008). Watershed vulnerability predictions for the Ozarks using landscape models. *Journal of Environmental Quality* 37 (5): 1769–1780.

Lund, M., Christensen, T.R., Lindroth, A., and Schubert, P. (2012). Effects of drought conditions on the carbon dioxide dynamics in a temperate peatland. *Environmental Research Letters* 7 (4): 045704. https://doi.org/10.1088/1748-9326/7/4/045704.

Ma, Z., Ni, H., Zeng, H., and Wei, J. (2011). Function formula for first flush analysis in mixed watersheds: a comparison of power and polynomial methods. *Journal of Hydrology* 402 (3): 333–339.

Maillard, P. and Santos, N.A.P. (2008). A spatial-statistical approach for modeling the effect of non-point source pollution on different water quality parameters in the Velhas river watershed–Brazil. *Journal of Environmental Management* 86 (1): 158–170.

Martins, M.V.A., Mane, M.Â., Frontalini, F. et al. (2015). Early diagenesis and clay mineral adsorption as driving factors of metal pollution in sediments: the case of Aveiro Lagoon (Portugal). *Environmental Science and Pollution Research* 22 (13): 10019–10033.

May, L., House, W.A., Bowes, M., and McEvoy, J. (2001). Seasonal export of phosphorus from a lowland catchment: upper River Cherwell in Oxfordshire, England. *Science of the Total Environment* 269 (1): 117–130.

Meyer, J.L., Paul, M.J., and Taulbee, W.K. (2005). Stream ecosystem function in urbanizing landscapes. *Journal of the North American Benthological Society* 24 (3): 602–612.

Mitsch, W.J. and Gosselink, J.G. (2000). *Wetlands*, 3e. New York, USA: Wiley New York.

Munir, T.M., Perkins, M., Kaing, E., and Strack, M. (2015). Carbon dioxide flux and net primary production of a boreal treed bog: responses to warming and water-table-lowering simulations of climate change. *Biogeosciences* 12: 1091–1111. https://doi.org/10.5194/bg-12-1091-2015.

Nash, D.M., Watkins, M., Heaven, M.W. et al. (2015). Effects of cultivation on soil and soil water under different fertiliser regimes. *Soil and Tillage Research* 145: 37–46.

Neris, J., Jiménez, C., Fuentes, J. et al. (2012). Vegetation and land-use effects on soil properties and water infiltration of Andisols in Tenerife (Canary Islands, Spain). *Catena* 98: 55–62.

Nie, W., Yuan, Y., Kepner, W. et al. (2011). Assessing impacts of landuse and landcover changes on hydrology for the upper San Pedro watershed. *Journal of Hydrology* 407 (1): 105–114.

Nugent, K.A., Strachan, I.B., Strack, M. et al. (2018). Multi-year net ecosystem carbon balance of a restored peatland reveals a return to carbon sink. *Global Change Biology* 24 (12): 5751–5768. https://doi.org/10.1111/gcb.14449.

Olefeldt, D., Euskirchen, E.S., Harden, J. et al. (2017). A decade of boreal rich fen greenhouse gas fluxes in response to natural and experimental water table variability. *Global Change Biology* 23 (6): 2428–2440. https://doi.org/10.1111/gcb.13612.

Panagopoulos, I., Mimikou, M., and Kapetanaki, M. (2007). Estimation of nitrogen and phosphorus losses to surface water and groundwater through the implementation of the SWAT model for Norwegian soils. *Journal of Soils and Sediments* 7 (4): 223–231.

Panagopoulos, Y., Makropoulos, C., and Mimikou, M. (2011). Diffuse surface water pollution: driving factors for different geoclimatic regions. *Water Resources Management* 25 (14): 3635–3660.

Planas-Clarke, A.M., Chimner, R.A., Hribljan, J.A. et al. (2020). The effect of water table levels and short-term ditch restoration on mountain peatland carbon cycling in the Cordillera Blanca, Peru. *Wetlands Ecology and Management* 28 (1): 51–69. https://doi.org/10.1007/s11273-019-09694-z.

Poor, C.J. and McDonnell, J.J. (2007). The effects of land use on stream nitrate dynamics. *Journal of Hydrology* 332 (1): 54–68.

Poorter, H., Niklas, K.J., Reich, P.B. et al. (2012). Biomass allocation to leaves, stems and roots: meta-analyses of interspecific variation and environmental control. *New Phytologist* 193 (1): 30–50. https://doi.org/10.1111/j.1469-8137.2011.03952.x.

Qaiser, K., Ahmad, S., Johnson, W., and Batista, J. (2011). Evaluating the impact of water conservation on fate of outdoor water use: a study in an arid region. *Journal of Environmental Management* 92 (8): 2061–2068.

Radu, D.D. and Duval, T.P. (2018). Precipitation frequency alters peatland ecosystem structure and CO 2 exchange: contrasting effects on moss, sedge, and shrub communities. *Global Change Biology* 24 (5): 2051–2065. https://doi.org/10.1111/gcb.14057.

Ratcliffe, J.L., Campbell, D.I., Clarkson, B.R. et al. (2019). Water table fluctuations control CO_2 exchange in wet and dry bogs through different mechanisms. *Science of the Total Environment* 655: 1037–1046. https://doi.org/10.1016/j.scitotenv.2018.11.151.

Reichenberger, S., Bach, M., Skitschak, A., and Frede, H. (2007). Mitigation strategies to reduce pesticide inputs into ground-and surface water and their effectiveness; a review. *Science of the Total Environment* 384 (1): 1–35.

Ren, W., Zhong, Y., Meligrana, J. et al. (2003). Urbanization, land use, and water quality in Shanghai: 1947–1996. *Environment International* 29 (5): 649–659.

Robroek, B.J., Albrecht, R.J., Hamard, S. et al. (2016). Peatland vascular plant functional types affect dissolved organic matter chemistry. *Plant and Soil* 407 (1–2): 135–143. https://doi.org/10.1007/s11104-015-2710-3.

Sahu, M. and Gu, R.R. (2009). Modeling the effects of riparian buffer zone and contour strips on stream water quality. *Ecological Engineering* 35 (8): 1167–1177.

Salimi, S., Almuktar, S., and Scholz, M. (2021a). The impact of climate change on wetland ecosystems: a critical review of experimental wetland mesocosms. *Journal of Environmental Management* 286: 112160. https://doi.org/10.1016/j.jenvman.2021.112160.

Salimi, S., Berggren, M., and Scholz, M. (2021b). Response of the peatland carbon dioxide sink function to future climate change scenarios and water level management. *Global Change Biology* 27: 5154–5168. https://doi.org/10.1111/gcb.15753.

Samson, M., Słowińska, S., Słowiński, M. et al. (2018). The impact of experimental temperature and water level manipulation on carbon dioxide release in a poor fen in northern Poland. *Wetlands* 38 (3): 551–563. https://doi.org/10.1007/s13157-018-0999-4.

Schilling, K.E. and Spooner, J. (2006). Effects of watershed-scale land use change on stream nitrate concentrations. *Journal of Environmental Quality* 35 (6): 2132–2145.

Schimelpfenig, D.W., Cooper, D.J., and Chimner, R.A. (2014). Effectiveness of ditch blockage for restoring hydrologic and soil processes in mountain peatlands. *Restoration Ecology* 22 (2): 257–265. https://doi.org/10.1111/rec.12053.

Schmidt, M.W., Torn, M.S., Abiven, S. et al. (2011). Persistence of soil organic matter as an ecosystem property. *Nature* 478 (7367): 49–56. https://doi.org/10.1038/nature10386.

Scholz, M. (2003a). Performance predictions of mature experimental constructed wetlands, which treat urban water receiving high loads of lead and copper. *Water Research* 37 (6): 1270–1277.

Scholz, M. (2003b). Sustainable operation of a combined flood-attenuation wetland and dry pond system. *Water and Environment Journal* 17 (3): 171–175.

Scholz, M. (2005). Hydraulics, water quality and vegetation characteristics of groundwater-fed open ditches. *Water and Environment Journal* 19 (1): 8–16.

Scholz, M. (2006). *Wetland Systems to Control Urban Runoff.* Amsterdam, Netherlands: Elsevier.

Scholz, M. (2010). *Wetland Systems: Storm Water Management Control.* Berlin: Springer.

Scholz, M. and Xu, J. (2002). Comparison of constructed reed beds with different filter media treating simulated mine drainage water. *Ecological Engineering* 18 (3): 385–390.

Schrautzer, J., Asshoff, M., and Müller, F. (1996). Restoration strategies for wet grasslands in northern Germany. *Ecological Engineering* 7 (4): 225–278.

Sen, P.K. (1968). Estimates of the regression coefficient based on Kendall's tau. *Journal of the American Statistical Association* 63 (324): 1379–1389.

Shaver, G.R., Billings, W.D., Chapin, F.S. et al. (1992). Global change and the carbon balance of arctic ecosystems. *BioScience* 42: 433–441. https://doi.org/10.2307/1311862.

Shrestha, S., Kazama, F., and Newham, L.T. (2008). A framework for estimating pollutant export coefficients from long-term in-stream water quality monitoring data. *Environmental Modelling & Software* 23 (2): 182–194.

Singh, A., Jakubowski, A.R., Chidister, I., and Townsend, P.A. (2013). A MODIS approach to predicting stream water quality in Wisconsin. *Remote Sensing of Environment* 128: 74–86.

Sliva, L. and Williams, D.D. (2001). Buffer zone versus whole catchment approaches to studying land use impact on river water quality. *Water Research* 35 (14): 3462–3472.

Smart, R., Soulsby, C., Cresser, M. et al. (2001). Riparian zone influence on stream water chemistry at different spatial scales: a GIS-based modelling approach, an example for the Dee, NE Scotland. *Science of the Total Environment* 280 (1): 173–193.

SMHI (2009). Dataserier med normalvärden för perioden 1961-1990 (Data Series with Normal Values for the Period 1961-1990), Sveriges Meteorologiska och Hydrologiska Institut (Swedish Meteorological and Hydrological Institute), Norrköping, Sweden.

Smith, A.P., Western, A.W., and Hannah, M.C. (2013). Linking water quality trends with land use intensification in dairy farming catchments. *Journal of Hydrology* 476: 1–12.

Sokal, R.R. and Rohlf, F.J. (1995). *Biometry*, 3e. New York: WH Freeman and Co.

Somura, H., Takeda, I., Arnold, J. et al. (2012). Impact of suspended sediment and nutrient loading from land uses against water quality in the Hii River basin, Japan. *Journal of Hydrology* 450: 25–35.

Stanley, E.H. and Maxted, J.T. (2008). Changes in the dissolved nitrogen pool across land cover gradients in Wisconsin streams. *Ecological Applications* 18 (7): 1579–1590.

Straková, P., Penttilä, T., Laine, J., and Laiho, R. (2012). Disentangling direct and indirect effects of water table drawdown on above-and belowground plant litter decomposition: consequences for accumulation of organic matter in boreal peatlands. *Global Change Biology* 18 (1): 322–335. https://doi.org/10.1111/j.1365-2486.2011.02503.x.

Sullivan, P.F., Arens, S.J., Chimner, R.A., and Welker, J.M. (2008). Temperature and microtopography interact to control carbon cycling in a high arctic fen. *Ecosystems* 11 (1): 61–76. https://doi.org/10.1007/s10021-007-9107-y.

Sulman, B.N., Desai, A.R., Saliendra, N.Z. et al. (2010). CO_2 fluxes at northern fens and bogs have opposite responses to inter-annual fluctuations in water table. *Geophysical Research Letters* 37 (19): https://doi.org/10.1029/2010GL044018.

Swenson, M.M., Regan, S., Bremmers, D.T. et al. (2019). Carbon balance of a restored and cutover raised bog: implications for restoration and comparison to global trends. *Biogeosciences* 16 (3): 713–731. https://doi.org/10.5194/bg-16-713-2019.

Tavares, A., Pato, R., and Magalhães, M. (2012). Spatial and temporal land use change and occupation over the last half century in a peri-urban area. *Applied Geography* 34: 432–444.

Tong, S.T. and Chen, W. (2002). Modeling the relationship between land use and surface water quality. *Journal of Environmental Management* 66 (4): 377–393.

Tota-Maharaj, K. and Scholz, M. (2010). Efficiency of permeable pavement systems for the removal of urban runoff pollutants under varying environmental conditions. *Environmental Progress & Sustainable Energy* 29 (3): 358–369.

Trepel, M. and Kluge, W. (2002). Ecohydrological characterisation of a degenerated valley peatland in northern germany for use in restoration. *Journal for Nature Conservation* 10: 155–169.

Trepel, M., Holsten, B., Kieckbusch, J., Otten, I. and Pieper, F. (2003). Influence of macrophytes on water level and flood dynamics in a riverine wetland in northern Germany. *Proceedings of the International Conference "EcoFlood – Towards Natural Flood Reduction Strategies"* (Warshaw, 6–13 September 2003). Institute for Land Reclamation and Grassland Farming, Raszyn, Poland.

Tu, J. (2009). Combined impact of climate and land use changes on streamflow and water quality in eastern Massachusetts, USA. *Journal of Hydrology* 379 (3): 268–283.

Tu, J. (2011). Spatially varying relationships between land use and water quality across an urbanization gradient explored by geographically weighted regression. *Applied Geography* 31 (1): 376–392.

Tu, J. (2013). Spatial variations in the relationships between land use and water quality across an urbanization gradient in the watersheds of northern Georgia, USA. *Environmental Management* 51 (1): 1–17.

UMWELTBUNDESAMT (2000). *Daten zur Umwelt: Der Zustand der Umwelt in Deutschland*, vol. 380, 2001. Berlin, Germany: Erich Schmidt Verlag (in German).

Van Herpe, Y. and Troch, P.A. (2000). Spatial and temporal variations in surface water nitrate concentrations in a mixed land use catchment under humid temperate climatic conditions. *Hydrological Processes* 14 (14): 2439–2455.

Van Vuuren, D.P., Edmonds, J., Kainuma, M. et al. (2011). The representative concentration pathways: an overview. *Climatic Change* 109 (1): 5–31. https://doi.org/10.1007/s10584-011-0148-z.

Voldoire, A., Sanchez-Gomez, E., Salas y Mélia, D. et al. (2013). The CNRM-CM5.1 global climate model: description and basic evaluation. *Climate Dynamics* https://doi.org/10.1007/s00382-011-1259-y.

Vymazal, J., Brix, H., Cooper, P.F. et al. (1998). *Constructed Wetlands for Wastewater Treatment in Europe*. Leiden, The Netherlands: Backhuys.

Walker, M.D., Wahren, C.H., Hollister, R.D. et al. (2006). Plant community responses to experimental warming across the tundra biome. *Proceedings of the National Academy of Sciences* 103 (5): 1342–1346. https://doi.org/10.1073/pnas.0503198103.

Walker, T.N., Ward, S.E., Ostle, N.J., and Bardgett, R.D. (2015). Contrasting growth responses of dominant peatland plants to warming and vegetation composition. *Oecologia* 178 (1): 141–151. https://doi.org/10.1007/s00442-015-3254-1.

Walsh, C.J., Roy, A.H., Feminella, J.W. et al. (2005). The urban stream syndrome: current knowledge and the search for a cure. *Journal of the North American Benthological Society* 24 (3): 706–723.

Wang, R., Xu, T., Yu, L. et al. (2013). Effects of land use types on surface water quality across an anthropogenic disturbance gradient in the upper reach of the Hun River, Northeast China. *Environmental Monitoring and Assessment* 185 (5): 4141–4151.

Ward, S.E., Bardgett, R.D., McNamara, N.P., and Ostle, N.J. (2009). Plant functional group identity influences short-term peatland ecosystem carbon flux: evidence from a plant removal experiment. *Functional Ecology* 23 (2): 454–462. https://doi.org/10.1111/j.1365-2435.2008.01521.x.

Ward, S.E., Ostle, N.J., Oakley, S. et al. (2013). Warming effects on greenhouse gas fluxes in peatlands are modulated by vegetation composition. *Ecology Letters* 16 (10): 1285–1293. https://doi.org/10.1111/ele.12167.

Watanabe, M., Suzuki, T., O'Ishi, R. et al. (2010). Improved climate simulation by MIROC5: mean states, variability, and climate sensitivity. *Journal of Climate* https://doi.org/10.1175/2010JCLI3679.1.

Weltzin, J.F., Pastor, J., Harth, C. et al. (2000). Response of bog and fen plant communities to warming and water-table manipulations. *Ecology* 81 (12): 3464–3478. https://doi.org/10.1890/0012-9658(2000)081[3464:ROBAFP]2.0.CO;2.

Weltzin, J.F., Bridgham, S.D., Pastor, J. et al. (2003). Potential effects of warming and drying on peatland plant community composition. *Global Change Biology* 9 (2): 141–151. https://doi.org/10.1046/j.1365-2486.2003.00571.x.

Williams, M., Hopkinson, C., Rastetter, E. et al. (2005). Relationships of land use and stream solute concentrations in the Ipswich River basin, northeastern Massachusetts. *Water, Air, and Soil Pollution* 161 (1–4): 55–74.

Wilson, D., Farrell, C.A., Fallon, D. et al. (2016). Multiyear greenhouse gas balances at a rewetted temperate peatland. *Global Change Biology* 22 (12): 4080–4095. https://doi.org/10.1111/ gcb.13325.

Wisskirchen, R. and Häupler, H. (1998). *Standartliste der Farn- und Blütenpflanzen Deutschlands*. Verlag Eugen Ulmer. Stuttgart, 765: Bundesamt für Naturschutz (in German).

Woli, K.P., Nagumo, T., Kuramochi, K., and Hatano, R. (2004). Evaluating river water quality through land use analysis and N budget approaches in livestock farming areas. *Science of the Total Environment* 329 (1): 61–74.

Woodhouse, C. and Duff, J.B.S. (2004). Treatment of Log Yard Runoff in an Aerobic Trickling Filter. *Water Quality Research Journal of Canada* 39 (3): 230–236.

Wu, L., Long, T., Liu, X., and Guo, J. (2012). Impacts of climate and land-use changes on the migration of non-point source nitrogen and phosphorus during rainfall-runoff in the Jialing River Watershed, China. *Journal of Hydrology* 475: 26–41.

Wu, J. and Roulet, N.T. (2014). Climate change reduces the capacity of northern peatlands to absorb the atmospheric carbon dioxide: the different responses of bogs and fens. *Global Biogeochemical Cycles* 28 (10): 1005–1024. https://doi.org/10.1002/2014GB004845.

Yin, Z., Walcott, S., Kaplan, B. et al. (2005). An analysis of the relationship between spatial patterns of water quality and urban development in Shanghai, China, Computers. *Environment and Urban Systems* 29 (2): 197–221.

Yin, J., Yin, Z., Zhong, H. et al. (2011). Monitoring urban expansion and land use/land cover changes of Shanghai metropolitan area during the transitional economy (1979–2009) in China. *Environmental Monitoring and Assessment* 177 (1–4): 609–621.

Yu, Z.C. (2012). Northern peatland carbon stocks and dynamics: a review. *Biogeosciences* 9 (10): 4071. https://doi.org/10.5194/ bg-9-4071-2012.

Zhang, H.X. and Yu, S.L. (2004). Applying the first-order error analysis in determining the margin of safety for total maximum daily load computations. *Journal of Environmental Engineering* 130 (6): 664–673.

Zhang, X., Shao, M., and Fu, H. (2000). The relationship of mineral nitrogen leaching in perennial ryegrass land with its runoff loss. *Acta Agrestia Sinica* 8 (2): 82–87.

Zhang, N., Yu, Y., Hong, B., and Chen, J. (2003). Factors influencing phosphorus loss by runoff process from farmlands in the dianchi watershed. *Chinese Journal of Environmental Science* 24 (3): 155–157.

Zhang, X., Cao, W., Guo, Q., and Wu, S. (2010). Effects of landuse change on surface runoff and sediment yield at different watershed scales on the Loess Plateau. *International Journal of Sediment Research* 25 (3): 283–293.

Zhang, P., Liu, Y., Pan, Y., and Yu, Z. (2013). Land use pattern optimization based on CLUE-S and SWAT models for agricultural non-point source pollution control. *Mathematical and Computer Modelling* 58 (3): 588–595.

Zhao, J., Peichl, M., and Nilsson, M.B. (2017). Long-term enhanced winter soil frost alters growing season CO_2 fluxes through its impact on vegetation development in a boreal peatland. *Global Change Biology* 23 (8): 3139–3153. https://doi.org/10.1111/gcb.13621.

Zhong, Y., Jiang, M., and Middleton, B.A. (2020). Effects of water level alteration on carbon cycling in peatlands. *Ecosystem Health and Sustainability* 6 (1): 1806113. https://doi.org/10.1080/20964129.2020.1806113.

2

Urban Water and Sustainable Drainage Systems

2.1 Full Silt Traps Discharging into Watercourses

2.1.1 Introduction

The dimensioning of silt traps is predominantly based upon the settlement properties of the particles and the maximum flow-through velocity. Generally, the outflow should be located on the same axis as the inflow, so that continuous flow ensures particle settlement over the full length of the silt trap (Schmitt et al. 1999). The structure should be small and simple to keep capital costs low.

Various international and British guidelines (e.g., British Standards, British Research Establishment and Construction Industry Research and Information Association) exist for the design of outflow structures (Abwassertechnische Vereinigung e. V. 1994, 1996; Pontier et al. 2001; Schmitt et al. 1999). In practice, designs vary considerably. However, silt traps are usually longer than wider by a factor of up to five. Silt trap depths vary considerably but are commonly approximately 0.40 m (Abwassertechnische Vereinigung e. V. 1994, 1996; Pontier et al. 2001; Salehi et al. 1997; Schmitt et al. 1999).

However, it is often the lack of maintenance concerning outflow structures such as silt traps that leads to problems years after the structure has been commissioned. Accumulated sediment must be removed from silt traps to retain their original design performance and to avoid re-suspension of accumulated pollutants during storms, causing pollution in the receiving water bodies (Abwassertechnische Vereinigung e. V. 1994, 1996; Pontier et al. 2001; Salehi et al. 1997).

In this case study, the Braid Burn, a semi-natural urban watercourse (Figure 2.1), which receives polluted surface water runoff, was studied. This watercourse is a small stream (discharge of usually less than 1 $m^3\ s^{-1}$) with a total catchment area of approximately 30 M m^2. The upstream reaches (urban streams: Howden Burn and Bonaly Burn) of the catchment area are steep and rural. Downstream of this rural upland area, the stream passes through suburban areas of Edinburgh; Hermitage (town area) and subsequently the golf course development near The King's Buildings (Science and Engineering Campus). The stream then flows within a culvert beneath a Shopping Centre at Cameron Toll (town area) and through Edinburgh, where it discharges into the Firth of Forth (North Sea) at Portobello (town area) according to Bullen Consulting (2002).

The overall aim of Chapter 2.1 is to analyse the treatment efficiency of a full silt trap. The objectives are to assess (i) the performance in removal of pollutants including suspended solids (SS), turbidity and heavy metals during dry- and wet-weather flow conditions; (ii) the distribution of sediment according to size within the full silt trap; (iii) the spatial distribution of metals within the full silt trap and the receiving urban watercourse as an indication of pollutant accumulation;

Sustainable Water Systems, First Edition. Miklas Scholz.

Figure 2.1 Outlet structure of the silt trap and the Braid Burn (urban stream) at The King's Buildings (Science and Engineering Campus, The University of Edinburgh) in winter 2003. See Section 2.1.2 for design details. River in the front and concrete-based outlet structure in the back. *Source:* Scholz and Zettel (2004)/with permission of John Wiley & Sons.

(iv) the water quality of the receiving watercourse; (v) the silt trap design and operation performance before and after emptying and (vi) alternative design and operation options. Chapter 2.1 is based on an original article by Scholz and Zettel (2004).

2.1.2 Study Site, Materials and Methods

The stormwater outlet (Figure 2.1) is located at The King's Buildings. All collected surface water runoff at The King's Buildings flows through the silt trap and discharges into the Braid Burn (Figure 2.1).

The diameter of the partially submerged (0.37 m) stormwater pipe leading into the silt trap chamber was 0.54 m. The dimensions of the rectangular silt trap chamber were as follows: length = 1.84 m, width = 1.84 m, chamber height = 1.14 m and silt trap outlet height = 0.42 m. The stormwater pipe inlet-to-outlet angle is 90°. Approximately 0.27 m of the silt trap was filled with potentially mature sediment (up to eight years old) that was removed on 12 March 2003.

The Braid Burn sampling sites 'before outlet' and 'after outlet' were located 13 m upstream and 6 m downstream, respectively, of the sample station 'at outlet' (Figures 2.1 and 2.2). To investigate the mixing properties of the stream, a further sample station located at the street 'Liberton Brae' was chosen 77 m downstream of the station 'at outlet'.

Sediment samples were dried at 105°C. Each sample was weighed, and wet-sieving procedures (British Standards 1990) were used to remove clay (and partially fine silt) from coarser sediment.

The remaining material was dried and weighed and the difference in weight was considered as a total estimation of the clay and silt fractions. A sediment size distribution curve (or grading curve) was established with the remaining material (British Standards 1990).

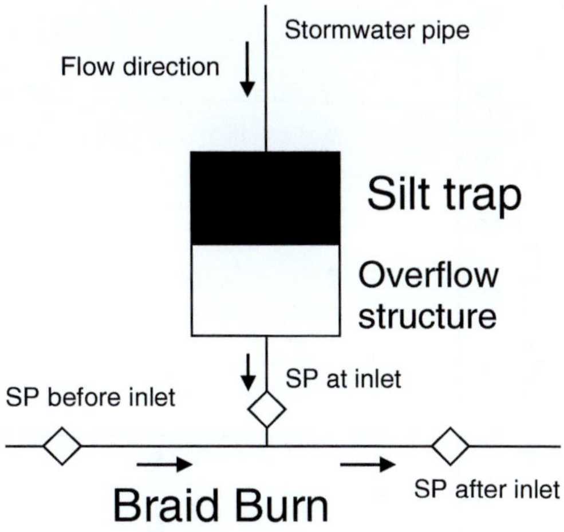

Figure 2.2 Diagram of the system and main sampling points (SP). Simple flow diagram showing sampling points before, at and after the inlet location.

Sampling took place between 15 January and 28 March 2003. The five days at 20°C ATU biochemical oxygen demand (BOD) was determined in all water samples with the OxiTop IS 12-6 system, a manometric measurement device, supplied by the Wissenschaftlich-Technische Werkstätten (Germany). Nitrification was suppressed by using N-Allylthiourea. A HACH 2100N turbidity meter was used.

Concerning the digestion methodology, 2 g of carefully sieved sediment samples (particles and sediment <0.002 m) were transferred to a 100-mL round bottom flask, and 21 mL of hydrochloric acid (strength of 37%) and 7 mL of nitric acid (strength of 69%) were added. The mixtures were heated on a Kjeldahl Digestion shelf for two hours to support the chemical reaction. After cooling, the solutions were filled into separate flasks and filtered through Number 41 Ashless Whatman filter papers. The filtered solutions were then filled up to 100 mL each with deionised water (Allen 1974). An Inductively Coupled Plasma Optical Emission Spectrometer (TJA IRIS instrument, ThermoElemental, USA) was used to determine total concentrations of metals in filtered (Whatman 1.2 μm cellulose nitrate membrane filter) water samples and the retrieved sediment extract.

Oxidised aqueous nitrogen was determined in selected water samples as the sum of nitrate-nitrogen and nitrite-nitrogen (generally low concentrations). Nitrate was reduced to nitrite by cadmium and determined as an azo dye at 540 nm (using a Perstorp Analytical EnviroFlow 3000 flow injection analyser) following diazotisation with sulphanilamide and subsequent coupling with N-1-naphthylethylenediamine dihydrochloride.

Ammoniacal nitrogen and ortho-phosphate-nitrogen were determined by automated colorimetry for selected water samples from reaction with hypochlorite and salicylate ions in solution in the presence of sodium nitrosopentacyanoferrate (nitroprusside), and reaction with acidic molybdate to form a phosphomolybdenum blue complex, respectively. The coloured complexes formed were measured spectrophotometrically at 655 and 882 nm, respectively, using a Bran and Luebbe autoanalyzer (Model AAIII) after Allen (1974).

2.1.3 Results and Discussion

2.1.3.1 Water and Sediment Quality of the Silt Trap

Concerning the outflow water quality of the silt trap before (33 samples per sampling location) and after emptying (5 samples per sampling location) on 12 March 2003, the standard deviations for all variables are high due to the elevated natural variability of the data (Butler and Davies 2000; Salehi et al. 1997; Schmitt et al. 1999). For example, the highly variable temporal distribution of SS is shown in Figure 2.3. The highlighted consent or threshold value of 30 mg L^{-1} indicates where the water quality exceeds the internationally accepted standard for secondary treated wastewater (e.g., treatment by an activated sludge system or trickling filter) according to Tchobanoglous et al. (2003).

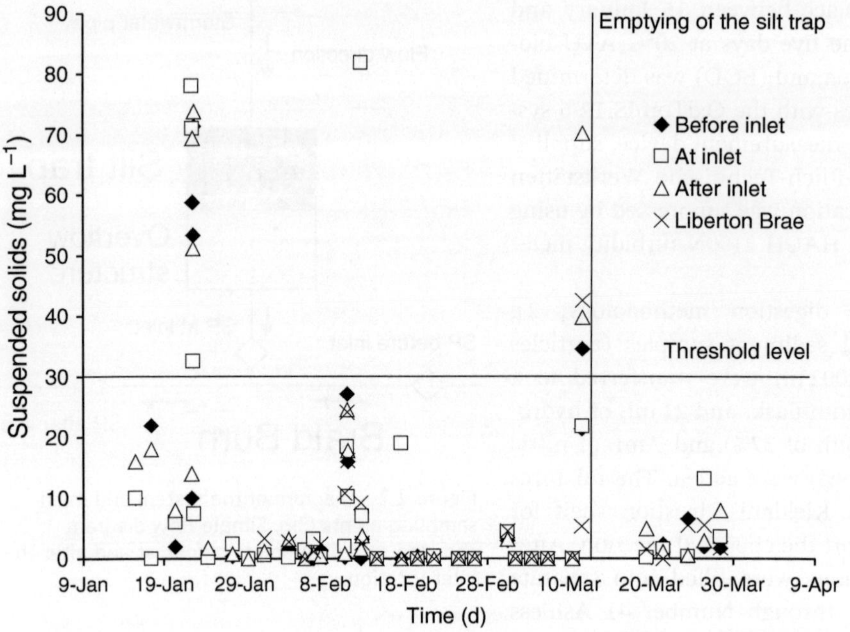

Figure 2.3 Distribution of suspended solids (SS) before, at and after the outlet where the silt trap discharges into the Braid Burn (urban stream), and further downstream at Liberton Brae (urban road). On 11 March 2003, the SS concentration at the outlet was 141.6 mg L^{-1}. This value has been omitted to aid visual clarity. The threshold of 30 mg L^{-1} is not met on three occasions before emptying the silt trap.

The emptying of the silt trap had a positive effect on the water quality despite the unusually short length of the chamber. SS and turbidity values were reduced by 76% and 82% to mean figures of 3.8 mg L^{-1} and 4.6 NTU, respectively.

The recovered sediment was contaminated with zinc, copper and nickel (approximately 940, 420 and 60 mg kg^{-1}, respectively). Therefore, the sediment was not suitable for local recycling as manure (Alloway 1995; Butler and Davies 2000).

The silt trap releases nitrate-nitrogen (8.69 mg L^{-1}) and ortho-phosphate-nitrogen (0.08 mg L^{-1}) into the Braid Burn (13 samples each). In contrast, ammoniacal nitrogen concentration of stormwater runoff is lower than the corresponding concentration of the stream water (4.59 mg L^{-1}). All detectable metals in the stream water are associated with low concentrations: barium (2.04 mg L^{-1}), boron (0.10 mg L^{-1}), calcium (32.90 mg L^{-1}), magnesium (6.18 mg L^{-1}), manganese (0.01 mg L^{-1}) and zinc (0.03 mg L^{-1}). However, the associated streamflow data sets are small (six samples per sample location).

Concerning the sediment properties, Figure 2.4 indicates the vertical sediment size distribution within the sediment of the full silt trap. The sediment is mixed in terms of its size distribution. However, the top layer of sediment (0.240–0.269 m above the bottom) contains less fine sediment. This relative loss can be explained by re-suspension of relatively fine sediment during wet-weather flow conditions (Pontier et al. 2001; Schmitt et al. 1999). Furthermore, re-suspension is enhanced by the unusual inflow to outflow angle of 90°. The ideal angle to minimise turbulence would be 180°, based on a common-sense approach (Abwassertechnische Vereinigung e. V. 1994, 1996; Butler and Davies 2000; Schmitt et al. 1999).

In addition to Figures 2.4 and 2.5 indicate that the sediment mixing during high flows leads to a random distribution of metals within the sediment. It follows that previously accumulated material

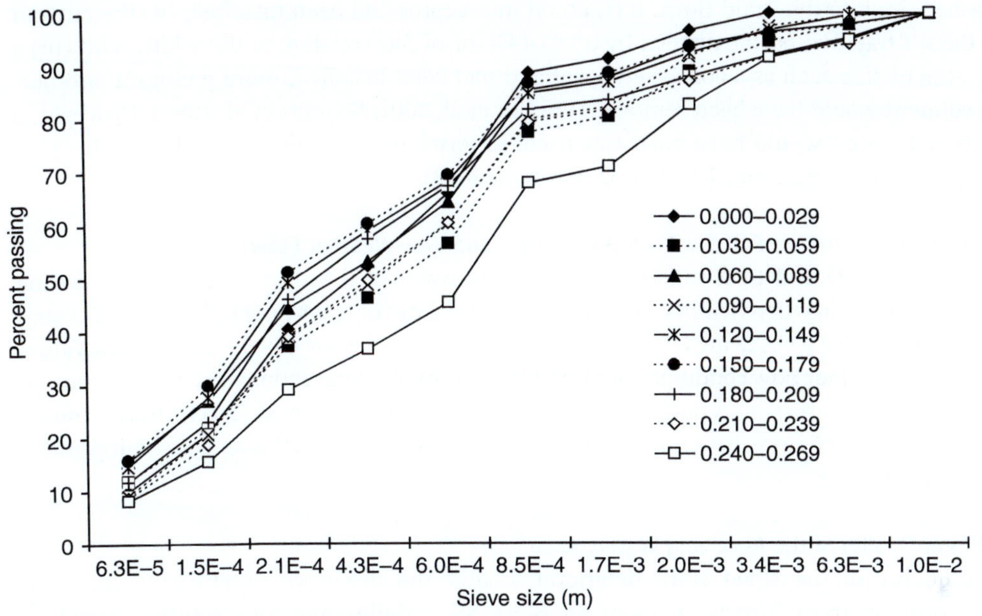

Figure 2.4 Sediment size distributions (or grading curves) within different layers (m) of the silt trap on 12 March 2003. Nine sieve curve lines that are only slightly overlaying each other on a few locations are shown.

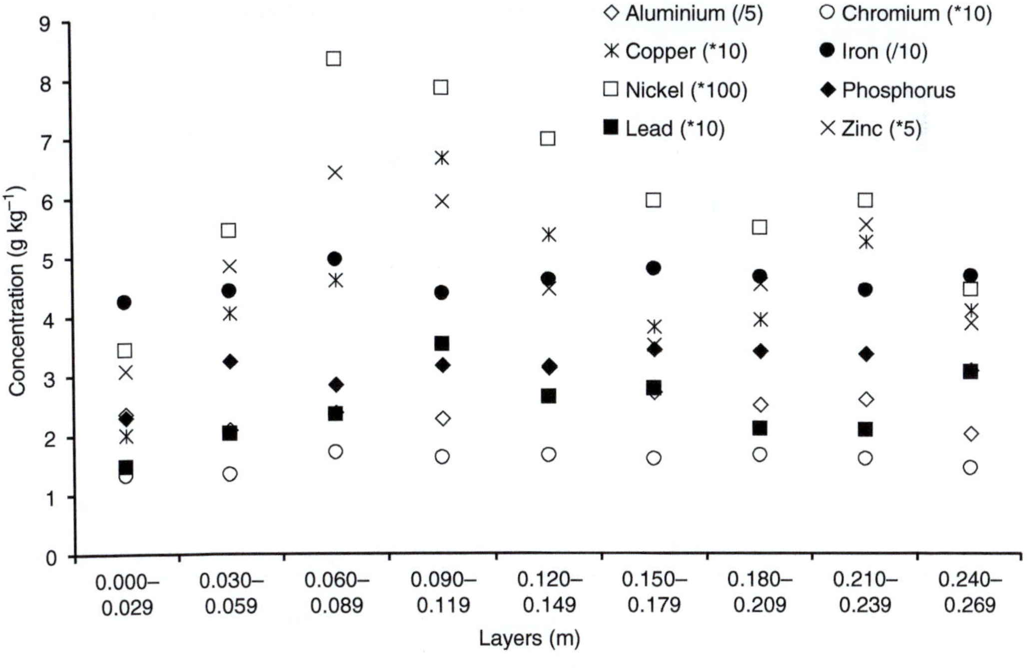

Figure 2.5 Metal concentrations within the different sediment layers of the silt trap on 12 March 2003. The sediment layers between 6 and 12 cm are most contaminated, but this statement is not very strong.

can be washed out into the Braid Burn. If (i) an oil interceptor had been installed; (ii) the current length of the silt trap had been extended by up to a factor of five (relative to the width) and (iii) a further treatment step such as a constructed wetland had been installed, more pollutant and particularly sediment would have been removed (Pontier et al. 2001; Schmitt et al. 1999). However, a more optimised design would have been expensive in terms of excavation costs due to the hilly terrain (Bullen Consulting 2002; Tchobanoglous et al. 2003).

2.1.3.2 Water Quality of the Full Silt Trap During Dry- and Wet-Weather Flow

The BOD, SS, turbidity and total solids for the outflow water of the silt trap are higher during wet-weather than during dry-weather flow conditions: 5.5 mg L^{-1}, 34.1 mg L^{-1}, 53.1 NTU and 597.0 mg L^{-1}, and 2.8 mg L^{-1}, 2.2 mg L^{-1}, 4.7 NTU and 300.9 mg L^{-1}, respectively. The outflow values for SS during wet-weather flow conditions are frequently above the internationally accepted consent value of 30 mg L^{-1} for treated secondary wastewater (Tchobanoglous et al. 2003). Furthermore, there is no obvious difference between dry- and wet-weather flow conditions concerning metal concentrations.

2.1.3.3 Water Quality of the Receiving Watercourse

The water quality of the Braid Burn deteriorated after the discharge of virtually untreated stormwater (full silt trap). This is apparent for BOD, SS, turbidity and total solids: 2.0 mg L^{-1}, 10.2 mg L^{-1}, 16.6 NTU and 200.6 mg L^{-1} before the outlet, and 2.2 mg L^{-1}, 12.7 mg L^{-1}, 19.1 NTU and 212.2 mg L^{-1} after the outlet, respectively (Figure 2.2). Furthermore, the mean SS concentration at Liberton Brae was only 5.5 mg L^{-1} indicating good mixing of particulate pollutants.

Concerning metal pollution, Figure 2.6 indicates the spatial distribution of metals identified in the sediment of the Braid Burn (Figure 2.1). Metals do not accumulate along the channelled stream

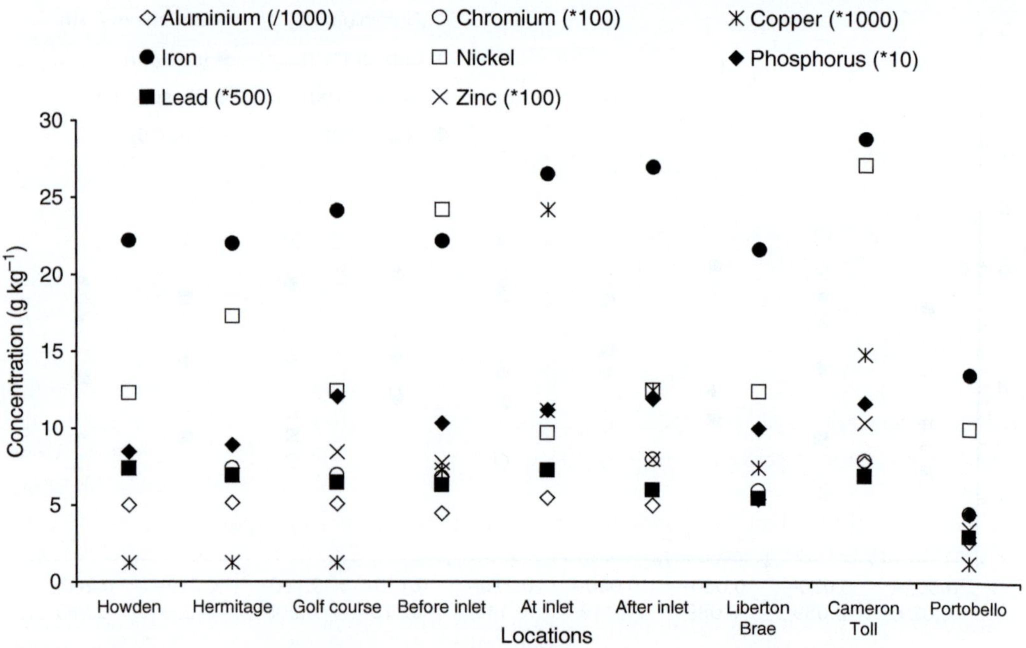

Figure 2.6 Metal concentrations within the sediment at different sampling locations of the Braid Burn (urban freshwater stream) on 13 February 2003 (after a prolonged period of rain). Howden (town area) is located near the spring and Portobello (town area) near the discharge location into the Firth of Forth (North Sea; salty water). The sample at Cameron Toll is most polluted, but this finding is not strong.

as previously indicated for other watercourses (Ciszewski 1998; Pontier et al. 2001), but are likely to be washed out into the Firth of Forth during storm events.

Furthermore, copper concentrations within the sediment of the Braid Burn immediately downstream of the silt trap are approximately 24 mg kg^{-1}. Figure 2.5 indicates that this relatively high concentration is likely to be linked to the surface water runoff from the Campus.

In order to reduce the risk of stream bed pollution, the author and his team recommend in principle the construction of a vertical flow (low footprint) constructed wetland after the silt trap (Figure 2.2) for the removal of SS and particularly heavy metals (Scholz 2004a, b; Scholz and Xu 2001). However, the current system layout (Figure 2.1) does not allow the retrofitting of a horizontal flow constructed treatment wetland due to lack of space. However, a small vertical-flow wetland might be able to treat the first foul flush after a storm event. An oil interceptor should also be located before the silt trap (Figure 2.2) that requires at least annual emptying. Regular cleaning of gully pots located upstream of the stormwater pipe would also help to reduce SS and associated BOD in the silt trap (Scholz 2004a).

2.1.4 Conclusions

The full silt trap was efficient in holding back SS during dry- but not wet-weather flow conditions. The accumulation capacity of pollutants in the silt trap was reduced due to high flow through velocities and an unconventional flow path design leading to re-suspension during periods of heavy rainfall.

The water quality of the receiving watercourse became polluted predominantly with SS (consent level >30 mg L^{-1}) due to the failing of the full silt trap. The sediment of the silt trap was contaminated with zinc, copper and nickel. Emptying of the silt trap resulted in significant water quality improvements despite a length-to-width ratio of 1 : 1 and a stormwater inlet to outlet angle of 90°.

A small vertical flow constructed treatment wetland should be installed after the silt trap to protect urban water from contaminants contained within the first flush of storm events. An oil interceptor could also be installed before the silt trap as an additional removal unit.

2.2 Filter Media, Plant Communities and Microbiology within Constructed Wetlands

2.2.1 Introduction

Reed beds are a cheap alternative to expensive treatment technology such as trickling filters and activated sludge processes. Vertical-flow and horizontal-flow wetlands based on soil, sand and/or gravel are used extensively to treat domestic and industrial wastewater (Adock et al. 1999; Bannister 1997; Cooper et al. 1996; Hammer 1989; Kadlec and Knight 1996; Robinson et al. 1999). They are also applied for passive treatment of diffuse pollution including mine wastewater drainage (Hammer 1989; Kadlec and Knight 1996; Mungur et al. 1997; Robinson et al. 1999) and motorway water runoff after storm events (McNeill and Olley 1998). Furthermore, wetlands serve as a wildlife conservation resource and can be seen as natural recreational places for the local community (Hawke and José 1996).

The functions of macrophytes within constructed wetlands have been reviewed intensively (Brix 1999). *Phragmites*, *Typha*, *Junctus* and other swamp plants are widely used in Europe and Northern America. A diverse microbial community is present in both the aerobic and anaerobic zones of wetland including a variety of bacteria, fungi, algae and protozoa (Cooper et al. 1996; Kadlec and Knight 1995; McNeill and Olley 1998).

Vertical-flow wetland technology can be compared to slow sand filters if cheap sand and gravel are used as traditional filter media. Moreover, micro-biological processes within wetlands and biological filters including biological activated carbon filters are expected to be similar (Hammer 1989). Controlled biological regeneration of activated carbon, slow sand and reed bed filter material due to sorption and subsequent metabolism of slowly biodegradable organic matter has been recognised by the water and wastewater treatment industry. Natural biological regeneration of exhausted porous media leads to an extended bed operation period for biological filters.

The diverse microbial ecology and plant community structure within complex wetland ecosystems has not yet been fully reported. It can be expected that microbial prey–predator relationships within freshwater wetlands are like those within seagrass communities in shallow coastal lagoons (Donelly and Herbert 1999). However, further research into wetland microbiology is required to perform sound modelling of microbial pollution indicators.

Chapter 2.2 aims to characterise experimental wetland filters with different design parameters including artificially planted macrophytes and microbial communities as well as filter media. The removal efficiencies for contaminants including copper and lead are compared and a cost-benefit description (based on the filter material used) performed. Scholz et al. (2001) provided the scientific basis for this chapter.

2.2.2 Materials and Methods

2.2.2.1 Environmental Conditions and Operation

A row of filter buckets was placed in the centre of a large unheated laboratory hall (windows at two sides and at the top), which was subject to sufficient direct and indirect variable natural (day-length) and artificial light (ambient light (Brix 1999): approximately between 700 and 1100 PAR (μmol m^{-2} s^{-1})). To approximately convert a natural daylight reading (waveband between 400 and 700 nm [PAR]) expressed in μmol m^{-2} s^{-1} to W m^{-2} divide by 4.6. However, the ambient light was not measured regularly with a PAR Quantum Sensor (SKP 200 and 215).

The climate was controlled to reduce parameters with high variations and potentially destructive effects on biota (freezing temperature, heavy rain and strong wind). For example, the temperature range (determined with Checktemp 2) within the laboratory during daytime (10 : 30) was between 14.2 and 27.8°C. However, the temperature in the laboratory was higher but more stable than outside. Therefore, biological growth rates were elevated, but better controlled.

The experiment was operated on a tidal flow system with inflow water from the Westbrook (a local brick-lined urban beck), which is subject to diffuse pollution including drainage water and sewage from combined sewer overflows during and after storm events. Copper and lead sulphate were added to the inflow water to give concentrations of dissolved copper and lead of 1.000 and 1.277 mg dm^{-3}, respectively, in order to simulate pollution including process water from mining (Read 1984). Natural Westbrook background concentrations of copper and lead were approximately 0.006 and 0.017 mg dm^{-3}, respectively.

The range of the theoretical hydraulic load per filter bucket was between 1.35 and 2.02 cm d^{-1} (mean: 1.91 cm d^{-1}). About 35% of the total water volume of each filter bucket was replaced on Monday, Wednesday and Friday during each week of operation. The water level was lowered carefully within one to two minutes until only approximately five millimetres of water remained on top of the submerged litter. The purpose of this procedure was not to destroy the anoxic biofilm structure. The inflow was distributed slowly and evenly into the *schmutzdecke* and litter zone of each filter. The effluent point was located at the centre of the bottom of the filter.

The average proportions of pure water evaporation, *Phragmites australis* evapotranspiration and *Typha latifolia* evapotranspiration accounted for approximately 0.08, 0.15 and 0.16 cm d^{-1}, respectively. For example, the average loss of water for Filter 6 accounted for approximately 0.39 cm d^{-1}.

The balanced controlled release fertiliser Osmocote (produced by Scotie Europe B. V., The Netherlands) was added to the top of Filter 6 (Table 2.1) during week 1 (21 January) and week 15 (28 April). Real-time indication is given because biological processes depend on seasonal changes. Approximately 50 g Osmocote was distributed within the charcoal media (Table 2.1). This N-P-K Fertiliser 14-13-13 contains 14% (w/w) total nitrogen, 13% (w/w) phosphorus pentoxide (P_2O_5) and 13% (w/w) potassium oxide (K_2O). Approximately 9.1% (w/w) of P_2O_5 (4.0% (w/w) total phosphorus) and 13% (w/w) of K_2O (10.8% (w/w) total potassium) were actually soluble in water.

Pure Simple Soap produced by Smith and Nephew (Birmingham) was dissolved in deionised water, and the solution was used as a means of pest control to reduce aphids (predominantly black flies) on plant leaves (predominantly *Typha latifolia*) according to March (1991). The soap consists of the environmentally friendly ingredients sodium tallowate, sodium palm kernelate, aqua (water), glycerin, sodium chloride, disodium phosphate, pentasodium pentetate and tetrasodium etidronate. Between 0.04 and 0.10 mg of dissolved soap was sprayed per cm^2 wetland area up to twice weekly after 15 March 2000 (week 9).

Table 2.1 Packing order of vertical-flow filter buckets simulating wetlands (before the start of the experiment).

Height (cm)	Filter 1	Filter 2	Filter 3	Filter 4	Filter 5	Filter 6
56–58	(Water/air)	(Water/air)	(Water/air)	(Water/air)	(Water/air)	(Water/air)
49–55	Water + C	Water + C	Water + C	Water + C	Water + C	Water + C
47–48	6	6 + A	6 + A + B	8 + A + B	8 + A + B	9 + A + B + Fs
41–46	6	6 + A	6 + A + B	8 + A + B	8 + A + B	8 + A + B
37–40	5	5 + A	5 + A + B	6 + A + B	7 + A + B	7 + A + B
35–36	5	5	5 + B	6 + B	7 + B	7 + B
33–34	4	4	4 + B	6 + B	6 + B	6 + B
29–32	4	4	4	5	6	6
25–28	4	4	4	5	5	5
21–24	3	3	3	4	5	5
17–20	3	3	3	4	4	4
15–16	3	3	3	3	4	4
9–14	2	2	2	3	3	3
0–8	1 + 2	1 + 2	1 + 2	1 + 2	1 + 2	1 + 2

1, cobblestones; 2, coarse gravel; 3, fine gravel; 4, pea-gravel; 5, coarse sand; 6, fine sand; 7, Filtralite; 8, activated carbon; 9, charcoal; A, *Phragmites australis*; B, *Typha latifolia*; C, marginal, floating and submerged plants; Fs, Osmocote.

2.2.2.2 Filter Media Composition

Wetland habitats were simulated on a laboratory scale with filter buckets of the following dimensions: height (58 cm), upper diameter (40 cm) and lower diameter (32 cm). The empty bucket volume was 59.2 dm^3. Six packing orders of filter media and plant roots were used in six vertical-flow wetland buckets (Table 2.1). The following aggregates were used (Table 2.1):

1) Large, washed Canterbury cobblestones (near the outlet) provided by a local aggregate supplier without a specification sheet.
2) Washed round Kent beach gravel (max. 40 mm).
3) Washed round gravel (approx. 8–20 mm).
4) Washed pea-gravel (approx. 4–10 mm). Aggregates 2–4 were provided by Hanson Aggregates Ltd. (Bradford).
5) Washed Ingleton sand (approx. 0.3–4.0 mm) provided by Hanson Aggregates Ltd. (Ingleton) and tested by Celtest Ltd. (Bangor).
6) Washed Derbyshire medium sand (approx. 0.15–2.36 mm) provided by Hanson Aggregates Ltd. (Bradford).
7) High-strength crushed Filtralite (1.5–2.5 mm) was supplied by Optiroc (Oslo, Norway). Filtralite is a light expanded clay aggregate which contains inert ceramic particles with dense shells surrounding porous cores; it is an expanded clay product with high porosity (Optiroc 1998).
8) Granular Activated Carbon Grade 207 EA (12 × 40 US Mesh) was supplied by Waterlink Sutcliffe Speakman Carbons Ltd., Ashton-in-Makerfield (Scholz and Martin 1998a, b).
9) BRIMAC 216 Charcoal (US mesh size: 20 × 60), supplied by Brimac Ltd. (Greenock), is a carbonaceous adsorbent derived from the carbonisation of appatite and is used primarily for liquid phase filtration. The adsorption capacity of 100 g BRIMAC 216 is 1340 mg lead and 540 mg copper.

2.2.2.3 Analytical Procedures

For heavy metal determination, sample water (5 cm^3) from the inflow (before adding metals), outflow and from the water layer on top of the litter zone was directly analysed with a Palintest Scanning Analyzer SA-1000. The performance range of this system was 0.005–0.100 (+/− 5%) mg dm^{-3} for lead and 0.700–2.000 mg dm^{-3} (+/− 10%) for copper. Sample water dilution for lead or concentration for copper (evaporation procedure used) was sometimes required.

If metal concentrations were below the detection limit, element concentrations were determined (for selected sample waters only) with the simultaneous ion detection system PlasmaQuad3 High Performance Quadrupole Inductively Coupled Plasma Mass Spectrometry (ICP-MS) produced by VG Elemental. Five samples (5 cm^3 each) were taken randomly from the sampling area or the inflow or outflow water tanks and mixed with each other to reduce bias and to increase accuracy. Lead and copper concentrations were based on a mean of three sub-samples (standard deviations <0.1 mg dm^{-3}).

A Hanna HI 8519N micro-processor bench-top pH meter, a Hanna HI 9142 portable waterproof dissolved oxygen (DO) meter and a Hanna C 102 multiparameter turbidity and ion specific meter were used to determine pH, DO and turbidity as well as low-range iron (HI 93746-01), respectively.

All analytical procedures including biochemical oxygen demand (BOD), suspended solids, total solids, DO and pH were performed according to the American standard methods for the examination of water and wastewater (Water Environment Federation 1995) except counting of microorganisms (Chapter 2.2.2.4). No replicates were taken unless stated otherwise in the standard methods or indicated below. Standard statistical methods like standard deviations,

Shannon–Weaver Diversity Index and Analysis of Variance (ANOVA) were applied (Fowler and Cohen 1998). Outliers were excluded from the analysis.

2.2.2.4 Micro-biological and Plant Examinations

Five water samples (5 cm^3 each) were taken randomly from the inflow water tank, outflow water tank or the top layer of filter media (mixed with plant litter) and mixed with each other to reduce bias and to increase accuracy and validity of the subsequent analysis of sub-samples. The analysis took place twice per week (predominantly on Tuesdays and Thursdays). Organisms within samples from the liquid phase of the top layer of filter media included free living, crawling and attached forms.

Immediately after mixing, micro-organisms including protozoa were counted in a first sub-sample of each filter by using a Fuchs-Rosenthal Haemacytometer (counting chamber: 0.20 mm× 0.25 mm× 0.25 mm) and a Wang Biomedical Research Microscope 6000 (bright field and phase-contrast). One count represents 313 organisms per cm^{-3} (if no mean was taken). Sub-samples (two replicates) from the inflow, outflow and top water layer of each filter were taken twice a week for analysis of micro-organisms including protozoa, actinopoda (AC) and diatoms (DT). Protozoa were identified based on their mode of movement (life action; e.g., movement with flagella or cilia) and morphology according to the methodology applied for the general subdivision of protozoa in British sewage treatment processes (Scholz and Martin 1998a, b).

Invertebrates including rotifers, water fleas, nematodes and segmented worms within a second sub-sample were counted in a Sedgewick-Rafter Cell S50 (counting chamber: 1 mm × 50 mm × 20 mm) at the same time as the micro-organisms described above. One count represents 10 organisms per cm^{-3} (if no mean was taken). The abundance of all organisms is expressed in numbers cm^{-3}. Furthermore, large macro-invertebrates including freshwater shrimps and hog lice were counted per dm^{-3} using a strainer to catch them (mesh size: 0.5 mm). This operation was performed only monthly.

A third sub-sample of 5 cm^{-3} was taken, and selected bacteria were cultured according to the standard methods summarised in the Oxoid Manual (Bridson 1998). Bacterial counts were obtained for water samples from the tops and outflows of the filter as well as the inflow. The colony numbers (colony-forming units; CFU) of total coliforms (TC; Bridson 1998) and total heterotrophic bacteria (THB; Bridson 1998) were estimated after growth on violet red bile glucose agar and nutrient agar, respectively. The incubation temperatures were 30°C. An average of three replicates was taken for each data point. Data were considered representative, if the difference between each replicate was less than 50% of the median. Results were expressed as CFU or colonies cm^{-3}. An increase of replicates and sample frequency for all microbiological data would have been desirable to decrease the variance. However, project resources were limited.

Mature *Phragmites australis* plants were taken from a natural University-based pond (Woodhall). *Typha latifolia* were separated from mature plants grown in a nursery from seeds. Mid-leaf colour (Pavey 1978) and plant growth indicated by height (Hendry and Grime 1993) to the nearest centimetre were determined twice a month for each plant. All plant heights were measured from the water surface (maximum water level) to the top of the plant including top leaf. These investigations were undertaken by one researcher only to reduce variability.

The spatial location of each stem was recorded to investigate edge effects of the filter buckets at the end of the growing season (research on-going). The root sizes have not yet been measured, because further research investigations are in progress. Dead plant material was removed once a month and placed in the litter zone of the corresponding filter. Further studies investigating the release of metals from decaying plant litter are currently in progress. However, this investigation is based on a different data set that includes the end of the growing season (autumn).

2.2.3 Results and Discussion

2.2.3.1 Filter Media Costs

A brief description of material prices is shown in Table 2.2 to compare setting-up costs for all wetland constructions. The chosen filter configurations (Table 2.2) have the following cost ratio in West Yorkshire (spring 2000); Filters 1–6 relate to the standardised set-up cost ratios 1 : 2 : 3 : 37 : 41 : 42. It follows, for example, that the costs for Filter 6 in comparison to Filter 1 are about 42 times greater (Table 2.2).

2.2.3.2 Comparison of Treatment Efficiencies

Heavy metals including copper and lead are adsorbed and absorbed well by porous filter material. Past experience with biological filters (used predominantly in the tertiary treatment stage) showed that expensive media including activated carbon and charcoal have greater adsorption surface areas than cheap sand or gravel (Table 2.1) according to Scholz and Martin (1998a, b). Copper and lead were removed well by the high-cost filters (Filters 4–6) containing additional adsorption media including activated carbon. In comparison, low-cost sand and gravel filters (Filter 1–3) showed an early breakthrough of metals. However, all filters performed equally well in terms of copper and lead reduction after nine weeks of operation (Figures 2.7 and 2.8).

Table 2.3 summarises the inflow water quality after addition of dissolved lead and copper sulphate which resulted in relatively low numbers of bacteria. The standard deviations for all inflow parameters (except for copper and pH) were high due to natural and artificial water quality variations (predominantly irregular sewage input) of the Westbrook. Nevertheless, the differences between the mean and median values were small (except for TC), which indicates adequate normal data distributions and consequently a stable wastewater inflow. Moreover, the *schmutzdecke* and/or

Table 2.2 Comparison of maximum capital costs (£ sterling) between all wetlands (Filters 1–6) for a filter volume of 1.00 m^3 × 1.00 m^3 × 0.48 m^3.

Material	Filter 1	Filter 2	Filter 3	Filter 4	Filter 5	Filter 6
1	2.00	2.00	2.00	2.00	2.00	2.00
2, 3 and 4	2.15	3.15	3.15	2.00	2.00	1.60
5 and 6	1.80	1.80	1.80	2.00	1.70	1.70
7	0.00	0.00	0.00	0.00	40.00	40.00
8	0.00	0.00	0.00	304.00	304.00	228.00
9	0.00	0.00	0.00	0.00	0.00	60.00
A	0.00	10.00	6.00	6.00	6.00	6.00
B	0.00	0.00	12.00	12.00	12.00	12.00
C	2.00	2.00	2.50	2.50	2.50	2.50
Fs	0.00	0.00	0.00	0.00	0.00	18.00
Total	8.95	18.95	27.45	330.50	369.80	371.80
Ratio	1	2.1	3.1	36.9	41.3	41.5

1, cobblestones; 2, coarse gravel; 3, fine gravel; 4, pea-gravel; 5, coarse sand; 6, fine sand; 7, Filtralite; 8, activated carbon; 9, charcoal; A, *Phragmites australis*; B, *Typha latifolia*; C, marginal, floating and submerged plants (optional; developed naturally in this case study); Fs, Osmocote.

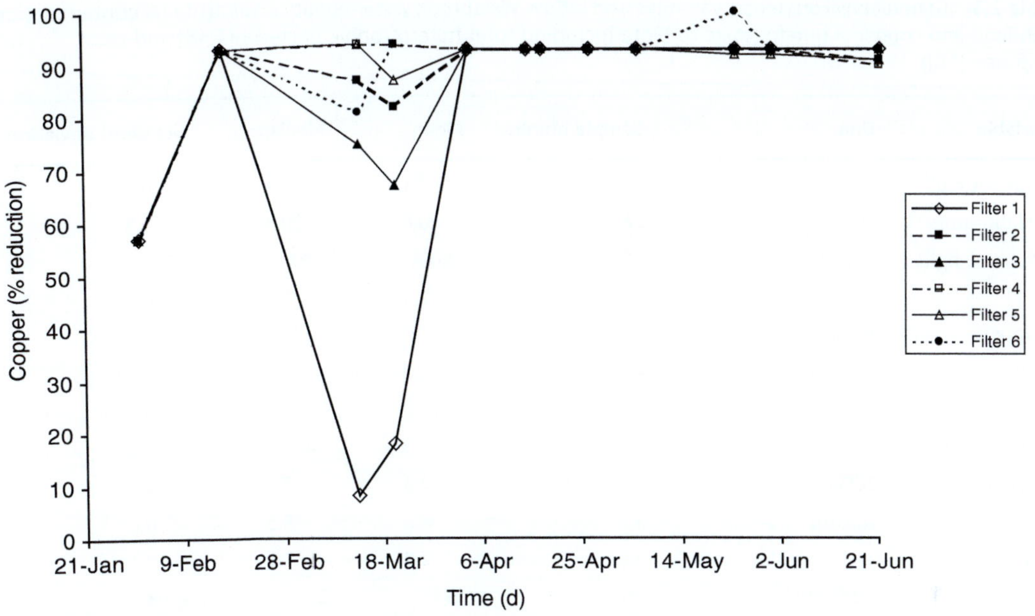

Figure 2.7 Relative % reduction of dissolved copper (mg dm^{-3}) for the outflow of Filters 1–6 (excluding concentrated samples, except ICP-MS result for Filter 6 during week 18; May 24). Note: reduction $(\%) = \frac{(in - out) \times 100\ (\%)}{in}$, where in = inflow value and out = outflow value. After a maturation period of about two months, the removal rate stabilised at about 90%.

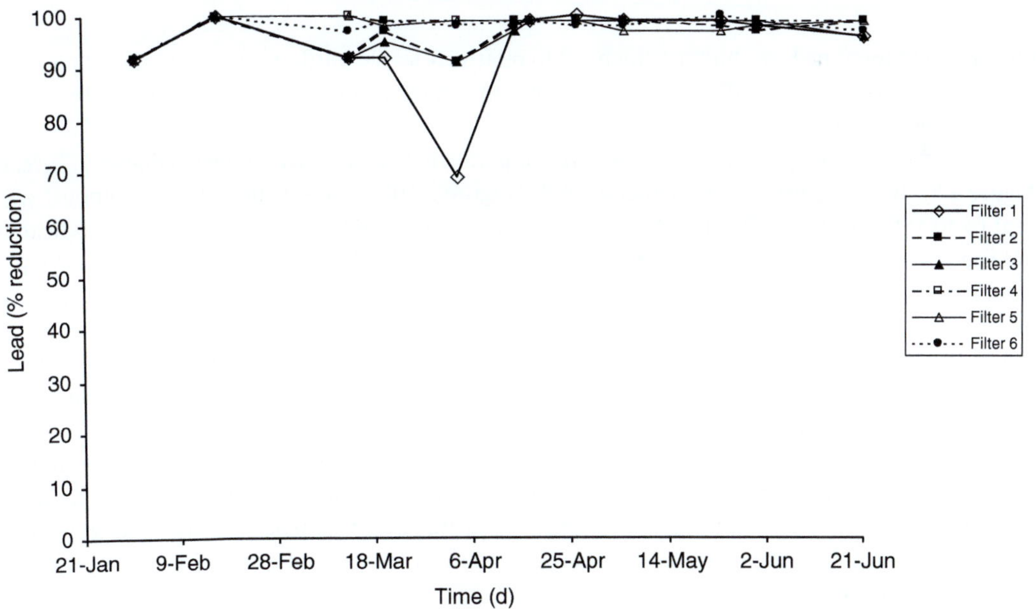

Figure 2.8 Relative % reduction of dissolved lead (mg dm^{-3}) for the outflow of Filters 1–6 (including diluted samples). Note: reduction $(\%) = \frac{(in - out) \times 100\ (\%)}{in}$, where in = inflow value and out = outflow value. After a maturation period of about three months, the removal rate stabilised at about 95%.

Table 2.3 General environmental variables and inflow Westbrook water quality after artificial contamination with lead and copper sulphate (toxic to biota including total heterotrophic bacteria [THB] and total coliforms [TC]).

Variable	Unit	Sample number	Mean	Median	Standard deviation
T (at source)[b]	°C	27	13.1	11.4	6.2
T (laboratory)[b]	°C	28	20.6	20.4	3.3
Ambient light	PAR (μmol m^{-2} s^{-1})	5	800	700	173
Lead	mg dm^{-3}	12	1.4	1.3	0.1
Copper	mg dm^{-3}	12	1.0	1.0	0.0
BOD	mg dm^{-3}	47	2.2	2.1	1.8
SS	mg dm^{-3}	28	17.0	14.9	13.8
Total solids	mg dm^{-3}	49	449.9	452.1	159.6
Turbidity	NTU	12	2.3	2.2	1.3
DO	mg dm^{-3}	52	8.5	8.5	0.6
pH	—	46	7.8	7.9	0.1
THB[a]	CFU cm^{-3}	16	2948	2908	2108
TC[a]	CFU cm^{-3}	17	368	201	479

[a] Data were considered if the differences between each of the three replicates were less than 50% of the median.
[b] Temperature was measured each morning at about 10:30 (local time).
BOD, biochemical oxygen demand; CFU, colony-forming unit; DO, dissolved oxygen; TC, total coliforms; THB, total heterotrophic bacteria; SS, suspended solids.

litter zone(s) functioned as a buffer, which led to relatively stable outflow water quality values; e.g., the standard deviations of BOD, suspended solids and turbidity were 0.9 mg dm^{-3}, 0.9 mg dm^{-3} and 0.1 NTU, respectively.

After five months of operation, the overall mean performance in terms of copper (Figure 2.7), lead (Figure 2.8), BOD (Figure 2.9), suspended solids (Figure 2.10), turbidity and bacteria removal was great (>40%) for all filters (Table 2.4). Moreover, an ANOVA (Fowler and Cohen 1998) indicated that the reductions of copper ($\log_{10}$ transformed data), suspended solids, total solids, lead ($\log_{10}$ transformed data) and BOD were statistically significantly similar. An ANOVA undertaken with the software package Microsoft Excel indicated that the corresponding levels of significance were 0.98, 0.71, 0.51, 0.46 and 0.36, respectively. However, Filter 3 had a lower BOD, turbidity and TC reduction in comparison to the overall average because of high plant decay indicated in advance by early mid-leaf colour transformation (after 24 March) (Fowler and Cohen 1998).

After initial strong growth, *Typha latifolia* in Filter 3 were reduced (after 21 April) by aphids despite spraying a solution of Pure Simple Soap. Total numbers of *Typha latifolia* were reduced between 38% and 75% at the end of June. Dead plant material was placed in the litter zone and contributed therefore to the increase of carbon, nutrients, trace minerals, turbidity and subsequent microbiological activity.

Potentially critical variables in terms of biomass efficiency and filtration performance were DO, suspended solids (top only) and total solids (out only) (Table 2.4) as well as plant growth: DO was reduced to <0.5 mg dm^{-3} for the Filters 4–6 during up to 7% of the running time of the experiment. Dissolved oxygen is the primary control variable critical for aerobic metabolism. Furthermore,

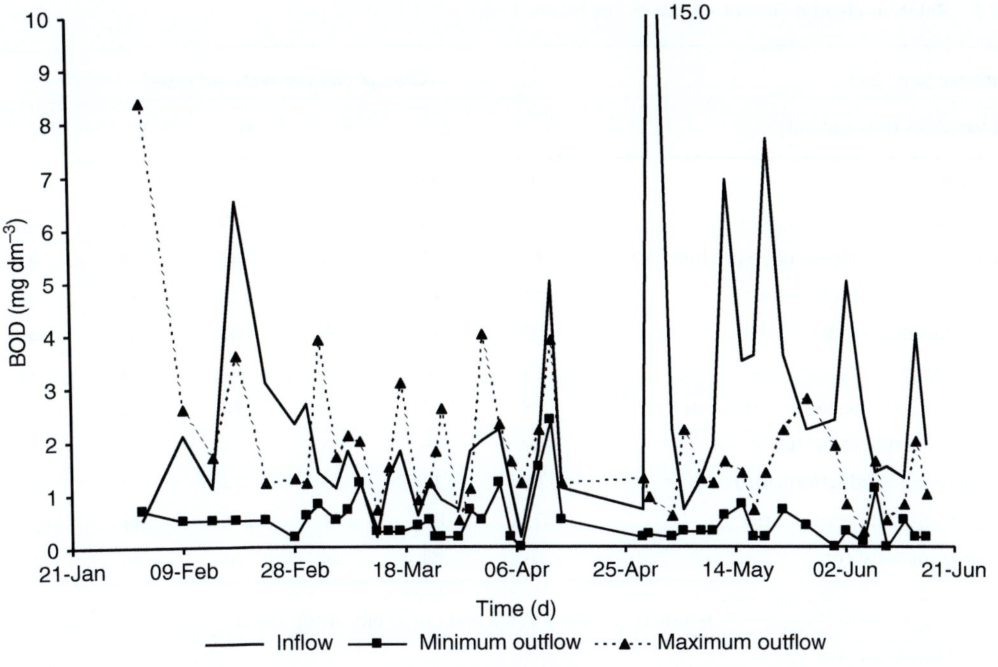

Figure 2.9 Inflow biochemical oxygen demand, BOD (mg dm^{-3}), in comparison to both the minimum and maximum outflow BOD (mg dm^{-3}) concentrations for Filters 1–6 (range of data for all filters). Each point represents one data point (outliers excluded). After a maturation period of about three months, the outflow concentrations were usually below the inflow ones.

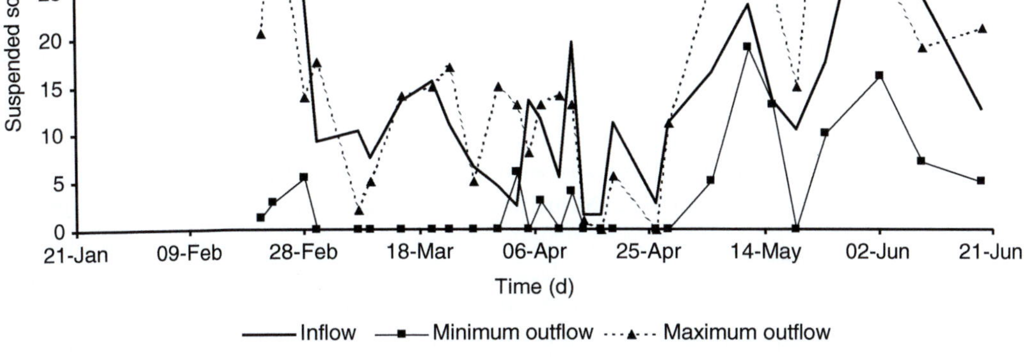

Figure 2.10 Inflow suspended solids, SS (mg dm^{-3}), in comparison to both the minimum and maximum outflow SS (mg dm^{-3}) concentrations for Filters 1–6 (range of data for all filters). Each point represents one data point (outliers excluded). Both inflow and outflow values are highly variable.

Table 2.4 Relative change (%)[a] of variables for filters 1–6.

Performance (out) and critical variables (top and out)	Change (%) per wetland filter[a]					
	1	2	3	4	5	6
Lead reduction (out)[b]	98	99	99	99	99	99
Copper reduction (out)[c]	96	98	97	99	98	99
Biochemical oxygen demand reduction (out)	60	57	41	45	53	41
Suspended solids reduction (out)	55	42	50	53	51	33
Turbidity reduction (out)	95	87	68	80	97	99
Dissolved oxygen reduction (out)	46	68	74	77	72	78
Total heterotrophic bacteria reduction (out)[d]	88	98	92	94	91	88
Total coliforms reduction (out)[d]	100	98	69	89	98	96
Dissolved oxygen reduction (top)	ns	32	76	74	84	87
Suspended solids increase (top)	371	188	171	276	241	100
Total solids increase (out)	ns	ns	ns	ns	ns	23

[a] change (%) $= \frac{(\text{in} - \text{out}) \times 100\ (\%)}{\text{in}}$, where in = inflow value and out = out or top value
[b] including diluted samples.
[c] including concentrated samples; top, water on top of the sediment; out, outflow water; ns, statistically not significant (The change (%) was only calculated if there was a statistically significant difference between inflow and outflow).
[d] Data were considered if the difference between each replicate was less than 50% of the median.

suspended solids increase in the outflow after week 13 (Figure 2.10) was high but often below recommended levels (25–100 mg dm^{-3}). The increase of solids is likely to have been due to the recycling of dead plant matter.

Furthermore, the standardised total plant heights of *Phragmites australis* for Filters 2–6 are 1.2, 1.0, 1.4, 1.1 and 1.7, respectively. The numerical value of one for Filter 3 represents a sum of 1698 cm. Only Filter 6 was different because it showed an increased biomass production rate indicated by plant growth which was probably the result of early fertiliser application (Table 2.1). This resulted in a relative increase in litter in Filter 6 after the growing season because dead plant matter was recycled. Moreover, the recycling of dead *Typha latifolia* plant material after aphid infestation resulted in low overall BOD reductions for Filters 3–6 (Table 2.4). In contrast to *Typha latifolia*, *Phragmites australis* was less targeted by aphids because the stems and leaves are hard and rigid.

2.2.3.3 Metal Analysis

An additional ICP-MS analysis for Filter 6 was performed to investigate the biomass performance in more detail (Table 2.5). Results obtained during week 18 suggested that Cu, As, Sb, Lu, Hf, Pb, Th and U accumulated as colloidal and/or particulate compounds in the *schmutzdecke* and charcoal zones (Table 2.1). For example, copper and lead could be present as copper and lead sulphide. Therefore, the toxicity effect of the remaining dissolved fraction of copper and lead on micro-organisms within the liquid phase of the litter zone was reduced. Table 2.5 indicates that lead in contrast to copper accumulated in the *schmutzdecke* of Filter 6.

A diverse invertebrate community (see below) including rotifers (250 cm^{-3}), crustaceans (18 cm^{-3}), segmented worms (29 cm^{-3}) and freshwater hog lice (2 dm^{-3}) formed the top of the

Table 2.5 Reed bed water analysis of filter 6 (week 8; 24/05/00) with the VG elemental PlasmaQuad3 high performance quadrupole ICP-MS. All units are expressed in mg dm^{-3}.

Element	Inflow	Top water	Sediment[a]	Outflow	Problem[b]
Silicon (Si)	0.000	0.000	0.000	6.010	w
Phosphorus (P)	0.000	0.026	0.000	0.012	gl
Sulphur (S)	0.000	0.000	0.000	26.250	w
Manganese (Mn)	0.311	0.077	0.184	1.061	hi, w
Iron (Fe)	0.370	0.000	0.000	0.004	hi, gl
Copper (Cu)	1.000	0.052	0.073	0.002	hi, t, sa
Zinc (Zn)	0.000	0.000	0.000	0.156	w
Arsenic (As)	0.005	0.004	0.059	0.004	sa
Yttrium (Y)	0.072	0.038	0.000	1.289	w
Antimony (Sb)	0.042	0.040	2.314	0.003	hi, sa
Barium (Ba)	0.037	0.081	0.070	0.317	w, hi
Lutetium (Lu)	0.008	0.008	0.444	0.000	sa
Hafnium (Hf)	0.039	0.038	2.239	0.003	sa
Lead (Pb)	1.292	0.052	1.909	0.002	hi, t, sa
Thorium (Th)	0.018	0.017	1.025	0.001	sa
Uranium (U)	0.026	0.025	1.446	0.002	sa

[a] Liquid phase surrounding the granules and litter within the top sediment.
[b] Potential problem with specific element or its concentration: w, subject to washout; gl, potentially growth limiting to biota; hi, high inflow concentration; t, potentially toxic to biota; sa, accumulates in the top sediment layer.

aquatic food chain in the liquid phase of the litter zone of Filter 6. This suggests that metals were predominantly present in a form that did not seem to harm the animals, or a part of the invertebrate community was resistant. This observation was supported by simultaneously performed THB plate count tests (Bridson 1998) indicating bacterial numbers between 215 and 800 CFU cm^{-3} in the top bulk water zone of all filters containing either sand or an adsorption medium as the top filter layer (Table 2.1). This range of CFUs may be considered as low, but plate count tests do only indicate a fraction of the real bacterial population and predation by protozoa reduce bacterial numbers considerably (Bridson 1998; Decamp and Warren 2000; Decamp et al. 1999; Ottová et al. 1997; Scholz 1997). However, further eco-toxicology research is necessary to investigate the resistance of various micro-organisms to heavy metal pollution.

Furthermore, compounds including Si, S, Mn, Zn, Y and Ba were washed out of the filter media and contributed to elevated total solids values in the outflow (Tables 2.4 and 2.5). The mean total solids outflow concentration was about 580 mg dm^{-3}.

The filters containing plants (in particular, *Typha latifolia*) contributed to the inflow BOD. The real BOD of the inflow was, therefore, the sum of the natural inflow BOD (mean: 2.2 mg dm^{-3}; 10–40%) and the BOD resulting from the decay of submerged dead plant material (60–90%). The BOD resulting from plant decay was greatest for filters containing *Typha latifolia*, because of early aphid infestation. Furthermore, the addition of fertiliser (Filter 6 only) increased the degradation rate of the plant material and therefore contributed to the decrease of oxygen content due to the respiration of aerobic bacteria and the breakthrough of total solids including mineralised plant matter (Table 2.4).

2.2.3.4 Water Quality and Micro-biology

The effluent of Filter 1 in comparison to the effluents of Filters 2–6 can be associated with a lower mean BOD (Table 2.4). This phenomenon occurred, because Filter 1 did not contain plants contributing to the inflow BOD. Moreover, the litter zone of Filter 6 contained large amounts of debris originating from rapidly decomposing plant leaves and stems (Table 2.1).

Heterotrophic bacteria, coliforms and protozoa distributions within reed beds have recently been studied (Decamp and Warren 2000; Decamp et al. 1999; Ottová et al. 1997). High numbers of bacteria within the inflow water seemed to be reduced effectively by protozoa and other organisms within the litter zone (Table 2.4). The investigation of other processes of heterotroph attenuation (toxic influent, low temperatures, anoxic conditions and so on) was beyond the scope of Chapter 2.2.

Figures 2.11 and 2.12 suggest a sufficient reduction of numbers of THB and TC after filtration for all filters. TC were significantly more abundant in mature carbon (869 CFU cm^{-3}) than in the sand-based sediment (6 CFU cm^{-3}). TC abundance was related to the presence of plant litter and correlated negatively with progressive plant matter mineralisation and a subsequent reduction of nutrient availability. It follows that TC numbers are expected to be low if plant decay is completed. However, data for all seasons (different temperatures and ambient light availabilities) would need to be analysed to justify a generalisation of the previous statement (research currently in progress).

The Shannon–Weaver Diversity Index (H; Fowler and Cohen 1998; Scholz and Martin 1998a) was calculated for the averages of individual counts per protozoa, algae or macro-invertebrate group for the inflow as well as the top and bottom (protozoa only) filter media layer of each filter (Table 2.6). The diversity indices were below $H = 1.85$, which is low for natural wetlands, but frequently reported for constructed wetlands (Kadlec and Knight 1995).

Considering protozoa and macro-invertebrate groups only, filters containing activated carbon (Filters 4–6) were associated with relatively high H values in contrast to low values for sand-based

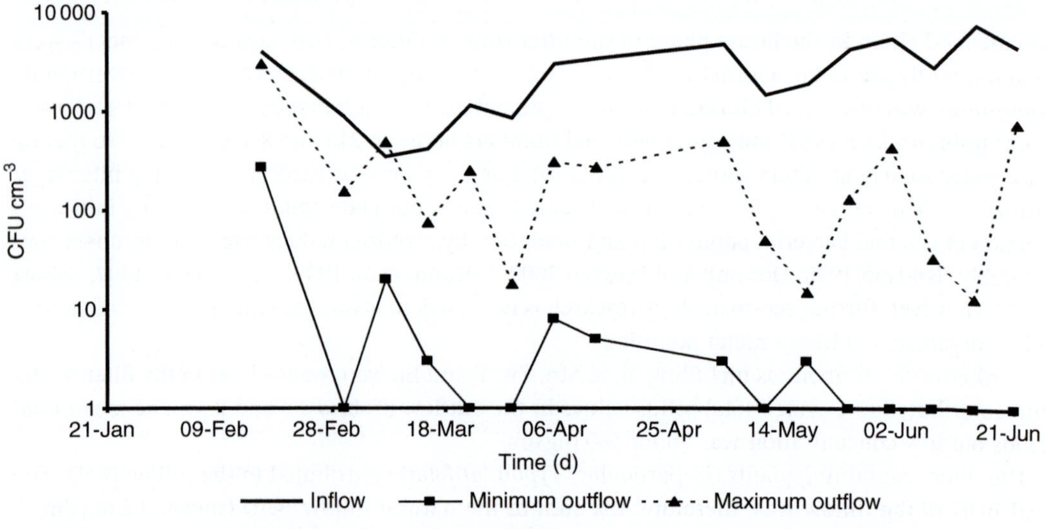

Figure 2.11 Inflow numbers (average of three replicates) of total heterotrophic bacteria, THB (CFU cm^{-3}) on a log scale distribution, in comparison to both the minimum and maximum outflow THB (CFU cm^{-3}) concentrations for Filters 1–6 (range of data for all filters). Bacteria numbers are expressed as colony-forming units (CFU). Data were considered if the difference between each replicate was less than 50% of the median. The statistical standard errors for the fluctuating temporal data series for the inflow, minimum outflow and maximum outflow of THB are 398, 13 and 132, respectively. Bacteria counts of the outflow are usually one or two orders of magnitude below the inflow one.

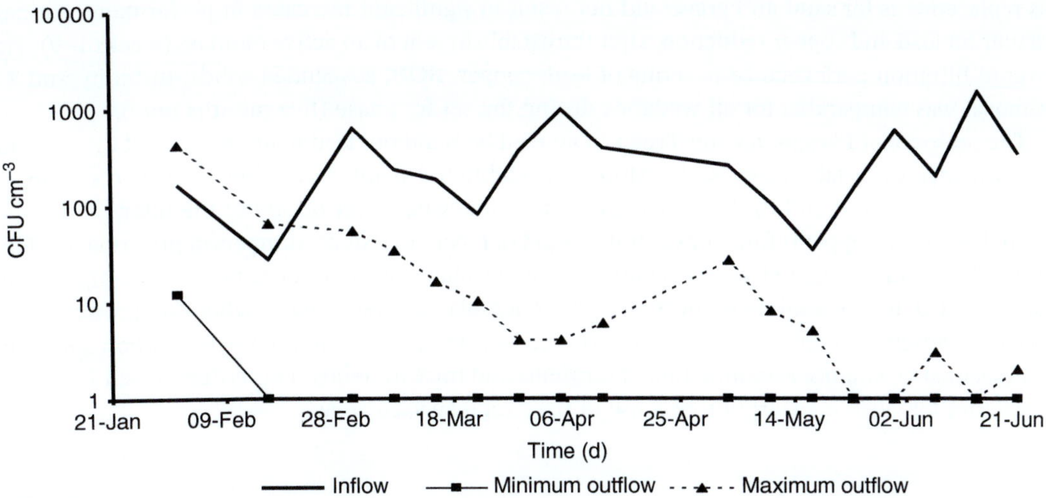

Figure 2.12 Inflow numbers (average of three replicates) of total coliforms, TC (CFU cm^{-3}) on a log scale distribution, in comparison to both the minimum and maximum outflow TC (CFU cm^{-3}) concentrations for Filters 1–6 (range of data for all filters). Bacteria numbers are expressed as colony-forming units (CFU). Data were considered if the difference between each replicate was less than 50% of the median The statistical standard errors for the fluctuating temporal data series for the inflow, minimum outflow and maximum outflow of TC are 95, 1 and 20, respectively. Coliform counts of the outflow are usually one or two orders of magnitude below the inflow one.

Table 2.6 Shannon–Weaver Diversity Index (*H*) for protozoa, algae and macro-invertebrate groups.

Micro-organism Groups	*H* of Inflow	*H* of filter					
		1	2	3	4	5	6
Protozoa (top)	0.63	0.33	0.71	0.80	0.98	1.03	1.01
Protozoa (out)	0.63	0.60	0.64	0.58	1.11	0.91	0.90
Algae (top)[b]	0.00[a]	1.85	1.40	1.52	1.01	1.14	1.26
Macro-invertebrates (top)	0.00[a]	0.00[a]	0.69	0.37	1.03	0.95	0.63

[a] very low higher micro-organism density within the sample water.
[b] predominantly diatom groups.
top, water on top of the sediment; out, outflow water.

filters (Filters 1–3). It may be the case that an increase in filter media adsorption capacity correlates positively with a slight increase in protozoa and macro-invertebrate diversity. Mature absorption media in comparison to sand are richer in nutrients and offer a greater surface area for growth and attachment. In contrast, this statement needs to be reversed for algae groups. Algae require light and a free water zone without plant litter for optimal growth.

2.2.4 Conclusions

The constructed wetlands studied containing additional adsorption filter media were approximately 26 times more expensive (material costs only) than traditional ones (£ 9 m^{-2}). Moreover, expensive adsorption filter material including granular activated carbon, charcoal and Filtralite as replacements for sand and gravel did not result in significant increases in performance, in particular for lead and copper reduction, after the establishment of an active biomass (weeks 1–9). The overall filtration performance in terms of lead, copper, BOD, suspended solids, turbidity and TC removal was comparable for all wetlands during the set-up phase (five months only).

The presence of *Phragmites* and *Typha* in all reed beds did not lead to an increase of the wetland performance on a laboratory scale. Moreover, additional plant decay appeared to result in an increase of the BOD and the bacterial load within the water layer on top of the litter.

High lead and copper inflow concentrations did not seem to reduce numbers of protozoa, rotifers, water fleas and segmented worms within the liquid phase of the litter zone, because the metals accumulated in the sediment. Furthermore, microbial diversity was higher for protozoa and macro-invertebrate groups within adsorption media compared to sand. Activated carbon and charcoal seemed to enhance accumulation of nutrients and trace minerals. This is due to their increased capacities for surface adsorption because of their high surface areas.

2.3 Vertical Subsurface Flow-Constructed Wetlands with Different Substrates

2.3.1 Introduction

The release of phosphorus-rich wastewater effluent into receiving watercourses results in environmental challenges such as eutrophication. Phosphorus removal has mainly been undertaken by using expensive wastewater purification systems such as activated sludge and enhanced biological phosphorus removal technologies (Sriwiriyarat and Randall 2005). In contrast, constructed wetlands are low-tech and low-cost alternatives for removing phosphorus from municipal, industrial and agriculture wastewater (Scholz 2006a).

Phosphorus removal in constructed wetlands takes place due to plant uptake (Greenway and Woolley 1999), accretions of wetland soils (Kadlec 1997), microbial immobilization (Reddy et al. 1999), adsorption and retention by substrates (Tanner et al. 1998) and precipitation in the water column (Diaz et al. 1994). Although macrophytes, soil, detritus and biomass are important sinks for phosphorus in the short term, substrate is the main sink for phosphorus in the long term (Sakadevan and Bavor 1997). In surface flow wetlands, most of the phosphorus within wastewater is in contact with the surface layer of substrates (Pant et al. 2001), whereas in sub-surface flow wetlands, wastewater is in direct contact with the substrates. Adsorption and retention are therefore the main pathways of phosphorus removal (Scholz 2006a).

Because of the great importance of substrate in phosphorus removal, various media including limestone, zeolite, shale, slag and gravel have been studied to assess phosphorus removal in constructed wetlands (Drizo et al. 1997; Mann 1997; Scholz and Xu 2001; Zhu et al. 1997). Johansson (1997) indicated that shale-based constructed wetlands with or without reeds (*Phragmites australis*) have between 98% and 100% phosphorus removal capacity. Blast furnace-slag (a by-product of limestone, coke and iron ore), zeolite, pelleted-clay (either alone or in combination with soils; Sakadevan and Bavor (1997)), alum, calcite and dolomite (Ann et al. 1999) improve the phosphorus-retention capacity of constructed wetlands.

Drizo et al. (1999) undertook regression analyses of the theoretical maximum phosphorus adsorption values (obtained from the Langmuir equation) plotted against physical properties of substrates (pH, electric conductivity, hydraulic conductivity, porosity, surface area and bonding capacity), and found that none of these properties had a significant effect on the phosphorus adsorption capacity.

The first criterion in comparing the phosphorus removal performance of different substrates is usually the maximal phosphorus adsorption capacity, which can be obtained from batch adsorption experiments (Drizo et al. 1999; Scholz and Xu 2001; Seo et al. 2005). Discrepancies in phosphorus removal performances for various substrates have been identified because of the differences in substrate grain sizes and quantities (Drizo et al. 1999, 2002; Pant et al. 2001; Sakadevan and Bavor 1997; Seo et al. 2005). Therefore, phosphorus batch adsorption experiments should be coupled with the assessment of material performance in columns to compare phosphorus removal efficiencies and retention capacities by different materials (Brooks et al. 2000; Scholz and Xu 2001; Seo et al. 2005).

According to Kadlec and Knight (1996), a modified first-order model can be used to describe the relationship between phosphorus removal and hydraulic retention time (HRT) in constructed treatment wetlands (Eq. 2.1). The background concentration C^* is often depleted (Kadlec and Knight 1996), which simplifies Eq. (2.1), leading to Eq. (2.2).

$$C_e = (C_i - C^*)\exp\left(-\frac{k\text{HRT}}{h}\right) + C^* \tag{2.1}$$

$$C_e = C_i \exp\left(-\frac{k\text{HRT}}{h}\right) \tag{2.2}$$

where C_e and C_i are concentrations of phosphorus (mg L^{-1}) in effluent and influent, respectively, k is the removal rate constant (m d^{-1}), *HRT* is the hydraulic residence time (d), h is the effective depth of the wetland and C^* is the background concentration (mg L^{-1}).

The main assumption for using Eq. (2.2) is that the removal rate constant k does not depend on the variables (HRT) and the influent phosphorus concentration C_i. There is, however, anecdotal evidence that the removal rate constant k is not a real constant, but changes with the HRT and influent concentrations of phosphorus.

The aim was to assess phosphorus removal performances for different unplanted vertical-flow constructed wetlands containing different substrates. The corresponding objectives were to (i) evaluate phosphorus adsorption capacities of different substrates; (ii) select shale to assess the effect of grain size on phosphorus adsorption capacity; (iii) compare phosphorus removal performances for different substrates in column experiments; (iv) assess the effect of HRT on phosphorus removal; (v) evaluate the influence of HRT on the first-order model removal rate constant k and (vi) assess factors affecting phosphorus removal applying linear regression. Chapter 2.3 is based on an original article by Tang et al. (2009).

2.3.2 Materials and Methods

Column experiments (Figure 2.13) with unplanted wetlands and supplementary batch adsorption experiments were conducted indoors to compare the phosphorus removal performances of hornblende, gravel, shale and ironstone. The mean elemental composition of shale was determined previously: 49.53% O, 15.13% Si, 10.37% C, 11.76% Ca, 4.87% Al and 3.02% Fe (Zhang et al. 2006). The composition of ironstone was 886.41 g kg^{-1} Fe_2O_3, 24.62 g kg^{-1} Al_2O_3, 19.70 g kg^{-1} MgO, 29.55 g kg^{-1} CaO, 24.62 g kg^{-1} TiO_2, 9.85 g kg^{-1} Mn_3O_4 and 1.73 g kg^{-1} K_2O. Measurements of selected physical properties are given in Table 2.7.

Equilibrium isotherm experiments were performed for all substrates with grain sizes between 0.5 and 1 mm. Moreover, three different ranges of grain sizes for shale (0.0–0.5 mm for shale A; 0.5–1.0 mm for shale B, 1.0–2.0 mm for shale C) were assessed to study the effect of substrate grain size on phosphorus adsorption capacity, supplementing previous work by Drizo et al. (1999) and Pant et al. (2001).

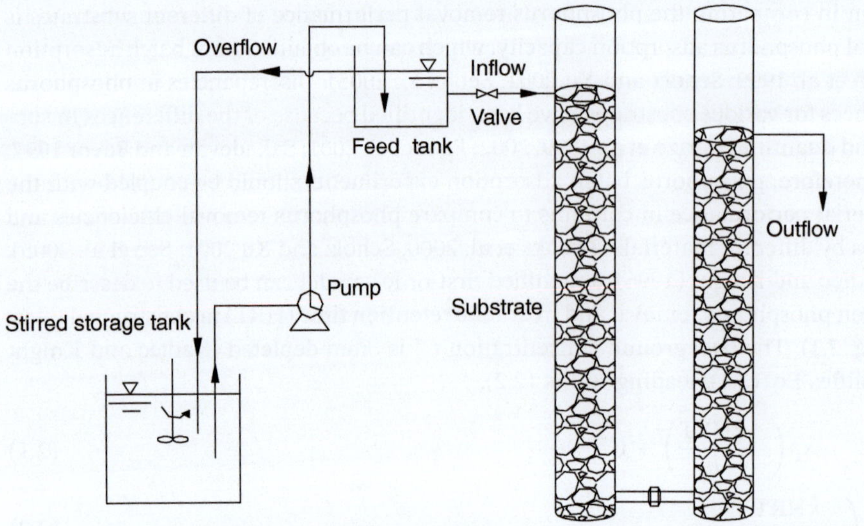

Figure 2.13 Indoor filter beds packed with either shale, gravel, ironstone or hornblende. The system comprises a storge tank, a pump, a feed tank and two columns with filter media. *Source:* Tang et al. (2009)/with permission of John Wiley & Sons.

Table 2.7 Selected physical characteristics of the materials tested for phosphorus removal.

Materials	Porosity (%)	Bulk density ($g\ cm^{-3}$)	d_{10} (mm)	d_{60} (mm)	d_{60}/d_{10} (-)	K_s ($cm\ s^{-1}$)
Gravel	48.4	1.48	9.8	18.1	1.85	48
Hornblende	35.8	1.38	0.9	2.2	2.40	6
Ironstone	41	1.65	2.3	14.6	6.35	35
Shale	46	1.40	1.6	11.5	11.50	8

d_i, particle size diameter corresponding to *i*% passing on cumulative particle-size distribution; K_s, hydraulic conductivity.

Nine phosphorus solutions ranging from 0 to 200 mg L^{-1} (i.e., 0, 1, 2.5, 5, 10, 25, 50, 100 and 200 mg L^{-1} phosphorus, using KH_2PO_4, which was used as the only phosphorus source in adsorption isotherm experiments) were tested for shale, gravel, ironstone and hornblende. For each test, 1 g substrate was placed in a 100-mL Erlenmeyer glass flask containing 20 mL of phosphorus solution. The remaining compound was deionized water. One drop of toluene was added to each flask to inhibit microbial activity. Blanks containing no substrate were included in each experiment; all experiments were conducted in triplicates.

Each solution was continuously shaken at 175 rpm within a water bath at 25°C for 24 hours. Suspensions were then centrifuged for 15 minutes at a speed of 5000 rpm. After adequate dilution, phosphorus remaining in the supernatant was determined according to APHA (1998). Adsorption isotherms such as the linear Eq. (2.3) were subsequently fitted to the data points.

$$N = pC_e + q \tag{2.3}$$

where N is the adsorbed phosphorus concentration (mg kg^{-1}), p is the slope (−), which was used to study the dependence of phosphorus adsorption capacity on the equilibrium phosphorus concentrations (C_e, mg L^{-1}) and intercept q (mg kg^{-1}) was the background phosphorus concentration of the substrate.

The Freundlich adsorption isotherm is shown in Eq. (2.4). The constant K may be considered as a hypothetical index of phosphorus sorbed from a solution with a unit equilibrium phosphorus concentration (Sakadevan and Bavor 1997) and may thus be taken to provide a measure of relative phosphorus adsorption capacity of the substrate being used. Equation (2.4) can also be written as Eq. (2.5), which can be used for a linear test plot with ln K as intercept and $1/n$ (where n is a constant) as the slope of the equation.

$$N = KC_e^{\frac{1}{n}} \tag{2.4}$$

$$\ln N = \ln K + \frac{1}{n} \ln C_e \tag{2.5}$$

where N is the adsorbed phosphorus concentration (mg kg^{-1}), K is a constant (mg kg^{-1}), C_e is the equilibrium phosphorus concentration (mg L^{-1}) and n is a constant (kg L^{-1}).

The Langmuir adsorption isotherm is shown in Eq. (2.6). The constant a is calculated from the slope obtained by plotting the equilibrium phosphorus concentration C_e against C_e/N (where N is the adsorbed phosphorus concentration) and can be viewed as the Langmuir adsorption maximum. The constant b relates to the binding energy (Eq. 2.6).

$$\frac{C_e}{N} = \frac{1}{ab} + \frac{C_e}{a} \tag{2.6}$$

where C_e is the equilibrium phosphorus concentration (mg L^{-1}), N is the adsorbed phosphorus concentration (mg kg^{-1}) and a (mg kg^{-1}) and b (L mg^{-1}) are constants.

Phosphorus removal processes by various substrates (see above) were studied in vertical-flow columns of 1.2 m height with 15 cm diameter polyvinylchloride tubes, cut to 1.2 m lengths (Figure 2.13). All experiments were run in a temperature control room at 25°C. Eutrophic Jin River water was pumped continuously into the stirred storage tanks. The hydraulic residence time was approximately 2.2 days, and mean water quality figures were as follows: chemical oxygen demand, 106.0 mg L^{-1}; ammonia-nitrogen, 5.74 mg L^{-1}; nitrate-nitrogen, 1.19 mg L^{-1}; total nitrogen, 7.34 mg L^{-1}; soluble reactive phosphorus (SRP), 0.398 mg L^{-1}; total phosphorus (TP), 0.516 mg L^{-1}; dissolved oxygen, 3.4 mg L^{-1}; pH, 7.7 mg L^{-1}; water temperature, 24.6°C.

Influent and effluent samples from all columns were taken every two days for analysis. The effluent flow rate was measured each time samples were taken. Samples were analysed for SRP using the ascorbic acid method (APHA 1998) on the date they were taken.

The HRT was increased to 3.1 days after 5 weeks for nearly 4 weeks to monitor changes in the SRP and TP effluent concentrations. Other operational conditions were not changed. The HRT was based on measured effluent flow rates using Eq. (2.7).

$$\text{HRT} = \frac{\pi r^2 H \varphi}{Q} \tag{2.7}$$

where HRT is the hydraulic residence time (days), r is the radius (cm) of the column, H is the height (cm) of the substrate in the column, ϕ is the porosity (%) of the substrate and Q is the measured flow rate (mL d^{-1}).

2.3.3 Results and Discussion

2.3.3.1 Adsorption Isotherm Experiments

The phosphorus concentrations in the equilibrium solution were related to phosphorus adsorption (Figure 2.14). The findings of the phosphorus adsorption isotherm experiments showed that adsorption capacities of different substrates varied considerably (Table 2.8); e.g., 619.7 mg kg^{-1} for shale in comparison to 89.05 mg kg^{-1} for gravel.

The results were comparable with those reported in literature: Gravel has a low phosphorus adsorption capacity of <50 mg kg^{-1} (Breen 1990), and shale has a maximum phosphorus adsorption capacity of 650 mg kg^{-1} (Drizo et al. 1999). However, some industrial byproducts had an extremely high phosphorus adsorption capacity; e.g., >44 000 mg kg^{-1} for blast furnace slag (Sakadevan and Bavor 1997). Furthermore, light expanded clay aggregates (produced by LECA) tested by Reddy et al. (1999) had highly variable phosphorus adsorption capacities (37–2900 mg kg^{-1}), depending on the soil of origin.

The ranking order of the maximum phosphorus adsorption capacity (Table 2.8) for the substrates was as follows: shale > ironstone > hornblende > gravel. Shale had the smallest grain size distribution and therefore the highest surface area, indicating a high potential for direct reaction with phosphate (Drizo et al. 1999).

Results show that the maximum phosphorus adsorption capacities (obtained by the Langmuir equation) were 953.8 mg kg^{-1} for shale A, 619.7 mg kg^{-1} for shale B and 383.8 mg kg^{-1} for shale C. The maximum phosphorus adsorption capacity increased as the grain size of the substrate decreased (Figure 2.15).

These findings compared well with reports in the literature: Xu et al. (2006) assessed the phosphorus adsorption capacity of nine different substrates. Their results indicate that the phosphorus adsorption capacities of soil and bentonite are higher than for sand, because their corresponding grain sizes are finer. An advantage of the current research is that the relationship between substrate grain size and phosphorus adsorption capacity has been derived from shale only and not from a mixture of different materials.

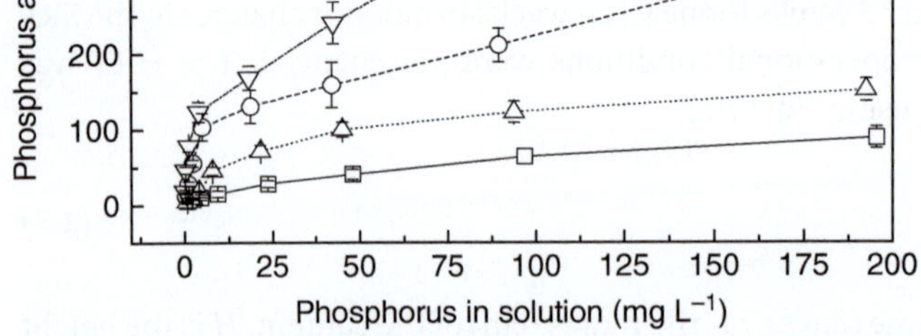

Figure 2.14 Relationship between the equilibrium solution of phosphorus (mg L^{-1}) and adsorbed phosphorus (mg kg^{-1}) for different substrates at 25°C. Only shale and ironstone adsorb phosphorus reasonably well.

Table 2.8 Statistics (mean ± standard deviation) based on three triplicate data sets for linear, Freundlich and Langmuir adsorption isotherms concerning shale, gravel, hornblende and ironstone with grain sizes between 0.5 and 1.0 mm.

	Linear			Freundlich			Langmuir		
	p	q	r^2	K	n	r^2	a	b	r^2
Material	**(L g^{-1})**	**(mg kg^{-1})**	**(–)**	**(mg kg^{-1})**	**(–)**	**(–)**	**(L g^{-1})**	**(mg kg^{-1})**	**(–)**
Shale	74.87 ± 2.17	3.29 ± 0.54	0.97 ± 0.06	69.16 ± 3.98	2.63 ± 0.46	0.97 ± 0.04	0.54 ± 0.07	619.70 ± 23.41	0.98 ± 0.04
Gravel	12.08 ± 0.89	0.43 ± 0.11	0.94 ± 0.04	5.29 ± 1.26	1.89 ± 0.14	0.99 ± 0.03	3.60 ± 0.23	89.05 ± 5.27	0.85 ± 0.05
Ironstone	63.28 ± 1.78	1.52 ± 0.37	0.89 ± 0.05	29.72 ± 2.05	2.10 ± 0.37	0.94 ± 0.05	5.57 ± 0.11	324.90 ± 14.98	0.99 ± 0.03
Hornblende	31.94 ± 1.24	0.74 ± 0.25	0.80 ± 0.07	7.75 ± 1.47	1.58 ± 0.18	0.95 ± 0.04	14.97 ± 0.72	153.10 ± 7.86	0.98 ± 0.04

a, constant; *b*, constant; *K*, constant; *n*, constant; *p*, slope; *q*, intercept; r^2, coefficient of determination.

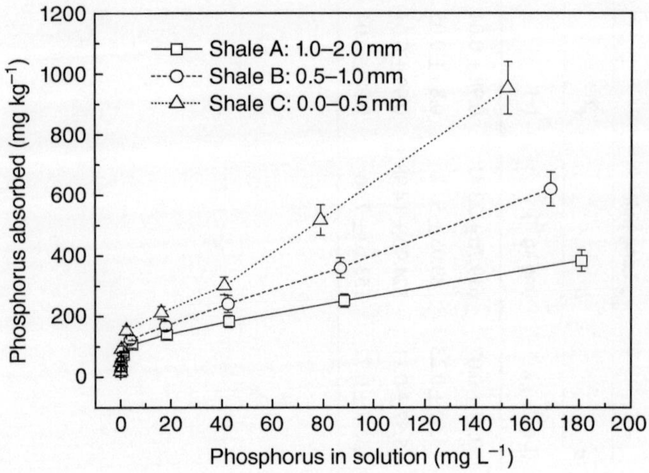

Figure 2.15 Relationship between the equilibrium solution of phosphorus (mg L^{-1}) and adsorbed phosphorus (mg kg^{-1}) for shale with different grain sizes at 25°C. A smaller diameter of shale leads to more phosphorus adsorption.

Data obtained from phosphorus adsorption experiments were fitted by linear, Freundlich and Langmuir adsorption isotherms. Statistics shown in Table 2.8 varied considerably, indicating a high variability of phosphorus adsorption characteristics for most substrates. After comparing the coefficient of determination r^2 values of the three adsorption isotherms (Table 2.8), it was found that linear equations were not appropriate for modelling the phosphorus adsorption capacities of shale, gravel, ironstone and hornblende.

For shale, ironstone and hornblende, Langmuir equations modelled phosphorus adsorption data much better (r^2 ranged between 0.98 and 0.99) than linear (r^2 ranged between 0.80 and 0.97) and Freundlich equations (r^2 ranged between 0.94and 0.97). This outcome compares well with several independent observations (Drizo et al. 1999; Pant et al. 2001; Sakadevan and Bavor 1997) showing that the Langmuir adsorption equation explained phosphorus adsorption better than the Freundlich equation.

An advantage of applying the Langmuir model for adsorption studies is that the corresponding equation enables the calculation of theoretical phosphorus adsorption maxima. Therefore, the Langmuir model is most frequently used. The Freundlich equation models phosphorus adsorption data for gravel better than linear and Langmuir equations ($r^2$of 0.99 compared to 0.85), if the phosphorus concentration ranges between 0 and 200 mg L^{-1}. This is because the Freundlich equation, although empirical, implies that the affinity for adsorption decreases exponentially with increasing saturation, which is more likely to be the case than the assumption of constant binding energy inherent in the Langmuir equation.

2.3.3.2 Column Experiments

The ranges of key water quality variables of the Jin River water were narrow during the experiment: chemical oxygen demand, 115–140 mg L^{-1}; total nitrogen, 7.8–9.5 mg L^{-1}; SRP, 1.2–1.3 mg L^{-1}; TP, 1.4–1.5 mg L^{-1}; pH, 8.32–8.64. Figures 2.16 and 2.17 indicate the relationships between HRT and phosphorus removal (SRP and TP) for the vertical-flow wetland columns. Findings show that 1.282 mg L^{-1} (86.27% of TP) was in the form of SRP. The removal of SRP (Figure 2.16) mirrored the pattern of TP removal (Figure 2.17).

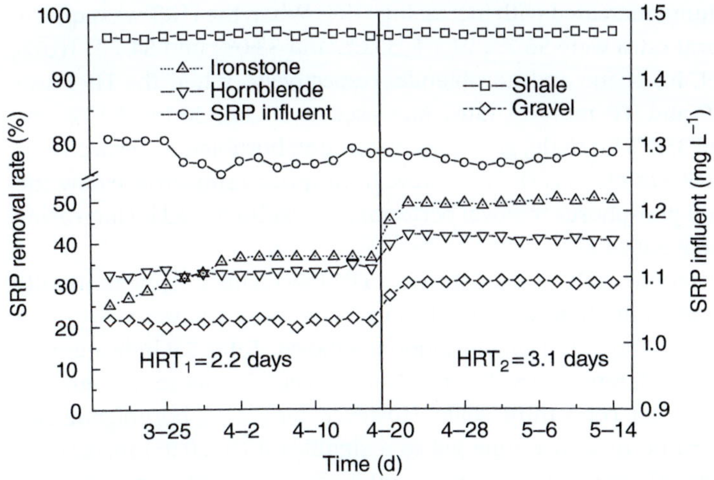

Figure 2.16 Soluble reactive phosphorus (SRP) removal rates at different hydraulic retention times (HRTs). The vertical line in the graph represents change of HRT. The removal rate increases with hydraulic retention rate for the media ironstone, hornblende and gravel. *Source:* Tang et al. (2009)/with permission of John Wiley & Sons.

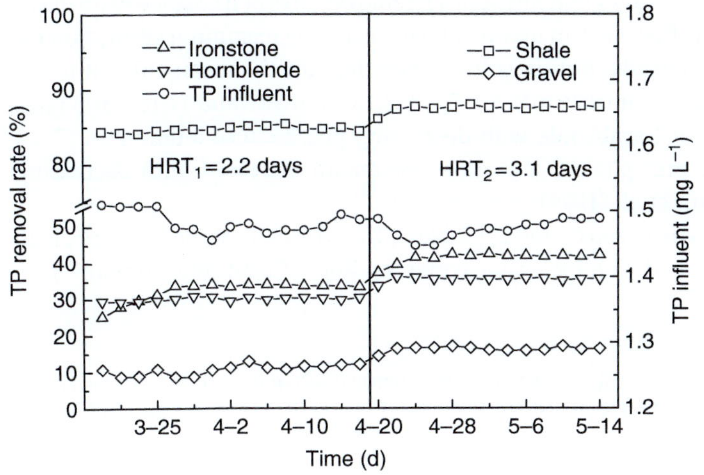

Figure 2.17 Total phosphorus (TP) removal rates at different hydraulic retention times (HRTs). The vertical line in the graph represents the change of HRT. The removal rate increases with hydraulic retention rate for the media ironstone, hornblende, gravel and shale. *Source:* Tang et al. (2009)/with permission of John Wiley & Sons.

It has previously been demonstrated that phosphorus removal from soil exhibited both rapid and slow phases (Nur Onar et al. 1996). However, this phenomenon was not observed for the entire data set of the column experiments (Figures 2.16 and 2.17). For phosphorus removal from soils, the initial rapid reaction was typically described as physical adsorption (Frossard et al. 1995).

Substrates used in constructed wetlands had much higher metal (Fe, Al and Ca) concentrations and larger grain sizes in comparison to soils (Drizo et al. 1999; Seo et al. 2005; Xu et al. 2006). Phosphorus removal by wetland substrate is therefore mainly due to chemical adsorption or precipitation, and physical adsorption is less dominant. Thus, the slow phase (see above) agreed with the data obtained from the column experiments.

Phosphorus removal in each column increased with increasing HRT. When the HRT was equal to 2.2 days, mean SRP and TP removal rates were 96.7%, 21.1%, 33.6% and 33.0%, and 84.7%, 10.6%, 32.4% and 30.0% for shale, gravel, ironstone and hornblende, respectively. When the HRT was increased to 3.1 days, mean SRP and TP removal rates increased to 97.1%, 30.1%, 50.2% and 40.6%, and 87.4%, 16.2%, 41.4% and 35.5% for shale, gravel, ironstone and hornblende, respectively. Experimental data indicate that SRP was much easier to remove from vertical sub-surface flow columns for all substrates than TP, and phosphorus removal performance was improved by increasing the HRT from 2.2 to 3.1 days (Table 2.9).

The inflow phosphorus concentrations (SRP: 1.2–1.3 mg L^{-1}; TP: 1.4–1.5 mg L^{-1}) were usually lower than the maximum phosphorus retention capacity of shale. A one-way analysis of variance was conducted to test the impact of different HRT on phosphorus removal. Table 2.9 indicates that there were no statistically significant differences in SRP and TP removal efficiencies for the selected short (2.2 days) and long (3.1 days) HRT concerning shale. However, for gravel, hornblende and ironstone, an increase of the hydraulic residence time led to a significant ($p < 0.05$) increase of the SRP and TP removal efficiencies. Moreover, the greatest combined removal efficiency increase of 49% was recorded for gravel, followed by hornblende (37%) and ironstone (21%).

Phosphorus adsorption onto iron and aluminium oxides (main composition of ironstone) are relatively high for acidic environments (Prochaska and Zouboulis 2006). In this study, the influent pH ranged between 8.32 and 8.64. Weak alkaline environments do not favour precipitation of phosphorus by iron and aluminium oxides. It follows that SRP and TP removal rates of ironstone were lower than those of hornblende during the first week of treatment. However, the maximum adsorption for ironstone was higher than for hornblende. But for column experiments, the mean effluent pH of columns filled with ironstone reduced from 8.26 to 8.02 (Figure 2.18), and SRP and TP removal rates exceeded the corresponding rate for hornblende with decreasing pH, because aluminium phosphate and ferric phosphate were precipitated, and their precipitate increased with decreasing pH as confirmed previously by Ann et al. (1999).

The significant positive correlation between C_e and C_i (Tables 2.10 and 2.11) indicates that the phosphorus effluent concentration increased with increasing corresponding influent concentration. This

Table 2.9 One-way analysis of variance for soluble reactive phosphorus (SRP) and total phosphorus (TP) removal efficiency (%) statistics (mean±standard deviation) at different hydraulic retention times (HRTs).

Filter Media	HRT (days)	*n* (–)	SRP (%)	TP (%)
Shale	2.2	14	96.65 ± 0.31^{g}	84.71 ± 0.35^{g}
Shale	3.1	13	96.84 ± 0.18^{g}	85.43 ± 0.45^{g}
Gravel	2.2	14	21.01 ± 0.69^{a}	10.48 ± 1.35^{a}
Gravel	3.1	13	30.31 ± 0.93^{b}	16.20 ± 0.68^{b}
Hornblende	2.2	14	34.20 ± 3.50^{d}	32.31 ± 2.81^{d}
Hornblende	3.1	13	49.53 ± 1.36^{f}	41.44 ± 1.42^{f}
Ironstone	2.2	14	33.02 ± 0.77^{c}	29.97 ± 0.51^{c}
Ironstone	3.1	13	41.13 ± 0.69^{e}	35.49 ± 0.59^{e}

n, sampling number; concerning SRP and TP removal efficiencies, letters a to g indicate values that are significantly different from each other ($p < 0.05$).

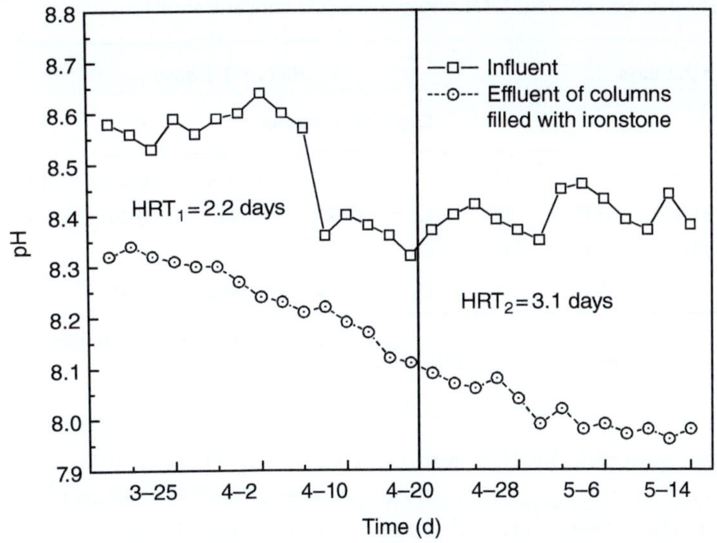

Figure 2.18 Effluent pH of filter beds filled with ironstone at different hydraulic retention times (HRTs). The vertical line in the graph represents the change of HRT. The pH value decreases with an increase in hydraulic retention time. *Source:* Tang et al. (2009)/with permission of John Wiley & Sons.

Table 2.10 Relationships between influent and effluent soluble reactive phosphorus (SRP) concentrations.

	HRT$_1$ = 2.2 days			HRT$_2$ = 3.1 days		
Material	**Empirical equation**	r^2	n	**Empirical equation**	r^2	n
Shale	$C_e = 0.1495C_i - 0.1489$	0.38	15	$C_e = 0.0201C_i + 0.0661$	0.00	13
Gravel	$C_e = 0.7714C_i + 0.0233$	0.70	15	$C_e = 1.4553C_i - 0.9710$	0.44	13
Ironstone	$C_e = 2.9494C_i - 2.9342$	0.69	15	$C_e = 0.7054C_i - 0.2570$	0.09	13
Hornblende	$C_e = 0.5776C_i + 0.1188$	0.53	15	$C_e = 0.7988C_i - 0.2689$	0.34	13

C_e, effluent SRP concentration (mg L^{-1}); C_i, influent SRP concentration (mg L^{-1}); HRT$_i$, hydraulic retention time (days) for the i part of the experiment; n, number of samples; r^2, coefficient of determination.

observation was also made by Pant et al. (2001), who found that the phosphorus removal efficiency of sub-surface flow constructed wetlands decreased with the increase of the phosphorus influent concentration and loading. The slope of the fitted equation and the corresponding r^2 for different substrates varied considerably both at low and high HRT (Tables 2.10 and 2.11). In general, the ranking order for the slope and r^2 for the substrates was as follows: shale < ironstone < hornblende < gravel. This order implied that adsorption was the predominant phosphorus removal process in this study, and phosphorus removal performance was more dependent on substrate phosphorus adsorption than on the influent concentration.

Additionally, the r^2 values in Table 2.11 were much higher than those in Table 2.10. It follows that under the same operational conditions in terms of HRT, TP removal in column experiments was more dependent on influent concentrations than SRP removal. Retention of particulate phosphorus (usually contained in TP) often occurred in constructed wetlands (Tanner et al. 1998). This process

Table 2.11 Relationships between effluent and influent total phosphorus (TP) concentrations.

	HRT$_1$ = 2.2 days			HRT$_2$ = 3.1 days		
Material	**Empirical equation**	r^2	n	**Empirical equation**	r^2	n
Shale	$C_e = 0.3183C_i - 0.2451$	0.64	15	$C_e = 0.2325C_i - 0.1574$	0.21	13
Gravel	$C_e = 1.1054C_i - 0.3136$	0.72	15	$C_e = 1.0700C_i - 0.3417$	0.49	13
Ironstone	$C_e = 2.5571C_i - 2.7913$	0.78	15	$C_e = 0.7121C_i - 0.1862$	0.19	13
Hornblende	$C_e = 1.0098C_i - 0.4597$	0.92	15	$C_e = 0.9209C_i - 0.4062$	0.55	13

C_e, effluent SRP concentration (mg L^{-1}); C_i, influent SRP concentration (mg L^{-1}); HRT$_i$, hydraulic retention time (days) for the i part of the experiment; n, number of samples; r^2, coefficient of determination.

also contributed to TP removal from eutrophic Jin river water in column experiments. It follows that r^2 values summarized in Table 2.11 are higher than those in Table 2.10. The r^2 values for all aggregates decreased when the HRT were increased from 2.2 to 3.2 days (Tables 2.8 and 2.10). This indicated that the increase of HRT prolonged the contact time between substrates and phosphorus added to the columns and improved the substrate adsorption performance.

The first-order model (Eqs. 2.1 and 2.2) could estimate the actual k value, since the phosphorus concentrations of the influent and effluent were known, in addition to the HRT. Results of mean k values are presented in Tables 2.12 and 2.13. The background concentration C^* was set to zero, which is based on a common assumption, which is, however, not always correct (Braskerud 2002; Kadlec 1997). According to the first-order model (Eqs. 2.1 and 2.2), the phosphorus effluent concentrations decreased with increasing hydraulic residence time. The SRP and TP retention efficiency was thus expected to increase as the HRT increased. Values of k increased with increasing

Table 2.12 Mean removal rate constant k (m d^{-1}) values for soluble reactive phosphorus removal (mg d^{-1}).

		Material			
HRT (days)	n	**Shale**	**Gravel**	**Ironstone**	**Hornblende**
2.2	15	1.542	0.107	0.186	0.182
3.1	13	1.115	0.116	0.221	0.191

HRT, hydraulic retention time (days); n, the number of the samples.

Table 2.13 Mean removal rate constant k (m d^{-1}) values for total phosphorus removal (mg d^{-1}).

		Material			
HRT (days)	n	**Shale**	**Gravel**	**Ironstone**	**Hornblende**
2.2	15	0.853	0.051	0.178	0.162
3.1	13	0.669	0.057	0.183	0.171

HRT, hydraulic retention time (days); n, the number of the samples.

HRT (Tables 2.12 and 2.13). However, the increase was not significance, probably due to the high content of SRP in the influent.

The study shows that mean *k* values (Tables 2.12 and 2.13) vary considerably but are still in sequence: shale > ironstone > hornblende > gravel. This sequence was the same for the phosphorus adsorption capacity of all substrates. It follows that the first-order rate constant *k* can also be used to evaluate the phosphorus retention performance of different substrates.

2.3.4 Conclusions

The maximal phosphorus adsorption capacities of shale, gravel, ironstone and hornblende were 619.7, 89.05, 324.9 and 153.1 mg kg^{-1}, respectively. The phosphorus adsorption capacity increased as shale size decreased.

Columns filled with shale showed the best SRP and TP removal performances. The relatively high concentration of SRP (86.27% of TP) mirrored the removal pattern of TP for both low and high HRTs. The SRP removal rates for each substrate were higher than those for TP.

The phosphorus adsorption capacities of shale, ironstone and hornblende could be interpreted better with Langmuir adsorption isotherms than Freundlich adsorption isotherms, while gravel data were fitted well with Freundlich adsorption isotherms. Mean values of the first-order removal rate constant *k* increased with increasing HRT.

Linear equations were inapposite in modelling the phosphorus adsorption capacity of all substrates. Phosphorus removal performance was more dependent on the phosphorus adsorption capacities of the substrates than on the influent concentrations.

2.4 Treatment of Gully Pot Effluent with Constructed Wetlands

2.4.1 Introduction

Horizontal-flow constructed wetlands consisting of soil, sand and/or gravel are used extensively to treat domestic and industrial wastewater (Cooper et al. 1996; Kadlec and Knight 1996), and have also been applied for passive treatment of mine wastewater drainage (Mays and Edwards 2001; Mungur et al. 1997). Vertical-flow systems can also be used for wastewater treatment, but there are only limited performance data available. Vertical-flow wetlands are flat and are intermittently flooded and drained, allowing air to refill the soil pores within the bed (Cooper et al. 1996; Gervin and Brix 2001; Green et al. 1998).

Macrophytes are widely used within treatment wetlands in Europe and Northern America (Cooper et al. 1996; Kadlec and Knight 1996). The functions of macrophytes within constructed wetlands have been reviewed in detail (Brix 1999; Scholz et al. 2002).

Metals occur in soluble or particulate forms. Heavy metals such as nickel and copper are most bioavailable when they are soluble, either in ionic or weakly complexed form (Cheng et al. 2002; Cooper et al. 1996; Wood and Shelley 1999). Metal bioavailability/removal is controlled by chemical processes including acid volatile sulphide formation and organic carbon binding/sorption in reduced sediments of constructed wetlands (Obarska-Pempkowiak and Klimkowska 1999; Wood and Shelley 1999). It follows that metals usually accumulate in the top layer (sediment and litter) of constructed wetlands (Cheng et al. 2002; Scholz et al. 2002).

Although constructed wetlands have been used for stormwater treatment since the 1980s (Cooper et al. 1996), there are insufficient system performance data currently available for the treatment of gully pot effluent from road drainage systems. Furthermore, the beneficial role of macrophytes and porous aggregates within the treatment process has frequently been questioned (Scholz et al. 2002).

Gully pots can be viewed as simple physical, chemical and biological reactors. They are particularly effective in retaining suspended solids (SS). Currently, gully pot effluent is extracted once or twice per annum from road drains and transported (often long distances) for treatment at sewage works (Butler et al. 1995; Memon and Butler 2002). A more sustainable solution would be to treat gully pot effluent locally in constructed wetlands, reducing transport and treatment costs.

Furthermore, the effect of toxic metals and varying loading rates on the biomass within constructed wetlands is unknown (Wood and Shelley 1999). Heavy metals within stormwater are associated with fuel additives, car body corrosion and tyre and brake wear. Common metal pollutants from cars include lead, zinc, copper, chromium, nickel and cadmium. Freshwater quality standards are most likely to be exceeded by copper (Cooper et al. 1996; Scholz et al. 2002; Tchobanoglous et al. 2003). It follows that an experimental study is required to establish design and management guidelines for constructed wetlands treating gully pot effluent.

The aim of Chapter 2.4 is to advance knowledge and understanding by formulating design and operation guidelines for constructed wetlands treating gully pot effluent water. The objectives are to assess (i) the performance of vertical-flow wetland filters, combined with a treatment pond (above the litter zone), as a more efficient process for gully pot effluent treatment; (ii) the removal efficiencies of experimental constructed wetlands treating gully pot effluent receiving high loads of nickel, copper, organic matter and salts; (iii) the impact of road gritting and salting on the filtration performance in cold climate; (iv) the function of macrophytes (with and without *P. australis*) and different aggregates (sand and gravel versus Filtralite, a light expanded clay product used to accumulate metals and nutrients) in wetland filters; and (v) the constructed treatment wetland performance in comparison with the corresponding alternative on-line treatment within the gully pots, stormwater pipeline network and stormwater sediment tank. Scholz (2004a) provided the scientific basis for Chapter 2.4.

2.4.2 Materials and Methods

2.4.2.1 Study Site and Filter Media Composition

Twelve wetland filters (Figure 2.19) were located outdoors for the six-month experiment at The King's Buildings campus (The University of Edinburgh, Scotland) to assess the system performance in cold climate (9 September 2002–28 February 2003). Inflow water (polluted by road run-off) was collected from randomly selected gullies on the campus and the nearby predominantly housing estates. Gully pot water was carefully obtained by manual abstraction with a 2 dm^3 beaker.

A round drainage pipe was used to construct the filters. All twelve vertical-flow wetland filters (Figure 2.19) were designed with the following dimensions: height = 83 cm and diameter = 10 cm. The calculated empty filter bed volumes (volume before adding aggregates) were approximately 6.2, 6.4, 4.0, 4.1, 3.8, 4.1, 3.8, 4.0, 3.8, 4.0, 4.0 and 4.0 dm^3 for Filters 1–12, respectively. Different packing order arrangements of filter media and plant roots were used in the wetland filters (Tables 2.14 and 2.15). The outlet construction of each filter comprised a simple hole in the bottom of each filter. Outflow velocities were altered with a mechanical valve.

The inflow waters of Filter 2 and Filters 7–12 were artificially contaminated with copper nitrate and nickel nitrate. Filters 1 and 2 are similar to a wastewater stabilisation pond or gully pot. In comparison, Filters 3, 5, 7 and 9 are like gravel and slow sand filters and Filters 4, 6, 8 and 10 are typical reed bed filters. The reed bed filters contain gravel and sand substrate and one native plant of Common Reed, *Phragmites australis* (Cav.) Trin. Ex Steud., all of similar total biomass

Figure 2.19 Constructed treatment wetland rig (The King's Buildings campus, The University of Edinburgh) on 9 September 2002. Picture showing twelve outside-located wetland filters in drainage pipes that are partly planted. *Source:* Scholz (2004a)/with permission of John Wiley & Sons.

Table 2.14 Systematic and stratified experimental set-up of filter content and operation.

Filter	Planted	Media type[a]	Contaminated[b]	Aerated	Increased loading[c]
1	No	1	No	No	No
2	No	1	Yes	No	No
3	No	2	No	No	No
4	Yes	2	No	No	No
5	No	3	No	No	No
6	Yes	3	No	No	No
7	No	2	Yes	No	No
8	Yes	2	Yes	No	No
9	No	3	Yes	No	No
10	Yes	3	Yes	No	No
11	Yes	3	Yes	Yes	No
12	Yes	3	Yes	Yes	Yes

[a] Media types: 1 = no media, 2 = standard, 3 = addition of Filtralite and barley straw (see also Table 2.15).
[b] addition of copper and nickel nitrate.
[c] Filter 12 received approximately 170% additional inflow load due to more frequent feeding (Table 2.16).

Table 2.15 Packing order of vertical-flow filters simulating wetlands.

Height (cm)	Filter 1	Filter 2	Filter 3	Filter 4	Filter 5	Filter 6
61–83 (top)	W	W	W	W	W	W
56–60	W	W	6	6 + P	7	7 + P
51–55	W	W	6	6 + P	6	6 + P
36–50	W	W	4	4 + P	5	5 + P
31–35	W	W	3	3	5	5
26–30	W	W	3	3	4	4
21–25	W	W	2	2	3	3
16–20	W	W	2	2	2	2
11–15	2	2	2	2	2	2
0–10 (bottom)	1	1	1	1	1	1
Height (cm)	**Filter 7**	**Filter 8**	**Filter 9**	**Filter 10**	**Filter 11**	**Filter 12**
61–83 (top)	W	W	W	W	W + A	W + A
56–60	6	6 + P	7	7 + P	7 + P	7 + P
51–55	6	6 + P	6	6 + P	6 + P	6 + P
36–50	4	4 + P	5	5 + P	5 + P	5 + P
31–35	3	3	5	5	5	5
26–30	3	3	4	4	4	4
21–25	2	2	3	3	3	3
16–20	2	2	2	2	2	2
11–15	2	2	2	2	2	2
0–10 (bottom)	1	1	1	1	1	1

1, stones (37.5–75.0 mm); 2, large gravel (10–20 mm); 3, medium gravel (5–10 mm); 4, small gravel (1.2–5.0 mm); 5, Filtralite; 6, sand (0.6–1.2 mm); 7, barley straw; A, aerated zone; P, Common Reed or *Phragmites australis* (Cav.) Trin. Ex Steud; W, water (if flooded).

weight and from the same source. However, Filters 5, 6, 9 and 10 also contain adsorption media. Additional natural adsorption media (Filtralite and barley straw) were used. Filtralite (containing 3% of CaO) with diameters between 1.5 and 2.5 mm is associated with enhanced metal and nutrient removal (Brix et al. 2001; Scholz et al. 2002). Furthermore, raw barley straw (natural product with varying physical properties) has a high adsorption area and is therefore associated with high heavy metal removal according to Frogmat International Holdings.

Finally, Filters 11 and 12 are more complex in their design and operation (Table 2.14). The top water layer of both filters is aerated (with air supplied by air pumps) to enhance oxidation and nitrification (Cheng et al. 2002; Green et al. 1998; Obarska-Pempkowiak and Klimkowska 1999). Filter 12 receives about 170% of Filter 11's inflow load (Table 2.14). The hydraulic regime of Filter 12 differs from that for Filters 1–11 (Table 2.16) to test the limits of filtration performance. A higher hydraulic load should result in greater stress on the plants and biomass.

Plant heights and diameters were measured regularly. The diameter was measured at a height of approximately 30 cm (avoiding nodes). On 2 December 2002, plant weights were estimated for stems greater than 50 cm of height after drying the plants at 105° for 24 hours to assess biomass development.

Table 2.16 Systematic regime for manually controlled filling and emptying (expressed as % drainage volume) the experimental filters (25 September 2002–9 October 2002). The first periodic cycle for Filters 1–11 is the duration from 25 September 2002–14 October 2002. The first periodic cycle for Filter 12 is the duration from 25 September 2002–4 October 2002. One and two complete cycles for Filters 1–11 and Filter 12 are shown, respectively.

No	Date	Day	Filter 1–11	Filter 12	No	Date	Day	Filter 1–11	Filter 12
1–16	9–24 September 2002	...	...	...	27	5 October 2002	Sat	100	100
17	25 September 2002	Wed	100	100	28	6 October 2002	Sun	100	100
18	26 September 2002	Thu	100	100	29	7 October 2002	Mon	100	100 → 50
19	27 September 2002	Fri	100	100 → 50	30	8 October 2002	Tue	100	50
20	28 September 2002	Sat	100	50	31	9 October 2002	Wed	100	50 → 100
21	29 September 2002	Sun	100	50	32	10 October 2002	Thu	100	100
22	30 September 2002	Mon	100	50 → 100	33	11 October 2002	Fri	100 → 0	100 → 0
23	1 October 2002	Tue	100	100	34	12 October 2002	Sat	0	0
24	2 October 2002	Wed	100 → 50	100 → 0	35	13 October 2002	Sun	0	0
25	3 October 2002	Thu	50	0	36	14 October 2002	Mon	0 → 100	0 → 100
26	4 October 2002	Fri	50 → 100	0 → 100	37	15 October 2002–30 January 2003	...	...	...

Fri, Friday; Mon, Monday; Sat, Saturday; Sun, Sunday; Thu, Thursday; Tue, Tuesday; Wed, Wednesday.

2.4.2.2 Environmental Conditions and Operation

The filtration system was designed to operate in batch-flow mode to save pumping costs. Manual flow control was therefore practised. All filters were periodically inundated with inflow water and partially drained (50%) or fully drained (100%) to encourage air penetration through the soils as indicated in Table 2.16 (Cooper et al. 1996; Gervin and Brix 2001; Scholz et al. 2002). The highly variable theoretical hydraulic residence time (about two to seven days) can be estimated with the information presented in Tables 2.14–2.16. The hydraulic conductivities were measured periodically for the experimental batch-flow filters, but findings were variable due to the small dimensions of the experimental set-ups. Moreover, hydraulic conductivity measurements would be more appropriate for large continuous flow-through systems.

Nickel and copper were selected as additional heavy metals for investigation because they are commonly occurring contaminants from cars and are not easily bio-available (Cooper et al. 1996; Scholz et al. 2002). It follows that these metals are likely to accumulate within the sediment of constructed wetlands. As the build-up continues, metal toxicity increases as does the risk of pollution due to leaching.

These metals do accumulate easily in constructed wetlands but may be released if environmental conditions change; e.g., road gritting in winter. Such transformation processes are not well understood.

Nickel nitrate and copper nitrate were added to the inflow water of Filter 2 and Filters 7–12 to give total concentrations of dissolved nickel and copper of approximately 1 mg dm^{-3} for each metal, comparable to figures reported for mine wastewater and urban water heavily contaminated with heavy metals (Cooper et al. 1996; Kadlec and Knight 1996; Mungur et al. 1997; Scholz et al. 2002).

In comparison, background concentrations of the raw gully pot water (before addition of heavy metals in the laboratory) were approximately 0.005 mg dm^{-3} for nickel and 0.033 mg dm^{-3} for copper. These background concentrations are low in comparison to the values measured after artificial contamination (Tables 2.14 and 2.17), but comparable to concentrations measured in Scotland. There are no clearly defined thresholds for metals released from constructed wetlands treating gully pot effluent in Scotland. However, the threshold concentrations for biochemical oxygen demand, SS and ammonia-N are 5, 25 and 1.3 mg L^{-1}, respectively.

The background concentration for nitrate-N (including nitrite-N) was approximately 2.8 mg dm^{-3}. Therefore, introduced nitrate-N contributed to only 25% (or approximately 0.9 mg dm^{-3}) of the overall nitrate-N (including nitrite-N) load. Nitrite-N concentrations were low. An alternative to adding the metals as nitrates would have been to add them as metal sulphates, but the latter do not dissolve as easily (Mungur et al. 1997; Scholz et al. 2002).

The filter volumes available for influent water differ among the filters due to different filter media compositions (Tables 2.14 and 2.15) and hydraulic regimes (Tables 2.14 and 2.16). The total loading rates for the contaminated filters were therefore between 51 and 92 mg for each metal.

2.4.2.3 Metal and Other Variable Determinations

Metal concentrations were determined in the raw gully pot effluent, sieved (pore size of 0.25 mm) gully pot effluent, contaminated (added metal nitrates) sieved gully pot effluent (actual inflow water for the rig) and the outflow waters from the experimental rig. Raw gully pot effluent was sieved to simulate preliminary and primary treated wastewater. This procedure is in line with common practice in industry (Cooper et al. 1996; Tchobanoglous et al. 2003). Sieving resulted in a reduction of five-day @ 20°C ATU biochemical oxygen demand (BOD) and SS by approximately 22% and 8%, respectively. Composite water samples were collected on Mondays, Wednesdays and Fridays from the inflow and outflow of all filters and stored frozen at −18°C until analysis.

To prepare acid digests of the vegetation, samples were dried at 105°C overnight in a drying oven (UM500, Memmert) prior to being ashed at 400°C for 12 hours in a muffle furnace (ELF 11/14, Carbolite). Ashed samples (0.7 g; standard deviation (SD): 0.28 g) were then digested under reflux in *aqua regia* for two hours, cooled, filtered through Whatman No. 5412 filter papers and made up to 100 mL with de-ionised water ready for analysis.

A Varian Spectr AA 400 Atomic Absorption Spectrophometer with a GTA-96 graphite furnace tube atomiser was used for the standard analysis of nickel and copper. Notched GTA partition tubes (coated) were applied, and the carrier gas was nitrogen for analyses before 9 November 2002 and argon thereafter. The change of the carrier gas (beyond the control of the author) had no significant influence on the results.

Samples were taken randomly from well-mixed waters and three sample replicates were taken from pre-filtered (applying Whatman 1.2 μm cellulose nitrate membrane filters) sub-samples. The spectrophotometer was calibrated with standards of 0, 0.01, 0.02, 0.03, 0.04 and 0.05 ppm nickel, and 0, 0.005, 0.01, 0.015, 0.020 and 0.025 ppm copper made up from 1000 mg dm^{-3} stock solutions

Table 2.17 Primary treated gully pot effluent: Water quality variables after contamination with nickel nitrate and copper nitrate (9 September 2002–28 February 2003).

Variable	Unit	Sample number	Mean	SD[a]	Mean (<13 December 2002)	Mean (≥13 December 2002)
Nickel	($mg\,dm^{-3}$)	21	1.005	0.149	1.032	1.090
Copper	($mg\,dm^{-3}$)	19	1.033	0.016	1.037	1.027
BOD[b]	($mg\,dm^{-3}$)	27	37.6	33.36	44.8	28.6
Nitrate-nitrogen[c]	($mg\,dm^{-3}$)	15	3.7	1.93	3.6	Not determined
Ammonia-Nitrogen	($mg\,dm^{-3}$)	15	1.4	1.58	1.1	Not determined
Phosphate	($mg\,dm^{-3}$)	15	0.2	0.32	0.2	Not determined
Suspended solids	($mg\,dm^{-3}$)	31	405.3	467.81	120.0	709.7
Total solids	($mg\,dm^{-3}$)	29	4783.90	9130.23	324.6	9561.70
Turbidity	(NTU)	31	321.40	453.84	117.0	539.50
Dissolved oxygen	($mg\,dm^{-3}$)	31	5.6	2.45	4.9	6.4
pH	—	31	6.69	0.238	6.59	6.80
Redox potential	(mV)	31	199	98.72	212	185
Conductivity	(μS)	31	7607.6	14570.60	361.7	15336.5
Temperature (air)	(°C)	66	6.7	4.067	8.2	5.3
Temperature (gully)	(°C)	29	5.8	2.760	7.5	4.0

[a] standard deviation.
[b] five-day ATU biochemical oxygen demand.
[c] includes nitrite-N.

(BDH Spectrosol). Some samples (e.g., contaminated inflow waters) were diluted accordingly. The calibration curves were calculated by least squares linear regression analysis. The coefficients of determinations (r^2) were generally above 0.99 for both nickel and copper standards.

An Inductively Coupled Plasma Optical Emission Spectrometer (ICP-OES) was used for selected samples only. The purpose was to economically screen the inflow and outflow to determine various trace concentrations such as zinc. There were no significant differences between data obtained by atomic absorption spectrophotometry and ICP-OES. Total concentrations of metals in filtered (Whatman 1.2 μm cellulose nitrate membrane filters) water samples were determined by ICP-OES using a TJA IRIS instrument (ThermoElemental, USA). Multi-element calibration standards with concentrations 0, 0.1, 1 and 10 $mg\,dm^{-3}$ were used and the emission intensity measured at appropriate wavelengths. For all elements, analytical precision (relative SD) was typically 5–10% for three individual aliquots.

The five-day @ 20°C ATU BOD was determined in all water samples with the OxiTop IS 12-6 system, a manometric measurement device, supplied by the Wissenschaftlich-Technische Werkstätten (WTW), Weilheim, Germany. The measurement principle is based on determining pressure differences estimated by piezoresistive electronic pressure sensors. Nitrification was suppressed by adding 0.05 mL of 5 g dm^{-3} N-Allylthiourea (WTW Chemical Solution No. NTH 600) solution per 50 mL of sample water.

Nutrients were determined with traditional chemical methods (Allen 1974). Oxidised aqueous nitrogen was determined in all water samples as the sum of nitrate-N and nitrite-N. Nitrate was reduced to nitrite by cadmium and determined as an azo dye at 540 nm (using a Perstorp Analytical EnviroFlow 3000 flow injection analyser following diazotisation with sulphanilamide and subsequent coupling with N-1-naphthylethylendiamine dihydrochloride (Allen 1974).

Ammonia and phosphate were determined by automated colourimetry in all water samples from reaction with hypochlorite and salicylate ions in solution in the presence of sodium nitrosopentacyanoferrate (nitroprusside), and reaction with acidic molybdate to form a phosphomolybdenum blue complex, respectively (Allen 1974). The coloured complexes formed were measured spectrophotometrically at 655 and 882 nm, respectively, using a Bran and Luebbe autoanalyser (Model AAIII).

Water samples were analysed for common anions including chloride and sulphate using a DX-500 ion chromatograph (DIONEX, Sunnyvale, CA, USA). Detection was via an ED40 electrochemical detector set at an output range of 30 μS. Instrument control and data collection were accomplished using PeakNet software.

The zeta potential was determined in the following water samples with a Zetasizer $3000HS_A$, using the Malvern zeta potential standard DTS0050: pre-treated inflow water plus metal nitrates (1 November 2002) and outflow waters for Filters 7–10 (8 November 2002 and 8 February 2003). Five replicates of each water sample were analysed at 25°C.

A Whatman PHA 230 bench-top pH meter (for control only), a Hanna HI 9142 portable waterproof dissolved oxygen (DO) meter, a HACH 2100N turbidity meter and a Mettler Toledo MPC 227 conductivity and pH meter were used to determine pH, DO, turbidity and conductivity, respectively. An ORP HI 98201 redox potential meter with a platinum-tip electrode HI 73201 was used. Composite water samples were analysed on Mondays, Wednesdays and Fridays. Usually, one replicate measurement was taken. All other analytical procedures were performed according to the American standard methods (APHA 1995).

Hydraulic residence times are not shown because of their deliberate high variability (method described in Table 2.16) to enhance biodegradation by introducing air into the system. It follows that the hydraulic conductivity is also highly variable due to frequent wetting and drying cycles.

2.4.3 Results

The capital and operational (air supply only) costs were calculated for each filter for the first year of operation. To scale up the filters to the likely operational size, material prices were requested for a volume of 100 m^3 per aggregate. The filter configurations have standardised set-up capital cost ratios of 1 (Filters 1 and 2): 6 (Filters 3, 4, 7 and 8): 7 (Filters 5, 6, 9 and 10): 9 (Filter 12): 13 (Filter 11), based on Edinburgh prices in January 2003. For example, the material costs for Filter 1 are approximately £550 per 100 m^2.

Table 2.17 summarises the inflow water quality after the addition of dissolved nickel nitrate and copper nitrate to sieved gully pot effluent water. The SD for all inflow parameters (except for pH) are high (Table 2.17) due to the random selection of gully pots (Butler et al. 1995).

The data set was divided into two sub-sets (before and after 13 December 2002) because road gritting and salting was practised from mid-December onwards. As a result, total solids (TS), SS, turbidity and particularly conductivity are high for the second data sub-set (Table 2.17).

Filtration performance figures are summarised in Tables 2.17 and 2.18 as well as Figures 2.20 and 2.21. The influence of metal addition (pre-treated inflow), the presence of plants (Filters 8 and 12),

Table 2.18 Relative reduction (%) of outflow variables (9 September 2002–28 February 2003).

	Change (%) per wetland filter[a]																	
	Filter 1			Filter 2			Filter 3			Filter 4			Filter 5			Filter 6		
Variables	A[b]	B[c]	C[d]	A[b]	B[c]	C[d]	A[b]	B[c]	C[d]	A[b]	B[c]	C[d]	A[b]	B[c]	C[d]	A[b]	B[c]	C[d]
Ni[e]	98	99	96	61	82	13	94	97	87	94	96	90	93	98	81	94	98	85
Cu[f]	98	nd	nd	97	nd	nd	98	nd	nd	98	nd	nd	98	nd	nd	98	nd	nd
BOD[g]	41	75	*−2*	*18*	67	*−22*	65	75	43	*16*	*8*	26	22	25	*16*	*13*	*−1*	37
NO_3[h]	94	94	Nd	77	73	nd	69	64	nd	97	96	nd	96	96	nd	79	76	nd
P[i]	59	59	Nd	60	52	nd	89	87	nd	93	92	nd	96	96	nd	96	95	nd
SS[j]	*39*	79	*40*	*33*	80	*29*	*57*	96	*59*	*56*	87	*59*	*62*	85	*60*	*54*	87	*57*
TS[k]	12	61	16	−21	48	−15	21	60	26	21	35	31	26	24	35	2	31	8
Cond[l]	−13	21	−11	−51	0	−48	−4	15	−1	1	−8	4	5	−5	8	−7	−19	−3
	Filter 7			Filter 8			Filter 9			Filter 10			Filter 11			Filter 12		
Variables	A[b]	B[c]	C[d]	A[b]	B[c]	C[d]	A[b]	B[c]	C[d]	A[b]	B[c]	C[d]	A[b]	B[c]	C[d]	A[b]	B[c]	C[d]
Ni[e]	88	97	66	80	93	48	87	97	65	82	94	53	84	95	57	78	92	47
Cu[f]	98	98	97	97	97	98	98	nd	nd	98	nd	nd	98	nd	nd	97	97	98
BOD[g]	97	79	52	52	64	42	24	−2	*55*	12	*6*	*20*	32	*11*	58	49	31	69
NO_3[h]	89	87	Nd	97	96	nd	97	97	nd	96	95	nd	97	97	nd	86	85	Nd
P[i]	93	92	Nd	94	93	nd	98	98	nd	94	93	nd	96	95	nd	94	93	Nd
SS[j]	*74*	96	*75*	*79*	95	*79*	*88*	93	*89*	*88*	93	*88*	*84*	97	*83*	*57*	93	*52*
TS[k]	49	50	53	58	57	62	83	18	86	75	19	79	69	36	72	17	22	10
Cond[l]	29	3	32	40	−20	93	65	−14	68	61	−9	69	56	−14	59	−11	−9	−19

[a] Change (%) = $\frac{(\text{in-out}) \times 100\ (\%)}{\text{in}}$, where in = inflow and out = outflow.
[b] mean.
[c] mean (<13 December 2002).
[d] mean (≥13 December 2002);
[e] nickel.
[f] copper.
[g] five-day ATU biochemical oxygen demand.
[h] nitrogen as nitrate (includes nitrogen as nitrite).
[i] phosphate-P.
[j] suspended solids.
[k] total solids.
[l] conductivity. In italics: BOD > 30 mg dm^{-3} and SS > 30 mg dm^{-3} (outflow values).

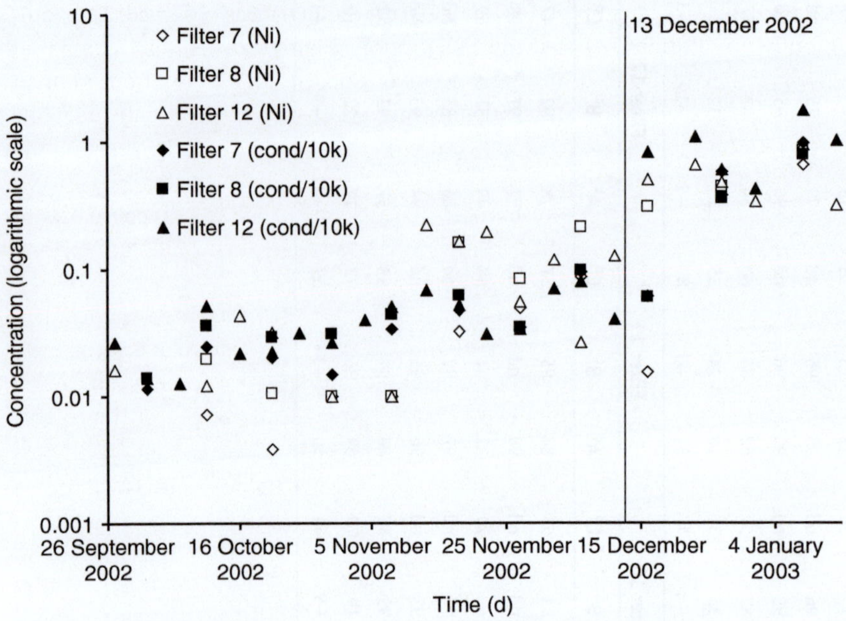

Figure 2.20 Nickel (mg dm^{-3}) and conductivity (cond (μS)) concentrations in the outflows of Filters 7 (unplanted), 8 (planted) and 12 (planted). Conductivity values have been divided by a factor of 10 000 (10k). The logarithmic scale has been chosen to indicate the good linear correlation. Road salting and gritting started on 13 December 2002. Graph showing increases in nickel and conductivity for all filters.

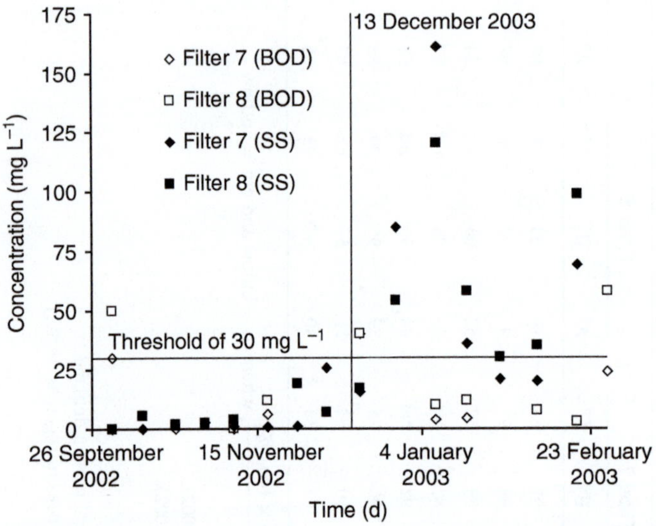

Figure 2.21 Five-day ATU biochemical oxygen demand (BOD (mg dm^{-3})), and SS (SS (mg dm^{-3})) concentrations for the outflows of Filters 7 (unplanted) and 8 (planted). A common threshold value for minimum American standards for the secondary treatment of effluent is 30 mg dm^{-3} (Tchobanoglous et al. 2003). Road salting and gritting started on 13 December 2002. Graph showing that BOD concentrations become more variable and SS concentrations increase.

filter aeration (Filter 12) and increased loading rate (Filter 12) on the BOD was observed. The addition of toxic metals to gully pot effluent reduced the BOD removal efficiency (Filter 1 compared with Filter 2; Table 2.18). Higher BOD concentrations occur usually in the outflow from planted filters, compared to unplanted gravel and sand filters. Concerning Filter 12, the potentially negative effect of a higher loading rate on the overall BOD removal efficiency was reduced by aerating the surface water (Tables 2.14, 2.17 and 2.18).

Furthermore, Table 2.18 indicates that the presence of toxic metals in Filter 2 (if compared to Filter 1) leads to a reduction in nitrogen (measured as nitrate) removal efficiency. Similar observations have been made, previously (Lim et al. 2003).

Anion measurements were undertaken before and after 13 December 2002. Chloride concentrations in the inflow and outflow waters were between 10 and 200 mg dm^{-3} before and frequently above 2000 mg dm^{-3} after mid-December. Sulphate concentrations were only between 1 and 200 mg dm^{-3} throughout the study. Other anions were around the detection limit of the instrument. It follows that some anions (including road salt anions like chloride) were not removed by all experimental filters (Kadlec and Knight 1996). However, an accumulation of salt in the filters was therefore not observed.

High conductivity values reflect the presence of salts (Butler et al. 1995; Kadlec and Knight 1996). Road salts seem to be linked with the leaching of nickel (Figure 2.20). It follows that high conductivity and nickel outflow concentrations for Filters 1, 2 and 5–12 were associated with the correlation coefficients 0.62, 0.68, 0.92, 0.72, 1.00, 0.95, 0.85, 0.97, 0.93 and 0.94, respectively.

Analysis by ICP-OES of selected inflow and outflow samples for a suite of cations showed that all waters generally contained low concentrations of heavy metals. Measured elemental concentrations were either low (barium, calcium, magnesium and manganese), close to the detection limit (iron) and for most heavy metals (including aluminium, copper and cadmium) below detection limit. Zinc was the pollutant measured in highest concentration. The mean inflow concentration for zinc was 0.15 mg dm^{-3} and the corresponding outflow concentrations were around 0.08 mg dm^{-3} (SD: 0.07 mg dm^{-3}).

Red *Tubifex tubifex* worms are indicators for polluted water including high BOD and low DO concentrations (Kadlec and Knight 1996). They were counted (by naked eye) in the outflow water from each filter as a rapid means of assessing water quality. Outflow water from filters planted with *P. australis* had a mean number of 2.2 (SD: 0.8) *T. tubifex* worms per litre. In contrast, unplanted gravel and sand filters contained on average 0.1 (SD: 0.1) *T. tubifex* worms per litre of outflow water. It follows that unplanted filters were less polluted with organic matter.

Furthermore, the dry weights of plants for Filters 4, 6, 8 and 10–12 were approximately 8.0, 4.7, 1.0, 1.1, 1.5 and 6.0 g, respectively. Corresponding stem diameters and heights were 3.7, 3.5, 2.7, 2.7, 2.6 and 3.5 mm, and 97, 74, 63, 62, 64 and 90 cm in this order. These data show that the dry weights, stem diameters and heights were greatest for Filter 4; the most simple uncontaminated planted filter.

Table 2.19 summarises the metal content within the stems and leaves of *P. australis*. Nickel concentrations are higher in the leaves than in the stems. This observation is also correct for copper except for Filters 4 and 10. These observations are high compared to previous recordings (Scholz et al. 2002) but still insignificant in terms of the overall metal loading rate.

The zeta potential values for the inflow and outflow waters were similar. For example, the corresponding zeta potentials for typical pre-treated inflow water with added nickel nitrate and copper nitrate, and the outflow waters for Filters 7–10 (8 November 2002) were as follows: −13.7 mV (SD: 1.1), −13.5 mV (SD: 3.4), −10.8 mV (SD: 1.6), −13.1 mV (SD: 0.4) and −11.9 (SD: 0.8), respectively. After the application of road gritting (including salt), the values for Filters 7–10 (8 January 2003)

Table 2.19 Metal concentrations within *Phragmites australis* (Cav.) Trin. Ex Steud. (19 December 2002).

	Nickel ($mg\ g^{-1}$)		Copper ($mg\ g^{-1}$)	
Filter	**Stems**	**Leaves**	**Stems**	**Leaves**
Filter 4	1.359	1.400	1.505	1.487
Filter 6	1.016	2.034	3.931	6.651
Filter 8	1.186	2.727	5.846	6.105
Filter 10	0.741	2.963	10.197	5.391
Filter 11	0.889	1.231	2.602	7.254
Filter 12	1.071	1.757	0.904	1.411

were −9.3 (SD: 0.4), −12.4 (SD: 0.4), −3.4 (SD: 0.6) and −2.2 (SD: 0.5) in this order. Except for Filters 9 and 10 (both containing Filtralite in contrast to Filters 7 and 8), the values were also similar.

Table 2.20 indicates the potential for treatment of surface water run-off by gully pots, extended storage within stormwater pipes and a small sedimentation tank as part of the associated outlet arrangement. After splitting the stormwater into dry and wet weather flow, the average SS content rose slightly above the permissible value of 30 $mg\ dm^{-3}$: 31.0 $mg\ dm^{-3}$ (SD: 33.06 $mg\ dm^{-3}$ for 10 measurements). However, this value is still lower than the outflow SS values of some experimental constructed wetlands (Tables 2.17 and 2.18).

Table 2.20 Comparison of untreated and treated gully pot water with the corresponding storm drainage outflow water after treatment within the gully pots, stormwater pipes and sedimentation tank on the University campus. The environmental pollution caused by the surface water run-off has also been indicated by summarising the water quality of the Braid Burn, an urban stream that receives the corresponding drainage water (15 January 2003–28 February 2003).

Type	Statistic	Biochemical oxygen demand ($mg\ dm^{-3}$)	Suspended solids ($mg\ dm^{-3}$)	Conductivity (μS)	pH (−)	Turbidity (NTU)	Total solids ($mg\ dm^{-3}$)
	Count	6	9	9	9	9	8
Gully pot effluent	Mean	52.7	1017.7	22351.7	6.77	724	141 350.8
	SD[a]	27.17	702	26732.02	0.183	703.77	15 259.94
Gully pot effluent	Count	7	9	9	9	8	9
filtered with a sieve	Mean	49.6	1061.3	22357.3	7.02	942.3	12 964
(diam.: 0.25 mm)	SD[a]	27.38	887.88	35 878.07	0.3	774.74	20 028.25
Wetland Filter 1	Count	3	5	5	5	5	5
(outflow)	Mean	40.7	592.6	23 180.6	6.84	116.3	12 073.4
	SD[a]	13.61	827.01	31 140.83	0.088	122.3	16 707.91
	Count	3	5	5	5	5	5

Table 2.20 (Continued)

Type	Statistic	Biochemical oxygen demand (mg dm^{-3})	Suspended solids (mg dm^{-3})	Conductivity (μS)	pH (–)	Turbidity (NTU)	Total solids (mg dm^{-3})
Wetland Filter 3							
(outflow)	Mean	13.2	408	21 413.6	6.97	21.1	10 733.3
	SD[a]	4.85	531.16	27 939.97	0.162	18.24	14 483.08
Wetland Filter 4	Count	4	5	5	5	5	5
(outflow)	Mean	31.5	403.6	20 131.6	6.8	42.7	9852.1
	SD[a]	16.46	506.82	25 568.09	0.104	31.76	12 979.06
Stormwater	Count	26	26	26	26	26	17
pipe	Mean	3.4	13.5	542.4	8.09	20.5	501.5
(outflow)	SD[a]	2.54	24.64	271.41	0.693	32.52	669.58
Braid Burn	Count	26	26	26	26	26	16
(before stormwater	Mean	1.9	10.8	399.5	7.98	17.4	230.1
outflow pipe)	SD[a]	0.78	20.10	158.82	0.196	23.99	116.14
Braid Burn	Count	16	16	16	16	16	16
(after stormwater	Mean	2.1	4.6	443.1	8.09	8.4	217.9
outflow pipe)	SD[a]	0.92	7.08	176.09	0.119	10.97	99.64

[a] standard deviation.

Furthermore, sediment samples were taken from the sediment tank for ICP-OES analysis on 12 March 2003. High concentrations of nickel (34–83 mg dm^{-3}), copper (202–667 mg dm^{-3}), zinc (621–1284 mg dm^{-3}), lead (148–353 mg dm^{-3}), iron (42 616–49 510 mg dm^{-3}) and phosphorus (2305–3426 mg dm^{-3}) were measured for nine sediment layers.

2.4.4 Discussion

2.4.4.1 Performance Comparison

The overall filtration performance in terms of copper was satisfactory (typically >96%) for all reed bed filters (Tables 2.17 and 2.18), compared to published performance data for similar reed beds (Kadlec and Knight 1996; Scholes et al. 1999; Shutes et al. 2001). However, the reductions in nickel, copper, BOD (Figure 2.21) and SS (Figure 2.21) were unsatisfactory for most filters, if compared to minimum American and European standards for the secondary treatment of effluent. Nickel, copper, BOD and SS concentrations should not exceed 0.0071, 0.0049, 30 and 30 mg dm^{-3}, respectively (Tchobanoglous et al. 2003). Some European standards are more stringent; e.g., BOD < 20 mg dm^{-3}.

In contrast to finding by Shutes et al. (2001), nickel instead of copper broke through the filter. While others (Shutes et al. 2001) speculate that the die-back of aquatic plants and consequently the release of metals resulted in the breakthrough of predominantly copper, the author assume that the breakthrough of nickel is predominantly linked to road grit (containing salt) (Figure 2.20). Nevertheless, *P. australis* containing high metal concentrations was harvested to avoid metal accumulation in the sediment (see also Table 2.19).

Furthermore, there is an indication (at least on the University campus) that some metals and phosphorus accumulate in the sediment of stormwater pipes and associated outlet structures like sedimentation tanks. This shows that metals including nickel and copper are not entirely retained by gully pots.

High levels of conductivity were recorded in the filter inflow and outflows due to road gritting (including salt) after 13 December 2002. Salt removal in the filters was low and lead to deterioration of the overall biological filtration performance. This is likely to have resulted in the breakthrough of nickel (see Table 2.18, Figure 2.20 and the correlation analysis).

Tables 2.17 and 2.18 indicate that even Filter 1 (control filter), which is like a gully pot, functions as a simple biological reactor (Butler et al. 1995). It has been suggested that mature and viable microbial biomass, in contrast to aggregates with high adsorption capacities, is responsible for the high overall filtration performances (Cooper et al. 1996; Scholz and Martin 1998a, b; Scholz et al. 2002).

However, it is difficult to classify objectively a biological system as mature without having undertaken extensive microbiological work. Nevertheless, the statistical set-up (Tables 2.14 and 2.15) and the treatment performance (Tables 2.18 and 2.20) show that extended storage leads to considerable reductions of nutrients, metals, SS and BOD.

2.4.4.2 Constructed Wetlands, Sustainability and Filterability

The application of constructed wetlands for treatment of wastewaters contaminated with toxic metals has been subjected to criticism (Cooper et al. 1996; Kadlec and Knight 1996; Scholz et al. 2002), as a potentially non-sustainable method, due to the nature of the non-biodegradable pollutants to be removed. The accumulation of metals may create a so-called chemical time-bomb: the sudden and unexpected release of removed metals after filter breakthrough or wastewater composition change like in this case study (occurrence of salts in the wastewater during winter). Constructed wetlands are therefore a more sustainable method for the treatment of biodegradable pollutants.

It is therefore not the purpose of Chapter 2.4 to promote constructed wetland technology for the treatment of industrial wastewaters contaminated with toxic metals from the manufacture industry. However, this applied research study shows how surface water run-off contaminated naturally with low metal concentrations form cars and high salt concentration from road gritting may pose a problem during the onset of winter, if treated with constructed wetlands.

According to previous findings by Cooper et al. (1996), the maximum salinity tolerance value for *P. australis* is about 45 g dm^{-3}. Considering that the measured conductivity and total solids inflow values are about 7600 μS and 4.8 g dm^{-3} (Table 2.17), respectively, the presence of salts should not significantly affect the macrophyte development.

Considering the systematic and stratified experimental set-up (Table 2.14), there seems to be no logical explanation for obvious differences in plant biomass (except for the possibility of random variation). Nevertheless, filter aeration (Filter 12, Table 2.16) is likely to lead to increased plant growth despite a high loading rate. However, a further experiment under controlled laboratory conditions with replicate filters would be required to confirm this observation.

The presence of *Tubifex tubifex* in planted filters indicates severely (most likely organically) polluted water, because this species is known to tolerate pollution (including high nitrate levels) and low oxygen levels by using a blood-like substance to help collect oxygen (Orten et al. 1995). This provides further evidence that *P. australis* adversely affected the quality of filter outflow water in this experiment.

An electrode double layer exists round each moving particle within the water body. Ions located beyond the double layer do not travel with the associated particle. The electronic potential at this boundary (surface of hydrodynamic shear) is the zeta potential that will be reduced if salt concentrations are high. This is due to compression of the double layer.

Particles with zeta potentials >30 mV or < -30 mV are stable. Measured zeta potentials were close to zero indicating that the waters were unstable (like tap water) and therefore difficult to filter efficiently. It is difficult to remove SS in unstable waters by filtration in cold climates because of the low biodegradation activity. However, filtration could be aided by lowering the zeta potential artificially, by increasing the pH or adding aluminium sulphate to increase the flocculation potential of suspended matter. A higher pH will also help to prevent the transformation of particulate metals into dissolved metals that are more toxic to the environment (Pontier et al. 2001; Scholz et al. 2002).

The presence of *P. australis* is not associated with a change in the zeta potential, despite the likely presence of negative ions from degrading plant material. In comparison, the presence of Filtralite (Tables 2.14 and 2.15) seems to be associated with a reduction of the zeta potential, despite the likely presence of positive ions from the aggregate (e.g., Ca^{2+} from CaO).

The zeta potential measurement technique was found to be partly inconclusive for this experiment although interesting results may be obtained in a controlled laboratory environment for a simplified system.

2.4.4.3 Alternative Treatment

Gully pot effluent from the University Campus would have been treated passively within the gully pots, stormwater pipeline network and sediment tank anyway, if it had not been abstracted for separate secondary treatment with constructed wetlands (Table 2.20). However, some data shown should be interpreted with caution; e.g., the obtained gully pot effluent is highly concentrated while the stormwater is diluted. Moreover, metals tend to accumulate within the entire stormwater network where they are impossible to control, if environmental conditions promote leaching change.

This study indicates that constructed wetlands are unsuitable for high toxic metal and salt loadings. However, they can be applied for gully pot effluent treatment if contamination is within normal concentration ranges (0–0.2 mg dm^{-3}) for most metals and road gritting (plus salt) is restricted or properly controlled.

Considering the present set-up, a polishing pond as a tertiary treatment step would help to further remove metals, BOD and SS. Alternative treatment technologies include traditional trickling filter and activated sludge secondary treatment steps. However, the application of active salt removal technologies is usually uneconomic.

2.4.5 Conclusions and Further Research

The sustainable constructed wetland showed signs of overloading in cold climatic conditions resulting in high five-day ATU @ 20°C BOD and SS outflow values. Waters were unstable in terms of their zeta potentials, and therefore difficult to strip of suspended solids. Copper concentrations were removed satisfactorily in comparison to values obtained from literature.

Furthermore, road salting led to a reduction in treatment performances due to high salt concentrations within the gravel and sand filters. A breakthrough of nickel correlated very well with high conductivity values because of road gritting and salting. Road run-off contaminated with road salts should therefore not be treated with constructed wetlands in the future.

The presence of Filtralite (adsorption filter media) as a replacement for aggregates such as sand and gravel did not result in an obvious reduction of metal and nutrient concentrations in outflow waters. Microbiological activity within each filter is likely to be responsible for the reduction of nickel and copper, rather than aggregates and macrophytes. This has implications for the future design and operation of constructed wetlands since costs can be saved on filter material and aquatic plants including their management.

Further research into the sustainability of constructed wetlands treating industrial wastewater contaminated with toxic metals should be undertaken. Moreover, more work on dynamic microbial community changes due to salt within road grit is required. It should be a further aim to assess the microbiological metal retention potential in comparison to physical and chemical processes.

2.5 Wetland and Dry Pond System

2.5.1 Introduction

Conventional stormwater and urban drainage systems are designed to dispose of runoff water as quickly as possible. This results in end-of-pipe solutions which often involve the provision of large interceptor or relief sewers, huge storage tanks in downstream locations and centralised sewage-treatment facilities. In contrast, wet and dry ponds (as a cost-effective end-of-pipe drainage solution) can be applied for local source control; e.g., the collection or diversion of roof drainage water. It is often possible to divert all drainage water for infiltration or storage and subsequent recycling; also, because the runoff from roofs is a major contributor to the quantity of surface water requiring disposal, this is a particularly beneficial approach where suitable ground conditions prevail.

The purpose of Chapter 2.4, which is based on the original article by Scholz (2003), is not only to investigate a case study of a sustainable urban drainage system, designed according to best management practice (Martin et al. 2000), but also to address the following key objectives: (i) to suggest water quality monitoring strategies; (ii) to identify difficulties and constraints associated with their design and operation; (iii) to investigate the cost-effectiveness and self-sustainability of pond systems and (iv) to investigate the compatibility of the primary needs of water removal and storage with amenity and wildlife enhancement.

2.5.2 Case Study and Methodology

A domestic property (Sandy Lane, Bradford, West Yorkshire, UK) was selected, comprising a wetland and dry pond for the storage, treatment, recycling and drainage of surface-water runoff (Figure 2.22). The most significant proportion of surface water came from a house roof and the roof of a double garage. In the original (1972) layout, rainwater drained into the public sewer. However, in May 2001, this layout was altered to feed a semi-natural attenuation (retention) wetland structure (Figure 2.23). The storage water could be used for watering garden plants in summer. If the attenuation wetland structure overflows, the water is transferred to a dry pond structure, which is designed to hold up to 1.9 m^3.

The system is based on a combined attenuation wetland and dry pond design. The area of enhanced ecological value (6.5 m × 8.5 m) comprises the pond structure as well as areas dominated

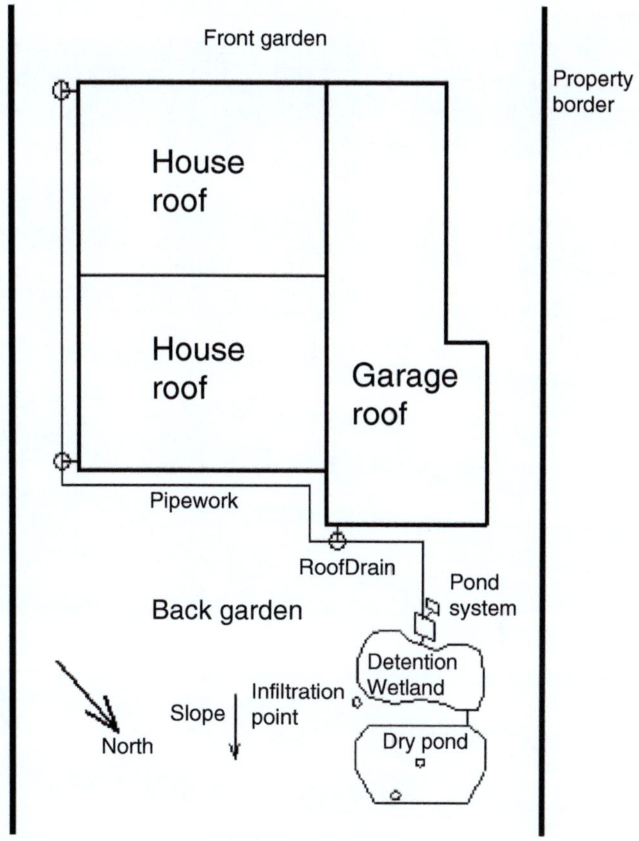

Figure 2.22 Approximate drawing of case study site showing roof areas, pipework and pond system. Roof runoff is channelled through pipes into a wet pond followed by a dry pond.

by lawn and shrubs. The distances between the garage and the attenuation wetland and dry pond are 1.5 and 5.0 m, respectively, and further details of the site are given in Table 2.21.

To determine the maximum water level of the semi-natural attenuation wetland during dry weather, the water depth was measured from the attenuation wetland centre (greatest depth) to the top water surface (outflow structure level) after artificial filling up with tap water until the attenuation wetland began to overflow. All water-level measurements were taken daily.

The pond was constructed with a strong plastic liner, and the attenuation wetland system was filled with stones and gravel (average depth: 40 mm). Most aquatic plants and animals were introduced in May and June 2001. The dominant macrophyte species were the common reed (*Phragmites australis*), reed-mace (*Typha latifolia*) and yellow iris (*Iris pseudochorus*). Canadian waterweed (*Elodea canadensis*) was the dominating floating plant. Two koi were introduced during construction of the pond, and 23 sticklebacks (*Gasterosteus aculeatus*) were added in September 2001 followed by 18 adult common grass frogs (*Rana temporaria*).

Water samples were analysed for biochemical oxygen demand (BOD), suspended solids (SS), total solids (TS), conductivity, turbidity, dissolved oxygen (DO) and pH (Table 2.22). Temporal (hourly, daily or weekly) and spatial sampling schemes were applied, and daily sampling took place at approximately 06 : 00 hours and/or 18 : 00 hours. Important (but easy to measure) and cost-effective

(a)

(b)

Figure 2.23 (a) Private ornamental attenuation wetland and (b) dry pond construction (winter 2002). A planted wet pond and a planted infiltration pond are shown. (a) *Source:* Scholz (2003)/with permission of John Wiley & Sons; (b) *Source:* Scholz (2003)/with permission of John Wiley & Sons.

water-quality parameters, including DO, pH and conductivity, were also sampled on a 0.4 m × 0.4 m grid. Infiltration rates were determined fortnightly or monthly by estimating the infiltration time of 200 mL tap water through a 240 mm long drainage pipe (diameter of 64.5 mm), which was buried in the ground at a depth of about 200 mm.

Table 2.21 Details of site.

Characteristic	Unit
Surface area of house roof (m^2)	29
Angle between roof and house (°)	23°
Theoretical horizontal area of the house roof (m^2)	53
Roof area of double garage (m^2)	33
Total horizontal area drained (m^2)	86
Length of plastic pipework (mean angle of 2°) (m)	19.6
Inner pipe diameter (mm)	64.5
Maximum horizontal dimensions of top pond (m)	0.3 × 0.4
Maximum horizontal dimensions of middle pond (m)	0.7 × 0.4
Maximum horizontal dimensions of bottom pond (attenuation wetland) (m)	3.2 × 1.7
Area of attenuation wetland (if completely filled with water) (m^2)	5.5
Maximum horizontal dimensions of dry pond (m)	3.7 × 2.5
Dry pond slope ratio towards each side (north and south)	1 : 1.9
Dry pond slope ratio towards lower garden (east)	1 : 2.4
Dry pond slope ratio towards the attenuation wetland (west)	1 : 1.6
Maximum depth of attenuation wetland (mm)	395
Maximum depth of dry pond (mm)	400

Table 2.22 Water quality of attenuation wetland (main sampling points MP, Figures 2.24–2.26) and infiltration rates for dry pond (April 2001–February 2002).

Variable and time of measurement	Unit	Overall sample number	Overall mean	Summer[a] mean	Autumn[b] Mean
Air temperature (06 : 00)	°C	30	9.8	13.3	9.1
Air temperature (18 : 00)	°C	198	12.0	16.9	8.9
Water temperature (06 : 00)	°C	29	9.0	14.1	8.4
Water temperature (18 : 00)	°C	178	11.0	15.4	8.6
BOD (06 : 00)	mg L^{-1}	37	4.2	4.2	3.1
Suspended solids (06 : 00)	mg L^{-1}	27	69	132	51
Total solids (06 : 00)	mg L^{-1}	26	276	239	294
Conductivity (06 : 00)	μS	41	93.2	154.3	72.7
Conductivity (18 : 00)	μS	134	45.5	75.1	37.5
Turbidity (06 : 00)	NTU	26	2.9	4.3	2.7
Dissolved oxygen (06 : 00)	mg L^{-1}	47	5.2	4.4	5.3
Dissolved oxygen (18 : 00)	mg L^{-1}	172	7.9	11.5	4.8

(Continued)

Table 2.22 (Continued)

Variable and time of measurement	Unit	Overall sample number	Overall mean	Summer[a] mean	Autumn[b] Mean
pH (06 : 00)	—	43	6.9	7.0	6.8
pH (18 : 00)	—	114	7.2	8.4	7.3
Redox potential (06 : 00)	MV	29	224	208	231
Algal cover (18 : 00)	%	146	51	61	50
Infiltration rate (Station 1)[a]	10^{-7} m s^{-1}	12	13	3	23
Infiltration rate (Station 2)[a]	10^{-7} m s^{-1}	12	1827	4270	95
Infiltration rate (Station 3)[a]	10^{-7} m s^{-1}	12	5810	8152	4778

[a] Summer: 21 June 2001–21 September 2001.
[b] Autumn: 22 September 2001–20 December 2001.

2.5.3 Results and Discussion

2.5.3.1 Water Quality

During the autumn of 2001, the BOD, SS, conductivity and turbidity were low because of dilution by rainwater runoff (Table 2.22). Large temperature datasets can be used to predict sparsely sampled variables such as DO and redox potential, and this would result in a reduction of sampling costs and effort. For example, correlations between air (water) temperature and DO and redox potential were 0.63 (−0.65) and 0.62 (0.52), respectively.

Figures 2.24–2.26 summarise spatial and temporal water quality findings with respect to pH. Spatially, pH was most stable at low water depths; i.e., less than about 200 mm (Figure 2.24). In contrast, conductivity and DO were only stable in areas, which were subject to shading by nearby buildings and trees. For example, pH and turbidity need to be sampled at least on a 1 m × 1 m sampling grid to stabilise the associated sample variances. However, this would increase sampling costs by a factor of approximately five. In contrast, conductivity and DO are spatially stable variables, if

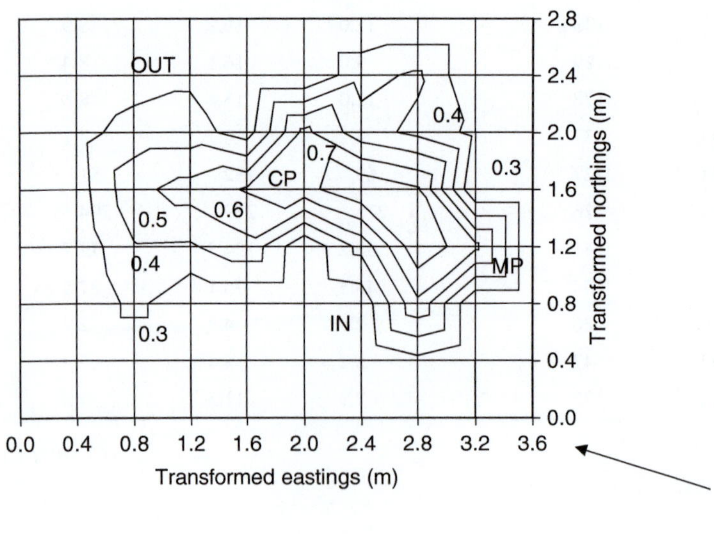

Figure 2.24 Attenuation Wetland: spatial distribution of the mean standard deviation for pH in autumn 2001 (17 samples). The arrow indicates geographical North. IN = inlet; OUT = outlet; CP = centre sampling point; MP = main sampling point for temporal sampling. Values range between 0.3 and 0.7 and increase towards the middle of the pond.

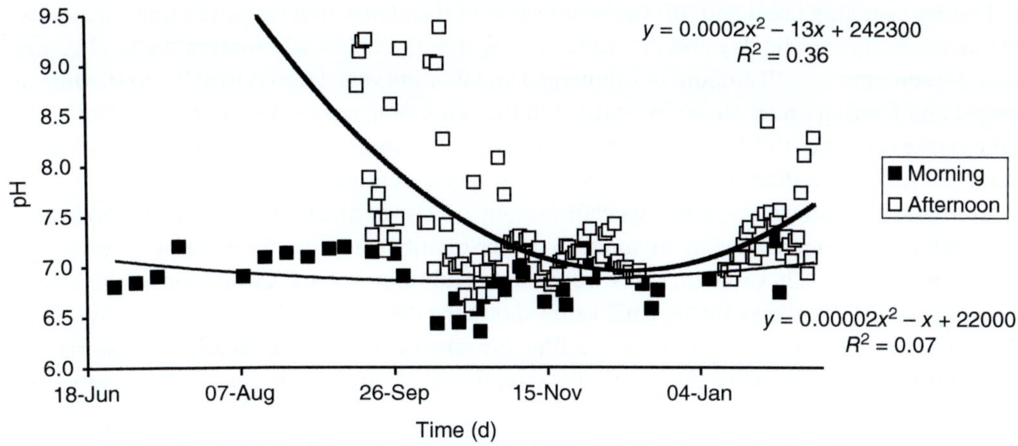

Figure 2.25 Attenuation Wetland: temporal distribution, morning and afternoon, for pH between summer 2001 and winter 2002. pH data are associated with the main sampling point (MP). Afternoon pH data points are usually higher and more variable compared to morning data.

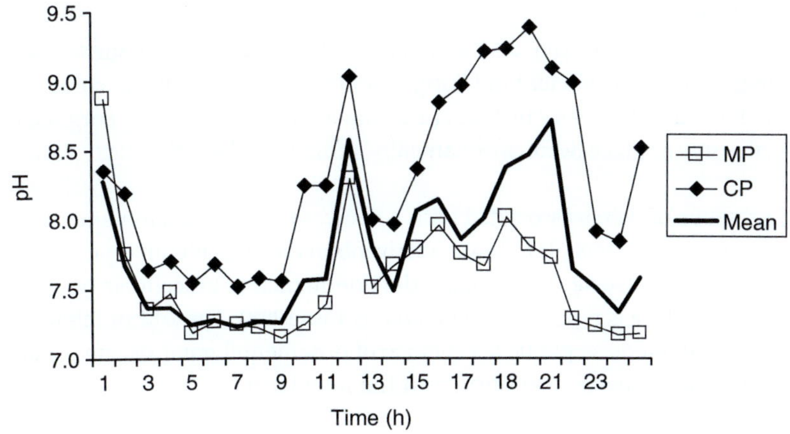

Figure 2.26 Attenuation Wetland: temporal distribution (24 hours) for pH on 15 March 2002. pH data are associated with the main sampling point (MP) for temporal sampling, the detention wetland centre sampling point (CP) and the mean value of six randomly selected detention wetland sampling points including MP and CP. CP data points are higher (except for the early beginning) than MP data points.

both sunshine intensity and temperature variations are low. Therefore, these variables require less dense sampling; e.g., on a 2 m × 2 m grid.

Temporally, the pH and DO of the attenuation wetland were generally high during the day and low during the night because of algal photosynthesis (Figures 2.25 and 2.26). High pH values during summer afternoons are not unusual because algal photosynthetic processes peak during the day (International Water Association Specialist Group on the use of Macrophytes in Water Pollution Control 2000). In contrast, conductivity was low during the afternoon and high in the morning.

Figure 2.26 is a random example for the daily temporal distribution of pH on 15 March 2002. pH values are associated with (i) the main sampling point for temporal sampling, (ii) the attenuation wetland centre sampling point and (iii) the mean value of six attenuation wetland sampling points. The pH increases during the day and decreases during the night. The attenuation wetland centre (Figure 2.24) contains a small amount of submerged and floating vegetation (virtually no shading of submerged and floating green algae) in contrast to the remaining sampling points. Therefore, the pH of the centre is frequently above the mean pH value for the attenuation wetland. The main sampling point of the attenuation wetland represents the mean trend (Figure 2.26).

The International Water Association Specialist Group on Use of Macrophytes in Water Pollution Control (2000) recommends weekly or monthly water quality sampling for treatment wetlands. However, Figures 2.24–2.26 show that this would not result in a realistic estimation of the actual water quality for pH. It follows that the pH value should be determined daily in the morning and weekly in the afternoon, assuming that its trendline is known or can be reasonably well estimated. Moreover, the location of the most representative sampling point must be determined in advance by spatial test sampling.

To remove seasonal trends, linear and second-order polynomial de-trending functions were applied to example datasets including pH, conductivity and DO. After the removal of outliers, trends were quantified and then subtracted from the corresponding sample series. However, no statistically significant (level of significance $P < 0.05$) increase of the coefficient of determination R^2 was achieved, if trend functions were fitted to the de-trended data. It follows that most temporal variation for pH and DO is random for large datasets. Therefore, frequent sampling is required to interpret data with high confidence.

An algal bloom (predominantly *Microspora spp.* >50 organisms/mL) was a problem due to low nutrient and trace mineral levels. Algae are efficient in taking up nutrients from the water column. However, the presence of small oat and barley straw bales (a commercial product called Forgmat) reduced this challenge. Furthermore, algae were mechanically removed when the algal cover exceeded about 60%.

It is well known that the presence of fish is associated with a decrease in zooplankton and consequently with an increase in algae (Moss 1998). It follows that the observed algal bloom is probably associated with the introduction of koi. However, normally the dominance of unicellular (green) algae in contrast to *Microspora*-like algae is reported in textbooks (Moss 1998). The author felt that the presence of koi was important to increase public acceptance of ornamental pond systems, and that sticklebacks help to provide an ecological equilibrium in the long term.

2.5.3.2 Capacity of the System

The roof, pipeline system and overflow structures can hold back a maximum of about 50 l rainwater during dry weather and about 300 L during a heavy storm. The semi-natural pond structure comprises three ponds having a total effective volume of about 1720 L during dry periods and 1900 L during storm events. The top pond (maximum 20 L) has only ornamental value, and rainwater drains into the middle pond (60 L maximum capacity). SS (e.g., decayed leaves and bird droppings) from the roof and pipework settle in this part of the structure, which acts as a silt trap.

Water from the middle pond drains and/or overflows into the bottom pond (volume 1670 L), which is designed as both an attenuation wetland and storage pond. Not more than 500 L of storage water can be used for irrigation without harming the aquatic-plant ecosystem at any time.

During heavy storm events, the bottom pond (Figure 2.23) can hold another 150 L. The dry pond can accommodate a maximum of 1870 L during heavy storm events before it overflows. However, at the onset of heavy rain, the overall system should have an additional capacity of no more than 480 L. Therefore, the overall system cannot cope with more than about 2500 L at any time. Indirect system capacity data are presented in Figure 2.27.

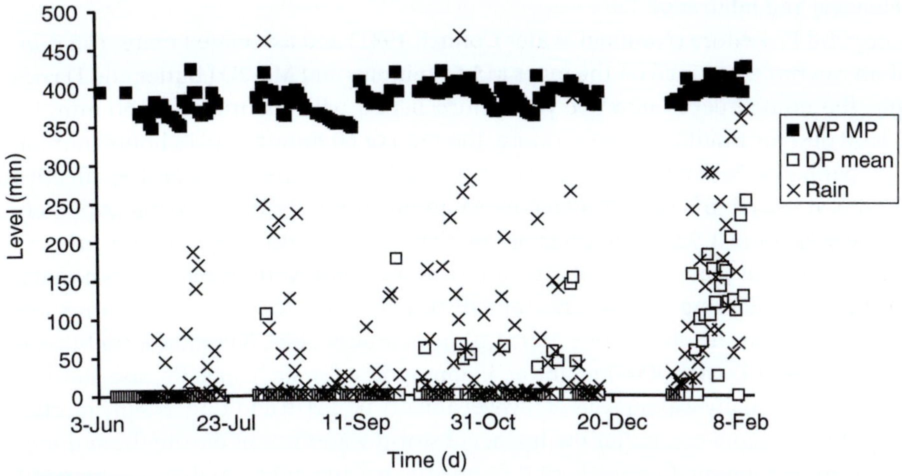

Figure 2.27 Daily water levels of the wet pond (WP; main sampling point; WP MP) and the dry pond (DP, mean of five stations; DP mean) as well as scaled-up daily rainfall data expressed in mm × 10 (Rain). High levels of rainfall on the following dates have been omitted from this graph: 84.4 mm on 27 September 2001 and 532 mm on 9 February 2002. MP, main sampling point for temporal sampling. The water level of the wet pond is stable while the water level of the dry pond depends strongly on the precipitation.

2.5.3.3 Economics

Yorkshire Water (regional water utility) gives domestic customers who are not connected to a sewer a rebate of £25. Therefore, an application to the local water utility to pay a lower sewerage charge for a domestic property where no surface water or groundwater drains to a public sewer was made successfully.

The four occupants (two adults and two children) of the detached house use a bath, shower, toilet and washing machine. No water was used for a dishwasher, water hose or sprinkler. The average annual water consumption is 150–230 m^3, and the associated bill is £226–385.

If half an hour's use of a hose pipe or sprinkler would consume 270–810 L, the cost would be £0.41–1.22. Furthermore, if the domestic property user recycled about 300 L of pond water 33 times a year to water plants during summer, he/she could save an additional £15. It follows that the property owner would have a reduced water bill of up to £40/annum.

Further savings of up to £30 are possible; e.g., if rainwater is recycled for toilet flushing. However, these savings have to be contrasted to capital and maintenance expenditure. The capital and labour costs (£800 and £1330, respectively) for the pond structure were approximately £2130. The capital costs of £500 for the construction of the attenuation wetland are the largest single item. Expenditure for the new pipeline layout (£120) and aquatic plants (£100) is also significant. However, the gain in property and amenity value is difficult to assess, and a discussion of these issues is beyond the scope of Chapter 2.4.

Cost figures are simplistic and are only based on the case study. A comparison with sites elsewhere in the UK and abroad would depend upon other variables including building and drainage regulations, water-consumption costs, water-metering service availability, cost of conventional systems, option to obtain a rebate for retrofitted systems, climatic conditions such as precipitation, soil-infiltration properties and material and labour costs. Therefore, a brief cost comparison is impossible to undertake and could be misleading.

2.5.3.4 Risk of Flooding and Infiltration

By using the Wallingford Procedure (National Water Council 1981) and associated maps, the rainfall data for Bradford can be determined on the maps M5-60 minutes and M5-2D (Butler and Davies 2000). Accordingly, the rainfall depth of a five-year return period with a duration of 60 minutes (M5-60 minutes) is 20 mm for Bradford. Furthermore, the ratio of 60 minutes to 48 hours' rainfall of a five-year return period for Bradford falls between 25% and 30% (25% considered). The effective area of the dry pond is at least 8 m^2. The estimated mean annual infiltration rate is approximately 10^{-4} m s^{-1} and a safety factor of 1.5 is used. Considering that the maximum depth is 400 mm, the structure has a high theoretical spare height of 270 mm. However, observations have shown that a depth of 4 m is justified considering that several rainfall-runoff events led to a maximum water depth of almost 260 mm (spare height of only 140 mm) in February 2002. Therefore, traditional design guidelines (Butler and Davies 2000) might be inappropriate for such specific case studies.

Figure 2.27 summarises the daily water levels of the wet and dry ponds in addition to daily rainfall data, and the attenuation wetland can buffer the impact of storm-water runoff during the summer and autumn. In contrast, the attenuation wetland is frequently full in winter, so that the dry pond receives excess water which must infiltrate into the ground. However, a system failure (e.g., flooding and structural damage) has not yet been recorded.

Table 2.22 summarises the mean infiltration rates which were calculated for the soil layers located 200 mm below the bottom of the dry pond centre (Station 1), the middle of the slope of the dry pond (Station 2) and the lawn (Station 3). The undisturbed unsaturated soil layer below the dry pond (>200 mm) has a hydraulic conductivity of 13×10^{-7} m s^{-1}; therefore, the seepage loss during storms is small. Hydraulic conductivity is the relative distance (m) of a fluid table travelling through a medium via gravity during a pre-defined period (s). The hydraulic conductivity is a function of both the porous medium (e.g., man-made soil) and the fluid (e.g., pretreated stormwater) flowing through the medium.

The recently disturbed man-made soil covering the slope and base of the dry pond has a hydraulic conductivity of 2×10^{-4} m s^{-1}, and the man-made lawn soil has a hydraulic conductivity of approximately 6×10^{-4} m s^{-1}. Both soil layers are capable of draining roof runoff water, which is produced during small moderate storms (Figure 2.27).

2.5.3.5 Sustainable Water Industry Asset Resource Decisions Analysis

The Sustainable Water Industry Asset Resource Decisions (SWARD) analysis support system was applied to assess the cost-effectiveness and sustainability of pond structures (Asley et al. 2001). The relevant primary criteria for the assessment of economic, environmental, social and technical categories are as follows:

Economic: life-cycle costs; willingness to pay and financial-risk exposure. Developers and householders should be willing to pay for pond structures, because the capital, operational and maintenance costs are low. The risk of loss associated with the investment is also low.

Environmental: resource utilisation and environmental impact. The resource utilisation (in terms of land and material use) could be substantial but is sustainable and cost-effective compared with traditional systems. The pond structure was designed to promote biodiversity by avoiding deep-water zones (>1 m), maximising shallow-water zones (<0.2 m: 45–55% of the total surface area) and avoiding steep slopes. Natural colonisation of vegetation was encouraged. However, planting was necessary in urban Sandy Lane. All dominant species were native, and preference was given to aquatic plants of local provenance, which are suited to local conditions in terms of water depth and quality.

Social: impact on human health; acceptability to stakeholders and public and local public participation and responsibility. The recreational and amenity value of pond structures is high, but provision needs to be made to make the pond structure safe for small children and pets.

Technical: performance of the system and flexibility. Flooding of gardens is the main potential problem of an unreliable and insufficiently designed system. However, permanent damage to property is unlikely with failing pond structures. Pond structures lack flexibility in terms of easy system removal.

2.5.4 Conclusions

The case study describes the successful design, operation and maintenance of a novel pond structure during the first 10 months of operation. The water quality of the attenuation wetland was acceptable for disposal and recycling (e.g., irrigation).

Frequent spatial and temporal water quality monitoring for pollution control has been suggested, because data series components are predominantly random and periodic. pH value and DO concentration show significant diurnal variations due to the photosynthetic activity of algae.

The pond structure is reasonably cheap, if labour costs are low. In the long term, a reduction of the sewerage charge and water recycling for irrigation helps to recover some capital costs.

A SWARD analysis identified low economic and technical risks as well as high environmental and social benefits. The risk of flooding was low, if high safety factors for traditional design formulae were applied.

2.6 Permeable Pavement and Ground Source Heating Pump Systems

2.6.1 Introduction

Traditional systems capture storm runoff and subsequently distribute it to nearby watercourses or sewer systems. Some of these systems have become ineffective (Schlüter and Jefferies 2002; Scholz 2006b). The need for modernising traditional drainage systems has become a priority for councils and water companies.

Sustainable (urban) drainage system (SuDS) treatment trains in the UK and best management practices (BMP) in the USA have become one of the key instruments to control urban runoff since 1990 (Ferguson 2005; Jefferies et al. 1999; Scholz 2006a). For most SuDS, runoff peak flows are reduced and delayed. Groundwater recharge is increased as SuDS are usually used as infiltration techniques. For example, permeable pavement systems (PPS) can be used both for water infiltration and storage and are considered being hard SuDS solutions (Ferguson 2005). PPS as well as geothermal heating and cooling systems are commercially available techniques and have been assessed independently in the past (Coupe 2004).

PPS are suitable for a wide variety of residential, commercial and industrial applications, yet are confined to light duty and infrequent usage, even though the capabilities of these systems allow for a much wider range of usage. The reason that PPS are traditionally used for light-duty surfaces is often because of geotechnical design considerations for structural loading, although heavy-duty application products are widely available. Furthermore, where there is any concern about the possible migration of pollutants into groundwater, PPS should be constructed with an impermeable layer, and the treated storm water should subsequently be discharged into SuDS (Wilson et al. 2003).

The general principle of PPS is simply to collect, treat and infiltrate freely any surface runoff to support groundwater recharge. In comparison to traditional drainage systems, storm water retention and infiltration is a sustainable and cost-effective process, which is suitable for urban areas (Andersen et al. 1999; Dierkes et al. 2000). Moreover, PPS have many potential benefits such as reduction of runoff, recharging of groundwater, saving water by recycling and prevention of pollution (Pratt et al. 1999).

A detailed review of PPS has been published by Scholz and Grabowiecki (2007). This reference led the groundwork for Chapter 2.6. Briefly, ground source heat pumps (GSHP) or geo-exchange systems are commonly used in North America, China, Japan and some European countries (Bi et al. 2002; Genchi et al. 2002; Hepbasli and Ozgener 2004; Nagano et al. 2006; Ozgeer and Hepbasli 2007). Most GSHP are using refrigerant to move unwanted energy (i.e., heat) out of buildings during summer and into them (if required) during winter (Bose 2005). They use constant temperatures of surrounding grounds, which are lower than the corresponding air temperatures during warm seasons (heat sinks) and higher during cold seasons (heat sources).

For ground connections, plastic pipes are installed within the soil. Various applications such as horizontal, vertical, looped or submerged designs can be used. The main thermal carrier within coils is a mixture between water and a de-icing agent. The length and width of the loops is determined by ground conductivity abilities. The most important variables are type of soil, geology and area of available land for such installations.

Various applications of PPS have been tested in the UK. The main pollutant used was mineral engine oils containing hydrocarbons. Furthermore, a specific geotextile incorporating polymer beads was designed to release nutrients for better microbial community growth and more efficient hydrocarbon removal (Spicer et al. 2006). This 'self-fertilising' geotextile demonstrates that nutrient sources can be incorporated into a polymer composite to influence positively microbial growth within PPS.

A different approach is to use gully pot liquor mixed with tap water and animal faeces in some experimental rigs as the main pollutants. Such mixtures would mimic the most extreme conditions, which may occur in practice. Gully pot liquor provides all possible pollutants available naturally.

A gully pot is a biochemical reactor where pollutants are released after acidic dissolution and sediment maturation. Also, various microbial degradation processes take place in the gully pot chamber (Morrison et al. 1995). Because gully pots operate under various regimes (e.g., wet or dry), concentrations of dissolved oxygen can drop rapidly during anaerobic conditions (Butler et al. 1995).

Because of their potential pathogenic nature, animal faeces (e.g., dog droppings) are not commonly used pollutants in academic SuDS research. Nevertheless, there are serious health concerns associated with PPS water (particularly if contaminated with faecal matter), which could potentially be recycled within buildings for toilet flushing and other applications as discussed by Grabowiecki and Scholz (2006).

Both PPS and GSHP are commercially available applications but are rarely used as combined systems in research projects. A combined system has the potential to capture, detain and treat runoff, and to either cool or heat a nearby building at the same time. The sub-base is only heated passively during the summer when hot air within an adjacent building needs to be cooled down by transferring access heat to below the permeable pavement. The heat is in fact a waste product that has the additional benefit of enhancing biodegradation within the sub-base. The rationale is to assess the impact of temperature changes induced by the geothermal heating and cooling system on water quality.

Table 2.23 Schematic layout of the experimental rigs.

	Inside rig						Outside rig					
Feature	**1**	**2**	**3**	**4**	**5**	**6**	**1**	**2**	**3**	**4**	**5**	**6**
Inbitex composite	√	√	√				√	√	√			
Inbitex geotextile				√	√	√				√	√	√
Cooling or heating	√	√		√	√		√	√		√	√	
Animal faeces	√			√			√			√		
Air thermometers	√				√	√	√				√	√
Vessel thermometers	√				√		√				√	
Carbon dioxide sampling	√				√		√				√	

The aim is to assess both rigs and to compare them with each other in terms of the treatment efficiencies and designs for the different PPS types (Table 2.23). The objectives are as follows: (i) to assess the combined PPS and GSHP performance; (ii) to characterise microbial activities under different temperature patterns; (iii) to assess the water quality at the bottom of tanked systems and (iv) to discuss the mobility of potentially pathogenic organisms. Chapter 2.6 is based on two original articles by Scholz and Grabowiecki (2009) and Tota-Maharaj and Scholz (2010).

2.6.2 Methodology

2.6.2.1 Environmental Conditions

As the two rigs were operated under controlled and uncontrolled conditions (i.e., internal and external rigs), there was a need for energy input to simulate GSHP workability. Aquarium heaters (VISI-THERM, Aquarium Systems NEWA, Loughborough, UK) were used to achieve temperature increases. In-line aquarium coolers (Titan 500, Aqua Medic, Bissendorf, Germany) were applied to decrease the temperatures within coils. No direct contact between heat transferring water and water stored within the PPS sub-base was allowed during the experiment.

Typical polyethylene 240-L wheelie-bins (Isoplastic, Warwickshire, UK) have been used as basic construction devices. Their dimensions (maximum height: 1070 mm; maximum length: 730 mm; maximum width: 580 mm) and maximum load weight (96 kg) were appropriate for this experiment. The bins mimicked impermeable tanked systems and provided suitable conditions for water collection.

The indoor PPS was composed of six bins and placed in a temperature-controlled room with a mean ambient temperature of 16°C. The outdoor rig was submerged within the ground and located outside where atmospheric temperature conditions prevailed. All bins were partly filled with the inflow water, and operated in batch-flow mode, which therefore simulates therefore 'real' car park and pavement runoff conditions better than a continuous-flow mode (Scholz 2006a).

2.6.2.2 Components of Rigs

Commercially available pre-washed aggregates were used for the construction of the sub-base. Their sizes were determined by the British Standards Institution (1992). Generally available materials were used; e.g., crushed gravel and rock with defined edges and sizes between 10 and

Table 2.24 Composition of layers within the experimental permeable pavement systems.

Depth (mm)	Type	Thickness (mm)	Aggregate diameter (mm)
80	Air	80	
>130	Clean stone	50	5
>132	Geotextile	2	
>232	Upper sub-base	100	5–20
>485	Upper part of lower sub-base	250	63–10
>735	lower part of lower sub-base	250	63–10

63 mm. A typical composition of layers comprises a lower sub-base (250 mm), an upper sub-base (100 mm), a geotextile, a clean 5-mm stone layer (50 mm) and finally paving blocks (80 mm) at the top, as shown in Table 2.24. Materials were free of any external organic or inorganic matter. The depth of the standardised sub-base was altered for research purposes to integrate GSHP. However, sub-base depths depend on the individual application and can therefore vary subject to the environmental conditions and the discretion of the design engineer. Usually, it is recommended to use a depth of 350 mm.

The depth of the lower sub-base was increased to 500 mm, because GSHP coils need to be fully submerged in water to operate appropriately. A system containing space filled with porous media under a constant head of water pressure delivers conditions similar to submerged systems containing GSHP installations and provides desirable conductivity to allow for sufficient heat transfers.

The second layer located in the upper sub-base contained stone with diameters between 5 and 20 mm to a depth of 100 mm. The porous Inbitex geotextile made of polyethylene and polypropylene fibres was placed in the top part of the upper base, as this is the area where considerable microbial degradation of pollutants is likely to take place. Either Inbitex on its own or Inbitex together with an impermeable layer (Terram Drainage Composite, Terram Geosynthetics, Gwent, UK), called composite, were used in the experiments (Table 2.23). The impermeable layers were overlapped with each other to allow water to percolate through the system but to avoid evaporation. Moreover, it is essential that the coils are always submerged.

On the top of the geotextile, an additional 5-mm layer of pea-gravel was placed and covered with permeable paving blocks provided by Hanson Formpave. Clean single-sized 3-mm pea-gravel was used to fill in the gaps between individual blocks. During installation of the experimental systems, reinforced 5-mm polypropylene tubes were placed within the lower sub-base. Their total length (seven to nine loops folded up) was 10 m. Both ends of each tube were located in a plastic water vessel. One of the ends was connected to the pump and the other end was used as an orifice for water discharge. This arrangement provided closed circulation of water (Figure 2.28).

2.6.2.3 Operational Conditions

In existing large-scale industrial applications, the source of energy for the heating or cooling system is provided via natural soil or a water reservoir. In the experimental rigs, however, it was not possible to create such environments without considerable additional costs. Therefore, a different heat transfer system was used, but the components of the GSHP remained similar. During the cold periods of the first year of research, heat was provided to the sub-base, resulting in a relative increase of

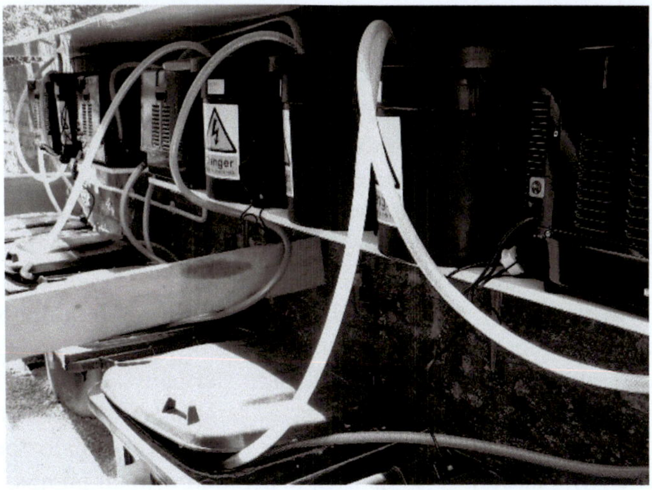

Figure 2.28 Composition of the rig located outdoors. Heat exchangers located above experimental bins and connected to each other with pipes and wires.

the temperature within the sub-base. This arrangement simulated 'real' site conditions of the outflowing water for the inside rig and increased relative temperature differences for the outside rig. During warm periods, the opposite arrangement was made, and the sub-base subsequently became relatively cold.

In general, in-door ventilation systems and GSHP are fully automated. For example, a required temperature decrease to reach a pre-programmed threshold would be undertaken automatically, if the external temperature (i.e., temperature outside the protected environment) would be too high; although there is a need for a regular maintenance provided by qualified technicians.

Concerning the experimental rigs, the heating and cooling run periods depend on the programmed duration of the set timers. The timers switch the required elements on or off. The elements heat up or cool down the water within the coils. The pumps run continuously. The coolers prevent the temperature to decrease by more than 5°C. The thermostats prevent overheating by switching the heaters off immediately when 22°C is reached. These settings were based on expert judgement.

Considering the needs in industry and the local temperature fluctuations, the following heating and cooling cycles were set to simulate a car park servicing a supermarket in Scotland: (i) Heating during winter: heating is switched on at 04 : 00, 09 : 00, 12 : 00, 16 : 00, and 19 : 00, and off at 08 : 00, 11 : 00, 15 : 00, 18 : 00 and 23 : 00 and (ii) Cooling during summer: cooling is switched on at 11 : 00, 13 : 00, 15 : 00 and 19 : 00, and off at 12 : 00, 14 : 00, 18 : 00 and 20 : 00.

The heating and cooling durations consider also equipment safety concerns such as overheating and economical constraints such as electricity supply costs, which are particularly high if the switching frequency is high. The on or off switching modules can be programmed up to six times within 24 hours.

Two main switching patterns exist in practice: domestic and commercial. Commercial sites such as large supermarkets usually prefer to have heating and cooling devices switched off as long as possible or to adopt a switching pattern with as few switches as possible to save energy. This type of running pattern was chosen for the experiment as this is the most likely application for the novel design on a large-scale application in practice; although the energy for heating would be provided from heat sinks such as the natural surroundings (e.g., sub-base of PPS, soil and lake).

2.6.2.4 Water Quality

Water was exchanged twice per week. The influent samples were prepared by collecting gully pot liquor and fresh dog faeces on the same day of analysis. Gully pot liquor (0.2 L) was mixed with tap water (2.0 L) in a plastic beaker. Approximately 3.1 g of fresh dog faeces were first weighed, and subsequently added and mixed properly to obtain a homogeneous mixture.

Approximately 2.2 L of sample water was slowly collected approximately 100 mm from the bottom of each bin by a hand lever pump. The temperature of the sample was immediately recorded at this stage. Samples were stored at a constant temperature (16°C) in a climate-controlled room for not longer than 0.5 hour before analysis to reduce data variability.

The five-day at 20°C N-Allylthiourea (nitrification inhibitor) biological oxygen demand (BOD) was determined using the OxiTop IS 12-6 system supplied by WTW (Wissenschaftlich-Technische-Werkstätten GmbH, Weilheim, Germany). The system uses a piezo-resistive measure of pressure differences. The amount of sample water used was usually 0.432 L for the outflows, and between 0.164 and 0.250 L for the inflows (i.e., usually the lower volume for systems contaminated with faeces). After 0.5 hour of aeration with air pumps, a nutrient inhibitor was added to the sample, and bottles were incubated at a constant temperature of 20°C for five days.

Nutrients were determined with traditional chemical methods based on procedures developed in the 1920s (Clesceri et al. 1998). Nitrate was reduced to nitrite by hydrazine in alkaline solution and determined at 550 nm by a colorimetric method using Bran Luebbe AA3 flow injection analyser (Bran and Luebbe 1999) following ISO standard methods (Bran and Luebbe 1999; Clesceri et al. 1998).

Ortho-phosphate-phosphorus was determined using ammonium molybdate, ascorbic acid and sulphuric acid. Ammonium molybdate and 0.5-mol bicarbonate of soda were also used in the analyses performed by automated colorimetry with an AA3 colorimeter for all water samples at 660-nm wavelength (Clesceri et al. 1998). Ammonia coloured blue compounds were measured at the same wavelength using a Bran Luebbe autoanalyser (model AA3; see above). Salicylate, dichloro isocynauric acid, sodium salt dehydrate and nitroprusside (catalyst) were also used in this method (Bran and Luebbe 1999).

American standard methods (Clesceri et al. 1998) were consulted for the examination of suspended solids, pH, conductivity and total dissolved solids. A Hanna HI 991 300 m was applied for this examination. Dissolved oxygen was measured with a WTW Oxi 315i m and the redox potential (ORP) reading was taken with a Hanna HI 98 201 m.

One of the objectives (see above) was to determine the effect of various variables such as temperature on the populations of indicator bacteria such as *Shigellae*, *Salmonella*, *Enterococci* and total heterotrophs in the effluent. The microbiological analysis was carried out according to American standard methods (Clesceri et al. 1998) applying plate count techniques. Microorganisms were grown on agars supplied by Oxoid (Basingstoke, UK) and prepared according to manufacturer's instructions.

MacConkey Agar Number 3 allows for the detection and isolation of *Salmonellae* and *Shigellae* species occurring frequently, for example, in pathological and food specimens (Clesceri et al. 1998). The Slanetz and Bartley Agar was designed to favour the growth of faecal *Enterococci* including *E. faecalis*, *E. faecium*, *E. avium*, *E. casseliflavus*, *E. cecorum*, *E. durans*, *E. gallinarum*, *E. hirae* and *E. malodoratus* (Tejedoret al. 2001). Nutrient Agar is a non-selective growth medium and allows for the growth of a wide variety of oxygen-tolerant genera. This agar promotes the growth of total heterotrophic bacteria providing information on calculating the percentage of pathogens as part of all bacteria present in rigs.

Duplicate samples were appropriately diluted before transferring them into Petri dishes (counting plates) containing suitable growth media. The plates were incubated at 37°, and the numbers of colonies present was subsequently counted (Clesceri et al. 1998). Good laboratory practice guidelines were adopted to avoid cross-contamination of microbiological samples.

2.6.3 Results and Discussion

2.6.3.1 Water Quality

The inflow water quality is summarised in Table 2.25. Approximately 3.1 g of dog faeces were added to selected bins twice per week resulting in a corresponding total of approximately 288 g. Although all bins received gully pot liquor diluted with tap water, only two bins per rig received additional dog faeces. Because of the strong variability of the chosen pollutants, the standard deviation for suspended solids was 411.4 and 180.8 mg L^{-1} for the inflow water with and without additional pollutants, respectively. Similar observations were made for conductivity and ammonia-nitrogen. Relatively stable and uniform values were only recorded for pH, DO and nitrate-nitrogen.

The mean bin intakes for ortho-phosphate-phosphorus were 1.9 and 26.2 mg L^{-1} without and with dog faeces, respectively. The corresponding concentrations for ammonia-nitrogen were 14.7 and 39.3 mg L^{-1}, respectively (Table 2.25).

Tables 2.26 and 2.27 show the summary statistics of the outflow water quality variables for the inside and outside rigs, respectively. The variability of pH is relatively low. In contrast, the maximum standard deviations for conductivity ranged between 25 and 80 µS for the inside and between 8 and 124 µS for the outside rig, respectively. Outdoor variability was usually greater than indoor (controlled environment) variability as expected. However, the most significant ($p < 0.002$) variability has been recorded for suspended solids with respect to the outside bin 5 and inside bin 2: standard deviations reached concentrations of 208 and 269 mg L^{-1}, respectively (Tables 2.26 and 2.27).

Mean dissolved oxygen concentrations were similar for both rigs: they were between 5 and 6 mg L^{-1} for the inside rig and between 5.5 and 7 mg L^{-1} for the outside one. Overall reductions in dissolved oxygen for both systems ranged between 22% and 46%. The BOD reductions were between

Table 2.25 Summary statistics of the water quality for the inflow (IN) without and with additional pollutants (dog faeces, P) for the permeable pavement systems (n = 55).

	IN	IN + P	IN	IN + P
Variable	**Mean**		**Standard deviation**	
Five-day @ 20°C biochemical oxygen demand (mg L^{-1})	7.5	92.7	8.49	78.59
Suspended solids (mg L^{-1})	227.5	133.3	411.37	180.76
Total dissolved solids (mg L^{-1})	62.8	98.0	34.74	29.86
Dissolved oxygen (mg L^{-1})	9.8	9.4	0.86	1.09
pH (−)	7.0	7.1	0.42	0.44
Conductivity (µS)	126.6	196.7	69.83	59.91
Ammonia-nitrogen (mg L^{-1})	14.7	39.3	63.63	117.60
Nitrate-nitrogen (mg L^{-1})	1.1	1.1	0.84	0.56
Ortho-phosphate-phosphorus (mg L^{-1})	1.9	26.2	6.46	81.45

Table 2.26 Summary statistics of the outflow water quality for the rig located indoors (n = 55).

		Bin numbers					
Variable	**Statistics**	**1**	**2**	**3**	**4**	**5**	**6**
Five-days @ 20°C biochemical oxygen demand	Mean	0.7	1.0	1.2	1.2	1.1	1.1
	SD	0.30	0.67	0.70	0.67	0.32	0.30
Suspended solids (mg L^{-1})	Mean	166.7	199.0	260.0	131.1	138.2	174.2
	SD	179.98	269.09	256.73	136.42	166.32	179.77
Total dissolved solids (mg L^{-1})	Mean	202	213	187	217	201	202
	SD	19.0	20.1	29.9	12.8	17.5	29.8
Dissolved oxygen (mg L^{-1})	Mean	5.0	5.3	6.5	5.4	4.6	5.6
	SD	1.52	1.71	2.20	1.34	1.38	2.08
pH (−)	Mean	7.4	7.5	7.5	7.5	7.4	7.5
	SD	0.26	0.21	0.26	0.28	0.21	0.22
Conductivity (μS)	Mean	403	428	374	435	403	416
	SD	38.4	40.7	59.5	25.4	35.1	80.4
Ammonia-nitrogen (mg L^{-1})	Mean	0.15	0.11	0.11	0.13	0.10	0.11
	SD	0.219	0.146	0.123	0.154	0.102	0.114
Nitrate-nitrogen (mg L^{-1})	Mean	3.2	1.0	2.7	1.4	0.4	2.2
	SD	3.69	1.35	3.10	1.35	0.49	2.78
Ortho-phosphate-phosphorus (mg L^{-1})	Mean	0.89	0.48	0.47	0.82	0.43	0.47
	SD	0.460	0.481	0.491	0.559	0.549	0.502

SD, standard deviation.

98% and 100%. Negative removal efficiencies for total dissolved solids (TDS) ranging between −272% and −75% were observed. It follows that the corresponding systems were sources and not sinks of TDS.

Even though the aggregates were washed in the quarry before delivery, some outflow parameters were influenced by the material of the sub-base composition (granite and sandstone). This is the reason why SS and TDS concentrations within the effluent were initially relatively high in comparison to the inflow.

Concerning nutrients (Figures 2.29–2.34), high reductions in ortho-phosphate-phosphorus (Figures 2.31 and 2.34) and ammonia-nitrogen (Figures 2.29 and 2.32) were observed. For the inside system, ortho-phosphate-phosphorus concentrations were less than 1 mg L^{-1} (corresponding reduction of 95%). For the outside system, concentrations fluctuated between 0.2 and 0.6 mg L^{-1} with corresponding reduction rates of approximately 95%. Ammonia-nitrogen reductions were up to 100% for both systems and the corresponding concentrations ranged between 0.1 and 0.2 mg L^{-1} for the inside and between 0.03 and 0.14 mg L^{-1} for the outside system, respectively.

An increase in nitrate-nitrogen (Figures 2.30 and 2.33) had been recorded. Figure 2.33 indicates the rate and start time of the increase. For the inside and outside bins, which received dog faeces, the highest releases of nitrate-nitrogen have been recorded. This is likely due to the additional load of nitrogen. There is also a slight increase for the inside bin 6 and outside bin 6 (no faeces, and no

Table 2.27 Summary statistics of the outflow water quality for the rig located outdoors (n = 30).

		Bin numbers					
Variable	**Statistics**	**1**	**2**	**3**	**4**	**5**	**6**
Five-days @ 20°C biochemical oxygen demand	Mean	1.2	0.1	0.3	0.1	0.3	0.4
	SD	1.72	0.30	0.65	0.30	0.65	0.67
Suspended solids (mg L^{-1})	Mean	90.0	77.3	130.0	71.3	96.0	42.0
	SD	112.03	146.70	275.56	97.02	208.44	78.00
Total dissolved solids (mg L^{-1})	Mean	202	213	187	217	201	202
	SD	19.0	20.1	29.9	12.8	17.5	29.8
Dissolved oxygen (mg L^{-1})	Mean	5.1	6.1	6.8	5.5	6.4	7.3
	SD	1.15	0.90	1.17	0.87	0.81	1.41
pH (–)	Mean	7.2	7.4	7.5	7.4	7.5	7.5
	SD	0.28	0.14	0.15	0.16	0.12	0.15
Conductivity (μS)	Mean	469	388	354	400	382	347
	SD	124.1	16.2	8.4	16.6	10.0	8.7
Ammonia-nitrogen (mg L^{-1})	Mean	0.14	0.03	0.03	0.04	0.03	0.03
	SD	0.103	0.027	0.022	0.036	0.020	0.027
Nitrate-nitrogen (mg L^{-1})	Mean	1.7	0.3	2.6	1.0	0.7	1.8
	SD	2.27	0.25	1.26	1.56	0.72	0.99
Ortho-phosphate-phosphorus (mg L^{-1})	Mean	0.35	0.22	0.22	0.60	0.14	0.23
	SD	0.428	0.336	0.253	0.298	0.120	0.236

SD, standard deviation.

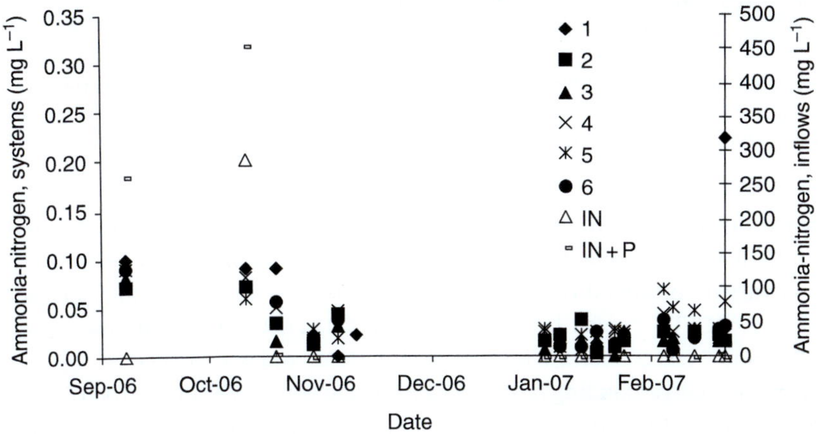

Figure 2.29 Ammonia-nitrogen distribution for the permeable pavement bins 1–6 located outdoors, and the inflow (IN) and the artificially polluted inflow (IN + P). Inflow and outflow concentrations are usually below 150 and 0.1 mg L^{-1}, respectively.

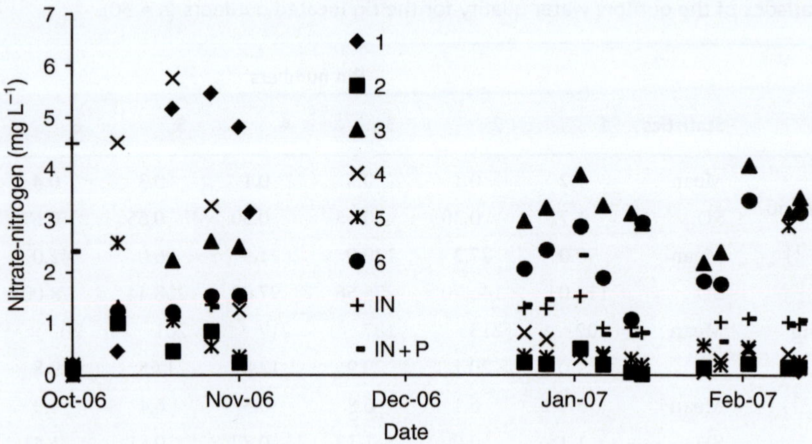

Figure 2.30 Nitrate-nitrogen distribution for the permeable pavement bins 1–6 located outdoors, and the inflow (IN) and the artificially polluted inflow (IN + P). Inflow and outflow concentrations are very variable and are usually below 6 mg L^{-1}.

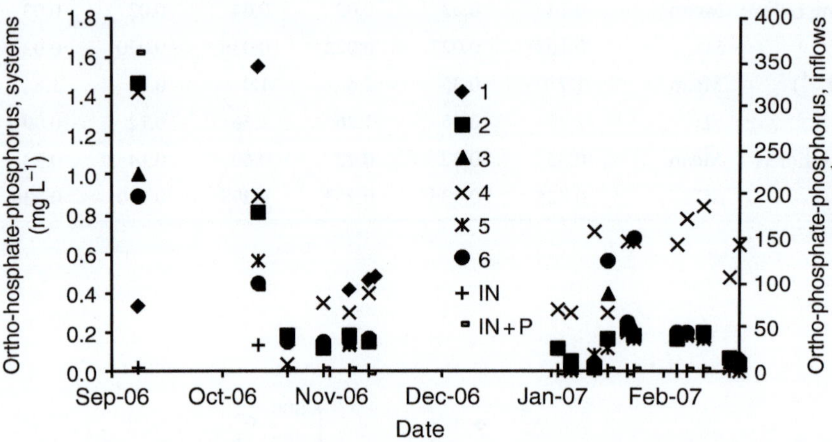

Figure 2.31 Ortho-phosphate-phosphorus distribution for the permeable pavement bins 1–6 located outdoors, and the inflow (IN) and the artificially polluted inflow (IN + P). Inflow and outflow concentrations are commonly below 400 and 1.6 mg L^{-1}, respectively.

heating or cooling). For the outside rig, nitrate-nitrogen reductions were observed for bins 2, 4 and 5. For the outside bins 1, 3, 6 and all inside bins, removal efficiencies were negative (−440–62%).

Microorganism counts varied considerably (Table 2.28). For the outside system, total heterotrophic colony-forming units (CFU) ranged between 38 500 and 121 000 for the outflow water, and with means of 101 250 and 7 605 000 for the inflow water without and with dog faeces, respectively. For the inside system, the corresponding numbers ranged between 51 500 and 180 000 for the outflow water and had means of 171 750 and 378 250 for the inflow without and with faeces, respectively.

Concerning the presence of *Salmonellae*, between 68 and 498 CFU for the outdoor rig and between 371 and 8931 CFU for the indoor rig were recorded. In comparison, *Shigellae* counts

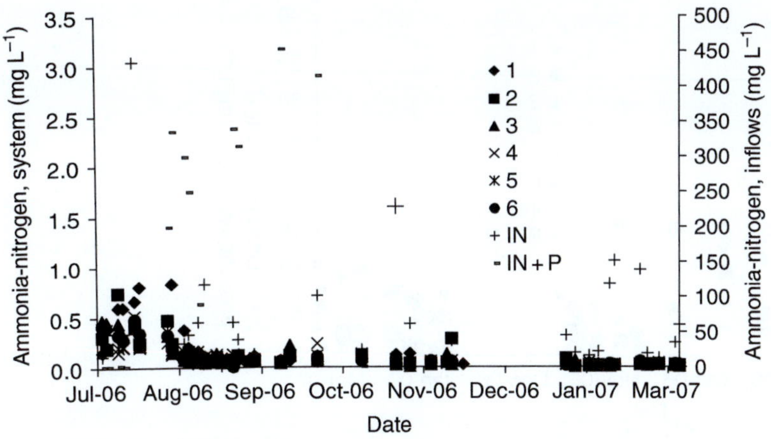

Figure 2.32 Ammonia-nitrogen distribution for the permeable pavement systems 1–6 located indoors, and the inflow (IN) and the artificially polluted inflow (IN + P). Inflow and outflow concentrations are usually below 500 and 1 mg L^{-1}, respectively.

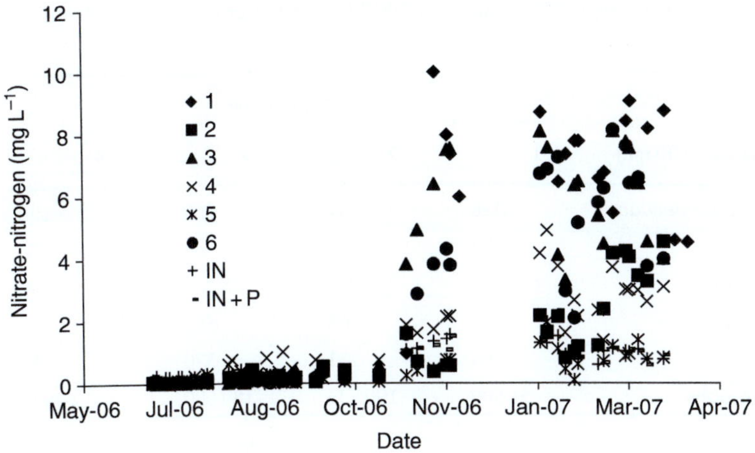

Figure 2.33 Nitrate-nitrogen distribution for the permeable pavement systems 1–6 located indoors, and the inflow (IN) and the artificially polluted inflow (IN + P). Inflow and outflow concentrations are very variable after October and are normally below 10 mg L^{-1}.

for the inflow ranged between 78 and 160 CFU for the inside rig and between 636 and 920 CFU for the outside rig.

Between 58 and 215 *Enterococci* CFU for the outside, and between 95 and 704 CFU for the inside rig were recorded. The input counts varied between 178 and 598 for the inside and between 880 and 3900 for the outside rig.

Figures 2.35 and 2.36 characterise the microbial activities within the systems, as they mirror the corresponding carbon dioxide concentrations. The highest concentrations were recorded close to the first two (i.e., the shallowest) sampling points for each bin. This indicates that the most intensive microbial activity takes place around the geotextile.

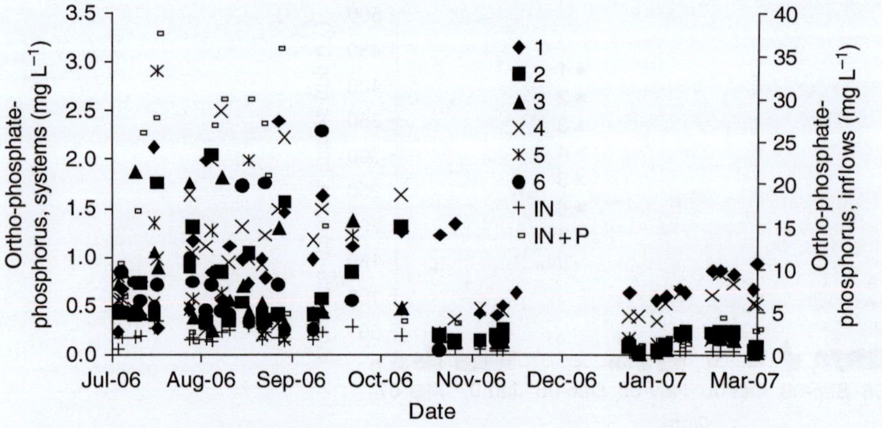

Figure 2.34 Ortho-phosphate-phosphorus distribution for the permeable pavement systems 1–6 located indoors, and the inflow (IN) and the artificially polluted inflow (IN + P). Inflow and outflow concentrations are almost always below 35 and 3 mg L^{-1}, respectively, and their variability decreases over time.

Table 2.28 Mean bacteria colony-forming counts for the inside and outside bins, inflow and inflow contaminated with dog faeces between November 2006 and April 2007 (inside system: n = 288; outside system: n = 192).

		Bin number			
Rig location	**Colony counts (per 100 mL)**	**1**	**2**	**3**	**4**
Inside	*Shigellae* and *Salmonellae*	498	441	260	195
	Enterococci	173	215	68	578
	Total Heterotrophs	121 000	56 750	38 500	74 000
Outside	*Shigellae, Salmonellae*	8931	368	598	371
	Enterococci	704	251	370	198
	Total Heterotrophs	51 500	100 833	180 000	12 9800
		Bin number		**Inflow**	
Rig location	**Colony counts**	**5**	**6**	**− faeces**	**+ faeces**
Inside	*Shigellae* and *Salmonellae*	198	68	636	920
	Enterococci	59	185	883	3900
	Total Heterotrophs	63 500	78 750	101 250	7 605 000
Outside	*Shigellae* and *Salmonellae*	708	485	160	78
	Enterococci	95	98	178	598
	Total Heterotrophs	77 667	76 833	171 750	378 250

2.6.3.2 Prevention of Water-related Diseases

Previous research was concerned with different geotextile materials, microbial communities within geotextiles and hydrocarbon removal by PPS using various oils (Pratt et al. 1999; Newmann et al. 2002, 2004). In contrast, this study focuses on the prevention of potential water-related diseases by

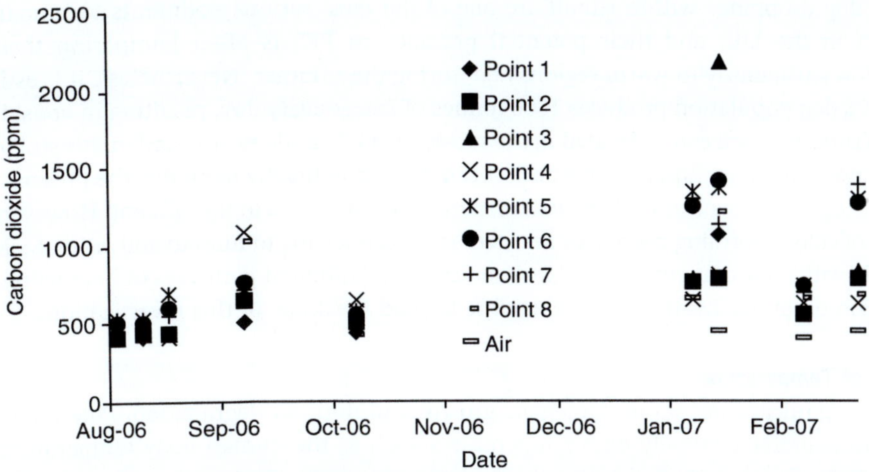

Figure 2.35 Carbon dioxide distribution within the permeable pavement system located indoors. Points 1–4 represent bin 1 and points 5–9 represent bin 4. The lower the point in the bin range, the shallower the level of sample collection. The concentration is below 2500 ppm and its variability decreases over time.

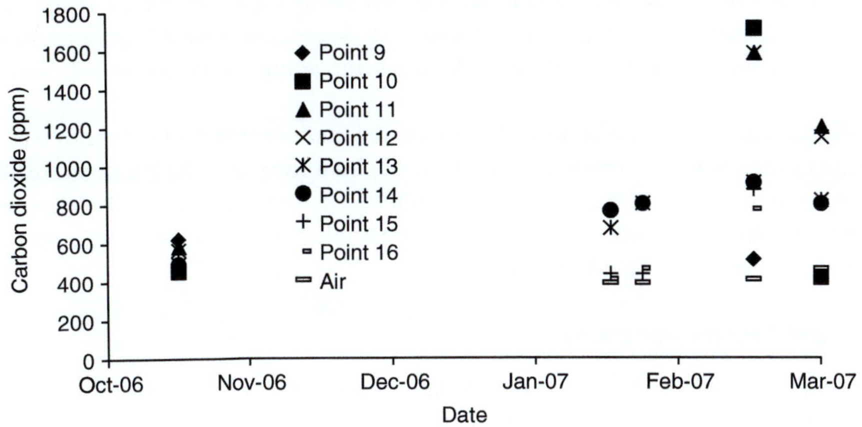

Figure 2.36 Carbon dioxide distribution within the permeable pavement system located outdoors. Points 9–12 represent bin 1 and points 13–16 represent bin 4. The lower the point in the bin range, the shallower the level of sample collection. The concentration is below 2500 ppm and its variability increases over time.

treating runoff contaminated with organics and faecal organisms within PPS. Common symptoms associated with these diseases may include abundant and watery diarrhoea, weight loss, nausea and low-grade fever (Abed 1996; Hunter 1997).

A synthetic cocktail of pollutants was not chosen to maintain the high variability of various pollutants within the runoff and to simulate real conditions. The main contaminant in this experiment is dog faeces, which is the second (after oils) most critical pollutant in car parks and driveways, and one the most frequent reasons of complaint concerning avoidable health and safety risks by the public to local authorities in the UK.

The remains of dog droppings within runoff are one of the most serious pollutants to human health within PPS in the UK, and their potential presence in PPS is often hampering their domestic installation particularly in warm regions and during the summer. Nevertheless, it is estimated that the UK's dog population produces 1000 tonnes of faeces every day, resulting in considerable pollution of runoff, which can be treated in the sub-base of PPS as demonstrated in this study.

The indicator microorganisms chosen in this project can be used indirectly to predict the presence of other pathogenic organisms such as *Escherichia coli*, parasites and viruses in the effluent. However, the specific values of colony-forming counts for selected bacteria referring to human and animal tolerances vary greatly with environmental boundary conditions and different pathways of interaction. A detailed discussion on public health risks is therefore beyond the scope of this investigation.

2.6.3.3 Influence of Temperature

Temperature has a significant impact on biological growth and nutrient degradation. Potentially pathogenic organisms prefer relatively high temperatures such as the human body temperature (Eke and Scholz 2008; Scholz 2006a; Scholz and Grabowiecki 2007). Temperature fluctuations due to the presence of the heating or cooling coils have therefore a major impact on the treatment performance of the PPS and the associated microbial community compositions within the sub-bases (see above). During the heating period, temperatures in the bins were between 20 and 25°C. During the cooling period, the temperature was approximately 5°C. These temperature recordings contrast air temperatures between 11 and 20°C during summer and between 3 and 13°C during winter. Considering that the temperatures are well below the human body temperature of approximately 37°C, regrowth of pathogens is limited. However, the elevated temperatures are likely to result in an enhanced degradation of biomass including nutrients (Abed 1996; Scholz 2006a).

The settings for all equipment were the same in the indoor and outdoor rigs, and the mean temperature for the temperature-controlled room was 14.7°C throughout the year. The temperature of the inflow water and the temperature of the outflow water differed between 0.3°C and 1.3°C during cooling. These temperature fluctuations are small and therefore unlikely to accelerate the growth of pathogens (Abed 1996; Hunter 1997; Scholz 2006a).

2.6.4 Conclusions and Recommendations

Chapter 2.6 compared the treatment efficiencies and designs of an inside with an outside PPS rig. The data variability of the inside rig was reduced by applying controlled environmental conditions such as a relatively stable temperature. The combined PPS and GSHP performance for both rigs was relatively good. The water quality at the bottom of the tanked systems was below secondary wastewater treatment standards for the BOD (<25 mg L^{-1}) and well above the threshold of 30 mg L^{-1} for suspended solids. Suspended solids values for the outside rig are on average 100% higher than those for the inside rig. This situation is caused by high dilutions with tap water and suspended solids are washed out of the sub-base. It is assumed though that higher SS concentrations in the input would be reduced in the same level as other parameters measured.

The ortho-phosphate-phosphorus and ammonia-nitrogen removal rates were very high (up to 95%), and the corresponding absolute concentrations fulfilled European urban wastewater treatment standards. Although an increase of nitrate-nitrogen has been observed, the concentrations were well within European standards.

The microbial activities during high temperature durations were high leading to better treatment performances, although the occurrence of a potential transfer of pathogens onto humans cannot be excluded. However, it has to be noted that the contaminated systems were over-polluted to simulate the worst-case scenario.

The elevated carbon dioxide concentrations and corresponding reductions in biochemical oxygen demand are evidence for the increased microbial activity within the sub-base, especially on the geotextile. The strong microbial community may increase the risk of potential transfer of pathogens to humans, although this requires additional assessments of large-scale systems, as the experimental rigs were intentionally overloaded with pollutants.

Further research on *Escherichia coli* (an important indicator for faecal pollution in aquatic environments) growth patterns involving molecular microbiological techniques should be undertaken to assess public health risks if the treated runoff is recycled for domestic use. Furthermore, heat transfer modelling within the PPS sub-base containing GSHP coils would be beneficial for a deeper understanding of spatial temperature distributions impacting on water quality. Future work should also evaluate correlations between temperature and different microorganism growth patterns, hopefully referring to certain temperature thresholds significantly reducing the risk of regrowth.

2.7 Permeable Pavement and Photocatalytic Titanium Dioxide Oxidation System

2.7.1 Introduction

Storm water collection, treatment and reuse are important activities that can contribute to sustainable development (Brown 2005; Scholz 2006a, b). It is accepted that storm water recycling is feasible and can contribute significantly towards sustainable urban water management (Chocat et al. 2007). Storm water treatment and reuse can reduce the pressure on freshwater resources and improve downstream water quality (Scholz 2006a).

There are serious health concerns associated with recycled effluent from permeable pavements, particularly if contaminated with faecal matter from animals such as dogs, cats and birds (Scholz and Grabowiecki 2007). However, this potential water resource could be reused in buildings (Scholz 2006a, b; Scholz and Grabowiecki 2007). In one gram of dog faeces, approximately 23 000 000 Coliforms can be found. Roughly 95% of faecal Coliforms within urban storm water comes from domestic pets such as dogs (Alderiso et al. 1996). Therefore, a combined application of permeable pavements and photocatalytic disinfection processes may enhance the use of storm water for domestic water recycling.

Chapter 2.7 assesses an experimental design comprising a combined permeable pavement and photocatalytic system treating and disinfecting diluted urban runoff obtained from concentrated urban runoff mixed with faecal matter. The objectives are to assess (i) the application of an experimental batch flow combined permeable pavement and photocatalytic disinfection system for the recycling of storm water runoff constants for indicator microbes; and (ii) the efficiencies of suspended and immobilised forms of titanium dioxide (TiO_2) in terms of disinfecting storm water. This chapter is based on an original article by Tota-Maharaj and Scholz (2013).

2.7.2 Materials and Methods

2.7.2.1 Permeable Pavement Engineering

Permeable pavement engineering aims to minimise the risk of surface flooding while meeting the requirements for hard structures withstanding traffic loads. The permeability, drainage rate and retention properties of permeable pavements have been found to be largely dependent on local surface area characteristics (Bean et al. 2007; Pratt et al. 1999). Permeable pavements minimise direct urban runoff discharge to receiving watercourses by allowing the infiltration of water through a load-bearing surface where it can be stored in an underground reservoir (Dierkes et al. 2002). The filtered storm water from the permeable pavements can be used to facilitate groundwater recharge. If the total runoff also includes storm water flows from roofs, the volume of water for recharge could be substantial. The principal mechanism for storm water quality control by permeable pavements includes (i) sedimentation, filtration and adsorption of pollutants, (ii) biological assimilation and biodegradation of contaminants by microbial activity within the porous media and (iii) volatilisation (conversion of chemical pollutants from liquid and solid states to a gaseous states) and water transport (quantity and quality control via gravitational settling) according to Pratt et al. (1999) as well as Scholz and Grabowiecki (2007, 2009).

Laboratory studies by Scholz and Grabowiecki (2009) have confirmed that permeable pavements reduce runoff and improve water quality in terms of removing phosphates, nitrates and ammonia. When compared to traditional, impervious asphalt, permeable pavements can reduce the runoff quantity, lower peak runoff rates and delay peak flows due to their high surface infiltration rates, water absorption into the sub-base materials and the retention time as storm water flows through the system (Bean et al. 2007; Brattebo and Booth 2003; Pratt et al. 1999; Scholz and Grabowiecki 2009).

Permeable pavements have also been shown to degrade pollutants such as metals and automotive oil (Brattebo and Booth 2003; Rushton 2001). Booth and Brattebo (2004) stated that various tested pavement systems reduced the concentrations of copper and zinc. Coupe et al. (2003) describe permeable pavements as efficient hydrocarbon traps and powerful in situ aerobic bioreactors, which have demonstrated that they remove hydrocarbons up to 98.7%. Geosynthetics within permeable pavements further support biomass growth (Pratt 2003). Bio-degradation processes supported by bacteria and fungi within permeable pavements are enhanced when prolonged aerobic, sulphate-reducing and denitrifying conditions occur (Coupe et al. 2003; Lei et al. 2005; Pratt 2003).

Borgwardt (2006) confirmed that the infiltration capacity of experimental permeable pavements decreases as the amount of oil, grease and fine organic and inorganic matter accumulates within the gravel filling the drainage openings. Permeable pavements can last for 20–25 years once properly constructed and maintained (Bean et al. 2007; Pratt 2003). Maintaining surface infiltration and resulting exfiltration is key to long-term pavement performance. With regular maintenance (vacuum sweeping and/or pressure washing) clogging of permeable pavements is reduced improving the infiltration capabilities (Bean et al. 2007).

2.7.2.2 Photocatalytic Processes with Titanium Dioxide

A heterogeneous photocatalytic system consists of semiconductor particles (photocatalyst), which are in close contact with a liquid (or gaseous) medium (Fujishima et al. 2000; Shannon et al. 2008), exposing the catalyst to light initiates processes such as redox reactions and molecular transformations (Herrmann 2005). Heterogeneous photocatalysts can carry out substrate oxidations and reductions concurrently. Ultraviolet light of long wavelengths (200–400 nm) or solar energy originating from sunlight can be used (Soares et al. 2007).

Titanium dioxide is known to have excellent properties such as high ultraviolet (UV) absorption and high stability, which allows it to be used in applications such as the photocatalytic degradation of chemicals and organic pollutants in soils, water and air (Fujishima et al. 2000). Photocatalytic processes gradually break down contaminant molecules resulting in no residue of the original material; i.e., no sludge is produced (Soares et al. 2007). The photocatalyst is unchanged during the disinfection process and no consumables (chemicals) are required, resulting in economic savings (Fujishima et al. 2000; Herrmann 2005; Soares et al. 2007). Rincón and Pulgarin (2003) discuss the key photochemical reactions involving TiO_2.

Heterogeneous photocatalytic processes (Salih 2002) use specific metal oxides that can readily generate hydroxyl radicals on the surface of particles when absorbing UV light. The anatase crystalline form of TiO_2 has a low band-gap energy (approximately 3.2 eV), which is almost equivalent to the wavelength of light (200–400 nm). Moreover, this catalyst behaves as a classical semiconductor (Salih 2002), where UV light is absorbed to promote electrons across the energy gap into the conduction band. Hence, the most important heterogeneous photocatalytic process is combined TiO_2 and UV radiation. Illumination of this photocatalyst in water generates electrons in the conduction band; these electrons form hydroxyl radicals, which can decompose organic compounds.

A simplified mechanism for the photo-activation of a semiconductor catalyst is presented in Figure 2.37. The photocatalytic process with UV light and titanium dioxide (TiO_2) absorbs ultraviolet radiation from either sunlight or the illuminated source (i.e., UV lamp) and produces the two oxidation reactants hydroxyl radicals and superoxide anions, which decompose toxic organic substances via oxidation. Note that the large vertical arrow on the left-hand side of Figure 2.37 represents the change in photochemical energy from a lower to a higher level. The valence band hole is strongly oxidising, and the conduction band electron is strongly reducing (Herrmann 2005). At the external surface, the excited electron and the hole can take part in redox reactions with adsorbed species such as water, organic compounds or oxygen (Herrmann 2005). The charges can react directly with adsorbed pollutants, but reactions with water are far more likely since the water

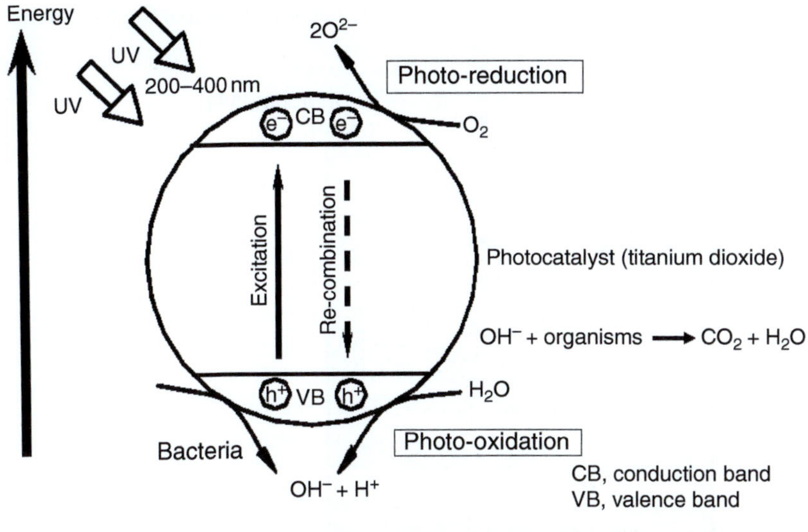

Figure 2.37 The photocatalytic process with ultraviolet (UV) light and titanium dioxide (TIO_2) after Wist et al. (2002) and Herrmann (2005). A drawing explaining the photo-reduction and photo-oxidation processes and the role of UV light and bacteria. *Source:* Wist et al. (2002) and Herrmann (2005).

molecules are more populous than contaminant molecules. Oxidation of water or OH^- by the hole produces the hydroxyl radical (•OH), an extremely powerful and indiscriminate oxidant (Wist et al. 2002). The electron and hole can recombine, releasing the absorbed light energy as heat, with no chemical alterations (Wist et al. 2002). The charges can participate in redox reactions with adsorbed organic pollutants in the photo-induced process. Photo-oxidation, photoreduction and adsorption occur on or near the surfaces of microbial organisms.

2.7.2.3 Experimental Permeable Pavement Set-up

Each experiment consisted of two phases: diluted gully pot liquor (obtained from randomly selected gully pots in Edinburgh) was treated and filtered by a permeable pavement system (PPS) acting as an aerobic reactor (Figure 2.38), and the effluent was subsequently treated by photocatalytic disinfection (Figure 2.39). The blank system (control; Figure 2.39) was in a dark room at 25°C. The first part of the experiment has been described in detail by Scholz and Grabowiecki (2009). Gully pot liquor was diluted using de-chlorinated tap water at a ratio of 1 : 10, to create a highly polluted and synthetic urban runoff, mimicking extreme worst-case scenarios onto the pavement surfaces.

In 2006, a field survey was conducted for the Scottish cities Glasgow and Edinburgh. The number of domestic dog droppings was recorded at a concrete pavement pathway loading rate of 4.67 g m^{-2} for Glasgow and 3.32 g m^{-2} for Edinburgh. Fresh faecal samples (still warm) were collected from domestic dogs (*Canis familiaris*) in Edinburgh, and 3.1 g m^{-2} were added weekly to each experimental set-up, which were operated over a 10-month period (May 2009–March 2010).

The construction of the permeable pavement reactor consisted of a 240-L wheelie-bin (height: 1370 mm; length: 730 mm; width: 580 mm), which was predominantly filled with aggregates (Figure 2.38). Aggregate sizes were selected based on expert opinion taking account of British Standards (British Standards Institution 1992) and internationally recognised good practice guidelines. The layers, and corresponding layer thicknesses (mm) and aggregate diameters (mm, if applicable) shown in brackets are as follows: permeable pavement blocks (80; –), bedding layer comprising pea-gravel (250; 3), geotextile (5; –), sub-base consisting of gravel (500, 63–100) and the saturated water zone (500; –). The Inbitex geotextile layer, which is made of polyethylene and polypropylene fibres, was placed in the top part of the upper base as shown in Figure 2.38. The top 30 mm of the pavement

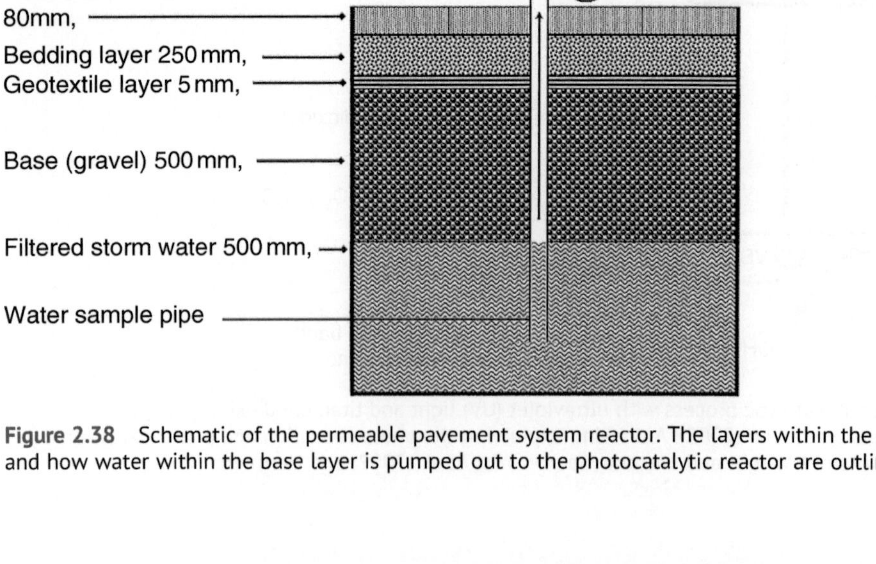

Figure 2.38 Schematic of the permeable pavement system reactor. The layers within the pavement reactor and how water within the base layer is pumped out to the photocatalytic reactor are outlined.

system and the geotextile zone is responsible for most of the microbial degradation of pollutants (Tota-Maharaj 2011). Permeable pavement blocks were used in the reactor with dimensions of (100 mm width × 200 mm length × 80 mm depth). Any gaps at the margins of the permeable pavement block area were filled with pea-gravel. All materials for the permeable pavement reactor were designed, manufactured and supplied by Hanson Formpave (Gloucestershire, United Kingdom).

2.7.2.4 Experimental Set-up of the Photochemical Experiments

Titanium dioxide was used because of its low cost (£2.50 per kg) and supplied via Fisher Scientific UK, Loughborough, United Kingdom. This photochemical is readily available and conservative in nature as a photocatalyst. Anatase, which is one of the three mineral forms of TiO_2 (small, isolated and sharply developed crystals), was used in the experiments. Pyrex glass beakers of 500 mL were applied as batch reactors.

Figure 2.39 shows the set-up of the photocatalytic experiment. A UV lamp (Ultra-Violet Products Ltd, Cambridge, United Kingdom) was fitted over the top centre of the beaker with a total distance of 5 cm from the surface of the water to enhance total incident irradiance. The irradiation experiments were performed at a fixed temperature of 16.5 ± 2°C and at a controlled humidity of 40 ± 3%. All external lighting was controlled, whilst a control experiment (blank) was placed in the dark at the same temperature, with no applications of UV radiation or photocatalyst disinfection.

Titanium dioxide catalyst was added to selected beakers in concentrations of 200, 400 and 600 mg L^{-1} for both experiments (i.e., immobilised (fixed on glass) and suspended). Previous studies (Cho et al. 2004; Rincón and Pulgarin 2003) have shown that concentrations of photocatalysts in this range have a high kill rate for aquatic protozoa, fungi and bacteria. Immobilisation of TiO_2 was achieved by evenly smearing anatase stuck by silicone adhesive onto a glass plate.

The photocatalytic treatment efficiency was tested by using artificial UV irradiation in a lab-scale reactor treating approximately 0.5 L of urban runoff. The photocatalytic reactor is constituted of

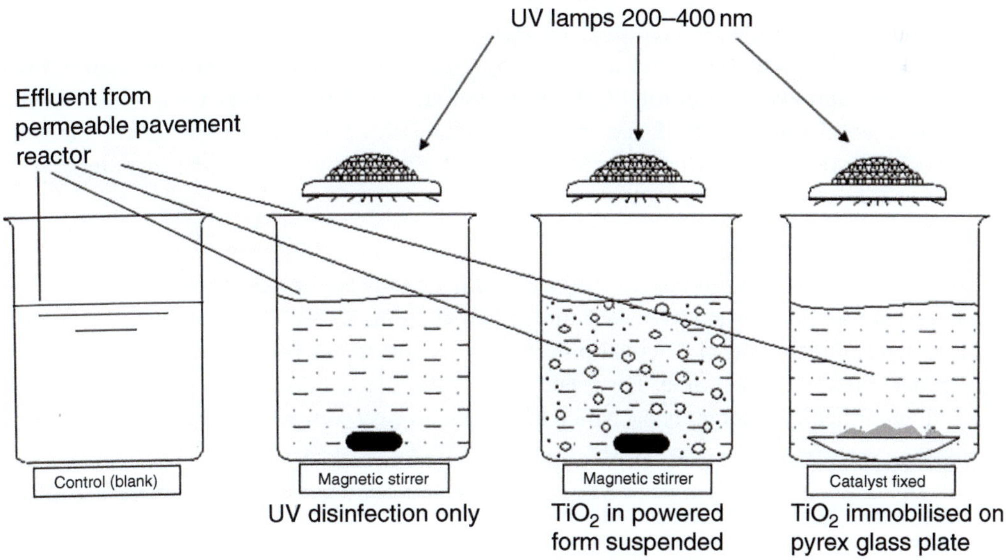

Figure 2.39 Schematic of the photocatalytic experimental apparatus for the photochemical reactors. Drawing showing that the permeable pavement reactor effluent is being treated by three different procedures that are compared with each other.

three parts: a Pyrex glass beaker (30 cm in height and 24 cm in diameter), UV lamp and support brace (25 cm in length) and a magnetic stirrer (Anachem Ltd, Bedfordshire, United Kingdom). The filtered urban wastewater was collected from the permeable pavement rig and placed in the photochemical reactor (Figure 2.39) where photocatalytic oxidation took place under continuous stirring with speeds of 300 rpm to allow complete mixing and application of UV irradiation.

After UV irradiation, the UV lamp was turned off, and the sample was left at a controlled room temperature of 21°C. After a total of 24 hours, total Coliforms, *E. coli* and faecal Streptococci colonies were counted.

2.7.2.5 Analytical and Microbiological Procedures

Permeable pavement influent (gully pot liquor spiked with dog faeces) and effluent (pre-treated urban runoff) water was tested. Samples were assessed for COD, five-day N-Allylthiourea biochemical oxygen demand (BOD), SS, colour, turbidity, total Coliforms, *E. coli* and faecal Streptococci. The COD was determined by common photometric tests using COD test tubes and a Nanocolor® 100D (Macherey-Nagel, Düren, Germany) photometer. The photometer's wavelength was adjusted for determination of colour. The turbidity of the samples was measured using a Hach 2100 turbidity meter (Hach Chemical Co, Colorado, USA) after 30 minutes of settling.

The BOD was determined using the OxiTop IS 12-6 system supplied by WTW (Wissenschaftlich-Technische-Werkstätten GmbH, Weilheim, Germany). The examination of SS, temperature, pH, electrical conductivity, dissolved oxygen and total dissolved solids were carried out following American standard methods (Clesceri et al. 1998).

The microbial degradation rates for total Coliform, *E. coli* and faecal Streptococci were assessed by applying standard culturing techniques. Samples of permeable pavement influent and effluent water were taken at regular intervals of 20 minutes, and directly used for the photocatalytic disinfection experiments. The water samples, including those obtained after disinfection with UV and UV-TiO_2, were collected in 2-L sterile bottles and analysed within two hours. The samples were diluted serially and cultured in triplicates to reduce the error margin.

The indicator microbes were identified by standard pour-plate and spread-plate techniques (Cappuccino and Sherman 1996) and described by Tota-Maharaj (2011). Nutrient Agar allowed for the detection and isolation of total Coliforms. Slanetz and Bartley Agar were used to favour the growth of faecal Enterococci. Eosin methylene blue (EMB) agar was used for the detection of *E. coli*. All culture media were ordered from Oxoid Ltd, Cambridge, United Kingdom.

2.7.2.6 Photocatalytic Disinfection Model

The Chick–Watson disinfection model outlined in Eq. (2.1) is applicable to photocatalytic treatment of wastewater (Cho et al. 2004; Fernández-Ibáñez et al. 2009). The purpose of this simple model is to determine the rate of inactivation of a microorganism as a function of the concentration of the disinfectant and contact time. The pseudo-first-order inactivation rate constant k was calculated for all identified microorganisms, which allows for an easy comparison of the effectiveness of disinfection.

$$\ln \frac{N}{N_o} = -kCT \tag{2.8}$$

where N_o is the initial microbial population (CFU/mL); N is the final microbial population (CFU/mL); k represents the pseudo-first order inactivation rate constant (/minute); C represents the OH radical concentration (mg L^{-1}) and T is the inactivation time (minute) of the microorganisms.

The efficiency of disinfection is given by ($N_t/N_0 \times 100$), which refers to the concentration ratio of the microbial pollutants in the water measured at any time t (N_t) compared to the initial concentration (N_0). The higher the calculated value, the more pronounced is the reduction in the microbial contaminant level, indicating a better photocatalytic inactivation efficiency (Fernández-Ibáñez et al. 2009). The efficiency of disinfection was determined to estimate the survival of different microorganisms in the presence of ultraviolet light radiation only, and in the presence of immobilised and suspended powdered forms of TiO_2 at different concentrations (see below).

2.7.3 Results and Discussion

2.7.3.1 Permeable Pavement Contaminant Removal Efficiency

Table 2.29 shows the mean water quality parameters for the PPS influent and effluent liquid. As a result of the influent containing gully pot liquor and dog faeces, it strongly resembled that of some domestic wastewaters in terms of BOD, COD, SS and turbidity as shown in Table 2.29.

Particulate matter from the urban runoff was trapped within the upper zone (joints between the pavement blocks) of the PPS, reducing the turbidity and SS of the inflow, and improving its colour (Tota-Maharaj 2011). The geotextile layer should also act as an additional buffer zone and absorb

Table 2.29 Water quality characteristics (n = 60) of the treated storm water obtained from the permeable pavement reactor between May 2009 and March 2010.

Parameter	Unit	Urban runoff influent (mean ± SD)	Permeable pavement reactor effluent (mean ± SD)	Removal rate (%)
Temperature	°C	16.5 ± 0.74	18.3 ± 1.08	—
pH	—	7.0 ± 0.15	7.6 ± 0.23	—
Electrical conductivity	μS	214 ± 71.3	78 ± 16.7	—
Total dissolved Solids	mg L^{-1}	188.2 ± 44.35	61.0 ± 27.17	67.6
Dissolved oxygen	mg L^{-1}	12.1 ± 1.14	3.1 ± 0.64	—
Turbidity	NTU	115.9 ± 18.92	8.0 ± 4.35	86.1
Suspended solids	mg L^{-1}	230.5 ± 58.63	19.1 ± 5.43	91.7
Colour	Hazen *units* (HU)	195 ± 6	13 ± 2	93.3
COD	mg L^{-1}	97.3 ± 4.50	12.0 ± 2.18	88.7
BOD	mg L^{-1}	65.6 ± 2.70	3.7 ± 0.57	95.4
Escherichia coli	CFU (100 mL)$^{-1}$	$4.40 \times 10^6 \pm 4900$	$7.54 \times 10^4 \pm 230$	98.1
Total Coliforms	CFU (100 mL)$^{-1}$	$1.50 \times 10^7 \pm 9800$	$9.54 \times 10^5 \pm 2000$	99.3
Faecal Streptococci	CFU (100 mL)$^{-1}$	$6.87 \times 10^5 \pm 3400$	$3.90 \times 10^3 \pm 170$	99.4

BOD, five-day biochemical oxygen demand; CFU, colony-forming unit; COD, chemical oxygen demand; SD, standard deviation.

contaminants (Coupe et al. 2003; Newman et al. 2002; Pratt 2003). Microbial parameters such as *E. coli*, total Coliforms and faecal Streptococci were subject to aerobic decomposition. Corresponding BOD and COD reductions were 95% and 89%, respectively (Table 2.29). As a result of the anaerobic conditions within the pavement structure, dissolved oxygen was reduced by approximately 75%. The PPS acted as a biological reactor (microbes digest organic matter), and thereby reduced oxygen levels. Aeration of the water in the sub-base was, however, not necessary for water reuse (e.g., gardening and washing cars). The suspended solids, BOD and COD concentrations met the European Commission Environment directive 91/271/EEC for urban wastewater treatment (EEC 1991).

2.7.3.2 Microbial Photochemical Deactivation

The effect of UV radiation on the kill rate of total Coliforms for UV combined with TiO_2 (both suspended and immobilised) and UV radiation only is presented in Figure 2.40. After 140 minutes, the surviving fraction of total Coliforms as a function of the time of UV exposure (without TiO_2) was negligible (Watts et al. 1995). Figure 2.40 illustrates that after just 80 minutes, heterogeneous photocatalysis at a concentration of 400 mg L^{-1} TiO_2 (suspended) was responsible for the reduction of approximately 99.99% of the initial concentration of total Coliforms. In comparison, the use of immobilised TiO_2 had a much longer reaction rate, almost as long as treatment with UV radiation individually.

Figure 2.40 indicates that there was a sharp increase in the inactivation of the total Coliform bacteria within the first 20–40 minutes. After 80 minutes, the remaining total Coliforms stabilised in numbers. As shown in Table 2.30, the inactivation rate constant *k* for total Coliforms exposed to UV

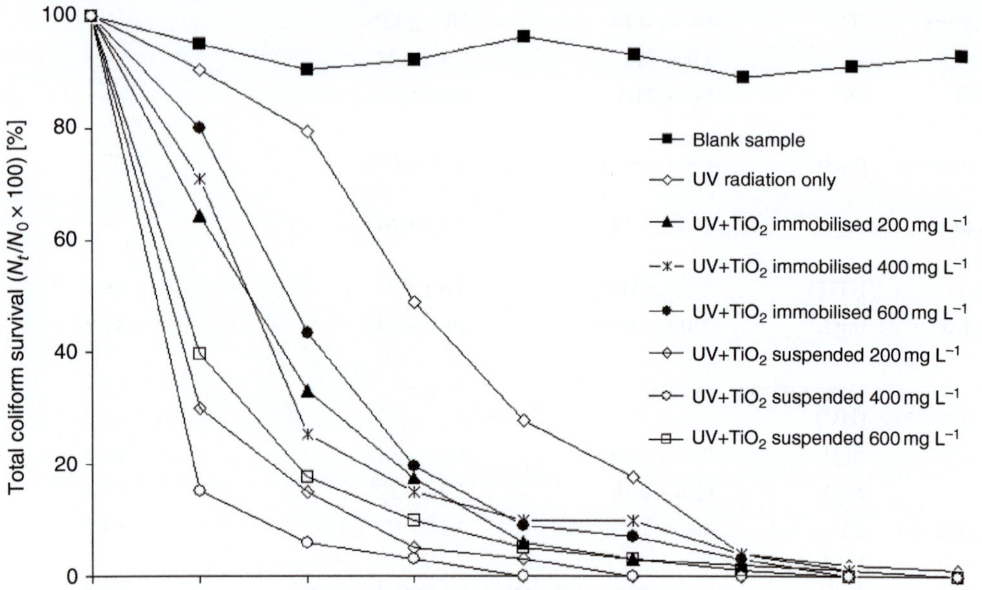

Figure 2.40 Survival of total Coliforms (CFU/100 mL) in the presence of ultraviolet (UV) light radiation only, and immobilised and suspended powdered forms of titanium dioxide (TiO_2). Graph showing a progressive reduction of total Coliforms with increasing treatment complexity and titanium dioxide concentration.

radiation only was the lowest (0.01563 per minute) in comparison to all other scenarios where TiO_2 was applied. Experiments with suspended TiO_2 at 400 mg L^{-1} had the highest inactivation rate (0.0744 per minute) followed by the same set-up but with a concentration of 200 mg L^{-1} TiO_2, where k was 0.05058 per minute. On average, the catalyst in suspension at a concentration of 400 mg L^{-1} was 64% faster than the catalyst in its immobilised form at the same concentration. This was followed by a 40% difference in the rates for both 200 and 600 mg L^{-1}, when compared again to the immobilised form at the same concentration. The photocatalytic disinfection of total Coliforms was a consequence of both the direct action of the UV and the photocatalytic degradation via the excited photon particles. As shown in Figure 2.40, there was a synergistic effect of UV and oxidative species generated by the photoactivation of TiO_2 (both suspended and immobilised). A mean turbidity of 4.2 NTU did not affect the rate of disinfection for the total Coliforms, indicating that the proposed treatment method works also for relatively visibly dirty urban runoff.

That ultraviolet light induced disinfection is well demonstrated in Figure 2.41 for *E. coli*. The efficiency of the process for the decontamination of *E. coli* can be seen by the steep reductions in the number of viable bacteria remaining after exposure between 80 and 160 minutes. Less than three hours of exposure to both UV and photocatalyst were sufficient to reduce *E. coli* numbers to an acceptable level. For experiments with the suspended form of TiO_2, there was an increase in the disinfection capacity like the one for the total Coliforms. At concentrations of 200 and 400 mg L^{-1}, the inactivation rate constants for suspended TiO_2 were 0.07470 and 0.07637 per minute, respectively (Table 2.30). When TiO_2 was immobilised, disinfection was less efficient because only microbes near TiO_2 crystals were killed as shown previously by Lonnen et al. (2005). This confirms why suspended TiO_2 was around 60% more effective than immobilised TiO_2 for the disinfection of *E. coli*. When *E. coli*-contaminated water was not exposed to irradiation (no photo-degradation), no change in *E. coli* numbers was observed. The inactivation kinetics of *E. coli* were consistent with those reported previously (Lonnen et al. 2005; Watts et al. 1995).

Figure 2.42 shows the reduction of faecal Streptococci following the Chick–Watson degradation kinetics. A lower efficiency in faecal Streptococci inactivation was observed for the immobilised

Table 2.30 Inactivation rate constant k (/minute) for total Coliforms, *Escherichia coli* and Faecal Streptococci with ultraviolet (UV) light and/or titanium dioxide treatment for immobilised and suspended growth.

System type	Treatment	Total Coliforms (min^{-1})	*E. Coli* (min^{-1})	Faecal Streptococci (min^{-1})
Immobilised (fixed on glass)	UV	0.01563	0.01209	0.01165
	UV + TiO_2 (200 mg L^{-1})	0.03027	0.02331	0.01855
	UV + TiO_2 (400 mg L^{-1})	0.02693	0.02003	0.02197
	UV + TiO_2 (600 mg L^{-1})	0.02437	0.03163	0.01884
Suspended	UV + TiO_2 (200 mg L^{-1})	0.05058	0.07470	0.18260
	UV + TiO_2 (400 mg L^{-1})	0.07440	0.07637	0.05504
	UV + TiO_2 (600 mg L^{-1})	0.04005	0.04450	0.04401

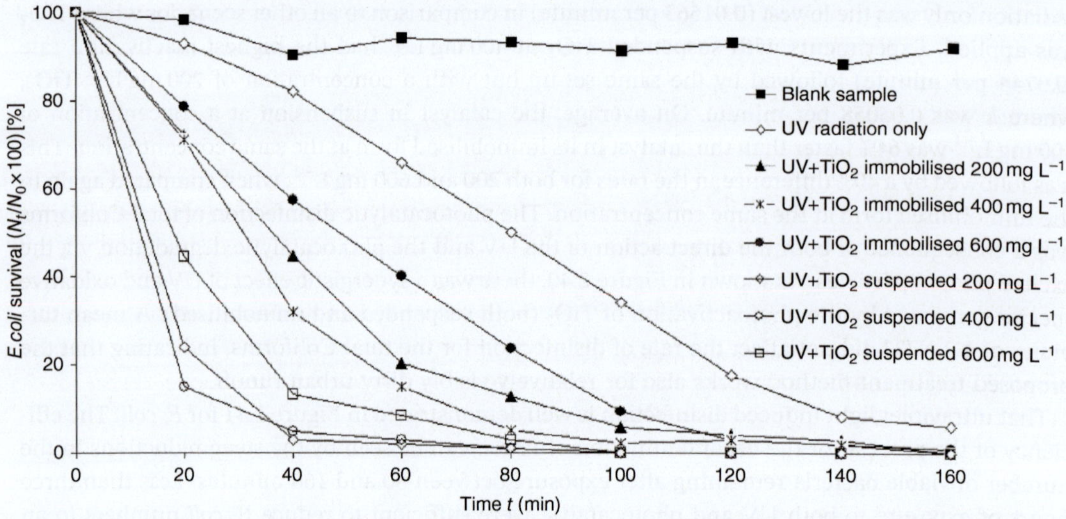

Figure 2.41 Survival of *Escherichia coli* (CFU/100 mL) in the presence of ultraviolet (UV) light radiation only, and immobilised and suspended powdered forms of titanium dioxide (TiO_2). Graph showing an almost perfect progressive reduction of *Escherichia coli* with increasing treatment complexity and titanium dioxide concentration.

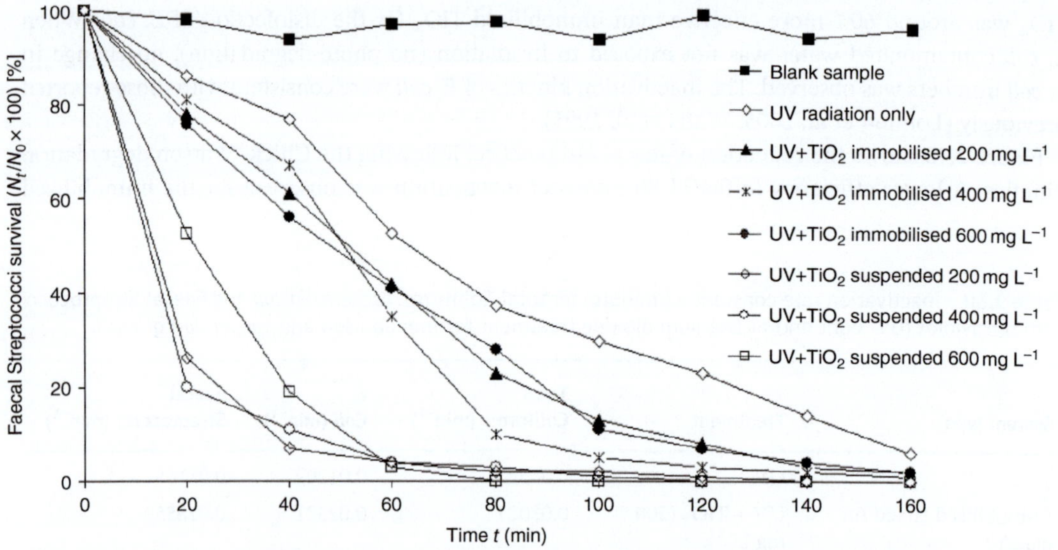

Figure 2.42 Survival of faecal Streptococci (CFU/100 mL) in the presence of ultraviolet (UV) light radiation only, and immobilised and suspended powdered forms of titanium dioxide (TiO_2). Graph showing an almost perfect progressive reduction of *faecal Streptococci* with increasing treatment complexity and titanium dioxide concentration.

form of TiO_2, which confirms the corresponding findings for both total Coliforms and *E. coli*. Combined UV and TiO_2 (suspended) treatment at TiO_2 concentrations of 400 and 600 mg L^{-1} had inactivation rate constants for faecal Streptococci of 0.05504 and 0.04401 per minute, respectively (Table 2.30). These reactions were 60–70% faster than those at the same concentrations for the immobilised form of TiO_2. There was increased disinfection efficiency at a dosage range between 400 and 600 mg L^{-1} for suspended TiO_2. A significant decrease ($p < 0.05$) of the initial faecal Streptococci counts at 400 mg L^{-1} TiO_2 in suspension of approximately 99.98% after 80 minutes was noted.

Particulate matter in suspension can negatively affect the process of water disinfection. In the case of heterogeneous photocatalysis, the active surface of the TiO_2 crystals can be reduced by the adsorption of particulate material (Herrera Melián et al. 2000). However, the photodegradation of faecal Streptococci was unaffected by colour (13 $\pm$2 Hazen units) and more importantly by turbidity (5.5 NTU).

Assuming that the treated effluent from the PPS could be brought into direct contact with sunlight, there is the potential to replace UV lamps with natural light particularly in warm countries. The combined use of solar radiation, TiO_2 photocatalysis and permeable pavements (Figure 2.43) could therefore reduce energy costs further, making the proposed technology more sustainable.

2.7.4 Conclusions

The PPS showed high removal efficiencies for BOD, COD, SS, colour, turbidity and microbiological parameters. Ultraviolet disinfection was efficient in reducing total Coliforms, *E. coli* and faecal Streptococci. However, the additional presence of TiO_2 enhanced the disinfection efficiency considerably.

The initial removal of organic matter within the PPS greatly improved the subsequent inactivation kinetics. The heterogeneous photocatalytic experiments indicate that the OH radical is the primary species responsible for the microbial inactivation. Experiments with suspended TiO_2 (400–600 mg L^{-1}) were more effective (40–60%) than those with immobilised TiO_2, totally removing

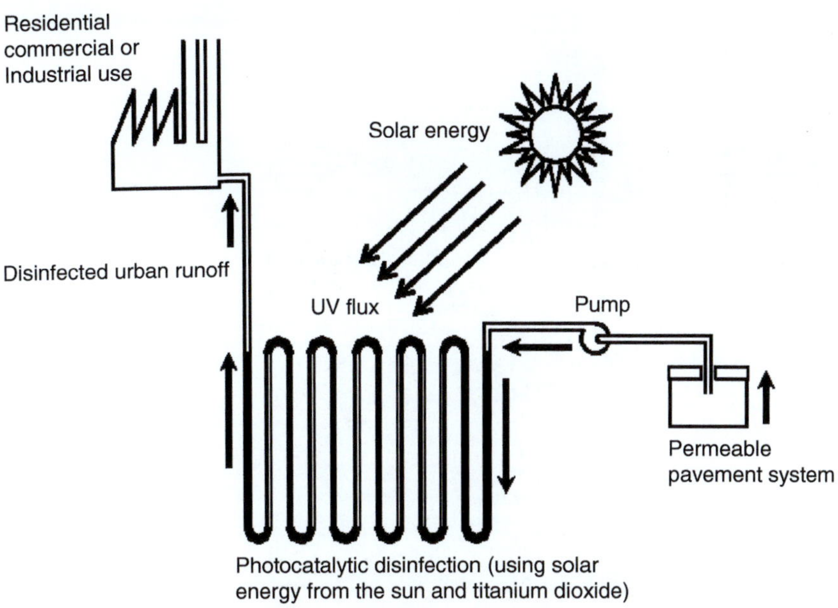

Figure 2.43 Schematic of an integrated permeable pavement system for initial filtration and treatment of urban runoff followed by photocatalytic decontamination using solar energy. Drawing showing the potential application of the new technology for used water (i.e., disinfected urban runoff) application in the urban environment.

the microbial pollutants within 80 minutes. Photocatalysis represents an attractive means to decontaminate urban runoff, particularly for countries with high natural UV radiation.

2.8 Refurbishment and Improvement of Screen Systems for Flood Control and Water Protection

2.8.1 Introduction and Background

2.8.1.1 Local Flooding Due to Heavy Rain

The frequency of flood events has increased and can be explained by climate change (Müller 2010; Scholz 2018). Forecasts suggest that the frequency of heavy rainfall events will continue to increase. Whole districts and streets of cities such as Oberursel (Taunus) near Frankfurt am Main are regularly affected by floods. The flood is triggered by heavy rainfall events as in January 2023 (Figure 2.44), August 2023 and May 2024.

However, the sewage system in Germany does not serve a flood protection purpose and therefore quickly reaches its stress maximum during heavy rainfall (Bayrisches Staatsministerium für Umwelt und Verbraucherschutz, and Bayrisches Staatsministerium für Wohnen, Bau und Verkehr 2019). Highly sealed urban areas and low grounds allow for flooding (Müller 2010; Scholz 2023).

Rainwater flows via open ditches, closed pipes and streams directly into the sewage system or into nearby watercourses. Streams and trenches are important receiving watercourses that also serve urban rainwater drainage. In the event of heavy rain events, these also receive combined sewer overflow water (Scholz 2018).

Both trenches and streams as well as separated and mixed water channels flood through too much surface drainage. In addition, screen systems (also called flotsam removal systems) can be

Figure 2.44 Example of flooding of streets, paths and property due to blockages in screens and pipes during heavy rain (January 2023) in Oberstedten, Oberursel (Taunus). Picture showing flooding close to a street within an urban area.

clogged due to rubble and organic material such as sticks and leaves. Then there may also be local floods associated with many screens in Oberursel (Figure 2.44).

Climate change leads to an increase in heavy rainfall events and floods (Hessisches Landesamt für Naturschutz, Umwelt und Geologie 2019). These changes require an adaptation of the water infrastructure, including the screen systems (including flotsam screens) to meet the new challenges (Müller 2010).

Figure 2.45 Example of a one-dimensional screen without sediment trap after heavy rain (May 2024) in Oberstedten, Oberursel (Taunus). Picture showing too much sediment located in front of a screen.

2.8.1.2 Existing Screen Systems

The municipality of Oberursel currently (summer 2024) maintains 56 screen systems. There are also two other urban screens in the planning phase. Thirty of the existing screen systems no longer adhere to the state-of-the-art, are damaged or suffer from inadequate maintenance. These systems are increasingly clogged and quickly reach their stress maximum during heavy rainfall (Figure 2.45).

Most of the screens in the community are one- and two-dimensional and have no sediment trap. This leads to frequent clogging from stones and sediment (Bhargava 2016). The maintenance teams for screens and streams are behind in terms of cleaning of clogged screens for several years due to a lack of workforce. The increased effort to avoid flood events costs time and money. Added to this is the increased water pollution by fine solids, which cannot be retained by screens and sediment traps (Scholz 2004b; Scholz and Zettel 2004).

2.8.1.3 Screen Maintenance

The entire construction, which includes the screen, is viewed as a coherent system. This involves, for example, elements such as the inlet area of the stream or ditch, a dead wood lock, a sediment and stone trap, the screen and the entrance area into the subsequent building element such as a rainwater or mixed water pipe.

Screen maintenance includes regular cleaning, functional control and repair. Due to increased heavy rainfall events, the maintenance of screens must be carried out more often. In addition, there are the necessary adjustments to the screen to new boundary conditions. This is a challenge for the maintenance teams and requires more efficient organisation and implementation of the cleaning work. The plans for the cleaning work must be time- and personnel-saving. If the screens are thoroughly cleaned, the duration until the next cleaning can be extended. Work equipment for the maintenance work must be suitable (Bayrisches Staatsministerium für Umwelt- und Verbraucherschutz, and Bayrisches Staatsministerium für Wohnen, Bau und Verkehr 2019; Scholz 2023).

2.8.1.4 Screen Design

The standards of the German Institute for Standardization provide the relevant design requirements for screen systems. In the worksheets and reports of the German Association for Water Management, Wastewater and Waste (DWA 2024), there are also many specifications on water-related structures. The DIN 19661 document deals with buildings in, on, above and among above-ground waters. This standard serves planning, construction, operation and maintenance (German Institute for Standardization 2022).

The accessibility of screen systems for maintenance work or rescue of people or animals must be upheld. Poor system accessibility leads to a reduction of the quality of maintenance work (Fang et al. 2020).

Screens are usually made of metal, while dead wood locks are often made of natural materials such as tree trunks (often hardwood; for example, obtained from oak trees). The screen can be one-, two- or three-dimensional. Depending on the area, location and flotsam, it can be decided which screen design is most appropriate. Some of the existing screens are often converted from one- to three-dimensional, because the screens can thereafter filter a greater capacity of water and then become more effective. The increasing heavy rainfall events require frequent checks of all screens concerning their function. Screen systems may have to be retrofitted by a finer or multi-dimensional screen and a sediment trap (Bhargava 2016; Lange and Lecher 2000).

2.8.1.5 Automated Screens

Screen systems can be operated with and without foreign energy to ensure cleaning. The drive can take place electrically, hydraulic, pneumatic or through hydropower. Cleaning can be done automatically, activated or carried out by maintenance work. However, screen systems for trenches and streams are simple and are, therefore, usually cleaned manually.

Automated cleaning requires a delimitation of the screen system from its surroundings to prevent external access and thereby minimise danger. However, since this is not always possible due to its location and the surrounding area, it must be avoided in many cases.

2.8.1.6 Supporting Structures

Additional structures such as dead wood locks, pre-screens as well as sand and sediment trap structures support the function of the main screen by holding larger amounts of flotsam and sediment back before they reach the main screen. This relieves the primary screen, which reduces the risk of flooding (Bhargava 2016; Lange and Lecher 2000; Müller 2010).

Most of the time, large items such as wooden trunks and shopping baskets are retained by pre-screens to avoid damage and clogging of the main screen. Depending on the location and nature of flotsam, the installation of pre-screens and sediment traps is an important step to improve the drain capacity of the streams (Lange and Lecher 2000).

Sand and stone traps have the task of holding back solids (mud, sand and stones) carried by the stream to prevent deposits in stretches further downstream (Bhargava 2016; Lange and Lecher 2000) and piped sections. However, they can also serve pre-treatment to relieve the main screen (Scholz 2004b; Scholz and Nettel 2004).

Many pollution parameters such as suspended solids, biological oxygen demand, nutrients and metals correlate well with sediment (Mustafa and Scholz 2011), which can be held back by a sediment trap (Fang et al. 2020). The federal guidelines for recognising ecologically critical water pollution by wastewater propose various solutions to improve the water quality of receiving watercourses such as streams (Fang et al. 2020; Hessisches Ministerium für Umwelt, Energie, Landwirtschaft und Verbraucherschutz 2024). It is therefore the responsibility of the municipalities to propose concepts for solutions such as improved screening systems, which the lower water authority and, if necessary, also the regional council must evaluate.

2.8.1.7 Objectives

The overarching goal is to optimise the local drainage system. The following objectives can be formulated: (i) protect pipes from clogging; (ii) prevent local floods; (iii) improve the water quality; (iv) improvement of preparations for storm events; (v) savings of personnel and material resources; (vi) get the screens faster cleaned; (vii) automation of screen system cleaning and processes; and (viii) improved screen monitoring.

2.8.2 Methodology

2.8.2.1 Locality

A detailed recording of the screen locations is required to carry out a systematic analysis and optimisation. In addition, other information is collected on monitoring and maintenance of the screen system including previously reported damage. Table 2.31 shows summary statistics of measured and estimated variables for all screen locations in the example municipality of Oberursel near Frankfurt.

Each calculation system in Oberursel can be clearly determined by the following entries that could be collected in a software such as Excel: screen number, screen name, street and house number, coordinates DD N and coordinates DD O. In this case study, Google coordinates were utilised.

Table 2.31 Summary statistics of measured and estimated variables.

Variable	Unit	Minimum	Median	Average	Standard deviation	Variance	Maximum
Measured (for 58 screens)							
Area	m^2	0.02	0.15	0.02	0.58	0.27	2.78
Rod distance[a]	cm	3	8.5	3	3.6	12.8	19
Water width	cm	0	156	0	167.3	27 846	1000
Water depth	cm	0	55	0	46.8	2236	200
Passage width[b]	cm	15	43	61.5	57.29	3051	343
Passage height[b]	cm	15	40	48.5	27.99	724	130
Estimated priorities							
Flotsam priority	—	1.0	4.0	3.4	1.3	1.7	5.0
Cleaning priority	—	1.0	3.5	3.0	1.5	2.4	5.0
Area priority	—	2.5	3.0	3.2	0.4	0.2	4.5
Water quality priority	—	1.0	3.0	2.9	1.0	1.1	5.0
Security priority	—	1.0	3.0	1.2	0.5	0.3	3.0
Renovation priority	—	1.0	1.0	3.2	1.2	1.5	5.0
Screen system renovation priority[c]	—	1.3	2.4	2.5	0.7	0.4	3.8

[a] Rod distance of the screen area, which is flowed head-on.

[b] The passage is the exit opening (e.g., a channel or a built-in stream) of the screen system that is located after the screen.

[c] This is a value that is composed of the upper six individual priorities after weighing up interests of stakeholder groups.

2.8.2.2 Digital Recording and Geo-systems

The digital recording of the screen locations in geographical information systems (GIS) such as Caigos offers several advantages such as the spatial analysis of data, which can identify and prioritise problematic areas. In addition, digital documentation facilitates long-term monitoring and maintenance of the screens by providing current information about the condition and performance of the screens.

2.8.2.3 Inspection of the Locality

The specific circumstances of each screen can be recorded and documented through local inspections. Parameters such as the passage width, passage height, shape of the screen, water width, water depth and floating angle are determined (Table 2.31). The screens are examined according to the type of flotsam and according to their dimension. The environment of the screen also plays a role to be able to assess the consequences of a local flood in advance and adapt the cleaning priorities accordingly.

2.8.2.4 Follow-up and Analysis

After cleaning the screen, a thorough quality control is crucial to check how well the system works. Outliers in the data need to be checked. The dimensions of the system are documented. Photos show the placement of the screen and explain its position, as well as the surrounding area before and behind it.

The DWA (2024) set of rules was adhered to during laboratory work. Water samples were determined photometrically with tests from Hach Lange.

2.8.2.5 Measures of Prioritisation

The screen system renovation priority (SSRP) was defined. The following parameters were determined by a team of local authority workers and the author: flotsam priority, cleaning priority, area priority, water quality priority, safety priority and renovation priority (Table 2.31). Each of the priorities mentioned is given a numerical value from 1 (very low) to 5 (very high).

The flotsam priority stands for the size and material nature of the flotsam and how likely it is that it causes clogging. The cleaning priority indicates how often a screen must be cleaned and which screen should be cleaned first in the event of a heavy rainfall. This priority is based on the experience of the local cleaning teams. The area priority indicates how important or vulnerable the area surrounding the screen is in terms of flooding. The water quality priority indicates how important the screen and a possible existing sediment trap are for the water quality of natural waters, which serve as a receiving watercourse. The security priority indicates how urgent a measure of public safety and health is. The renovation priority represents how much a renovation is necessary at the present time to ensure full screen functioning. Subsequently, the SSRP indicates the overall rating based on an interest group weighting of flotsam, cleaning, area, water quality, safety and renovation priority (Table 2.32).

The following measures were specified for screen system upgrades: locksmith, concrete, sediment and stones, safety, deadwood, ditch, early warning, screen creation and automatic screen cleaning. Metal processing is undertaken by the locksmith. The measures concrete, sediment and stones as well as dead wood and ditch are measures for road construction work. Early warning and automatic screen cleaning measures require external work with a special company for information technology. The measures and their costs should be described in Excel tables. The screens and the measures should be sorted according to the SSRP values of the corresponding screens. The higher the SSRP,

Table 2.32 Example of a stakeholder assessment as a basis for the screen system renovation priority.

Stakeholder group[a]	Weighing of the SG	Flotsam priority	Cleaning priority	Area priority	Water quality priority	Security priority	Renovation priority
Residents	10	2	5	1	2	5	4
Urban drainage	8	5	5	4	4	3	5
Road construction	4	2	3	3	1	4	2
Environment and ecology	5	1	1	1	5	1	3
Public funds[b]	8	3	5	5	4	4	3
Security and health	15	2	3	5	2	5	2
Screen system renovation priority considering different SG							
Relative part (Total = 1)		0.13	0.19	0.18	0.14	0.20	0.16

[a] Stakeholder group = SG.
[b] From the point of view of the municipality, which should provide the funds for screen system maintenance and renovation work.

the more important is the measure. The measures can then be processed from top to bottom by local workers or companies to be commissioned (e.g., locksmith).

One- and two-dimensional screens should be checked for their conversion to three-dimensional screens. This measure should be discussed with the responsible local team for screen construction. First, the individual priorities of the screen should be checked. If screens have a low cleaning priority, their conversion is in most cases superfluous. It must be determined whether a potential screen upgrade would have a benefit. There is the possibility that screens with a low SSRP value will still be upgraded, because this would make a higher contribution to flood protection.

Then the location of the screen must be considered. A three-dimensional screen may not be installed due to the specific nature of the inlet, stream or ditch. In this case, it should be checked whether a screen upgrade offers a good solution. The passage or inlet behind the screen must be considered. If the inlet is too small, it is usually not worth converting the screen, because the low load capacity of the passage or inlet is not optimised. Since the enlargement of an inlet is linked significantly to more effort and costs, it must be carefully examined whether this is worth it.

In addition, the conversion of screen systems should not lead to any additional difficulties for the cleaning teams. It must be considered which work equipment the teams use so that cleaning is not made more difficult. It is also important that a screen system upgrade (still) ensures the safety of the cleaning teams.

Similar criteria like for the conversion are used to decide on the construction of dead wood locks as well as sediment and stone traps. In addition, however, there is the type of flotsam. In forests, this measure offers a solution for retaining swimming tree trunks and large sticks.

Another measure is to install early warning systems such as surveillance sensors. Here, too, a cost–benefit assessment must determine for which screens this measure should be carried out. The weighing takes place based on the priority list and the experience made by cleaning teams during heavy rainfall events.

The most complex measure of all is the construction of an automated cleaning system. This measure can only be implemented for very few screens. Automatisation is only worthwhile for screens with high cleaning needs and can only be installed if the screens are large enough and if the surroundings of the screens offer sufficient space to install the system. In addition, the automated screens must be inaccessible to people and animals to prevent dangerous situations leading to accidents. Very few existing screens in the case study meet these criteria.

2.8.3 Results and Discussions

2.8.3.1 Characteristics of Screen Systems

Screens are intended to prevent blocking of pipes, streams and trenches as well as retain rubble, wood and plant parts. In the event of a heavy rain, screens are an important factor to prevent flood events and hold back large objects (Müller 2010). Without screens, many objects get into the openings of channels and cause blockages. The cleaning of pipes is much more expensive than the maintenance of screens (Bhargava 2016; Lange and Lecher 2000; Scholz 2018, 2023).

Screens play an important role in local flood protection by maintaining the optimal water flow rate by preventing blockages. With the increase in heavy rain events, screens are becoming increasingly important for the water infrastructure and for local disaster protection (Müller 2010). The construction of dead wood screens is very helpful here because it intercepts much greater material and thereby relieves the primary screen. As a result, this has a lower risk of flooding, because the main screens can ensure the design drainage.

The flow angle influences the efficiency of the screen. Flow-optimised construction can compensate for flow restrictions such as partial clogging. The greater the inclination between the screen and the horizontal, the higher the pressure loss. The hydraulic calculation of these screens is crucial to ensure their efficiency and functionality (Scholz 2023).

The flow rate is determined by a flow exit coefficient that depends on the construction of the screen and blockage. The rod width and space of the screen determine the size of the flotsam that can be held back.

The inclination of the screen is the angle between the screen and the horizontal. A greater inclination can make cleaning easier, but also influences the hydraulic pressure loss.

The pressure or energy loss caused by the screen can be calculated to understand the effects on the hydraulic performance of the system. However, a correct calculation can only be done theoretically, because the loss of pressure differs significantly in reality and changes quickly.

The flow velocity is the speed of the water through the screen. It influences the forces acting on the dead wood lock, the screen and the drift material. Here, too, the calculation is only possible theoretically, because the speed is not constant in reality.

The classification of the type of screen according to the rod spacing (gap width) is as follows: protective screens (60–200 mm), coarse screens (20–100 mm) and fine screens (8–20 mm). In Oberursel, there are 43 coarse screens (two more in planning) and 13 protective screens. The rod spacing influences the type of drift material that is held back by the screen. A too large distance between the rods can lead to blockage in the further course of the pipe (Heydarzadeh et al. 2019). In comparison, a distance that is too small leads to increased cleaning effort. The rod spacing of each system is adapted to the type of flotsam and the area of the passage behind the screen (Bhargava 2016; Scholz 2018, 2023).

Pre-screens have a large rod spacing to relieve the main screen by intercepting large items. Fine screens, on the other hand, have low bar distances, because they are designed to intercept small objects. They are more effective in some cases but have a greater risk of blockage.

With a change in weather conditions, the optimal rod distance of a screen system can change, because the drift material type and size may alter. In the event of heavy rain, much larger material such as wooden trunks and garbage cans can be expected.

A reduction in the flow speed can increase the efficiency of the screen and reduce the risk of clogging. Screens are exposed to shock loads. Depending on the stability of the screens, it can be helpful to reduce the flow speed through a natural stream design to keep the shock loads low. Personnel costs can be saved with the right flow speed. However, it must be noted that in the case of stone and sand screens, higher velocities are necessary to prevent large amounts of deposits to enter the screen systems in the case of hydraulic maximum loads causing a fault (Lange and Lecher 2000; Scholz 2023).

In the case of inappropriate screens located within streams, fish passage can be negatively affected. However, fish ladders can help here. A fish protection system can be used to guide fish away from screens. They do not enter the danger zone, so do not penetrate the main screen, and are guided by the construction arrangement to a bypass. The bypass leads the fish harmlessly to the underwater (Lange and Lecher 2000). In Oberursel, however, the construction of such protective systems is not necessary, because there are no fine screens on flowing waters with fish stock.

In the case of automated cleaning systems, a screen is installed. The screen system filters the water and leads the flotsam to a collection unit, which must be tailored to the screen area. In the German case study, there are two automatic screen systems in planning.

A sole jump is the difference in height for which a screen system can be installed. Depending on where it is, its impact changes on the rake. In principle, the sole jump counteracts deposits and thus prevents clogging. If the sole jump is at the base of the screen or behind it, its positive effects are at their best. Especially behind the screen, the sole jump usually offers the best support against clogging. For the case study in Oberursel, 19 rakes are known, where the sole jump is located behind the screen. For nine other one-dimensional screens, the sole jump is situated directly below the screen.

Often clogged screen systems can be supplemented by an emergency relief. If the amount of water exceeds the resilience of the screen system, the excess water can continue to flow through the emergency relief. The emergency relief must be installed in such a way that water can only flow through sideways from or over the rake in the event of an overload. There is the possibility of a short-term and automatic opening of the screen in the event of an overload, which leads to relief. The screen is often adapted by its height so that the water can continue to flow over the screen if an increase in the water level takes place. This cannot lead to an increase in the backstream level within the canal, ditch or stream, and the flood risk is reduced accordingly (Müller 2010). In Oberursel, there are currently six screens with built-in emergency relief options for increased water levels.

Sediment traps relieve screens and prevent functional disorder (Lange and Lacher 2000). They also prevent clogging of pipes and help to save time and staff effort. Twenty-one screen systems in Oberursel have a sediment trap. Six more are planned.

Sediment basins also contribute to better water quality and ecology. By intercepting sediment and rubble, streams are also relieved ecologically. The water quality priority in Oberursel has an average of 3.3 for screen systems with current and future sediment traps. For screen systems with no sediment basin, the average is only 2.7.

Sediment traps are usually dug with a depth between 10 and 100 cm and a length of 100–700 cm in front of or behind the screen. Sediment basins form a lowering (often in front of the screen), where stones, sand and sediments are deposited. The stones and sand are deposited at the bottom of the trap when the water flows through and continues to the screen or pipe (opening). The pool usually consists of concrete or natural stone (Bhargava 2016; Lange and Lecher 2000).

In the case of trenches and streams that run through a forest, a dead wood lock is very effective. A lot of dead wood gets through the forest into ditches and streams. This type of flotsam is mostly large and would clog the channel and possibly damage the screen. It remains in front of the screen and clogs it. In the event of heavy rainfall events, large wood is washed-up, and it only takes a few hours for the screen to clog and create a localised flood. This can best be prevented by a dead wood lock. In this way, clogging during heavy rainfall events can be prevented and the cleaning teams are relieved. There are no dead wood locks in Oberursel at the present time, but three are being planned. As expected, the flotsam priority for the three screen systems, where a dead wood lock is planned, is 5.0.

The screen systems in Oberursel are in public spaces. Accordingly, they must ensure people's security. If there is a high risk of falling, railings must be attached. All risks that arise from a screen system must be eliminated.

In addition, with large pipes and channels, there is a risk that children and animals enter them. For this reason, the channel must be made inaccessible from the outside. The safety priority in connection with the pipe has been assessed. The larger the area, the greater is also the security priority.

2.8.3.2 Screen System Maintenance

Regular functional controls and cleaning are necessary to ensure the effectiveness of the screen (Lange and Lecher 2000). In the optimal case, small items of flotsam are not removed when cleaning, but they are led to the sewage treatment plant with the mixed wastewater. Depending on the screen system, cleaning can also be automated, but mostly an external energy source is required.

The following cleaning methods can be used: ejector beam (air-water mixture), rotating stripe comb, broom, cleaning bar, cleaning top spiral, optimised room teeth, cleaning brushes, snail-wear with brush replacement and fan glands.

If water flows through the entire screen, it must be cleaned in continuous operation. Fine screens must be cleaned more often due to their relatively small rod distances. Cleaning can be automated for fine screens at suitable boundary conditions to save time and staff effort.

Screen systems are operationally safe and often also low in maintenance (dependent on the weather). Optical clogging control can often succeed. For this reason, it is worthwhile to apply surveillance sensors in many cases. Much of the control and maintenance can then succeed in the surveillance systems. Together with an automated cleaning system, local inspections rarely need be made. This saves time and personnel costs, increases the effectiveness and enables more screens to be cleaned and checked in the same period.

In the event of flood events, the cleaning team must clean the rakes in optimal order based on a priority list, which is a function of both cleaning priority and travel times. The locations of the screens must be included in the planning. This concept for cleaning builds on the analysis of the screen.

Personnel resources play a major role in maintaining and cleaning screens. In the event of heavy rainfall events, many screens must be cleaned quickly due to their high priority rating. In such cases, a large cleaning team would be very helpful. This would make it possible to divide the team into several groups to clean or free the screen from flotsam and debris more efficiently.

In addition to maintaining the screen system, there is also the treatment of the collected flotsam. Before this is removed, it must be pressed to save space during transport. Piston presses, roller presses and snail presses are three relevant press systems with different functions. In addition, the weight and volume of the screen is reduced by drainage. Before transportation, the flotsam is temporarily stored in collecting containers. The collected waste is often disposed of by landfill, combustion or composting.

When it comes to screen maintenance, the screen must be protected in winter from freezing. This may best be realised by a total enclosure.

The sediment basin is cleaned by removal of the rubble. Sedimental cleaning is part of the screen plant care. Regular cleaning is important to ensure the function of the silt trap. Due to the weight of the sediment and rubble, cleaning is often carried out by an excavator. Since the width of the sediment trap usually does not exceed one meter and the height usually does not exceed half a meter, the excavation of the trap must be performed carefully to prevent damage to the concrete or stone of the structure. Cleaning causes higher effort and costs, but it is essential to ensure the function of the sediment trap and thereby also the task of the screen system.

Sediment and rubble pools can also be placed behind the screen. As a result, cleaning can often only succeed manually. In this case, the system comprises mostly a coarse screen. The bar distances are then large enough that rubble and sediment can flow through. So that rubble and sediment do not clog the entrance area of the pipe, it is deposited in the sediment trap (Bhargava 2016).

Table 2.33 shows an overview of the water quality parameters of representative sediment traps. Sediment tanks with and without many fine components contribute to water pollution of receiving watercourses due to their high chemical oxygen demand. Fine materials such as organic silt often also have an increased nutrient value.

Generally, screens have to be repaired very rarely. The rods of screens are mostly made of metal and are, therefore, difficult to break or bend. Damage is rarely caused by drift material or impact load. Much damage is caused by humans, either by the maintenance teams or by vandalism. By acting heavily from the outside, screens can become loose in their installation place or get fractures in some places, which affects their function.

Damaged screens must be repaired promptly to maintain their functionality. If the same defect is independent of humans and occurs repeatedly, optimisation of the screens must be considered. As part of the repair, an optimisation or expansion of the system (for example, by a pre-screen, a dead wood lock or a rubble collection trap) can also be carried out to save time. Figure 2.46 shows a successful screen renewal.

Table 2.34 provides an overview of the most common options of original and future compositions of screen systems. A future composition is based on the recommendations for action by a local team

Table 2.33 Overview of quality parameters (after filtering) of representative sediment traps on 16 September 2024: Screen 17 with a high content of fines (99% of the sample below 63 μm) and screen 21 with a normal sieve size distribution.

Variable	Unit	Number 17	Number 21	Threshold
pH value	—	7.7	7.6	—
Chemical oxygen demand	$mg\,L^{-1}$	66.2	68.1	35
Ammonium-nitrogen	$mg\,L^{-1}$	0.243	0.694	1.2
Nitrate-nitrogen	$mg\,L^{-1}$	11.80	3.06	—
Total nitrogen	$mg\,L^{-1}$	16.50	6.48	12
Ortho-phosphate-phosphorus	$mg\,L^{-1}$	0.434	0.099	—
Phosphate	$mg\,L^{-1}$	1.11	0.584	0.7
Organic compounds (according to weight share)	%	25	3	—

Remarks: 2 L of tap water were added to each sediment test of 4-L to get a water sample after one minute of stirring. Threshold values refer to the discharge of cleaned sewage water directly into the Urselbach (receiving watercourse).

(a) (b)

Figure 2.46 Representative example: screen number 28 (a) before and (b) after renewal. Pictures comparing a two-dimensional with a three-dimensional screen.

Table 2.34 Overview of the original and future screen system composition. Number of systems: 58.

Option	Screen: None	Screen: 1-D	Screen: 2-D	Screen: 3-D	Ditch	Sediment trap (old)	Sediment trap (new)	Early warning system	Number of group members
Current situation (no individual cases)									
1			✓		✓				18
2			✓		✓	✓			17
3		✓			✓				12
4		✓			✓	✓			4
5				✓	✓				3
6	✓				✓				2
7			✓						2
Future situation (plus eight individual cases)									
1			✓		✓				13
2			✓		✓	✓			13
3		✓			✓				6
4		✓			✓	✓			2
5				✓	✓				6
8			✓		✓	✓	✓		3
9				✓	✓		✓		3
10			✓		✓		✓		2
11				✓	✓			✓	2

of experts in this case study. Clear developments (trends) can be seen for options 1–5. These options contain at least 69% of the group members (specific type of screen system).

Option 1 shows a change by five group members, which can be justified by optimising the screen systems with a sediment trap or upgrading the screen dimension. Some of the group members of options 1 and 5 in the state of the future are in the actual state in option 3. The number of group members for option 2 is also reduced, because a new sediment trap is planned, or the screen dimension is increased.

Option 3 shows a noticeable decline in the number of group members by 50%. These are screen systems that consist only of a one-dimensional screen. This means that there are many optimisation options or that they are essential. The compensation dimension is mainly optimised, which supports options 1 and 5 in the state of the future.

The only option that has a significant increase of group members is option 5. The members of this option increase by 100% to six members in the future scenario. Most group members of the screen systems of options 8–11 differ from the other screen systems by their relatively high SSRP value. The individual cases can be attributed to installations of dead wood locks, early warning systems and automated cleaning systems.

Protective walls or flood relief structures for some screen structures may be necessary to improve their stability and efficiency. Protective walls could make sense to achieve a more efficient functioning of the screen. Furthermore, flood relief structures are sometimes necessary to prevent blockages.

If a data analysis indicates that some geometric conditions between the screen system and the water flow direction are inappropriate, a protective wall can be the solution. Especially due to heavy rainfall events, some screens are no longer suitable for the current situation. In addition, increasing technical advances result in more efficient screen systems and methods, which means that older screens will have to be modified or replaced. Through systematic before and after comparisons (before and after installation of the screen or before and after the screen wall change), someone can make statements about the effectiveness of the system and make recommendations for future renovations. By observing the screen usage and functioning under different conditions, someone can determine whether a screen system is dimensioned too small. From a flotsam cover of well over 60%, an increase of the screen dimensioning must be considered. For example, a one-dimensional screen could be replaced by a multi-dimensional one.

The distance between the dead wood lock and the screen must not be too long to prevent deadwood from causing clogging of the screen from damaged trees between the dead wood lock and the screen. Depending on the location and environment, it must be decided to which side the dead wood should be guided by the dead wood lock. Ideally, there is a meadow, field or forest on one side of the lock. The dead wood does not cause any major problems at such locations. In some cases, the dead wood must be redirected to a path or a small road because the other side is blocked or inaccessible.

In general, and especially for deadwood locks, the total height must be significantly higher than that of the nearby area to avoid the drifting away of the already collected flotsam. The static stress on the structure must also be considered. This very simple type of coarse screen used as a pre-screen is particularly helpful during heavy rainfall events, because it intercepts dead wood from trees and thereby relieves the main screen. However, cleaning must be carried out extensively with suitable equipment.

2.8.3.3 Recommendations for Action

In view of the challenges of climate change and the increasing heavy rain events, it is essential to create a comprehensive database of all screen locations and to maintain it continuously. The load stress and the stress intensity of individual screen systems should be examined. To make decisions, functional analyses of screens must be made regularly.

The construction of a new screen, the conversion of an existing one or the relocation of a screen to a nearby more suitable area should be considered. The installation of sedimentation traps and rubble catches as well as dead wood locks at suitable locations should be checked. Screening automation in terms of cleaning should also be considered.

It is necessary to create sensible maintenance and cleaning plans, which are both time- and personnel-saving as well as effective. It must be determined which screen can be maintained and/or repaired and where external help should be used. Locations, where the use of sensors is worthwhile in connection with an early warning system, should be identified.

State funds can be often applied to comprehensively renovate or expand screens to control flooding and diffuse pollution. There are also funding programs on different administrative levels in Germany.

2.8.4 Conclusions and Outlook

Repairing or enlarging screens often optimises their functionality and reduces the local flood risk. Screens should be provided with additional structures such as sand and debris traps. A special focus should also be placed on increasing the screen dimension and positioning.

The safety priority increases with the passage area (often a pipe) behind the screen. This also applies to the bar spacing of the screen. These screens should therefore be prioritised during maintenance.

Regular cleaning of sediment traps reduces the risk of pipe blockages and improves the water quality of receiving waters that are polluted by stormwater pipes. Fine particles in the sediment often have too high nutrient and oxygen demand values.

The installation of remote monitoring systems can significantly improve the efficiency of maintenance teams. Real-time monitoring of the screens allows problems to be identified early and quickly resolved. This reduces personnel costs and enables a faster response time in the event of blockages or damage. Such systems can make maintenance much easier, especially in hard-to-reach or remote areas.

Installing pre-screens can also increase the efficiency of the main screens. Pre-screens catch larger debris before they reach the main screens, thus reducing the load on the main screens. This leads to a reduction in cleaning intervals and a higher functionality of the entire screen structure.

Optimising travel routes for maintenance teams can be achieved by analysing locations and strategically planning maintenance routes. This saves time and resources and ensures that the screens are regularly and efficiently maintained. The combination of digital recording, remote monitoring and optimised travel routes forms the basis for effective and sustainable screen maintenance.

A long-term strategy for adapting and maintaining the debris screens to climate change is essential. This should include regular inspections, timely repairs and adjustments to the screen structures, and continuous training of maintenance teams.

References

Abel, P.D. (1996). *Water Pollution Biology*. London: Taylor and Francis.

Abwassertechnische Vereinigung e. V. (1994). *Richtlinien für die hydraulische Dimensionierung und Leistungsnachweis von Regenwasser-Entlastungsanlagen in Abwasserkanälen und –leitungen. Guideline A 111*. Hennef, Germany: Abwassertechnische Vereinigung e. V. (in German).

Abwassertechnische Vereinigung e. V. (1996). *Richtlinien für die hydraulische Dimensionierung und Leistungsnachweis von Regenwasser-Entlastungsanlagen in Abwasserkanälen und –leitungen. Guideline 112*. Hennef, Germany: Abwassertechnische Vereinigung e. V. (in German).

Alderiso, K., Wait, D., and Sobsey, M. (1996). Detection and characterization of make-specific RNA coliphages in a New York city reservoir to distinguish between human and nonhuman sources of contamination. *Proceedings of a Symposium on New York City Water Supply Studies*.

Allen, S.E. (1974). *Chemical Analysis of Ecological Materials*, 1e, 89. Oxford, UK: Blackwell Scientific Publications Ltd.

Alloway, B.J. (1995). *Heavy Metals in Soils*, 2e. Suffolk, UK: Chapman and Hall Ltd.

Andersen, C.T., Foster, I.D.L., and Pratt, C.J. (1999). Role of urban surfaces (permeable pavements) in regulating drainage and evaporation: development of a laboratory simulation experiment. *Hydrological Processes* 13 (4): 597–609.

Ann, Y., Reddy, K.R., and Delfino, J.J. (1999). Influence of chemical amendments on phosphorus immobilization in soils from a constructed wetland. *Ecological Engineering* 14 (1–2): 157–167.

APHA (1998). *Standard Methods for the Examination of Water and Wastewater*, 20e, 4–112. Washington, DC: American Public Health Association.

APHA American Public Health Association (1995). *Standard Methods for the Examination of Water and Wastewater*, 19e. Washington DC, USA: APHA-AWWWA-WEF.

Ashley, R. M., Smith, H., Jowitt, P. W. et al. (2001). A multi-criteria analysis/risk management tool to assess the relative sustainability of water/wastewater systems: SWARD (sustainable water industry asset resource decisions). *Proc. of 1st First National Conf. on Sustainable Drainage*, Coventry University, Coventry, UK, p. 221.

Bannister, A.F. (1997). Lagoon and reed-bed treatment of colliery shale tip water at dodworth, south Yorkshire. *Proceedings of a CIWEM National Conference held 5th September 1997 at the University of Newcastle*, ed by Younger PL. CIWEM, UK, pp. 105–131.

Bayrisches Staatsministerium für Umwelt und Verbraucherschutz, and Bayrisches Staatsministerium für Wohnen, Bau und Verkehr (2019). Arbeitshilfe: Hochwasser- und Starkregen in der Bauleitplanung. München, Deutschland. https://www.stmuv.bayern.de/themen/wasserwirtschaft/hochwasser/doc/arbeitshilfe.pdf (accessed 16 August 2024).

Bean, E.Z., Hunt, F.W., and Bidelspach, A.D. (2007). Field survey of permeable pavement surface infiltration rates. *Journal of Irrigation and Drainage Engineering* 133 (3): 249–255.

Bhargava, A. (2016). Physico-chemical waste water treatment technologies: an overview (*Physikalisch-chemische Abwasserbehandlungstechnologien: ein Überblick*). *International Journal of Scientific Research and Education* 4 (5): 5308–5319. https://doi.org/10.18535/ijsre/v4i05.05.

Bi, Y.H., Chen, L.G., and Wu, C. (2002). Ground heat exchanger temperature distribution analysis and experimental verification. *Applied Thermal Engineering* 22: 183–189.

Booth, D.B. and Brattebo, B.O. (2004). Permeable pavement update. *Journal of the American Planning Association* 70 (1): 98–98.

Borgwardt, S. (2006). Long-term in-situ infiltration performance of permeable concrete block pavement. *Proceedings of the 8th International Conference on Concrete Block Paving* (6–8 November). San Francisco, California, USA.

Bose, J.E. (2005). Space conditioning: the next frontier, in international ground source heat pump association air innovations conference (24–26 August 2005), US Environmental Protection Agency, Office of Air and Radiation, Chicago

Bran and Luebbe (1999). *Bran Luebbe AA3 Autoanalyser Methods: G-109-93 and 94*. Norderstedt: Bran and Luebbe.

Braskerud, B.C. (2002). Factors affecting phosphorus retention in small constructed wetlands treating agricultural non-point source pollution. *Ecological Engineering* 19 (1): 41–61.

Brattebo, B.O. and Booth, D.B. (2003). Long-term stormwater quantity and quality performance of permeable pavement systems. *Water Research* 37 (18): 4369–4376.

Breen, P.F. (1990). A mass balance approach of assessing the potential of artificial wetlands for wastewater treatment. *Water Research* 24 (6): 689–697.

Bridson, E.Y. (1998). *The Oxoid Manual*, 8e, 2-161, 2-216 and 2-217. Basingstoke, UK: Oxoid Ltd.

British Standards (1990). *Methods of Test for Soils for Civil Engineering Purposes. BS 1377 – 1 to 9.* London, UK: British Standards Publishing Ltd.

British Standards Institution, BS 882 (1992). *Specification for Aggregates from Natural Sources for Concrete.* London, UK: British Standards Institution.

Brix, H. (1999). Functions of macrophytes in constructed wetlands. *Water Science and Technology* 29 (4): 71–78.

Brix, H., Arias, C.A., and del Bubba, M. (2001). Media selection for sustainable phosphorus removal in subsurface flow constructed wetlands. *Water Science and Technology* 44 (11–12): 47–54.

Brooks, A.S., Rozenwald, M.N., Geohring, L.D. et al. (2000). Phosphorus removal by wollastonite: a constructed wetland substrate. *Ecological Engineering* 15 (1–2): 121–132.

Brown, R.R. (2005). Impediments to integrated urban stormwater management: the need for institutional reform. *Environmental Management* 36 (3): 455–468.

Bullen Consulting (2002). Braid burn flood prevention scheme – environmental statement (Volume I). Reference 1018260, The City of Edinburgh Council. Edinburgh, UK.

Butler, D. and Davies, J.W. (2000). *Urban Drainage.* London, UK: E & FN Spon (an imprint of the Taylor and Francis Group plc).

Butler, D., Xiao, Y., Karunaratne, S.H.P.G., and Thedchananamoorthy, S. (1995). The gully pot as a physical, chemical and biological reactor. *Water Science and Technology* 31 (7): 219–228.

Cappuccino, J.G. and Sherman, N. (1996). *Microbiology, a Laboratory Manual*, 4e, 115–118. Suffern, New York, USA: Rockland Community College and 175.

Cheng, S.P., Grosse, W., Karrenbrock, F., and Thoennessen, M. (2002). Efficiency of constructed wetlands in decontamination of water polluted by heavy metals. *Ecological Engineering* 18 (3): 317–325.

Cho, M., Chung, H., Choi, W., and Yoon, J. (2004). Linear correlation between inactivation of E. coli and OH radical concentration in TiO_2 photocatalytic disinfection. *Water Research* 38 (4): 1069–1077.

Chocat, B., Ashley, R., Marsalek, J. et al. (2007). Towards the sustainable management of urban stormwater. *Indoor and Built Environment* 16 (3): 273–285.

Ciszewski, D. (1998). Channel processes as a factor controlling accumulation of heavy metals in river bottom sediments: consequences for pollution monitoring (Upper Silesia, Poland). *Environmental Geology* 36 (I–II): 45.

Clesceri, L.S., Eaton, A.D., Greenberg, A.E., and Franson, M.A.H. (1998). *Standard Methods for the Examination of Water and Wastewater*, 20e. Washington, District Council, USA: American Public Health Association, American Water Works Association and Water Environment.

Cooper, P.F., Job, G.D., Green, M.B., and Shutes, R.B.E. (1996). *Reed Beds and Constructed Wetlands for Wastewater Treatment.* Swindon, UK: WRc plc.

Coupe, S.J. (2004).Biodegradation and microbial ecology within permeable pavements. PhD thesis. School of Science and the Environment. Coventry University, Coventry.

Coupe, S.J., Smith, H.G., Newman, A.P., and Puehmeier, T. (2003). Biodegradation and microbial diversity within permeable pavements. *European Journal of Protistology* 39: 495–498.

Decamp, O. and Warren, A. (2000). Investigation of Escherichia coli removal in various designs of subsurface flow wetlands used for wastewater treatment. *Ecological Engineering* 14: 293.

Decamp, O., Warren, A., and Sanchez, R. (1999). The role of ciliated protozoa in subsurface flow wetlands and their potential as bioindicators. *Water Science and Technology* 40 (3): 91–98.

DEUTSCHES INSTITUT FÜR NORMUNG (2022). Bauwerke in, an, über und unter oberirdischen Gewässern (DIN 19661). https://www.din.de/de/mitwirken/normenausschuesse/naw/entwuerfe/wdc-beuth:din21:352419795 (accessed 16 August 2024).

Diaz, O.A., Reddy, K.R., and Moore, P.A. (1994). Solubility of inorganic P in stream water as influenced by pH and Ca concentration. *Water Research* 28 (8): 1755–1763.

Dierkes, C., Göbel, P., Benze, W., and Wells, J. (2000). Next generation water sensitive storm water management techniques. *2nd National Conference on Water Sensitive Urban Design* (2–4 September 2000). Brisbane.

Dierkes, C., Kuhlman, L., Kandasamy, J., and Angelis, G. (2002). Pollution retention capability and maintenance of permeable pavements. *Proceedings of the 9th International Conference on Urban Drainage* (ed. E.W. Strecker and W.C. Humber) (8–13 September), International Water Association: Portland, Oregon, USA.

Donelly, A.P. and Herbert, R.A. (1999). Bacterial interactions in the rhizosphere of seagrass communities in shallow coastal lagoons. *Journal of Applied Microbiology* 85: 151S–160S.

Drizo, A., Frost, C.A., Smith, C.A., and Grace, J. (1997). Phosphate and ammonium removal by constructed wetlands with horizontal subsurface flow, using shale as a substrate. *Water Science and Technology* 35 (5): 95–102.

Drizo, A., Frost, C.A., Grace, J., and Smith, K.A. (1999). Physico-chemical screening of phosphate-removal substrates for use in constructed wetland systems. *Water Research* 33 (17): 3595–3602.

Drizo, A., Comeau, Y., Frost, C.A., and Chapuis, P. (2002). Phosphorus saturation potential: a parameter for estimating the longevity of constructed wetland systems. *Environmental Science & Technology* 35 (21): 4642–4648.

DWA (2024). Das Deutschen Vereinigung für Wasserwirtschaft, Abwasser und Abfall e. V. (DWA) Regelwerk. https://de.dwa.de/de/die-dwa.html (accessed 3 Oktober 2024).

Eke, P.E. and Scholz, M. (2008). Benzene removal with vertical-flow constructed treatment wetlands. *Journal of Chemical Technology and Biotechnology* 83: 55–63.

European Commission Environment (EEC) (1991). European Commission urban wastewater treatment directive 91/271/EEC. http://ec.europa.eu/environment/water/water urbanwaste/index_en.html (accessed March 2010).

Fang, H., Huang, L., Zhao, H. et al. (2020). *Bedload Transport of Bio-sediment: Mechanics of Bio-Sediment Transport*. Berlin, Germany: Springer.

Ferguson, B.K. (2005). *Porous Pavements*. Boca Raton: Taylor and Francis.

Fernández-Ibáñez, P., Sichel, C., Polo-López, M.I. et al. (2009). Photocatalytic disinfection of natural well water contaminated by fusarium solani using TiO_2 slurry in solar CPC photo-reactors. *Catalysis Today* 144 (1–2): 62–68.

Fowler, J. and Cohen, L. (1998). *Practical Statistics for Field Biology*. West Sussex, UK: Wiley.

Frossard, E., Brossard, M., Hedley, M.J., and Metherell, A. (1995). Reactions controlling the cycling of P in soils. In: *Phosphorus in the Global Environment* (ed. H. Tiessen), 107–139. Chichester, UK: Wiley.

Fujishima, A., Rao, T.N., and Tryk, A.D. (2000). Titanium dioxide photocatalysis. *Journal of Photochemistry and Photobiology, C: Photochemistry Reviews* 1 (1): 1–21.

Genchi, Y., Kikegawa, Y., and Inaba, A. (2002). CO_2 payback-time assessment of a regional-scale heating and cooling system using a ground source heat-pump in a high energy-consumption area in Tokyo. *Applied Energy* 71: 147–160.

Gervin, L. and Brix, H. (2001). Removal of nutrients from combined sewer overflows and lake water in a vertical-flow constructed wetland system. *Water Science and Technology* 44 (11–12): 171–176.

Grabowiecki, P. and Scholz, M. (2006). Pavements: a new source of water and energy? *Proceedings of the Institution of Civil Engineers* 159: 54.

Green, M., Friedler, E., and Safrai, I. (1998). Enhancing nitrification in vertical-flow constructed wetlands utilizing a passive air pump. *Water Research* 32 (12): 3513–3520.

Greenway, M. and Woolley, A. (1999). Constructed wetlands in Queensland: performance efficiency and nutrient bioaccumulation. *Ecological Engineering* 12 (1–2): 39–55.

Hammer, D.A. (1989). *Constructed Wetlands for Wastewater Treatment – Municipal.* Chelsea, Michigan, USA: Industrial and Agricultural. Lewis Publishers.

Hawke, C.J. and José, P.V. (1996). *Reed Bed Management for Commercial and Wildlife Interests.* Sandy, UK: RSPB.

Hendry, G.A.F. and Grime, J.P. (1993). *Methods in Comparative Plant Ecology – A Laboratory Manual.* London, UK: Chapman & Hall.

Hepbasli, A. and Ozgener, L. (2004). Development of geothermal energy utilization in Turkey: a review. *Renewable and Sustainable Energy Reviews* 8: 433–460.

Herrera Melián, J.A., Doña Rodríguez, J.M., Viera Suárez, A. et al. (2000). The photocatalytic disinfection of urban waste waters. *Chemosphere* 41: 323–327.

Herrmann, J.-M. (2005). Heterogeneous photocatalysis: state of the art and present applications. *Topics in Catalysis* 34 (1–4): 49–65.

Heydarzadeh, R., Tabesh, M., and Scholz, M. (2019). Multiobjective optimization in sewer network design to improve wastewater quality. *Journal of Pipeline Systems Engineering and Practice* 10 (4): 04019037.

Hunter, P.R. (1997). *Waterborne Disease.* Chester: Wiley.

International Water Association Specialist Group on the Use of Macrophytes in Water Pollution Control (2000). Constructed Wetlands for Pollution Control. Scientific and Technical Report No. 8. IWA Publishing, London.

Jefferies, C., Aitken, A., McLean, N. et al. (1999). Assessing the performance of urban BMPs in Scotland. *Water Science and Technology* 39: 123–131.

Johansson, L. (1997). The use of leca (light expanded clay aggregates) for the removal of phosphorus from wastewater. *Water Science and Technology* 35 (5): 87–93.

Kadlec, R.H. (1997). An antibiotic wetland phosphorus model. *Ecological Engineering* 8 (2): 145–172.

Kadlec, R.H. and Knight, R.L. (1995). *Treatment Wetlands.* Boca Raton, Florida, USA: Lewis Publishers.

Kadlec, R.H. and Knight, R.L. (1996). *Treatment Wetlands*, 893. New York: Lewis Publishers.

Lange, G. and Lecher, K. (2000). *Gewässerregelung & Gewässerpflege – Naturnaher Ausbau und Unterhaltung von Fließgewässern*, 3e. Singhofen, Deutschland: Paul Parey Zeitschriftenverlag.

Lei, L., Khodadoustb, A.P., Suidana, M.T., and Tabakc, H.H. (2005). Biodegradation of sediment-bound PAHs in field contaminated sediment. *Water Research* 39 (2–3): 349–361.

Lim, P.E., Tay, M.G., Mak, K.Y., and Mohamed, N. (2003). The effect of heavy metals on nitrogen and oxygen demand removal in constructed wetlands. *The Science of the Total Environment* 301 (1–3): 13–21.

Lonnen, J., Kilvington, S., Kehoe, S.C. et al. (2005). Solar and photocatalytic disinfection of protozoan, fungal, and bacterial microbes in drinking water. *Water Research* 39 (55): 877–883.

Mann, R.A. (1997). Phosphorus adsorption and desorption characteristics of constructed wetland gravel and steelworks by-products. *Australian Journal of Soil Research* 35 (2): 375–384.

March, K. (1991). *The Environmentalist's Guide to Garden Pest Control* (ed. S. Forty). Shepperton, Surrey, UK: Ian Allan Ltd.

Martin, P., Turner, B., Waddington, K. et al. (2000). *Sustainable Urban Drainage Systems – Design Manual for England and Wales.* London: Construction Industry Research and Information Association.

Mays, P.A. and Edwards, G.S. (2001). Comparison of heavy metal accumulation in a natural wetland and constructed wetlands receiving acid mine drainage. *Ecological Engineering* 16 (4): 487–500.

McNeill, A. and Olley, S. (1998). The effects of motorway runoff on watercourses in south-west Scotland. *Water and Environment Journal* 12: 433–439.

Memon, F.A. and Butler, D. (2002). Assessment of gully pot management strategies for runoff quality control using a dynamic model. *The Science of the Total Environment* 295: 115–129.

Morrison, G.M., Revitt, D.M., and Ellis, J.B. (1995). The gully pot as a biochemical reactor. *Water Science and Technology* 31: 229–236.

Moss, B. (1998). *Ecology of Fresh Waters – Man and Medium, Past and Future*, 3e. Oxford, UK: Blackwell Science Ltd.

Müller, U. (2010). *Hochwasserrisikomanagement: Theorie und Praxis*. Wiesbaden, Germany: Vieweg und Teubner.

Mungur, A.S., Shutes, R.B.E., Revitt, D.M., and House, M.A. (1997). An assessment of metal removal by laboratory scale wetland. *Water Science and Technology* 35 (5): 125–133.

Mustafa, A. and Scholz, M. (2011). Nutrient accumulation in typha latifolia and sediment of a representative integrated constructed wetland. *Water, Air, and Soil Pollution* 219 (1): 329–341.

Nagano, K., Katsura, T., and Takeda, S. (2006). Development of a design and performance prediction tool for the ground source heat pump system. *Applied Thermal Engineering* 26: 1578–1592.

National Water Council (1981). *Design and Analysis of Urban Storm Drainage: The Wallingford Procedure*. UK: NWC.

Newman, A.P., Pratt, C.J., Coupe, S.J., and Cresswell, N. (2002). Oil bio-degradation in permeable pavements by microbial communities. *Water Science and Technology* 45 (7): 51–56.

Newman, A.P., Puehmeier, T., Kwok, V. et al. (2004). Protecting groundwater with oil-retaining pervious pavements: historical perspectives, limitations and recent developments. *Quarterly Journal of Engineering Geology and Hydrogeology* 37: 283–291.

Nur Onar, A., Balkaya, N., and Akyuz, T. (1996). Phosphorus removal by adsorption. *Environmental Technology* 17 (2): 207–213.

Obarska-Pempkowiak, H. and Klimkowska, K. (1999). Distribution of nutrients and heavy metals in a constructed wetland system. *Chemosphere* 39 (2): 303–312.

Optiroc (1998). Filtralite for Water Purification. Optiroc (formerly known as a. s. Norsk Leca), Oslo, Norway.

Orton, R., Bebbington, A., and Bebbington, J. (1995). *Freshwater Name Trial*. Shrewsbury, UK: Field Studies Council Publications.

Ottová, V., Balcarová, J., and Vymazal, J. (1997). Microbial characteristics of constructed wetlands. *Water Science and Technology* 35 (5): 117–123.

Ozgener, O. and Hepbasli, A. (2007). Modeling and performance evaluation of ground source (geothermal) heat pump systems. *Energy and Buildings* 39: 66–75.

Pant, H.K., Reddy, K.R., and Lemon, E. (2001). Phosphorus retention capacity of root bed media of sub-surface flow constructed wetlands. *Ecological Engineering* 17 (4): 345–355.

Pavey, D. (1978). *Methuen Handbook of Colour*, 3e. Norfolk, UK: Eyre Methuen Ltd., text by Cox & Wyman Ltd., colour plates by Sankt Jørgen Tryk Ltd. (with inks provided from Sadolin & Holmblad Ltd.).

Pontier, H., Williams, J.B., and May, E. (2001). Metals in combined conventional and vegetated road run-off control systems. *Water Science and Technology* 44 (11–12): 607–614.

Pratt, C.J. (2003). *An Application of Geosynthetics in Sustainable Drainage System*, 121–135. Loughborough, UK: Thomas Telford.

Pratt, C.J., Newman, A.P., and Bond, P.C. (1999). Mineral oil bio-degradation within a permeable pavement: long term observations. *Water Science and Technology* 39: 103–109.

Prochaska, C.A. and Zouboulis, A.I. (2006). Removal of phosphates by pilot vertical-flow constructed wetlands using a mixture of sand and dolomite as substrate. *Ecological Engineering* 26 (3): 293–303.

Read, H.H. (1984). *Mineralogy*, 26e. London, UK: George Allen & Unwin (Publishers) Ltd.

Reddy, K.R., Kadlec, R.H., Flaig, E., and Gale, P.M. (1999). Phosphorus retention in streams and wetlands: a review. *Critical Reviews in Environmental Science and Technology* 29 (1): 83–146.

Rincón, A.G. and Pulgarin, C. (2003). Photocatalytical inactivation of E. coli: effect of (continuous–intermittent) light intensity and of (suspended–fixed) TiO_2 concentration. *Applied Catalysis, B: Environmental* 44 (3): 263–284.

Robinson, H., Harris, G., Carville, M. et al. (1999). The use of engineered reed bed system to treat leachate at monument hill landfill site, southern England. In: *Constructed Wetlands for the Treatment of Landfill Leachates* (ed. G. Mulamoottil, M.B. EA, and F. Rovers), 27. Boca Raton, Florida, USA: Lewis Publishers.

Rushton, B.T. (2001). Low-impact parking lot design reduced runoff and pollutant loads. *Journal of Water Resources Planning and Management* 127 (3): 172–179.

Sakadevan, K. and Bavor, H.J. (1997). Phosphate adsorption characteristics of soils, slags and zeolite to be used as substrates in constructed wetland systems. *Water Research* 32 (2): 393–399.

Salehi, F., Lagace, R., and Pesant, A.R. (1997). Construction of a year-round operating gauging station for sediment and water quality measurements of small watersheds. *Journal of Soil and Water Conservation* 52 (VI): 431.

Salih, F.M. (2002). Enhancement of solar inactivation of Escherichia coli by titanium dioxide photocatalytic oxidation. *Journal of Applied Microbiology* 92: 920–926.

Schlüter, W. and Jefferies, C. (2002). Modelling the outflow from a porous pavement. *Urban Water* 4: 245–253.

Schmitt, F., Milisic, V., Bertrand-Krajewski, J.L. et al. (1999). Numerical modelling of bed load sediment traps in sewer systems by density currents. *Water Science and Technology* 39 (IX): 153.

Scholes, L.N.L., Shutes, R.B.E., Revitt, D.M. et al. (1999). The removal of urban pollutants by constructed wetlands during wet weather. *Water Science and Technology* 40: 333–340.

Scholz, M. (1997). Optimisation of biological activity in granular activated carbon beds. Ph.D. thesis. The University of Birmingham, UK.

Scholz, M. (2003). Sustainable operation of a small-scale flood-attenuation wetland and dry pond system. *Journal of the Chartered Institution of Water and Environmental Management* 17 (3): 171–175.

Scholz, M. (2004a). Treatment of gully pot effluent containing nickel and copper with constructed wetlands in a cold climate. *Journal of Chemical Technology and Biotechnology* 79 (2): 153–162.

Scholz, M. (2004b). Storm water quality associated with a silt trap (empty and full) discharging into an urban watercourse in Scotland. *International Journal of Environmental Studies* 61 (4): 471–483.

Scholz, M. (2006a). *Wetland Systems to Control Urban Runoff*. Amsterdam, The Netherlands: Elesevier.

Scholz, M. (2006b). Practical sustainable urban drainage system decision support tools. *Proceedings of the Institution of Civil Engineers–Engineering Sustainability* 159: 117–125.

Scholz, M. (2018). *Sustainable Water Management: Engineering Solutions for a Variable Climate*. Amsterdam, The Netherlands: Elsevier.

Scholz, M. (2023). *Wetlands for Water Pollution Control*. Amsterdam, The Netherlands: Elsevier.

Scholz, M. and Grabowiecki, P. (2007). Review of permeable pavement systems. *Building and Environment* 42: 3830–3836.

Scholz, M. and Grabowiecki, P. (2009). Combined permeable pavement and ground source heat pump systems to treat urban runoff. *Journal of Chemical Technology and Biotechnology* 84 (3): 405–413.

Scholz, M. and Martin, R.J. (1998a). Control of bio-regenerated granular activated carbon by spreadsheet modelling. *Journal of Chemical Technology and Biotechnology* 71: 253–261.

Scholz, M. and Martin, R.J. (1998b). Biological control in granular activated carbon beds. *International Review of Hydrobiology* 83: 657–664.

Scholz, M. and Xu, J. (2001). Comparison of vertical-flow constructed wetlands for treatment of wastewater containing lead and copper. *Water and Environment Journal* 15 (4): 287–293.

Scholz, M. and Zettel, S. (2004). Stormwater quality associated with a full silt trap discharging into an urban watercourse. *Water and Environment Journal* 18 (4): 226–229.

Scholz, M., Xu, J., and Dodson, H.I. (2001). Comparison of filter media, plant communities and microbiology within constructed wetlands treating wastewater containing heavy metals. *Journal of Chemical Technology and Biotechnology* 76 (8): 827–835.

Scholz, M., Höhn, P., and Minall, R. (2002). Mature experimental constructed wetlands treating urban water receiving high metal loads. *Biotechnology Progress* 18 (6): 1257–1264.

Seo, D.C., Cho, J.S., Lee, H.J., and Heo, J.S. (2005). Phosphorus retention capacity of filter media for estimating the longevity of constructed wetlands. *Water Research* 39 (2): 2445–2457.

Shannon, M.A., Bohn, P.W., Elimelech, M. et al. (2008). Science and technology for water purification in the coming decades. *Nature* 452: 301–310.

Shutes, R.E., Revitt, D.M., Scholes, L.N.L. et al. (2001). An experimental constructed wetland system for the treatment of highway run-off in the United Kingdom. *Water Science and Technology* 44 (11–12): 571–578.

Soares, E.T., Lansarin, M.A., and Moro, C.C. (2007). A study of process variables for the photocatalytic degradation of rhodamine B. *Brazilian Journal of Chemical Engineering* 24 (1): 29–36.

Spicer, G.E., Lynch, D.E., Newman, A.P., and Coupe, S.J. (2006). The development of geotextiles incorporating slow-release phosphate beads for the maintenance of oil degrading bacteria in permeable pavements. *Water Science and Technology* 54: 273–280.

Sriwiriyarat, T. and Randall, C.W. (2005). Performance of IFAS wastewater treatment processes for biological phosphorus removal. *Water Reseach* 39 (16): 3873–3884.

Tang, X., Huang, S.L., and Scholz, M. (2009). Comparison of phosphorus removal between vertical subsurface flow constructed wetlands with different substrates. *Water and Environment Journal* 23 (3): 180–188.

Tanner, C.C., Sukias, J.P.S., and Upsdell, M.P. (1998). Relationships between loading rates and pollutant removal during maturation of gravel-bed constructed wetlands. *Environmental Quality* 27 (2): 448–458.

Tchobanoglous, G., Burton, F.L., and Stensel, H.D. (2003). *Wastewater Engineering: Treatment and Reuse*, 4e. New York, USA: Metcalf & Eddy Inc., McGraw Hill Companies Inc.

Tejedor, J.M.T., Gonzalez, M.M., Pita, M.L. et al. (2001). Identification and antibiotic resistance of faecal enterococci isolated from water samples. *International Journal of Hygiene and Environmental Health* 203: 363–368.

Tota-Maharaj, K. (2011). *Permeable Pavements for Urban Stormwater Runoff Enhancement and Reuse.* Müller Saarbrücken, Germany: VDM Verlag Dr.

Tota-Maharaj, K. and Scholz, M. (2010). Efficiency of permeable pavement systems for the removal of urban runoff pollutants under varying environmental conditions. *Environmental Progress & Sustainable Energy* 29 (3): 358–369.

Tota-Maharaj, K. and Scholz, M. (2013). Combined permeable pavement and photocatalytic titanium dioxide oxidation system for urban runoff treatment and disinfection. *Water and Environment Journal* 27 (3): 338–347.

Watts, R.J., Kong, S., Orr, M.P. et al. (1995). Photocatalytic inactivation of coliform bacteria and viruses in secondary wastewater effluent. *Water Research* 29 (1): 95–100.

Wilson, S., Newman, A.P., Puehmeier, T., and Shuttleworth, A. (2003). Performance of an oil interceptor incorporated into a pervious pavement. *Proceedings of the Institution of Civil Engineers-Engineering Sustainability* 156: 51–58.

Wist, J., Sanabria, J., Dierolf, C. et al. (2002). Evaluation of photocatalytic disinfection of crude water for drinking-water production. *Journal of Photochemistry and Photobiology, A: Chemistry* 147 (3): 241–246.

Wood, T.S. and Shelley, M.L. (1999). A dynamic model of bioavailability of metals in constructed wetland sediments. *Ecological Engineering* 12 (3–4): 231–252.

Xu, D.F., Xu, J.M., Wu, J.J., and Muhammad, A. (2006). Studies on the phosphorus sorption capacity of substrates used in constructed wetland systems. *Chemosphere* 63 (3): 344–352.

Zhang, J., Zhang, X., Li, G.H., and Zhang, H.T. (2006). Physical and chemical effects and plant growth suitability of substrates in subsurface flow wetlands. *Environmental Sciences* 27 (5): 874–880.

Zhu, T., Jenessen, P.D., Maehlum, T., and Krogstad, T. (1997). Phosphorus sorption and chemical characteristics of lightweight aggregate (LWA)-potential filter media in treatment wetlands. *Water Science and Technology* 35 (5): 103–108.

3

Sustainable Flood Retention Basins including Integrated Constructed Wetlands

3.1 Sustainable Flood Retention Basin Management

3.1.1 Introduction

The European Union has introduced the Flood Directive 2007/60 EC (EU 2007), which requires each member state to develop flood risk management plans and flood defences. The concept of sustainability has been applied to flood defences, implying that all available infrastructures should be reassessed (Shih and Nicholls 2007). An assessment system that can be applied economically to a wide range of water bodies, wetlands and artificial impoundments would be of value in implementing the flood directive (EU 2007).

A sustainable flood retention basin (SFRB) is defined as an impoundment or wetland which has a pre-defined or potential role in flood defence and diffuse pollution control that can be accomplished cost effectively through best management practice, achieving sustainable flood risk management and enhancing sustainable drainage, pollution reduction, biodiversity, green space and recreational opportunities for society. The word 'sustainable' in SFRB means capable of being maintained at a steady level without exhausting natural resources, harming the environment or causing severe ecological damage (McMinn et al. 2010). To represent diverse functions of SFRB, six types of SFRB were named as follows (Scholz 2007; Scholz and Sadowski 2009): Hydraulic Flood Retention Basin (type 1), Traditional Flood Retention Basin (type 2), Sustainable Flood Retention Wetland (type 3), Aesthetic Flood Treatment Wetland (type 4), Integrated Flood Retention Wetland (type 5) and Natural Flood Retention Wetland (type 6).

Kriging is a group of geostatistical techniques designed to interpolate the value of a random field at an unobserved location derived from studies of its value at nearby sites. The spatial variation of a variable is quantified by a semi-variogram, which is a graph describing the characteristics of a regionalised variable (Rivoirard 1994). Various kriging techniques have also been applied successfully in hydrogeology and remote sensing (Webster and Oliver 2001), natural resource management (Oliver et al. 1996), environmental science (Rivoirard 1994) and agriculture (Liu et al. 2006). Geostatistics has also been applied in the wider area of water resource management; e.g., floodplain soil property mapping (Gallardo 2003), river water quality monitoring (Karamouz et al. 2009) and river network discharge mapping (Sauquet 2006). An effective and rapid geostatistical tool providing examples of flood management involving SFRB, which are characterised by clear and relevant characterisation variables, would support planning and communication among practitioners and planners.

Sustainable Water Systems, First Edition. Miklas Scholz.

The aim of Chapter 3.1 is to assess the potential role of SFRB in flood risk management with the help of geostatistical tools to improve communication between stakeholders. The key objectives are as follows: (i) to spatially analyse all relevant SFRB characterisation variables which could be used to assess flood control options; (ii) to assess attributes at unobserved sites using ordinary kriging with the support of a large and detailed example case study dataset; and (iii) to apply disjunctive kriging to assess the probability for individual variables to exceed specific management thresholds relevant for flood control. Yang et al. (2012) delivered the scientific basis for Chapter 3.1.

3.1.2 Methodology

3.1.2.1 Data

The survey of any SFRB is a two-stage process combining a desk study and a field visit during which 40 variables (Table 3.1) are assessed (McMinn et al. 2010). A guidance manual used to determine the 40 variables characterising water bodies including SFRB has been published by Scholz and Yang (2010). The variables include conventional hard engineering attributes such as *Dam Length* and *Dam Height*, together with more holistic attributes such as how *Engineered* the structure appears (McMinn et al. 2010; Scholz and Yang 2010).

A dataset of 371 SFRB sites, which are in the wider central Scotland area (Figure 3.1), has been developed. Figure 3.1 shows the locations of the different SFRB types. Each SFRB site is associated

Table 3.1 Classification variables used for the assessment of sustainable flood retention basins.

ID	Variable and unit	ID	Variable and unit
1	*Engineered* (%)	21	*Impermeable Soil Proportion* (%)
2	*Dam Height* (m)	22	*Seasonal Influence* (%)
3	*Dam Length* (m)	23	*Site Elevation* (m)
4	*Outlet Arrangement and Operation* (%)	24	*Vegetation Cover* (%)
5	*Aquatic Animal Passage* (%)	25	*Algal Cover in Summer* (%)
6	*Land Animal Passage* (%)	26	*Relative Total Pollution* (%)
7	*Floodplain Elevation* (m)	27	*Mean Sediment Depth* (cm)
8	*Basin and Channel Connectivity* (m)	28	*Organic Sediment Proportion* (%)
9	*Wetness* (%)	29	*Flotsam Cover* (%)
10	*Proportion of Flow within Channel* (%)	30	*Catchment Size* (km^2)
11	*Mean Flooding Depth* (m)	31	*Urban Catchment Proportion* (%)
12	*Typical Wetness Duration* ($d\ a^{-1}$)	32	*Arable Catchment Proportion* (%)
13	*Estimated Flood Duration* ($d\ a^{-1}$)	33	*Pasture Catchment Proportion* (%)
14	*Basin Bed Gradient* (%)	34	*Viniculture Catchment Proportion* (%)
15	*Mean Basin Flood Velocity* ($cm\ s^{-1}$)	35	*Forest Catchment Proportion* (%)
16	*Wetted Perimeter* (m)	36	*Natural Catchment Proportion* (%)
17	*Maximum Flood Water Volume* (m^3)	37	*Groundwater Infiltration* (%)
18	*Flood Water Surface Area* (m^2)	38	*Mean Depth of the Basin* (m)
19	*Mean Annual Rainfall* (mm)	39	*Length of Basin* (m)
20	*Drainage* ($cm\ d^{-1}$)	40	*Width of Basin* (m).

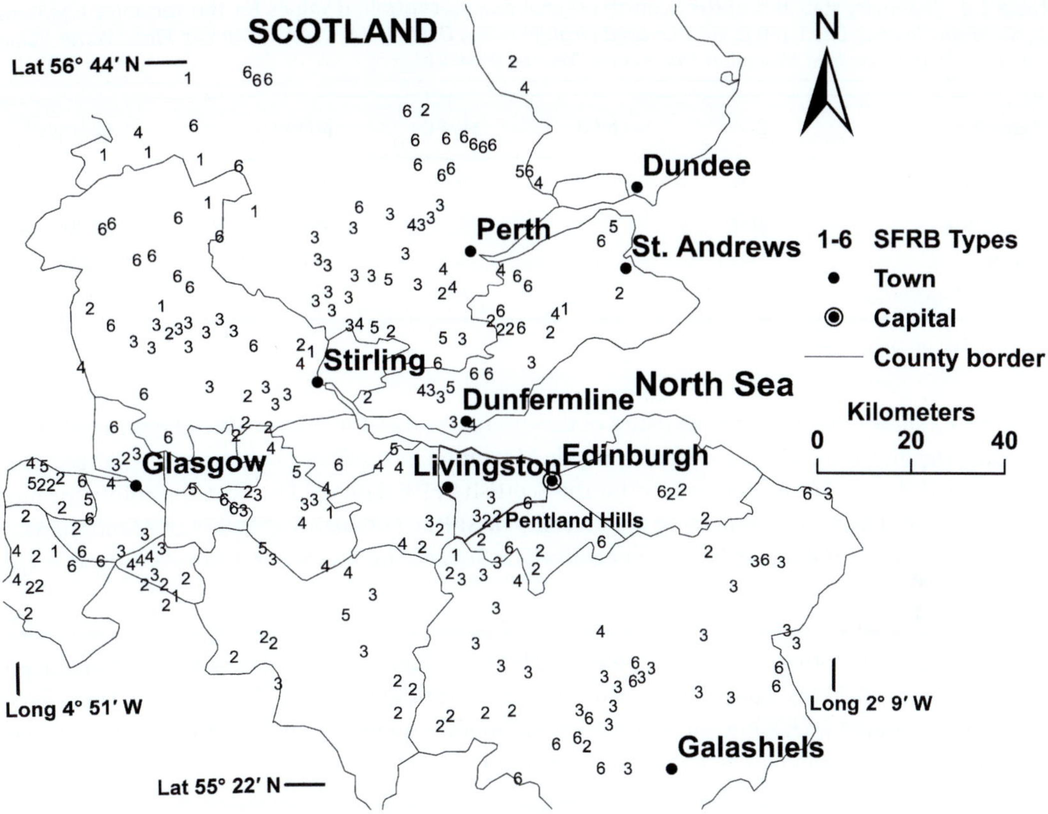

Figure 3.1 Study area, administrative boundaries and the 371 identified sustainable flood retention basin (SFRB) types in the wider central Scotland area (United Kingdom). A map showing the location of the assessed SFRB located roughly within a radius of 100 km around the town of Livingston situated about 20 km west of Edinburgh.

with 40 variables. At least 100 sites are required to produce reliable maps based on stable variograms (Webster and Oliver 2001). Thus, the selected number of sites is sufficient to apply geostatistical methods such as kriging.

Chapter 3.1 only focuses on flood-related variables such as *Engineered* (relative value, expressed in %, based on expert judgement of how engineered a basin appears in contrast to a natural water body), *Mean Flooding Depth* (average depth within the basin during flooding), *Maximum Flood Water Volume* (maximum volume of water within a basin during a typical flood) as discussed by Scholz (2007), McMinn et al. (2010), and Scholz and Yang (2010), and two compound variables: *Managed Mean Flooding Depth* and *Managed Maximum Flood Water Volume*. The first compound variable can be derived from the variables *Mean Flooding Depth* and *Mean Depth of the Basin*. *Managed Mean Flooding Depth* data for drinking water reservoirs are similar to the values collected for *Mean Flooding Depth,* while values for lakes are calculated by subtracting *Mean Depth of the Basin* from *Mean Flooding Depth*. The second compound variable *Managed Maximum Flood Water Volume* is derived by multiplying *Managed Mean Flooding Depth* by *Flood Water Surface Area*. For details on how to define and determine the values for the variables, readers may refer to Scholz and Yang (2010).

Table 3.2 Summary statistics of the Scottish original quality-controlled values for the variables *Engineered* (*E*, %), *Mean Flooding Depth* (MFD, m), *Managed Mean Flooding Depth* (MMFD, m), *Maximum Flood Water Volume* (MFWV, m^3) and *Managed Maximum Flood Water Volume* (MMFWV, m^3).

Statistic	*E*	MFD	MMFD	MFWV	MMFWV
Minimum	0	1	0	2450	0
Maximum	100	70	42	$260\,000 \times 10^4$	$14\,200 \times 10^4$
Mean	40.5	6.9	4.5	2312×10^4	462×10^4
Standard. deviation	33.52	8.13	6.01	$18\,801 \times 10^4$	1589×10^4

The number of samples was 371.

Table 3.2 shows the summary statistics of the original values for the variables *Engineered, Mean Flooding Depth, Managed Mean Flooding Depth, Maximum Flood Water Volume* and *Managed Maximum Flood Water Volume* determined for the Scottish SFRB dataset. The data variability for most variables is relatively high, reflecting the diversity of SFRB. The high variability of *Managed Maximum Flood Water Volume* is likely to negatively influence the prediction errors of kriged maps.

3.1.2.2 Variograms

In geostatistics, the variogram is a function describing the degree of spatial statistical dependence of a spatial random field (called spatial autocorrelation). A variogram interpolates a raster from a set of points using kriging with a known semi-variogram model and its parameters (Rivoirard 1994). The experimental variogram can be computed by using Eq. (3.1).

$$\gamma(h) = \frac{1}{2n}\sum_{i=1}^{n}\{z(x_i) - z(x_i + h)\}^2 \tag{3.1}$$

where $\gamma(h)$ is the sample semi-variance at lag h, which is a vector in both distance and direction, $z(x_i)$ and $z(x_i + h)$ are the values of $Z(x)$ at locations x and $x + h$, respectively, and n is the number of pairs of comparisons separated by a lag h for $i = 1, 2, ..., n$.

The parameters used to describe variograms are nugget, partial sill and range. The parameter nugget is the height of the jump of the semi-variogram at the discontinuity at the origin. The sill is the limit of the variogram tending to infinity lag distance, whilst the range is the distance in which the difference of the variogram from the sill becomes negligible. A partial sill is the sill minus the nugget.

The empirical variogram, which provides information on the spatial autocorrelation of the SFRB dataset, is fitted by a mathematical function, which describes the structure of variation and ensures validity of the variogram. Before the production of any kriged map, the most suitable model was fitted by applying the least squares method to the points forming the empirical semi-variogram.

3.1.2.3 Kriging

The key application of kriging in flood management with SFRB is the prediction of attribute values at unknown locations. Kriging forms weights from a semi-variogram based on surrounding measured values to predict figures for unmeasured sites. The measured values nearest to the unmeasured sites have the greatest influence. Predictions were made for each location in the wider central Scotland area based on the semi-variogram and the spatial arrangement of measured values that are located within the vicinity.

Ordinary kriging provides best linear unbiased estimations with minimum error variance and is the most used type of kriging. Furthermore, kriging weighs the surrounding measured values to derive a prediction for an unmeasured location. The general formula is formed as a weighted sum of the data (Eq. 3.2). The weight λ_i depends on a fitted model to the measured points, the distance to the prediction location and the spatial relationships among the measured values around the prediction location.

$$Z(S_0) = \sum_{i=1}^{N} \lambda_i Z(S_i) \tag{3.2}$$

where $Z(S_i)$ denotes the measured value at the ith location, λ_i is an unknown weight for the measured value at the ith location, S_0 is the prediction location and N means the number of measured values.

Disjunctive kriging is a nonlinear generalisation of kriging. This estimation technique allows for the conditional probability that the value of a spatially variable SFRB characterisation parameter is greater than a cut-off level yet to be calculated (Liu et al. 2006; Rivoirard 1994). The method can be used in flood risk management decision-making to help determine when some action, such as the construction of a new SFRB or a change in reservoir management, is necessary. Two input parameters are required to use the technique: a cut-off level and the critical probability level (Oliver et al. 1996). The relevant standard formulae for disjunctive kriging are shown in Eqs. (3.4–3.5). Sample values x in the original space A are transformed into y in a normal space B that has a standard normal distribution such that $x = \varphi(y)$. The function φ is written as a linear combination of Hermite polynomials as shown in Eq. (3.2) and discussed by Grad (2006).

$$\phi(y) = \sum_{k=0}^{\infty} C_k H_k(y) \tag{3.3}$$

where $\varphi(y)$ is the function of a linear combination of Hermite polynomials, C_k are coefficients to be calculated from the sample values x_i ($i = 1, 2, ..., n$) and $H_k(y)$ is a Hermite polynomial of the order k.

In the space B, the value y has the so-called coordinates $H_0(y)$, $H_1(y)$, ..., $H_k(y)$, ..., while in the space A, the corresponding value $x = \varphi(y)$ has the coordinates $C_0H_0(y)$, $C_1H_1(y)$, ..., $C_kH_k(y)$. Each sample value x_i is transformed to a value y_i ($i = 1, 2, ..., n$) with coordinates $H_0(y_i)$, $H_1(y_i)$, ..., $H_k(y_i)$. To obtain the disjunctive kriging estimate μ_D^*, which belongs to the space A, someone first calculates the corresponding value v_D^* in the space B. If H_{0v}^*, H_{1v}^*, ..., H_{kv}^*, are the coordinates of v_D^* in B, these coordinates are obtained by a linear combination of the corresponding coordinates of the sample values. The estimator H_{kv} is of the form shown in Eq. (3.5). To obtain μ_D^* from v_D^*, the function φ is used as shown in Eq. (13.5). The geostatistical analysis was performed within the ArcGIS 9.2 software environment (Liu et al. 2006).

$$H_{kv} = \sum_{i=1}^{n} b_i H_k(y_i) \tag{3.4}$$

where H_{kv} is the coordinate of the normalised block estimator v_D in space B, calculated from the surrounding sample values y_i ($i = 1, 2, ..., n$), and b_i are weights, optimally defined through the use of the autocorrelation function of the sample values.

$$\mu_D = \varphi(v_D) = \sum_{k=0}^{\infty} C_k H_{kv} \tag{3.5}$$

where μ_D represents the disjunctive kriging estimator and v_D is the corresponding value in normalised space B.

3.1.3 Results and Discussion

3.1.3.1 Findings Based on Ordinary Kriging

Ordinary kriging was applied for the key flood control variables *Engineered, Mean Flooding Depth, Managed Mean Flooding Depth, Maximum Flood Water Volume* and *Managed Maximum Flood Water Volume*. The statistics of these target variables are summarised in Table 3.2. The high variance of these variables reflects the diverse types of SFRB. Table 3.3 presents the ordinary kriging characteristics for these variables.

Figure 3.2a–3.2e shows map examples applying ordinary kriging for the variables *Engineered, Mean Flooding Depth, Managed Mean Flooding Depth, Maximum Flood Water Volume* and *Managed Maximum Flood Water Volume*, respectively. High numerical values for the variable *Engineered* generally indicate the likely necessity for high civil engineering investment to be made when planning for the construction of a new SFRB (Figure 3.2a).

The most engineered SFRB structures are likely to be found in the south-west of the study area, which coincides with the highest density of reservoirs and lakes used for water supply purposes. In contrast, for the study area in the north, relatively low investment for flood infrastructure is required. This variable is particularly useful when a decision has to be made on where old flood infrastructure needs to be upgraded or new SFRB constructed. It would be preferable to have areas prone to flooding being located within catchments associated with low values for the variable *Engineered* and high values for those such as *Managed Mean Flooding Depth* and *Managed Maximum Flood Water Volume*.

The spatial distribution for the variables *Mean Flooding Depth* and *Managed Mean Flooding Depth* is shown in Figure 3.2b, c, respectively. The *Mean Flooding Depth* is relatively high in the less populated upland areas of the north-west and south of the study area as well as within the Pentland Hills, a small area directly located south-west of Edinburgh (Figure 3.1). Low values for the variable *Mean Flooding Depth* are rare and patchy.

In comparison, the *Managed Mean Flooding Depth* variable is high only in the north-east and south-west of the study area. Moreover, elevated values have also been noted for some parts of the Pentland Hills. This area situated to the south-west of Edinburgh also has the highest density of reservoirs that could be used for hydraulic purposes such as flood control to protect the capital.

Table 3.3 Summary of ordinary kriging characteristics for the variables *Engineered* (*E*, %), *Mean Flooding Depth* (MFD, m), *Managed Mean Flooding Depth* (MMFD, m), *Maximum Flood Water Volume* (MFWV, m^3) and *Managed Maximum Flood Water Volume* (MMFWV, m^3).

Variogram and error	Parameter	*E*	MFD	MMFD	MFWV	MMFWV
Variogram	Model	exponential	Gaussian	spherical	spherical	Gaussian
	Range	125 534	69 788	96 548	167 522	167 522
	Partial sill	255.04	52.19	24.30	3.6297×10^{16}	2.5236×10^{14}
	Nugget	953.06	25.26	17.59	1.5532×10^{16}	1.5356×10^{4}
Mean standardised error for prediction		−0.0018	−0.0103	−0.0036	0.0044	0.0072

Number of samples: 371. Variograms: no data transformation.

(a)

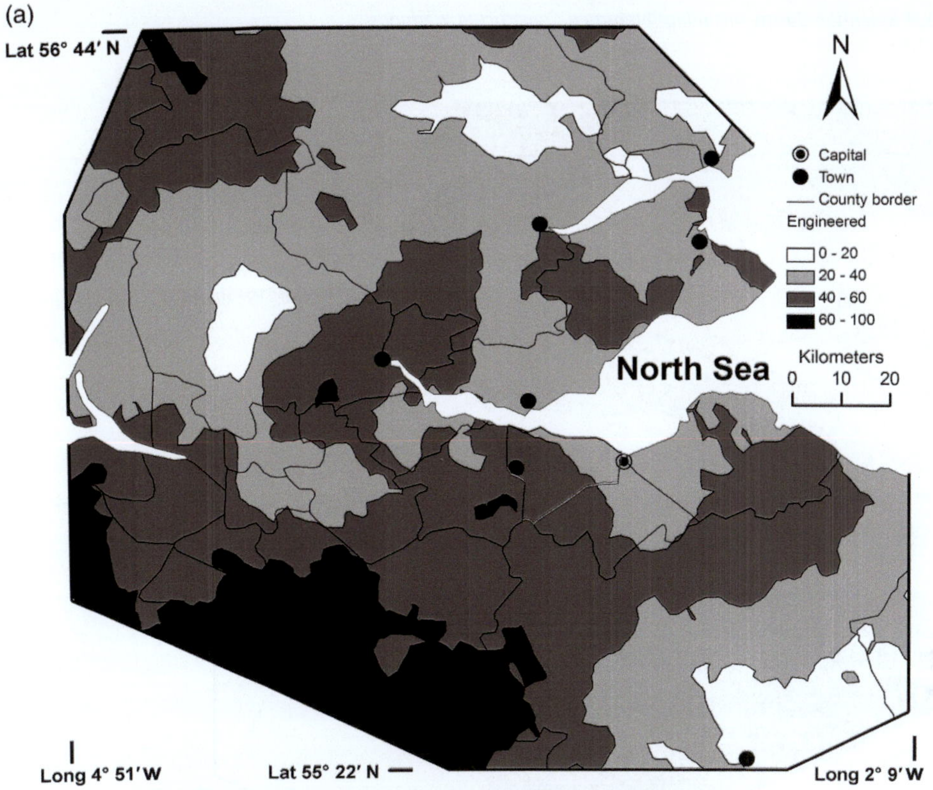

(b)

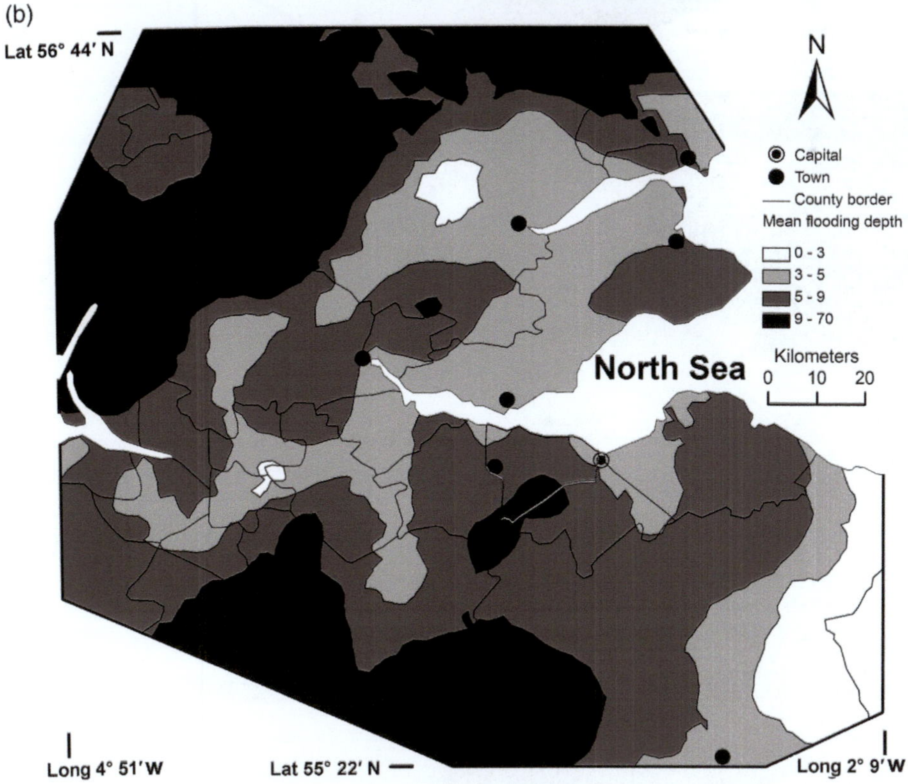

Figure 3.2 Map examples showing the application of ordinary kriging for (a) *Engineered* (%), (b) *Mean Flooding Depth* (m), (c) *Managed Mean Flooding Depth* (m), (d) *Maximum Flood Water Volume* (m^3) and (e) *Managed Maximum Flood Water Volume* (m^3). Higher values for most variables are usually found in the far North, far West and far South.

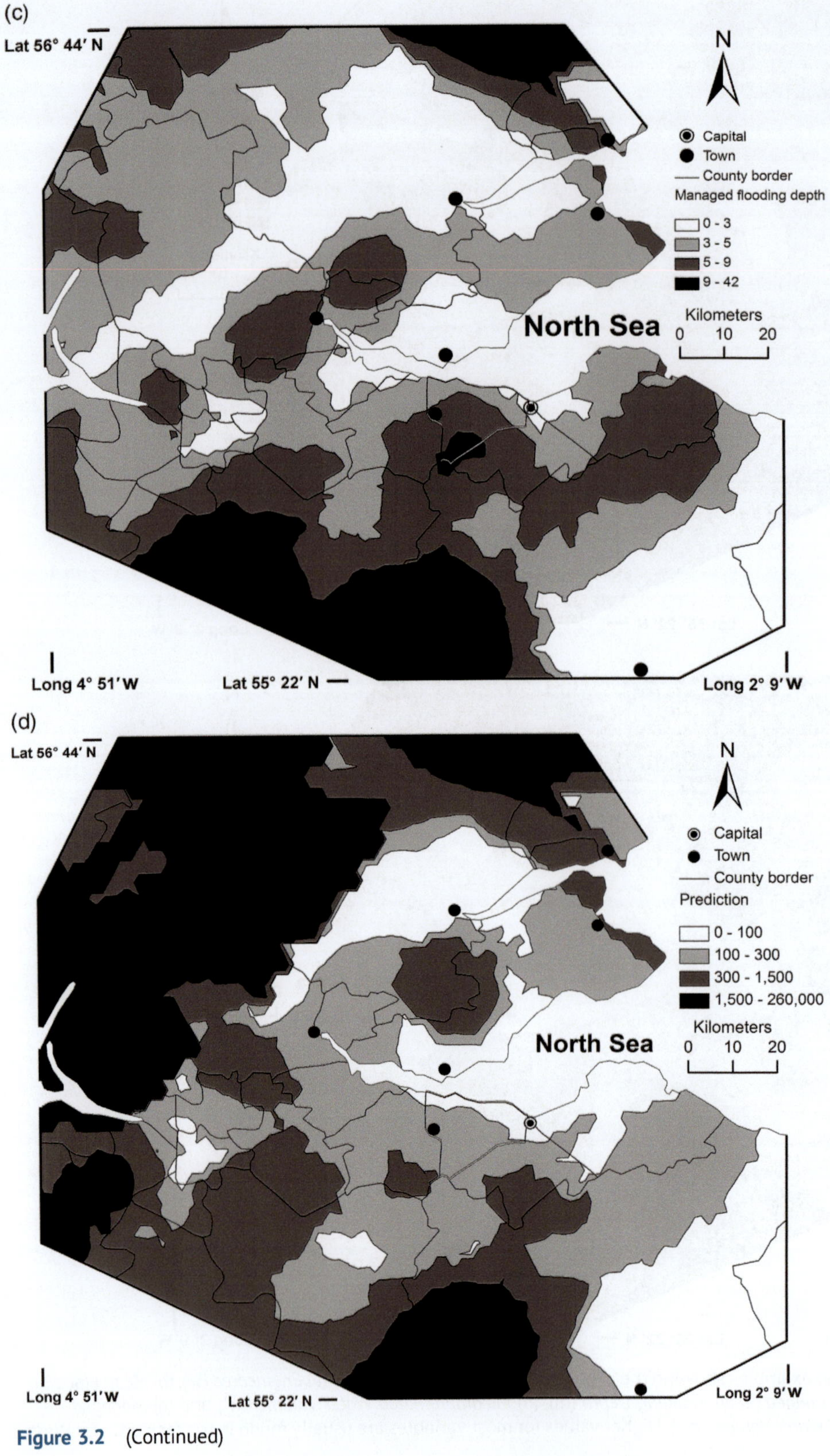

Figure 3.2 (Continued)

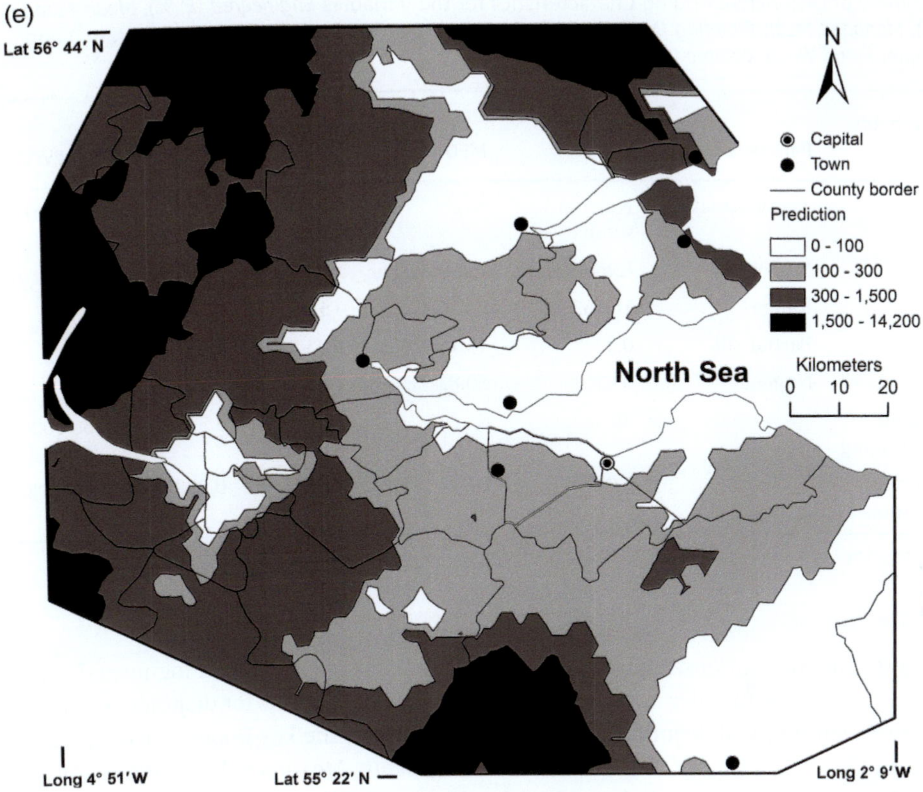

Figure 3.2 (Continued)

The south-east and north-central regions are dominated by shallow floods. The comparison indicates that the new variable *Managed Mean Flooding Depth* has much lower values than the old variable *Mean Flooding Depths*, because the permanent water contained in lakes has not been taken into account when calculating the former variable. The new variable is therefore a much better indicator for increased flood control potential.

Figure 3.2d, e shows the most likely values for the variables *Maximum Flood Water Volume* and *Managed Maximum Flood Water Volume*. These volume-based variables mirror the depth-based variables, indicating that higher depths relate to increased volumes, which is particularly the case for upland areas far away from major urban settlements.

Ordinary kriging has proved to be ideal for more than 90% of the SFRB ground surveys, because it results in maps which are easy to understand for practitioners (e.g., Figure 3.2). Maps appear to be smooth because most short-range noise was removed during kriging, which allows the flood risk manager to identify the key underlying patterns within complex data structures.

3.1.3.2 Findings Based on Disjunctive Kriging

The maps produced with ordinary kriging assist flood risk managers in identifying areas of low or high values for various SFRB variables. However, these maps should not be taken at face value since many are misleading to a greater or lesser extent. It is therefore necessary to use a technique such as disjunctive kriging, which provides estimates of the probability based on data given that the true

Table 3.4 Summary of Disjunctive Kriging Characteristics for the Variables *Engineered* (*E*, %), *Mean Flooding Depth* (MFD, m), *Managed Mean Flooding Depth* (MMFD, m), *Maximum Flood Water Volume* (MFWV, m^3) and *Managed Maximum Flood Water Volume* (MMFWV, m^3).

Variogram, thresholds and error	Parameter	*E*	MFD	MMFD	MFWV	MMFWV
Variogram	Transformation	Normal score	None	Normal score	Log	Normal score
	Model	Exponential	Exponential	Spherical	Spherical	Spherical
	Range	16 663	34 300	15 06	61 058	37 555
	Partial sill	0.32	15.16	0.41	0.79	0.22
	Nugget	0.61	50.83	0.48	4.36	0.74
Primary threshold for disjunctive kriging		30	3	3	350 000	350 000
Mean standardised error for estimation		0.0173	0.0422	0.0351	0.0988	−0.0098

Number of samples: 371.

values exceed a threshold at an unsampled location. It is necessary to perform a lognormal transformation for some variables in order not to violate underlying assumptions for disjunctive kriging.

Table 3.4 shows a summary of disjunctive kriging parameters for the key flood control variables *Engineered*, *Mean Flooding Depth*, *Managed Mean Flooding Depth*, *Maximum Flood Water Volume* and *Managed Maximum Flood Water Volume*. Based on data properties and expert judgement, the thresholds for the variable *Engineered* and any variables indicating flooding depth and flood water volume were set as 30%, 3 m and 350 000 m^3, respectively. Similar to ordinary kriging, the model that led to minimum mean standardised error was selected as the most suitable model fitting the variables.

Map examples showing the application of disjunctive kriging for the key variables are summarised in Figure 3.3a–e. Areas of low and high probabilities for the variable *Engineered* are relatively small and patchy (Figure 3.3a). The probability map shown in Figure 3.3a can be used in conjunction with Figure 3.2a and all maps indicating flooding depth and flood water volume to determine the areas of greatest investment potential, if flooding is likely to be a problem.

The maps showing probabilities of exceeding 3 m flooding depth associated with the variables *Mean Flooding Depth* and *Managed Mean Flooding Depth* should be used to estimate the likely return in flood infrastructure investment throughout the study area (Figure 3.3b, c). The higher the probability, the more likely it is that an existing or planned SFRB is making a positive impact on flood control. In contrast to the *Mean Flooding Depth*, the map for *Managed Mean Flooding Depth* indicates much lower probability values. Moreover, only those areas that are dominated by reservoirs, which could be used by Scottish Water (water utility of Scotland) and local authorities for flood control management, are shown. The greatest potential for active flood control is in areas situated to the south-west of the capital such as the Pentland Hills.

Figure 3.3d, e show that the areas with the greatest flood storage capacity are in upland catchments distant from populated lowland areas. The probabilities for the likely volumes that could be used for active flood control management by Scottish Water and local authorities are clearly shown in Figure 3.3e, because unmanageable storage volumes within natural water bodies have been

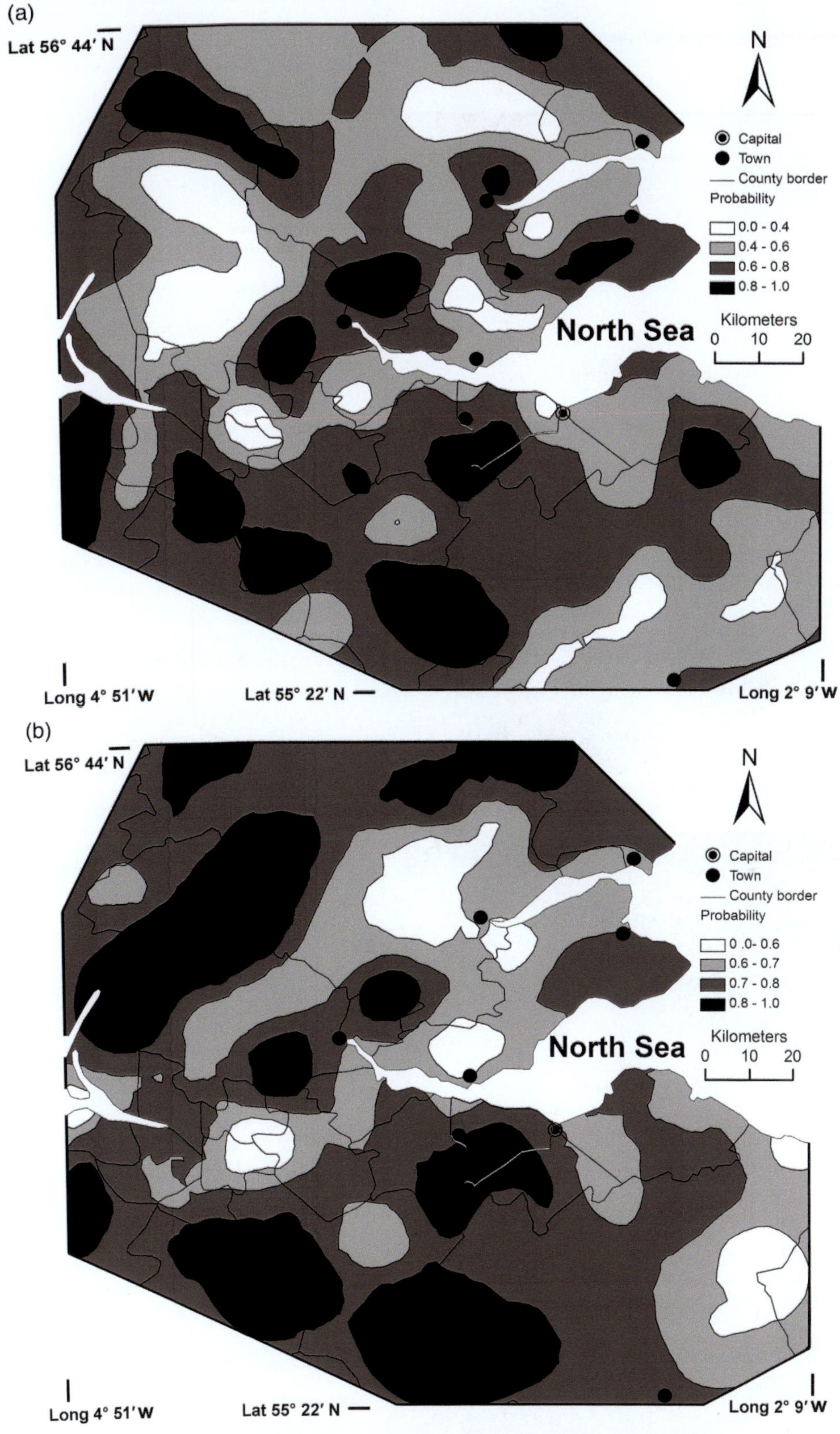

Figure 3.3 Map examples showing the application of disjunctive kriging for (a) *Engineered* (%; exceeding 30%), (b) *Mean Flooding Depth* (m, exceeding 3 m), (c) *Managed Mean Flooding Depth* (m, exceeding 3 m), (d) *Maximum Flood Water Volume* (m^3, exceeding 35×10^4 m^3) and (e) *Managed Maximum Flood Water Volume* (m^3, exceeding 35×10^4 m^3). No clear pattern emerges for the variables *Engineered*, *Mean Flooding Depth* and *Managed Mean Flooding Depth*. Higher values for *Maximum Flood Water Volume* and *Managed Maximum Flood Water Volume* are usually found in the far North, far West and far South.

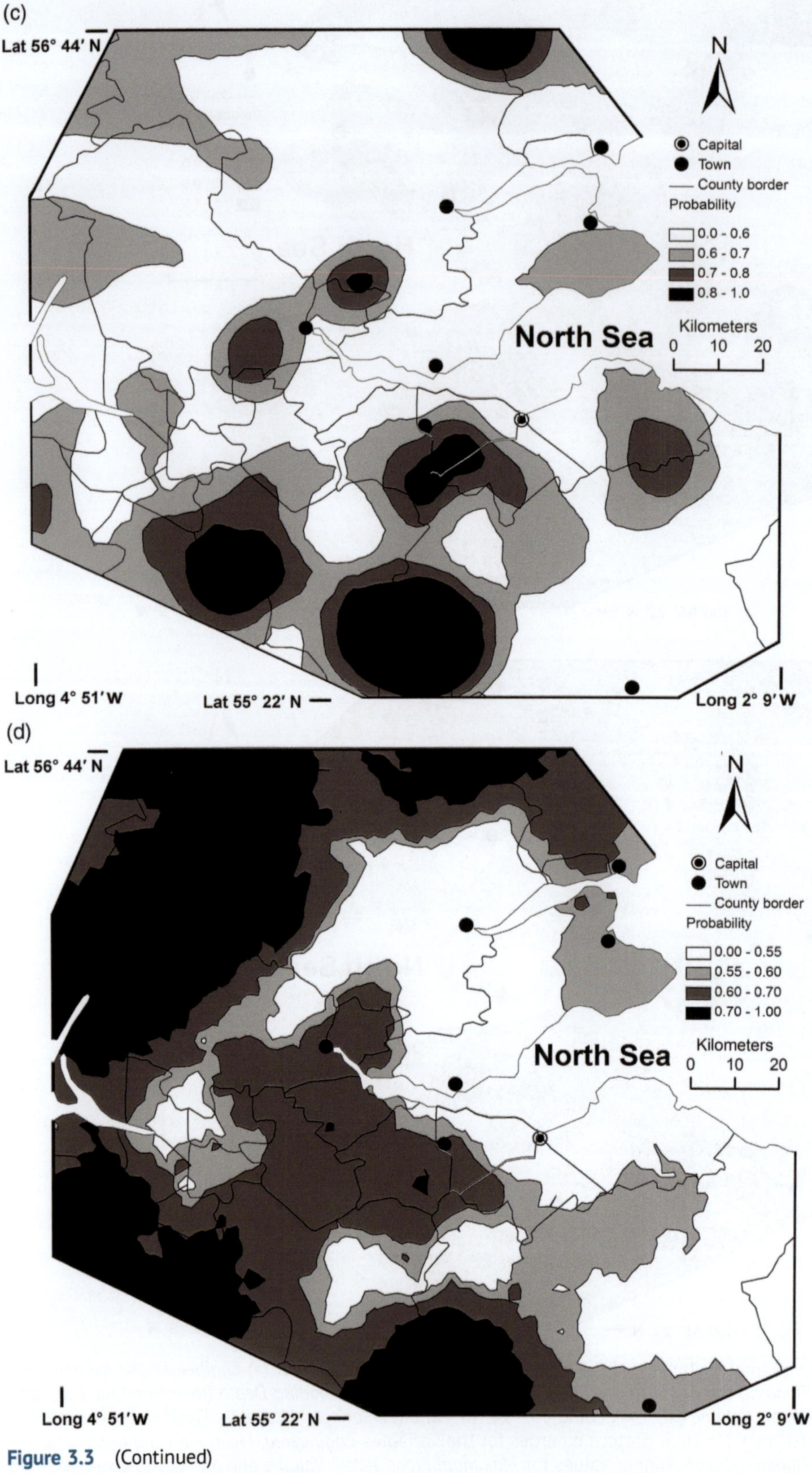

Figure 3.3 (Continued)

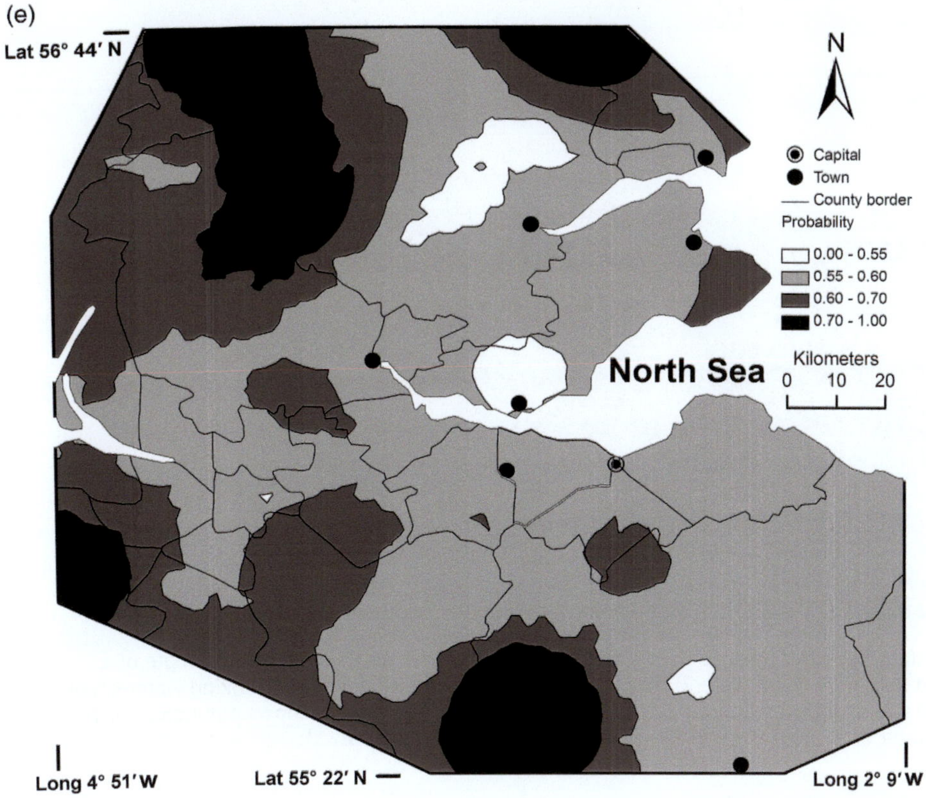

Figure 3.3 (Continued)

excluded from the probability map. A comparison of Figure 3.3a, e indicates that the north-west of the study area has the greatest potential for low-cost SFRB investment yielding a greater flood water storage volume return.

3.1.3.3 Consequences for Flood Risk Management

Traditional Flood Retention Basins (SFRB type 2) and Natural Flood Retention Wetlands (SFRB type 6) dominate the wider central Scotland area (McMinn et al. 2010). Figure 3.4 shows an example of a Traditional Flood Retention Basin that is, however, currently only used for drinking water supply purposes. There is great underutilised potential in using former and less important current potable supply reservoirs for flood control purposes. In comparison, Figure 3.5 is a representative example of a Natural Flood Retention Wetland, which is predominantly used for environmental protection and recreational purposes and has limited flood control potential.

The fieldwork program identified many former water supply reservoirs, which were predominantly identified as SFRB type 2. In the vast majority of cases, these structures now fulfil multiple roles providing opportunities for recreation, nature conservation and angling with many former drinking water reservoirs or industrial water supply structures being managed as fisheries.

A feature of these sites, based on the majority of current (SFRB type 1) and former (SFRB type 2; Figure 3.4) drinking water supply reservoirs surveyed, is that they are maintained at their maximum water retention volumes, and their corresponding spillways are continuously in operation. In this mode of operation, the extensive infrastructure is making very little contribution to water retention (i.e., flood control) in the upper catchments.

Figure 3.4 Glenfarg Reservoir (near to Rossie Ochill, County of Perth and Kinross), an example of a Traditional Flood Retention Basin (type 2) that is, however, currently used only for drinking water supply purposes (picture taken by Qinli Yang on 20 November 2008). The photograph shows a massive earth dam holding back a water body. *Source:* Qinli Yang taken on 20 November 2008.

Figure 3.5 Hare Myre (2.5 km South-east of Blairgowrie, County of Perth and Kinross), an example of a Natural Flood Retention Wetland (type 6) that is predominantly used for environmental protection, recreational and diffuse pollution control purposes (picture taken by Qinli Yang on 7 July 2009). A natural waterbody surrounded by wild vegetation can be seen on the photograph. *Source:* Qinli Yang taken on July 2009.

The proposed methodology could be used directly for planning purposes by classifying water bodies into SFRB types. For example, the SFRB concept could support the Water of Leith (located in the Pentland Hills) Flood Prevention Scheme to protect Edinburgh from flooding. A proper classification of the water bodies located within the Water of Leith catchment area that have flood control potential would clarify their individual planning status. Clarification of their current purpose (e.g., water supply, flood attenuation, recreation and environmental protection) would benefit communication between all stakeholders (e.g., local authorities, landowners and Scottish Water) involved with this case study to optimise their planning effort.

It follows that a change in management practice of reservoir-like structures by Scottish Water and local authorities could assist in sustainable flood risk management planning, leading to more sustainable reservoirs (SFRB type 3). Effectively, this would require some water to be released from the reservoirs prior to expected heavy precipitation. As the vast majority of former drinking water reservoirs have manual water level control, this would require on-site visits to manually release excess water, only closing the reservoir prior to a heavy rainfall event. This simple operation would enhance the reservoir capacity for water storage in the upper reach of the catchment and retard the peak flows from the upper catchment, which is likely to lead to a reduction of flooding downstream. Moreover, a geostatistical approach to river network management (attempted in France (Sauquet 2006) and Iran (Karamouz et al. 2009)) may also benefit flood risk management planning in Scotland.

3.1.4 Conclusions

The geostatistical analysis techniques of ordinary kriging and disjunctive kriging were successfully applied to SFRB management in the context of flood control. This novel approach provides efficient and rapid predictions relevant to flood management with respect to large areas comprising sites that have not been surveyed. The proposed geostatistical methodology will aid stakeholder communication by delivering information regarding the most favourable locations for SFRB development.

Traditional Flood Retention Basins consisting predominantly of former drinking water reservoirs are clearly a noticeable component of the Scottish landscape. These structures could be used for low-cost flood control purposes, if their water levels were actively controlled on a seasonal basis.

The key flood control variables are *Engineered*, *Mean Flooding Depth*, *Maximum Flood Water Volume*, *Managed Mean Flooding Depth* and *Managed Maximum Flood Water Volume*. The latter two compound variables are novel and essential in identifying the under-utilised flood control potential of former and current water supply reservoirs.

The spatial flood control assessment methodology ordinary kriging allowed for a clear interpretation of areas requiring further flood control investment. Disjunctive kriging was successfully used to assess the probability of individual variables to exceed specific management thresholds, which are relevant for flood control.

3.2 Nutrient Release from Integrated Constructed Wetland Sediment

3.2.1 Introduction

The integrated constructed wetland (ICW) concept was developed in the early 1990's to address water pollution and provide ecological and other wetland ecosystem services by reanimating wetland infrastructure (Harrington et al. 2005). The main functional features of ICW systems are shallow water depth, dense emergent vegetation, sequential flow through a series of wetland cells and the use of on-site materials and topography. The use of locally available soils is a key factor delivering adequate area at acceptable cost and thus there is a general absence of the use of synthetic (e.g., concrete or plastic) liners. The ICW utilises the same structures and biogeochemical processes that exist in natural wetlands to effectively treat wastewater (Scholz et al. 2007) and diffuse pollution (Harrington 2011). The ICW systems were originally designed and constructed to treat farmyard runoff and domestic wastewater within the Anne Valley in south County Waterford in Ireland.

Wetland sediments may retain or release accumulated contaminants over time and changing environmental conditions. The contaminant removal processes within wetland sediments can be abiotic (physical/chemical) or biotic (microbial/phytological). The primary abiotic processes include settling, sedimentation and sorption. Settling and sedimentation efficiently remove particulate matter and suspended solids (SS). Sorption, which includes the chemical processes of adsorption and absorption, can result in either short-term retention or long-term immobilisation of contaminants. Biotic processes such as biodegradation and plant uptake also influence contaminant removal. They can be affected by inflow quality, basin topography, hydrological features, amount of vegetation cover and land use management (Knox et al. 2006).

Wetland plants play a key role in treatment processes (Kadlec and Knight 1996). Nutrient removal by plants is predominantly achieved by two processes: uptake and microorganism activity around rhizome (Boon 1986). In the rhizosphere, interactions take place between contaminants and microorganisms. The removal efficiency can be enhanced through microbial growth and activities stimulated by the release of oxygen and organic compounds into the rhizosphere (Stottmeister et al. 2003). Furthermore, in the rhizosphere roots can increase the reactive areas and open wastewater passageways to counter clogging (Cooper 2005). In general, the key role of hydrophytes is to uptake pollutants, promote the microbial growth on media surfaces and assist the transport of oxygen through aggregates, subsequently increasing the treatment efficiency.

Although no diminution of overall performance has been found in the multi-cellular ICW systems over a period of 15 years (Harrington 2011), the long-term performance of individual cells and the cascading transfer of intercepted pollutants is less known. In general, there have been few studies conducted to evaluate the long-term treatment effectiveness of constructed wetlands. Although wetland aging through necro-mass accumulation is an understood challenge for their management, the loss of pollutant sequestration is seen as a perceived problem which may contribute to decrease in contaminant removal rates over time.

The specific objectives of this study are: (i) to assess the nutrient removal efficiency of ICW sediments accumulated in the first cell (most contaminated) of two studied ICW over nine years of full-scale operations; (ii) to compare the variation of two species of macrophytes – the helophyte common reed (*Phragmites australis*) and widely adaptive terrestrial common bent grass (*Agrostis stolonifera*) on nutrient reduction; and (iii) to evaluate exfiltration through the supporting soil structure that is commonly perceived as a potential risk of groundwater contamination. Dong et al. (2013) provided the scientific basis for Chapter 3.2.

3.2.2 Materials and Methods

3.2.2.1 Site and Experimental Set-up

The ICW systems (site 7 and site 11) studied in Chapter 3.2 was constructed as part of a series of 15 wetland systems to improve the water quality of an entire catchment area (Annestown-Dunhill watershed, Ireland). The site 7 (Figure 3.6) has a total area of 0.3 ha and was designed to treat local

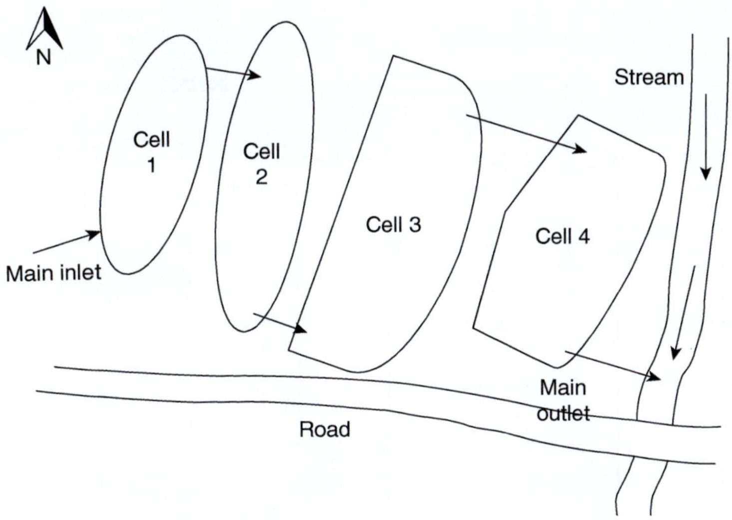

Figure 3.6 Sketch of the integrated constructed wetland site 7 layout. The drawing shows four wetland cells draining into a nearby river.

sewage. Domestic wastewater from nearby households is collected via a sewage system and then piped to the ICW system. The wastewater inflow and outflow are approximately 40 and 24 m^3 day^{-1}, respectively. In comparison, the site 11 (Figure 3.7) was constructed to treat farmyard runoff. The system has a total area of 0.76 ha. The influent comes from a dairy farm of 0.5 ha with 77 cows. Both wetland sites had four unlined cells in a row and the corresponding subsoil was reworked and used as a natural liner. The primary vegetation types planted were emergent species (helophytes).

Five mesocosms were set up in Edinburgh to stimulate contaminant removal and/or accumulation processes within mature sediments from full-scale horizontal surface flow ICW systems. Details of mesocosm design are illustrated in Figure 3.8. The laboratory room temperature was maintained at 15°C. Artificial light was provided by programmable controlled UV lights to simulate sunshine. The mesocosms were constructed using polyvinyl chloride drainage pipes with identical

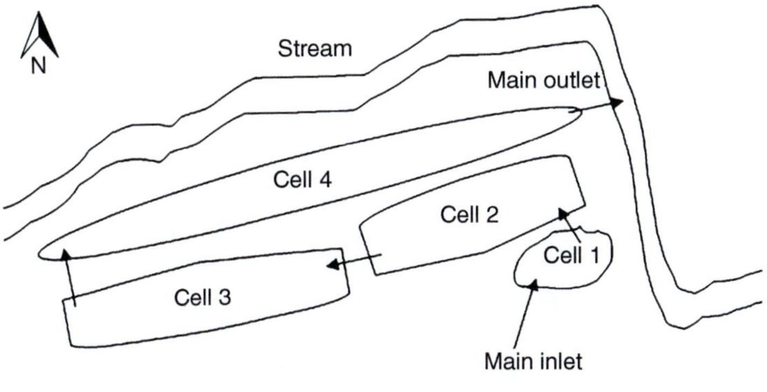

Figure 3.7 Sketch of the integrated constructed wetland site 11 layout. The drawing shows four wetland cells draining into a nearby river.

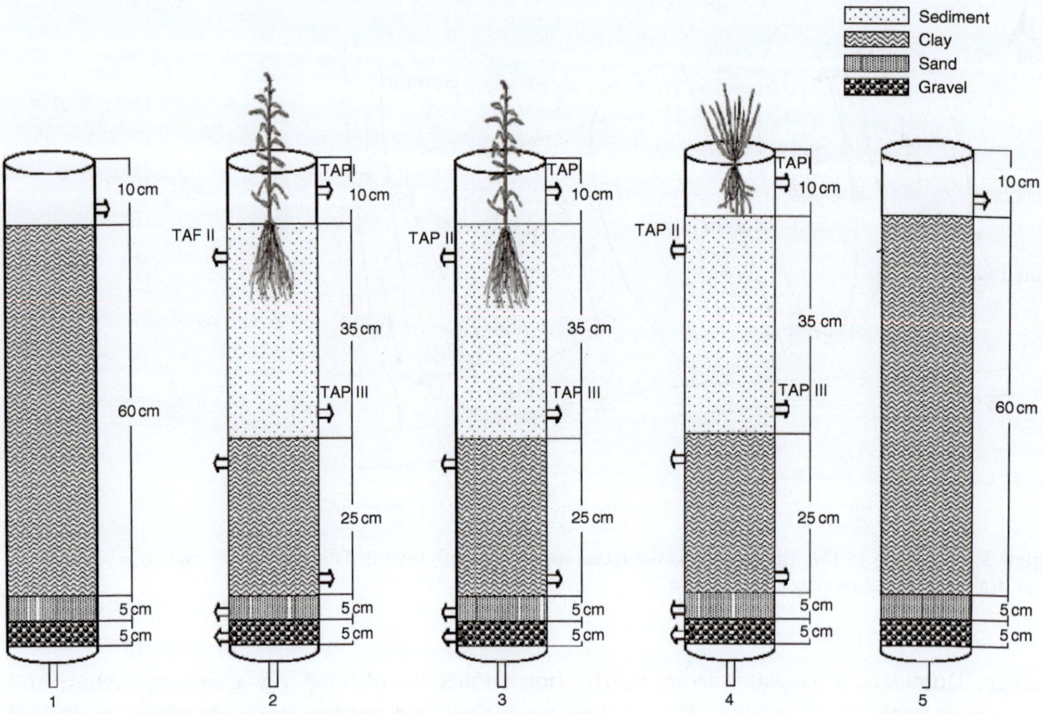

Figure 3.8 Schematic diagram of the integrated constructed wetland (ICW) mesocosms. The drawing shows five wetland columns of which columns 2–4 are planted.

dimensions (height: 83 cm; diameter: 10 cm). Seven plastic taps were evenly placed around the circumference of the pipe. The outlet valves were located at the centre of the bottom plate.

Of the five mesocosms, two were fed with lightly contaminated farmyard runoff (mesocosms 1 and 2), which was collected from an outdoor farmyard runoff drain at the Gorgie City Farm (Gorgie Road, Edinburgh). Three mesocosms (mesocosms 3, 4 and 5) were fed with pre-treated domestic wastewater taken from the Seafield Wastewater Treatment Plant (Marine Esplanade, Seafield Road, Edinburgh). The influents' water quality is summarised in Tables 3.5–3.7.

Mesocosms 2, 3 and 4 were packed with four successive layers of aggregates (from bottom to top): 50 mm of small gravel (1.2–5.0 mm), 50 mm of sand (0.6–1.2 mm), 250 mm of sodium bentonite clay (permeability of $10^{-}9$ m s^{-1}) and 350 mm of core sediment (incorporating plants if applicable). The arrangement of soil liner, type, depth and permeability was according to the full-scale ICW systems. Core sediment samples were taken directly from the surface of site 7 (cell 1) and site 11 (cell 1), respectively. These mesocosms were consistently flooded with inflow water up to the top (a depth of 100 mm). The hydraulic loading rate was approximately 9.14×10^{-3} m day^{-1}. Two control mesocosms were set up as benchmarks to ensure the same flow conditions and saturation ratios. The soils and aggregates (from bottom to top) contained 50 mm of small gavel (1.2–5.0 mm), 50 mm of sand (0.6–1.2 mm) and 600 mm of sodium bentonite clay (permeability of 10^{-9} m s^{-1}).

Mesocosms 2 and 3 were planted with *Phragmites australis* (Cav) Trin. ex Steud; whereas mesocosm 4 was planted with *Agrostis stolonifera* L. rhizomes. The control mesocosms were left unplanted. The *Phragmites australis* rhizomes were obtained from the outdoor wetland rig constructed for previous research studies at The King's Buildings, Edinburgh; *Agrostis stolonifera* rhizomes were obtained from site 7, Waterford.

Table 3.5 Water quality variables for mesocosms 1 and 2 (February 2009–October 2010).

								Mesocosm 2								
			Influent		Mesocosm 1 (control)			Tap II			Tap III					
Variable	Unit	*n*	Mean	SD	Mean	SD	Eff %	Mean	SD	Eff %	Mean	SD	Eff %	Mean	SD	Eff %
COD	mg L^{-1}	66	87.3	31.0	122.5	23.2	−40.3	98.1	28.1	−12.4	103.7	24.3	−18.8	137.4	41.3	−57.4
Ammonia-nitrogen	mg L^{-1}	61	1.2	1.3	0.2	0.1	83.3	3.1	2.3	−158.3	16.8	4.5	−1300.0	78.8	13.1	−6466.7
Nitrate-nitrogen	mg L^{-1}	57	0.9	1.5	21.2	4.8	−2255.6	0.4	0.4	55.6	0.7	0.7	22.2	1.1	0.9	−22.2
MRP	mg L^{-1}	61	0.9	0.5	3.6	1.3	−300.0	2.9	1.0	−222.2	3.8	1.0	−322.2	10.8	2.3	−1100.0
Chloride	mg L^{-1}	62	21.5	12.4	209.1	36.1	−872.6	56.1	15.1	−160.9	49.6	12.8	−130.7	156.4	74.0	−627.4
Suspended solids	mg L^{-1}	68	31	13	35	24	−12.9	59	45	−90.3	151	79	−387.1	44	42	−41.9
Conductivity	μS cm^{-1}	69	287	148	1312	236	–	509	141	–	691	137	–	2045	303	–
Redox	mV	67	−65	90	−93	96	–	−138	81	–	−190	120	–	−98	102	–
Water temperature	°C	70	18.3	2.4	16.9	1.7	–	17.3	1.6	–	17.6	2.0	–	17.0	1.3	–
pH	–	70	6.8	0.4	7.6	0.3	–	6.4	0.2	–	6.0	0.2	–	6.9	0.4	–

COD, chemical oxygen demand; Eff, removal efficiency; MRP, molybdate reactive phosphorus; *n*, sample number; SD, standard deviation.

Table 3.6 Water quality variables for mesocosms 3 and 5 (May 2009–October 2010).

								Mesocosm 3								
			Influent		Mesocosm 5 (control)			Tap I			Tap II			Tap III		
Variable	**Unit**	***n***	**Mean**	**SD**	**Mean**	**SD**	**Eff %**	**Mean**	**SD**	**Eff %**	**Mean**	**SD**	**Eff %**	**Mean**	**SD**	**Eff %**
COD	mg L^{-1}	64	70.3	44.4	82.4	39.3	−17.2	78.7	38.1	−11.9	119.0	48.0	−69.3	145.2	54.9	−106.5
Ammonia-nitrogen	mg L^{-1}	54	9.3	4.1	0.3	0.3	96.8	8.1	7.7	12.9	43.7	20.5	−369.9	42.6	29.5	−358.1
Nitrate-nitrogen	mg L^{-1}	37	0.3	0.1	7.2	5.8	−2300.0	0.6	0.3	−100.0	0.7	0.5	−133.3	1.0	0.7	−233.3
MRP	mg L^{-1}	57	1.2	0.7	3.5	1.1	−191.7	3.2	2.0	−166.7	3.8	1.8	−216.7	2.9	1.1	−141.7
Chloride	mg L^{-1}	48	108.7	46.2	419.0	81.0	−285.5	233.2	45.0	−114.5	222.0	46.7	−104.2	319.1	147.7	−193.6
Suspended solids	mg L^{-1}	62	34	16	18	16	47.1	11	10	67.6	12	15	64.7	25	20.2	26.5
Conductivity	μS cm^{-1}	62	617	121	2236	235	–	1056	179	–	1350	155	–	2273	620	–
Redox	mV	60	−78	87	−98	90	–	−180	69	–	−216	107	–	−170	104	–
Water temperature	°C	63	17.1	2.5	16.9	1.8	–	17.2	1.9	–	16.8	1.4	–	17.2	1.6	–
pH	–	63	6.7	0.3	7.5	0.2	–	6.7	0.2	–	6.6	0.2	–	7.6	0.7	–

COD, chemical oxygen demand; Eff, removal efficiency; MRP, molybdate reactive phosphorus; n, sample number; SD, standard deviation.

Table 3.7 Water quality variables for mesocosms 4 and 5 (May 2009–October 2010).

								Mesocosm 4								
			Influent		Mesocosm 5 (control)			Tap I			Tap II			Tap III		
Variable	**Unit**	***n***	**Mean**	**SD**	**Mean**	**SD**	**Eff %**	**Mean**	**SD**	**Eff %**	**Mean**	**SD**	**Eff %**	**Mean**	**SD**	**Eff %**
COD	$mg\ L^{-1}$	64	70.3	44.4	82.4	39.3	−17.2	81.0	32.7	−15.2	105.5	46.7	−50.1	117.3	44.7	−66.9
Ammonia-nitrogen	$mg\ L^{-1}$	54	9.3	4.1	0.3	0.3	96.8	12.6	10.2	−35.5	34.5	19.2	−271.0	43.6	13.2	−368.8
Nitrate-nitrogen	$mg\ L^{-1}$	37	0.3	0.1	7.2	5.8	−2300.0	0.5	0.3	−66.7	0.9	0.5	−200.0	1.2	0.7	−300.0
MRP	$mg\ L^{-1}$	57	1.2	0.7	3.5	1.1	−191.7	3.5	1.8	−191.7	4.4	2.4	−266.7	4.4	2.0	−266.7
Chloride	$mg\ L^{-1}$	48	108.7	46.2	419.0	81.0	−285.5	159.4	27.6	−46.6	165.0	30.2	−51.8	219.0	69.8	−101.5
Suspended solids	$mg\ L^{-1}$	62	34	16	18	16	47.1	23	19	32.4	8	10	76.5	17	18	50.0
Conductivity	$\mu S\ cm^{-1}$	62	617	121	2236	235	–	841	133	–	1056	152	–	1401	193	–
Redox	mV	60	−78	87	−98	90	–	−236	118	–	−224	113	–	−229	107	–
Water temperature	°C	63	17.1	2.5	16.9	1.8	–	17.3	1.6	–	17.5	2.0	–	16.8	1.3	–
pH	–	63	6.7	0.3	7.5	0.2	–	6.3	0.3	–	6.2	0.2	–	6.6	0.3	–

COD, chemical oxygen demand; Eff, removal efficiency; MRP, molybdate reactive phosphorus; n, sample number; SD, standard deviation.

3.2.2.2 Sampling and Analytical Methods

Water samples (up to 75 mL each) were collected weekly from top to bottom taps (Figure 3.8). Virtually no water infiltrated into deeper layers due to the presence of bentonite and biomass. Therefore, water samples from the remaining sampling points (below tap III) could not be collected.

Collected samples were analysed for pH, conductivity, total dissolved solids (TDS) and temperature by using the EC/TDS/pH/Temperature Meter. The redox potential was measured using the ORP measurement device. Dissolved oxygen (DO) was determined by the handheld Oxygen Meter. Samples of 50 mL were filtered by using glass fibre filter paper to determine SS. The chemical oxygen demand (COD) was determined by the Palintest COD standard measuring method. The above analyses were conducted at The University of Edinburgh, and all instruments were calibrated before testing. Meanwhile, frozen samples were shipped by overnight courier on dry ice (around −20°C) to Waterford County Council water laboratory where ammonia-nitrogen, nitrate-nitrogen, nitrite-nitrogen, total organic nitrogen, chloride (Cl) concentration and molybdate reactive phosphorus (MRP) were analysed using APHA (2005) unless mentioned otherwise.

All statistical analyses were carried out by using the Origin 8.0. A parametric analysis of variance was used to determine any significant ($p < 0.05$) differences in pollutant removal efficiencies. The univariate and multivariate statistical tests were performed to explore relationships between water quality parameters and pollutant concentrations. Variables from statistically significant univariate regressions were subsequently selected to be included into multiple regression models. To determine the contaminant loads, both oxidised and reduced sediment layers were simply assumed to be equal in height (i.e., 175 mm).

This experiment was only operated under tighter environment control that aimed to reduce unexplainable variability; therefore, seasonal variations cannot be assessed appropriately. It should also be emphasized that hydrological conditions were flexible in semi-natural and open systems. However, this variability was not captured in this study.

3.2.3 Results and Discussion

3.2.3.1 Comparison of Treatment Performances

Mesocosms 1 and 2 were assessed between February 2009 and October 2010. Whereas mesocosms 3, 4 and 5 were investigated between May 2009 and October 2010 (Tables 3.5–3.7). The first eight weeks for mesocosms 1 and 2 and first four weeks for mesocosms 3–5 were seen as the start-up periods, which are characterized by system instability; therefore, the data have not been included in Chapter 3.2.

Tables 3.5–3.7 summarise the variables measured at the inlet and outlets, and the corresponding contaminant removal efficiencies. The influent water quality varied considerably; probably due to highly variable farmyard runoff and raw domestic wastewater characteristics.

For mesocosm 2, generally a higher COD concentration was observed at sampling point III than at other sampling points. However, small differences appeared in COD concentrations among sampling points over the study period (Figure 3.9).

The COD concentrations in mesocosms 3 and 4 fluctuated initially due to system instability. In addition, it can be found that COD concentrations in the lower sediment layer (sampling point III) were higher than in the upper sediment (sampling point II) and water layers (sampling point I) as shown in Figure 3.10.

These results indicate that ICW mesocosms act as sources of COD rather than as sinks. The sediments released more organic matter in comparison to the amount of incoming organic matter that could be degraded. This finding agrees with other studies carried out in recent years. Hunt and

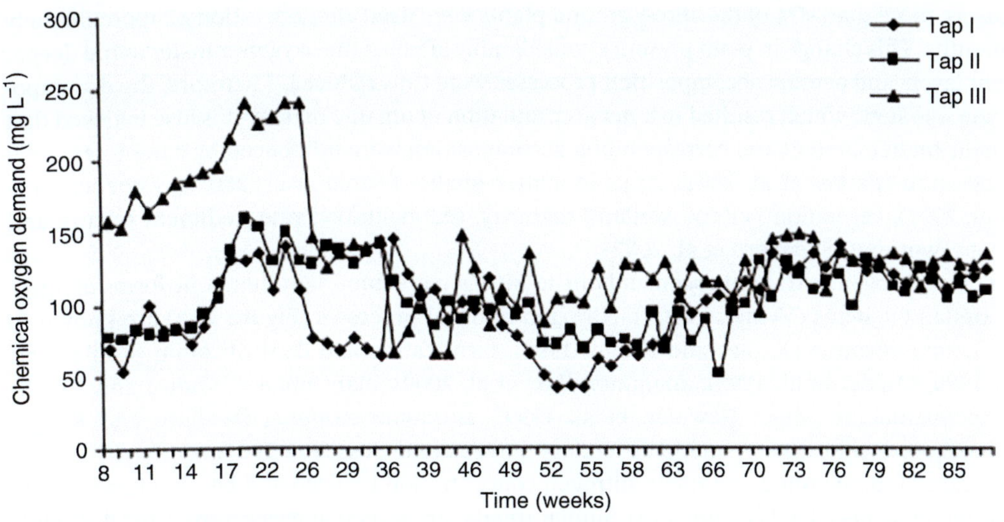

Figure 3.9 The concentrations of chemical oxygen demand (COD) for mesocosm 2. The COD usually increases with increasing tap number.

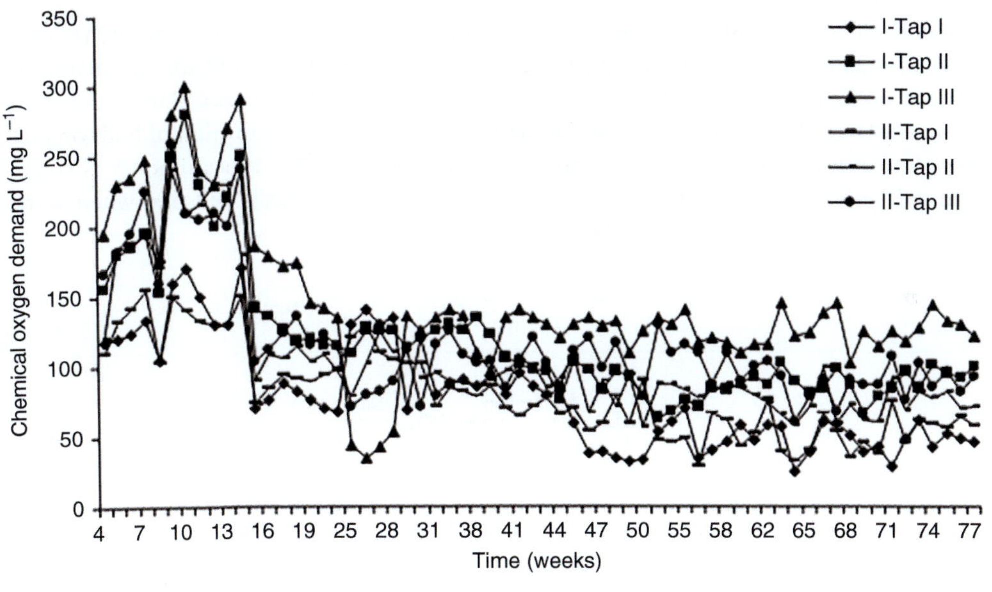

Figure 3.10 The concentrations of chemical oxygen demand (COD) for mesocosms 3 (I) and 4 (II). The COD commonly increases with increasing tap number.

Poach (2001) explain that constructed wetland systems cannot completely remove carbon and solid compounds because of the microbial and vegetative decay, which continuously adds carbon to the system.

The reduction of organic matter in constructed wetlands can be the result of aerobic heterotrophic bacteria (Vymazal 2007), and a considerable amount of oxygen is, therefore, required to decompose the organic pollutants. However, due to high concentrations of ammonia in mesocosms, more than 90% of the above-ground plants were dead after operation of approximately three months. This change in plant presence considerably affected the oxygen transfer to the deeper sediment layers and aerobic decomposition processes were thus reduced. Therefore, the decomposition rate was slow, which resulted in a net accumulation of organic matter. It is also believed that the organic matter content and corresponding decomposition were influenced by parameters such as temperature (Barber et al. 2001), organic matter quality (Turcq et al. 2002), residence time (Yu et al. 2002), vegetation pattern, wetland maturity, sedimentation rate, sediment texture and sediment reworking (Shepherd et al. 2007).

Ammonia-nitrogen, nitrate-nitrogen and nitrite-nitrogen are important inorganic forms of nitrogen in wetland systems (Vymazal 2007). Nitrogen removal processes (Figure 3.11) are known to involve ammonification (Kadlec and Knight 1996), nitrification and denitrification (Kadlec and Knight 1996; Mulder et al. 1995), anammox (Lee et al. 2009), plant uptake (Kadlec and Knight 1996), sedimentation (Majer Newman et al. 1999), ammonia stripping (Wallace and Knight 2006), breakpoint chlorination and ion exchange (Mitsch and Gosselink 2007).

In the case of mesocosm 2, ammonia-nitrogen concentration reduced and became less variable from sampling points III to I (Figure 3.12). Similar trends can be seen in mesocosms 3 and 4, regardless of the type of wetland plant (Figure 3.13).

The water samples from mesocosms 2 to 4 showed large variability in nitrate-nitrogen among sampling points. Nevertheless, an increase in nitrate-nitrogen at sampling point III was observed during most of the study period (Figures 3.14 and 3.15).

In general, the ammonia-nitrogen concentrations at all the sampling points were higher than those for the influent, in particular for tap III. This is probably because a large amount of ammonia-nitrogen accumulated at the sediment layer and was subsequently released during the study period. In a constructed wetland system, ammonia-nitrogen is subject to sorption of both organic and inorganic substrates through ion exchange. The ionized and adsorbed ammonia is subsequently bound loosely to the substrate such as sediment and can be easily released when water chemistry and other environmental factors change (Kadlec 2009).

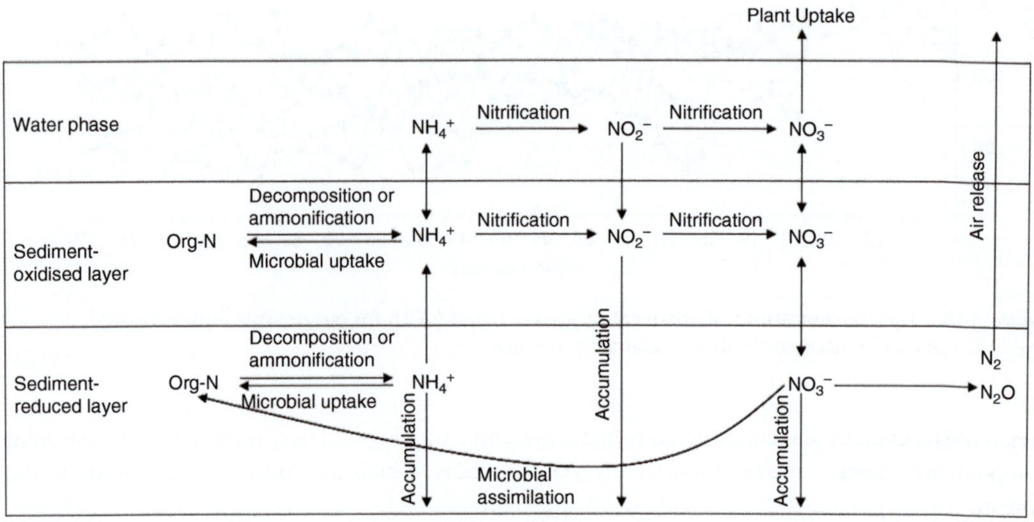

Figure 3.11 Nitrogen transformations in the water column and sediment layers. Processes such as plant update, air release, ammonification, nitrification, denitrification and accumulation of nitrogen compounds are illustrated in the drawing.

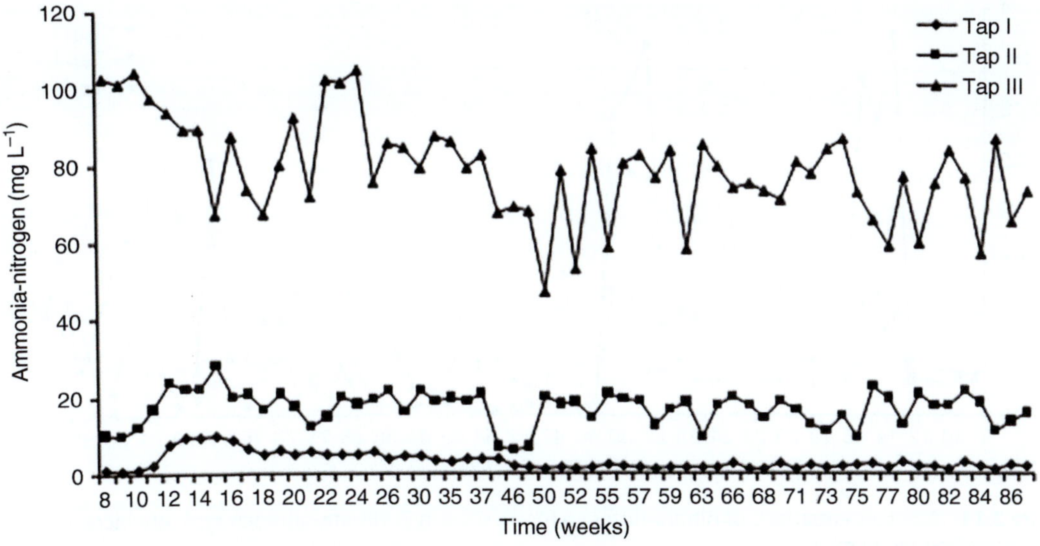

Figure 3.12 The concentrations of ammonia-nitrogen for mesocosm 2. Ammonia-nitrogen commonly increases with increasing tap number.

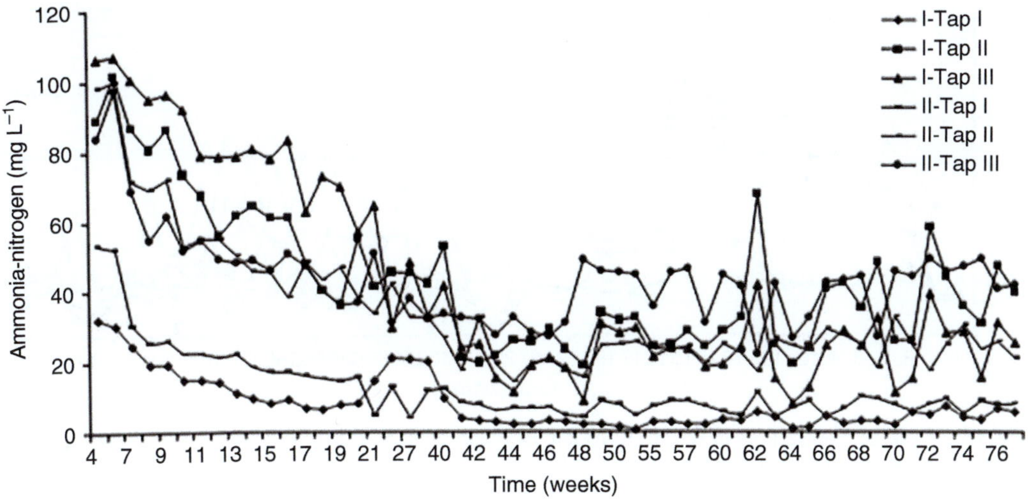

Figure 3.13 The concentrations of ammonia-nitrogen for mesocosms 3 (I) and 4 (II). Ammonia-nitrogen normally increases with increasing tap number.

Kinetic ammonification (mineralization) of organic nitrogen proceeds more rapidly than nitrification, thus creating the potential increase in ammonium concentration in the outlet (Kadlec and Knight 1996). Similar studies confirm these findings. For instance, Hammer and Knight (1994) reported that the nutrient removal efficiency, especially for ammonia-nitrogen, was low within most wetland systems. In addition, Reinhardt et al. (2006) assessed the nitrogen removal efficiencies in a small, constructed wetland located in Switzerland and revealed that the total nitrogen removal efficiency was 27%. In comparison, the ammonia removal rate was only −1%. On the other

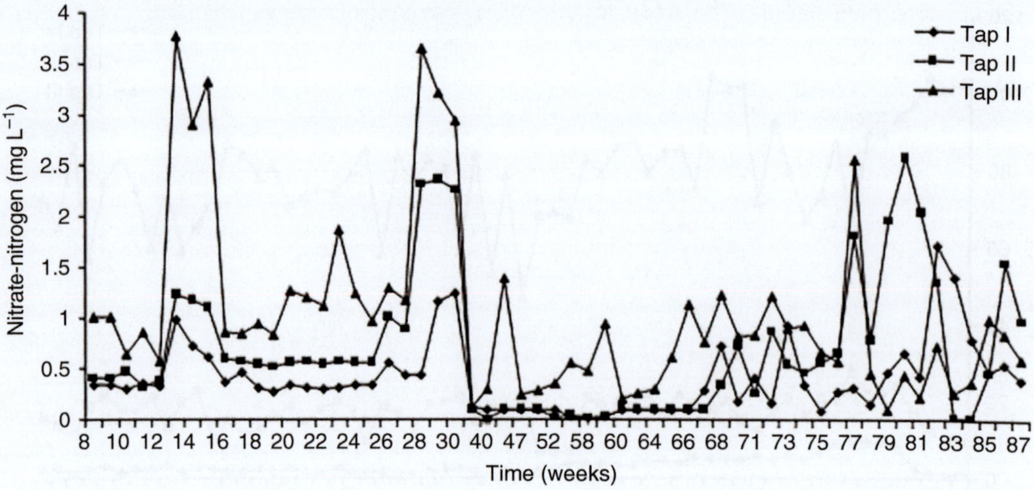

Figure 3.14 The concentrations of nitrate-nitrogen for mesocosm 2. Nitrate-nitrogen typically increases with increasing tap number.

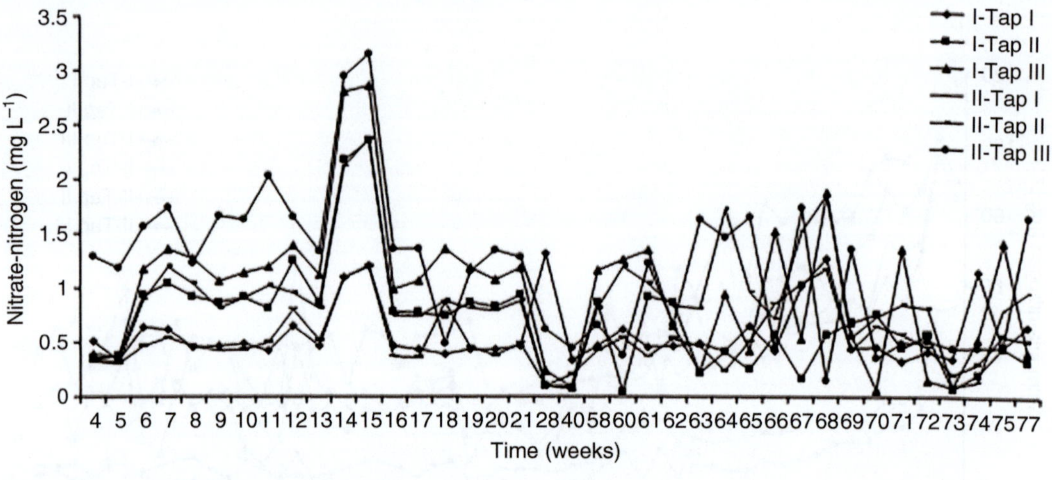

Figure 3.15 The concentrations of nitrate-nitrogen for mesocosms 3 (I) and 4 (II). Nitrate-nitrogen usually increases with increasing tap number.

hand, ammonia-nitrogen can be reduced by several processes, which include adsorption, plant uptake and volatilization. However, it is believed that the contribution of these processes is very limited (Lee et al. 2009).

The low nitrification rate could be caused by several factors including oxygen shortage, low pH values and low temperature. The nitrification process is very oxygen demanding. During the study period, plants did not grow well and resulted in a low rate of oxygen transformation to the sediment (DO < 4 mg L^{-1}). In addition, the pH value is also a key influencing factor. When the pH value drops to below 7, nitrification rates swiftly decline (Ahn 2006). As can be seen from Tables 3.5 to 3.7, all samples had low pH values (in particular tap II), which might adversely affect the nitrification process. However, pH values of the control systems were above 7. This difference in pH resulted in an enhancement of the nitrification processes in control systems where high concentrations of nitrate-nitrogen were present.

Moreover, temperature also plays a key role in the nitrification and denitrification processes of treatment wetlands (Langergraber 2008). Varying temperatures affect both microbial activity and oxygen diffusion rates in constructed wetlands. Vymazal (2007) reported that the optimum temperature range for nitrification is between 30°C and 40°C in soils. At low temperature, nitrification can be insufficient to prevent a net increase in ammonia concentration.

Chapanova et al. (2007) showed that ammonia transformation is sensitive to the wastewater salinity concentration. Ammonia conversion was significantly reduced after increasing the salinity of the influent. In addition, Dinçer and Kargi (1999) demonstrated that the salt content might reduce the performance of nitrification and denitrification if the salt content is above 2% and 1%, respectively. Of the studied mesocosms, the Cl concentrations of influent and effluent were high for all mesocosms (Tables 3.5–3.7). This might also explain the low nitrogen removal efficiency.

The mean concentrations of MRP of mesocosm 2 were 2.9 ± 1.0, 3.8 ± 1.0, and 10.8 ± 2.3 mg L^{-1} (Tables 3.5). The average effluent MRP concentrations of mesocosms 3 and 4 were as follows: 3.2 ± 2.0 and 3.5 ± 1.8 mg L^{-1} at tap I; 3.8 ± 1.8 and 4.4 ± 2.4 mg L^{-1} at tap II; 2.9 ± 1.1 and 4.4 ± 2.0 mg L^{-1} at tap III (Tables 3.6 and 3.7). For mesocosms 2, 3 and 4, the loads of MRP were higher at sampling points II and III than that at sampling point I (Figures 3.16 and 3.17).

These results indicate that sediments were saturated with MRP and were acting as a source for MRP. The alteration of environmental conditions triggered the phosphorus release process, resulting in a decrease of removal efficiency (Panswad and Anan 1999). Moortel et al. (2009) showed that surface flow and subsurface flow constructed wetlands initially remove phosphorus and later

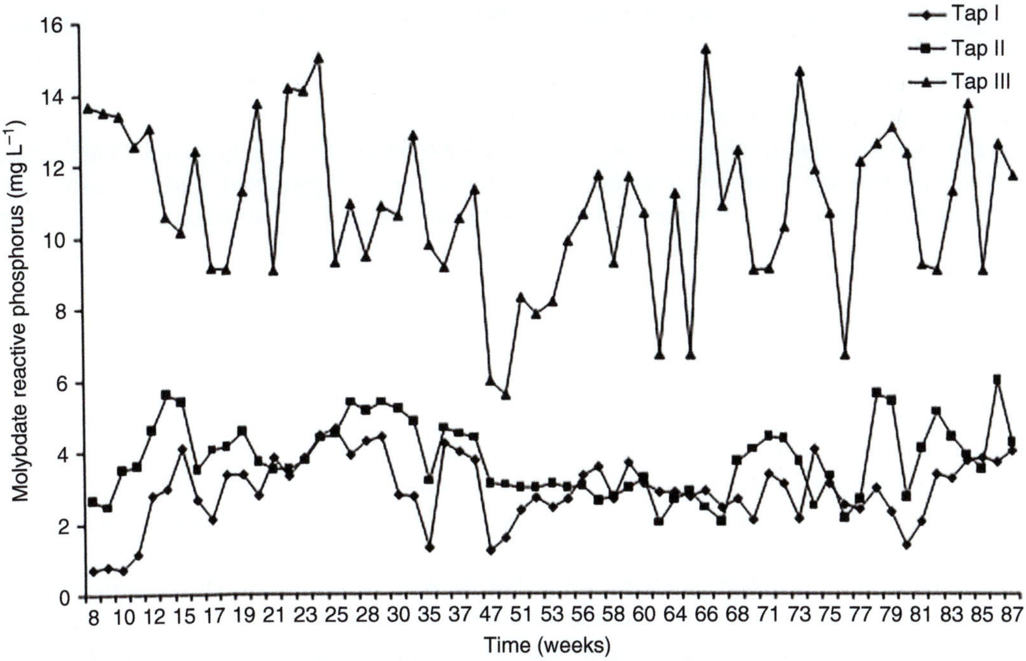

Figure 3.16 The concentrations of molybdate reactive phosphorus (MRP) for mesocosm 2. Nitrate-nitrogen typically increases with increasing tap number.

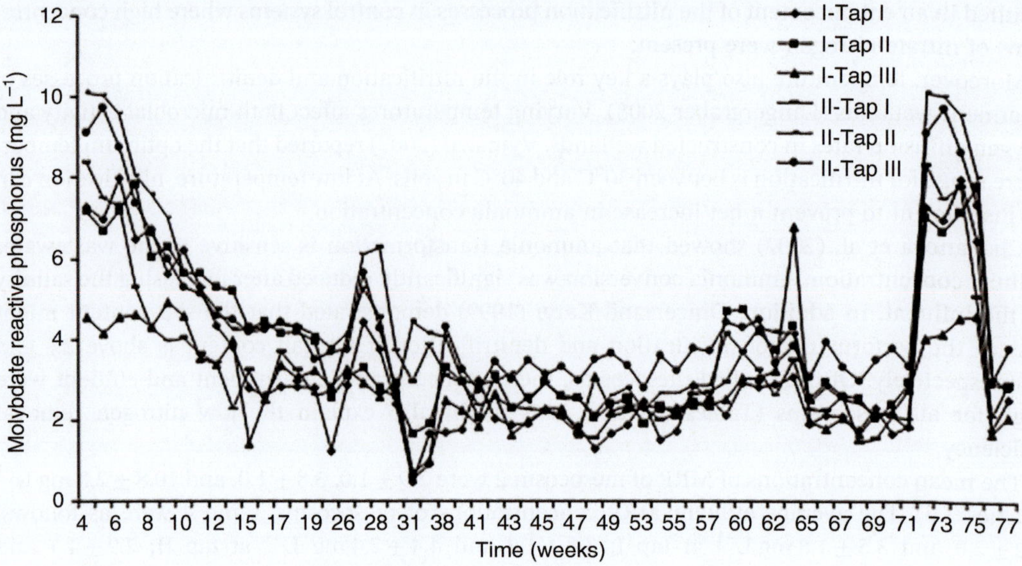

Figure 3.17 The concentrations of molybdate reactive phosphorus (MRP) for mesocosms 3 (I) and 4 (II). The MRP generally increases with increasing tap number.

release it again from the substrate. In addition, the carbon concentration may also influence the phosphorus removal efficiency either by blocking the adsorption sites or by competing for them with phosphates (Vohla et al. 2007).

A study by Mann and Bavor (1993) monitored the performance of a full-scale subsurface flow gravel wetland over two years and demonstrated that non-reactive sorbents can remove significant amounts of phosphorus, but that sites quickly become saturated. Similarly, Kayranli et al. (2010) reported that the MRP removal efficiency decreased during the research period due to blockage of the adsorption sites within wetland soils and sediments.

The phosphate-accumulating microorganisms are sensitive to high salinity (Scholz 2006). Zhang et al. (2008) evaluated the nutrient removal performance of constructed wetlands using self-organisation map models and reported that chloride and conductivity correlated well with MRP, indicating that increased salt concentrations had negative effects on MRP removal. The mean effluent chloride concentrations at all sampling points were higher than those of the corresponding influent concentrations. It is possible that the high chloride concentration supported the release of phosphorus from the mature sediments.

A simple regression analysis between contaminants and other water quality parameters were conducted for all ICW mesocosms. The pH, conductivity and Cl were significantly related to the COD (Table 3.8). The concentrations of ammonia-nitrogen correlate with conductivity and DO (Table 3.9), whereas MRP concentrations can be explained by the variations of redox and DO (Table 3.10). Fewer correlations exist between nitrate-nitrogen concentrations and water quality variables.

Based on simple regression analyses, the variables which were significantly correlated with contaminants are included in the multiple regression models (Table 3.11). Multiple regressions of COD concentrations for selected variables are highly significant. Conversely, the multiple regression models concerning ammonia-nitrogen for tap I of mesocosm 2, MRP for tap I of mesocosm 2 and tap I and tap II of mesocosm 4 are not statistically significant.

Table 3.8 Regressions estimates of chemical oxygen demand (COD) on water quality parameters.

	Mesocosm 2			Mesocosm 3			Mesocosm 4		
Variables	**Tap I**	**Tap II**	**Tap III**	**Tap I**	**Tap II**	**Tap III**	**Tap I**	**Tap II**	**Tap III**
Chemical oxygen demand (mg L^{-1}) versus pH	0.054 (0.059)	0.029 (0.169)	0.201 (<0.001)	0.224 (<0.001)	0.213 (<0.001)	0.170 (<0.001)	0.513 (<0.001)	0.148 (<0.001)	0.363 (<0.001)
Chemical oxygen demand (mg L^{-1}) versus conductivity (μS cm^{-1})	0.467 (<0.001)	0.029 (0.175)	0.236 (<0.001)	0.036 (0.138)	0.118 (<0.001)	0.100 (0.012)	0.324 (<0.001)	0.368 (<0.001)	0.007 (0.512)
Chemical oxygen demand (mg L^{-1}) versus chloride (mg L^{-1})	0.215 (<0.001)	0.101 (0.015)	0.210 (<0.001)	0.068 (0.048)	0.339 (<0.001)	0.142 (<0.001)	0.009 (0.487)	0.051 (0.090)	0.151 (<0.001)

p values (in brackets) > 0.05 indicate that a linear relationship between variables is not significant.

Table 3.9 Regressions estimates of ammonia-nitrogen on water quality parameters.

	Mesocosm 2			Mesocosm 3			Mesocosm 4		
Variables	**Tap I**	**Tap II**	**Tap III**	**Tap I**	**Tap II**	**Tap III**	**Tap I**	**Tap II**	**Tap III**
Ammonia-nitrogen (mg L^{-1}) versus conductivity (μS cm^{-1})	0.002 (0.723)	0.156 (<0.001)	0.219 (<0.001)	0.026 (0.252)	0.106 (0.018)	0.392 (<0.001)	0.106 (0.019)	0.404 (<0.001)	0.023 (0.280)
Ammonia-nitrogen (mg L^{-1}) versus dissolved oxygen (mg L^{-1})	0.025 (0.223)	0.005 (0.599)	<0.001 (0.953)	0.109 (0.015)	0.126 (<0.001)	0.119 (0.011)	0.025 (0.250)	0.144 (<0.001)	0.308 (<0.001)

p values (in brackets) > 0.05 indicate that a linear relationship between variables is not significant.

Table 3.10 Regressions estimates of molybdate reactive phosphorus on water quality parameters.

Variables	Mesocosm 2 Tap I	Mesocosm 2 Tap II	Mesocosm 2 Tap III	Mesocosm 3 Tap I	Mesocosm 3 Tap II	Mesocosm 3 Tap III	Mesocosm 4 Tap I	Mesocosm 4 Tap II	Mesocosm 4 Tap III
Molybdate reactive phosphorus (mg L^{-1}) versus Redox (mV)	0.060 (0.063)	0.004 (0.635)	0.195 (<0.001)	0.149 (<0.01)	0.008 (0.04)	0.148 (<0.001)	0.010 (0.477)	0.071 (0.055)	0.219 (<0.001)
Molybdate reactive phosphorus (mg L^{-1}) versus Dissolved Oxygen (mg L^{-1})	0.142 (<0.001)	0.048 (0.090)	0.064 (0.048)	0.003 (0.680)	0.102 (0.015)	0.116 (<0.001)	0.005 (0.619)	0.041 (0.133)	0.136 (<0.001)

p values (in brackets) > 0.05 indicate that a linear relationship between variables is not significant.

Table 3.11 Multiple regressions of chemical oxygen demand (COD), ammonia-nitrogen and molybdate reactive phosphorous (MRP) on water quality parameters.

Dependent variable		Regression	*n*	Adjusted R^2	*F*	*p*
Chemical oxygen demand						
Mesocosm 2	Tap I	COD = −36.021 + 9.187pH + 0.117 Conductivity + 0.275Chloride	58	0.452	16.653	<0.001[a]
	Tap II	COD = −77.633 + 24.268pH + 0 .012 Conductivity + 0.581Chloride	58	0.092	2.914	0.042[a]
	Tap III	COD = 316.857 − 6.376pH − 0.050 Conductivity − 0.194Chloride	57	0.321	9.831	<0.001[a]
Mesocosm 3	Tap I	COD = 661.544 − 86.532pH − 0.045Conductivity − 0.232 Chloride	55	0.257	7.230	<0.001[a]
	Tap II	COD = 677.437 − 77.531pH − 0.054Conductivity − 0.558Chloride	55	0.530	21.257	<0.001[a]
	Tap III	COD = 283.629 − 11.005pH − 0.012Conductivity − 0.085Chloride	55	0.132	3.733	0.017[a]
Mesocosm 4	Tap I	COD = 369.434 − 54.678pH − 0.089Conductivity − 0.110Chloride	55	0.584	26.296	<0.001[a]
	Tap II	COD = 289.258 − 45.716pH + 0.185Conductivity − 0.579Chloride	54	0.516	19.855	<0.001[a]
	Tap III	COD = 695.976 − 99.861pH − 0.079Conductivity − 0.147Chloride		0.426	14.332	<0.001[a]

Ammonia-nitrogen						
Mesocosm 2	Tap I	NH_3-N = 1.215 + 0.001Conductivity + 0.583Dissolved Oxygen	59	−0.004	0.874	0.423
	Tap II	NH_3-N = 6.898 + 0.013Conductivity + 0.330Dissolved Oxygen	59	0.130	5.322	0.008[a]
	Tap III	NH_3-N = 126.223 − 0.020Conductivity − 1.708Dissolved Oxygen	58	0.201	8.149	<0.001[a]
Mesocosm 3	Tap I	NH_3-N = −5.720 + 0.005Conductivity + 3.698Dissolved Oxygen	52	0.104	3.973	0.025[a]
	Tap II	NH_3-N = 22.042 + 0.043Conductivity − 14.767Dissolved Oxygen	52	0.186	6.842	0.002[a]
	Tap III	NH_3-N = 140.189 − 0.028Conductivity − 10.834Dissolved Oxygen	52	0.480	24.534	<0.001[a]
Mesocosm 4	Tap I	NH_3-N = −1.368 + 0.022Conductivity − 3.262Dissolved Oxygen	52	0.104	3.961	0.025[a]
	Tap II	NH_3-N = −23.939 + 0.070Conductivity − 7.368Dissolved Oxygen	51	0.430	19.850	<0.001[a]
	Tap III	NH_3-N = 67.917 − 0.001Conductivity − 8.122Dissolved Oxygen	52	0.266	10.238	<0.001[a]
Molybdate reactive phosphorus						
Mesocosm 2	Tap I	MRP = 3.617 − 0.002Redox − 0.496 Conductivity	58	0.144	5.799	0.005[a]
	Tap II	MRP = 3.253 + 0.001Redox + 0.264 Conductivity	58	0.020	1.574	0.217
	Tap III	MRP = 11.887 + 0.010Redox − 0.062Conductivity	58	0.166	6.683	0.003[a]
Mesocosm 3	Tap I	MRP = 4.600 + 0.014Redox + 0.618 Conductivity	53	0.148	5.526	0.007[a]
	Tap II	MRP = 7.124 + 0.006Redox − 0.808 Conductivity	53	0.147	5.476	0.007[a]
	Tap III	MRP = 3.975 + 0.003Redox − 0.142 Conductivity	53	0.124	4.689	0.014[a]
Mesocosm 4	Tap I	MRP = 3.231 + 0.002Redox + 0.400 Conductivity	53	−0.018	0.541	0.585
	Tap II	MRP = 7.439 + 0.005Redox − 0.821 Conductivity	52	0.048	2.290	0.112
	Tap III	MRP = 7.214 + 0.008Redox − 0.335Conductivity	53	0.198	7.420	0.002[a]

[a] Indicates a significant relationship between the variables in the regression and the $p \leq 0.05$ level.

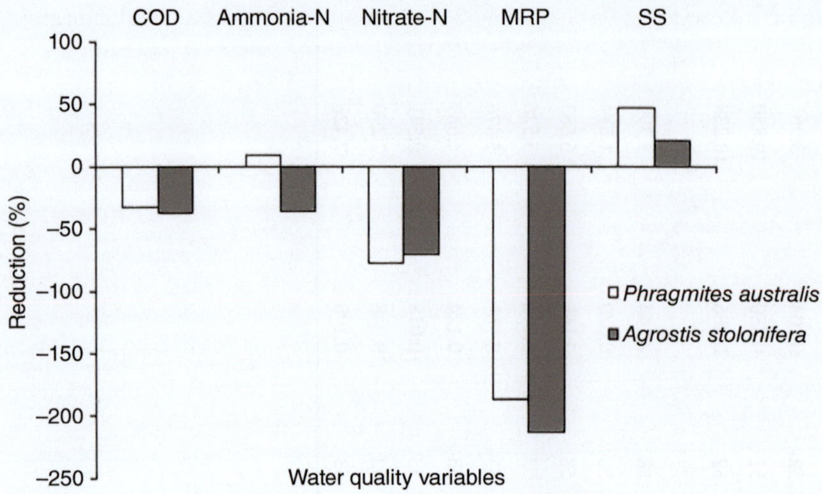

Figure 3.18 Comparison of contaminant removal efficiencies of mesocosms 3 (*Phragmites australis*) and 4 (*Agrostis stolonifera*). COD, chemical oxygen demand; Ammonia-N, ammonia-nitrogen; Nitrate-N, nitrate-nitrogen; MRP, molybdate reactive phosphorus; SS, suspended solids. Reductions were only noted for SS (both plants) and ammonia-N (*Phragmites* only).

3.2.3.2 Vegetation

There is a link between contaminant reductions and wetland vegetation. Concerning the two mesocosms treating domestic wastewater, mesocosm 3 obtained higher reduction removal efficiencies of COD, ammonia-nitrogen, MRP and SS compared to those of mesocosm 4 (Figure 3.18). This suggests that the presence of *Agrostis stolonifera*, a typical floating vegetation, limited the diffusion of oxygen (as electron reception) to the lower sediment layers. Relatively low oxygen concentrations can be identified as the main contributor to the inadequate removal of COD and ammonia-nitrogen.

3.2.3.3 Groundwater Contamination

The ICW was constructed using in situ soils. As the contaminated influent passes through the system, the suspended matter settles on the soil surface and subsequently slows the infiltration of contaminants through the wetland cells (Mustafa et al. 2009; Scholz 2006). An outlet valve was located on the base plate of each experimental mesocosm to gauge the possibility of groundwater contamination by infiltration of the polluted water. However, no leachate was collected throughout the study period. This can be explained by the presence of a compact bentonite clay layer, acting as an excellent barrier to prevent pollutants transferring to potential aquifers. Furthermore, biogeochemical processes are taking place within sediments. Some of these play an important role in, for instance, the clogging of the soil matrix, biomass accumulation and/or biogas (i.e., methane) formation through soil microbes (Tokida et al. 2005).

3.2.3.4 Sediment Management

Sediments serve as a reservoir of contaminants within a wetland system, and they were identified as a potential source for contaminants leaching after long-term operation (Kang et al. 1998). Previous sediment accumulation research indicated that the approximate sludge removal frequency of cell 1 of wetland site 11 would be nine years. The frequency is a function of the overall pond capacity. However, from the contaminant removal perspective, ICW cells might require desludging every five to six years. For some heavily loaded systems, more frequent desludging is necessary.

Sediments contain elements required for plant growth and can be put to beneficial use. The most appropriate solution would be land spreading on a farm in accordance with good farm management practice. Scholz et al. (2007) calculated that the phosphorus content of accumulated sediment would satisfy between 6% and 12% of the agricultural annual requirement at a relatively low fertilization rate of 10 kg ha^{-1} phosphorus, and between 4% and 7% of the agricultural annual requirement, if the fertilization rate increases to 15 kg ha^{-1} phosphorus.

Previous research indicated that treatment wetland sediment can be significantly contaminated with metals such as Cu, Zn and Cd (Domingo et al. 2006; Lesage et al. 2007). Therefore, the land-spreading of accumulated sediment requires good agricultural practice. Water UK and British Retail Consortium (BRC) have developed the Safe Sludge Matrix to guide the application of sewage sludge to agricultural land. Wetland managers and/or farmers should comply with those regulations to ensure land and environmental protection.

3.2.4 Conclusions and Recommendations

The study demonstrates the importance of mature wetland sediments (from the first cell of the ICW system, which receives the most contaminated pollutants) and plants on the nutrients removal performance. Five identical ICW mesocosms were constructed to duplicate full-scale ICW operations.

The contaminant removal performance of the sediment was poor, and the mesocosms acted as nutrient sources rather than nutrient sinks. Due to many years of operation and high nutrient loads entering the full-scale ICW systems, nutrients were trapped and accumulated at the studied sediment and remobilized when environmental conditions changed. In addition, the restricted oxygen transfer through vegetation adversely affected the removal efficiency of COD and ammonia-nitrogen. The univariate and multiple regression analyses provide useful insights into potential relations between pollutants removal and water quality variables.

The mesocosm planted with *Agrostis stolonifera* (floating vegetation) has less capacity to diffuse oxygen (as electron reception) to lower sediment layers as compared to a mesocosm planted with *Phragmites australis*. This resulted in an inadequate level of contaminant removal.

The compact bentonite layer was an excellent barrier to prevent contaminant transfer to potential aquifers. Furthermore, the biogeochemical processes taking place within sediments play a key role in clogging soil matrix and in providing impedance to infiltration.

The absence of a synthetic liner (e.g., concrete or plastic) makes ICW technology affordable. However, appropriate sediment management (in particular, the first cell) is paramount to maintain treatment performance and protect receiving watercourses though very little evidence of reduced ICW efficacy has been collected.

There is still limited understanding of key biotic and abiotic variables, their interactions at temporal and spatial scales and their effects on nutrient removal or accumulation within sediments. Chapter 3.2 suggests the following key areas for future research regarding wetland sediments: first, the nutrient removal capacity within sediments and the corresponding responsible parameters need to be studied at both pilot-scale and full-scale ICW under natural environmental conditions and second, the role of wetland plant species on nutrient removal should be fully assessed.

3.3 Groundwater Quality Impacts from an Integrated Constructed Wetland

3.3.1 Introduction

The concept of integrated constructed wetlands (ICW) has been developed to achieve water treatment, landscape fit and biodiversity enhancement targets through an innovative wetland design methodology (Scholz et al. 2007). Characterised by a series of interconnected free water surface flow constructed wetland cells, which incorporate the concept of restoration ecology (Jordan et al. 1987), ICW systems specifically mimic the structure and processes of natural wetlands (Scholz et al. 2007). In addition, ICW systems have shallow water depths (10–30 cm) and contain many plant species. This facilitates microbial and animal diversity (Jurado et al. 2010; Nygaard and Ejrnæs 2009), and generally, is aesthetically appealing, which enhances recreation and amenity values.

The ICW systems are traditionally built by utilising resources from the building site so that they will not require the installation of expensive artificial liners such as plastic or concrete to impede wastewater infiltration, and the consequent contaminants loading into the nearby groundwater system. The local on-site subsoil materials are reworked and compacted to form a low permeability liner for the ICW cells. Typically, ICW cells are underlain by at least 1.0 m of subsoil, with the upper 0.5 m enhanced where highly permeable subsoil is encountered, to hydraulic conductivity of 1×10^{-8} m s^{-1} (Carty et al. 2008; Dunne et al. 2005a, b; Harrington et al. 2007; Scholz et al. 2007).

Because the soil is a dynamic system that has been known to transmit water contaminants through its pore space, the question of wastewater infiltration from ICW cells as a potential source of groundwater pollution has often been raised. Particularly in Ireland, concerns are raised regarding the movement of nitrogen compounds into local drinking water supplies, where approximately 26% of all the drinking water supply is provided by groundwater (EPA 2008). Ultimately, groundwater flow can transport the nutrient-rich infiltration water to surrounding and nearby surface waters, where water quality degradation may occur (Hathaway et al. 2010).

In a previous study, Dzakpasu et al. (2012) noted infiltration rates of 4.3×10^{-9} m s^{-1}, 3.7×10^{-9} m s^{-1} and 1.0×10^{-8} m s^{-1}, respectively, through the cell liners of the first three ICW cells during the first 18 months of operation. Infiltration water originating from the ICW cells may potentially contain relatively high concentrations of organic carbon compounds, nutrients (mostly nitrogen and phosphorus) and pathogens. Such inputs to the groundwater may cause pollution or pose a risk to public health. Whilst excess organic carbon compounds could impact the groundwater quality negatively, its presence in groundwater, nonetheless, plays important roles in controlling geochemical processes by acting as proton donors/acceptors and as pH buffers, by affecting the transport and degradation of pollutants, and by participating in mineral dissolution/precipitation reactions. Dissolved and particulate organic carbon compounds may also influence the availability of nutrients and serve as a carbon substrate for microbially mediated reactions (Aiken 2002). Nitrogen compounds, on the other hand, may be microbially oxidised to form nitrate (NO_3-N), excessive amounts of which are of a great concern to public health (WHO 2008).

Nitrate often occurs naturally in groundwater, typically with a baseline not exceeding 2 mg L^{-1} (Wakida and Lerner 2005). Therefore, groundwater wells with NO_3-N concentrations of 3–10 mg L^{-1} may indicate input from external sources (Power and Schepers 1989); excessive concentrations mostly being associated with water quality degradation (Canter and Knox 1985; Koh et al. 2007). Furthermore, although typically non-toxic, phosphorus (P) compounds can considerably decrease the quality of water resource systems. Ortho-P and NO_3-N are often the limiting nutrients that could stimulate the growth of aquatic plants, most notably algae, leading to eutrophication problems and the consequent loss of recreational and commercial values of water systems (Akpor and Muchie 2011; Cloern 2001; Khan and Ansari 2005).

Several published studies such as Dunne et al. (2005a, b), Mustafa et al. (2009), Kayranli (2010), Dong et al. (2011) and Dzakpasu et al. (2011) have indicated that ICW systems can provide an effective method of wastewater treatment for both point and diffuse sources. Nevertheless, the question is whether the use of local soil materials to line ICW cells is capable of providing effective protection to the underlying and associated groundwater. Furthermore, limited studies have been carried out concerning the potential infiltration of water contaminants from full-scale constructed wetlands into groundwater systems.

Chapter 3.3 assesses the groundwater quality near a full-scale ICW system treating domestic wastewater. The specific objectives are (i) to evaluate the temporal and spatial extent of groundwater contamination near the ICW system; and (ii) to identify the possible sources of contamination in the groundwater and assess the contributions thereof, of contaminant sources on the groundwater quality parameters by using chemometrics models. Chemometrics is the field of extracting information from multivariate chemical data using mathematical and statistical methods. Dzakpasu et al. (2014, 2015) provided the scientific basis for Chapter 3.2.

3.3.2 Materials and Methods

3.3.2.1 Study Site Description

The ICW treatment system at the centre of the study is located within the walls of the Castle Leslie Estate at Glaslough in County Monaghan, Ireland (06°53′37.94″ W, 54°19′6.01″ N). The site is surrounded by woodland and, therefore, required sensitive development in terms of landscape fit, and biodiversity, amenity and habitat enhancement.

The ICW (Figure 3.19) comprises a small pumping station, two sedimentation ponds and a sequence of five shallow and vegetated wetland cells. Hydraulic characteristics of the wetland cells are presented in Table 3.12. The ICW system was commissioned in October 2007 to treat sewage from Glaslough village and to improve the water quality of the Mountain Water River, which flows through the site. The Mountain Water River is within the catchment area of the Blackwater River in Monahan.

The design capacity of the ICW system is 1750 p.e. (105 kg BOD d^{-1}). The functional water area of the ICW cells is 3.25 ha (HLR of 0.54 cm d^{-1}) within a curtilage area of 6.74 ha. The wetland cells have no artificial lining. Excavated local soil material was used to construct the base of the wetland cells and was compacted to a thickness of 500 mm to form a low permeability liner. A site investigation conducted in September 2005 indicated a soil coefficient of permeability of 9×10^{-11} m s^{-1} (IGSL Ltd., Business Park, Naas, Co. Kildare, Ireland). The soils at the study area comprise a mixture of coarse- and fine-grained materials, namely alluvium, organic soils, tills and gravel. Subsoil samples collected from the first 2 m near the sedimentation ponds were classed as sandy gravely clay; whereas those collected near cells 1 and 2 were classed as sandy silt and silty clay, respectively (IGSL Ltd., Business Park, Naas, Co. Kildare, Ireland). Typically, the mineral content of the soils is in the following range: Al^{3+}, 4.1–5.0%; Fe^{2+}, 2.01–2.5%; Ca^{2+}, 0.301–0.45%; Na^{+}, 0.501–0.65%; K^{+}, 0.751–1.00%; Mg^{2+}, 0.41–0.5%; Organic C, 6.1–10% and pH, 5.01–5.5 (Fay et al. 2007).The main ICW system is flanked by the Mountain Water River and the Glaslough Stream (Figure 3.19).

Influent primary domestic wastewater from the village is pumped directly into a receiving sedimentation pond. The system contains two sedimentation ponds that can be used alternately so that one can be desludged without interrupting the whole treatment process. The purpose of the sedimentation ponds is to retain the suspended solids contained in the influent wastewater. In this

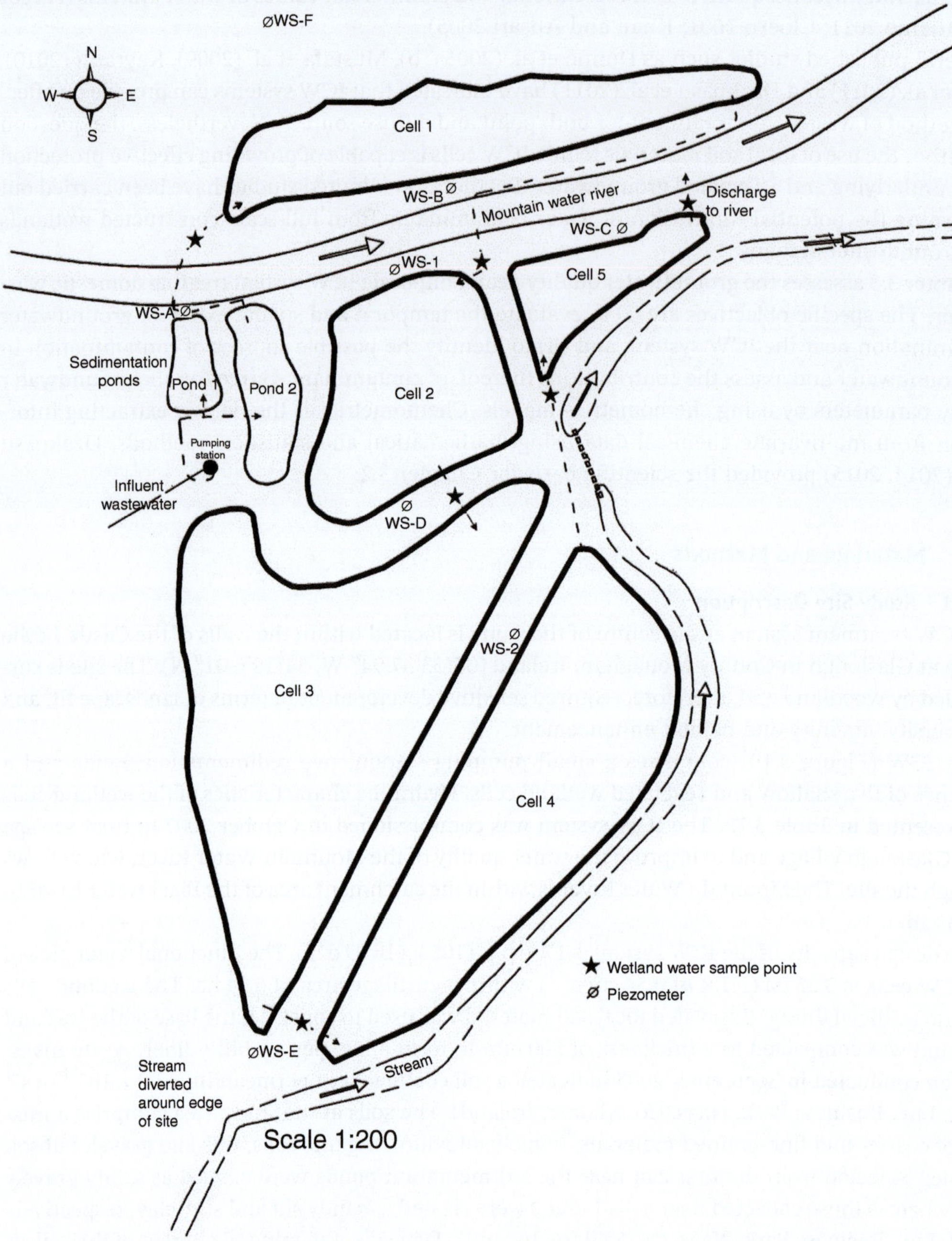

Figure 3.19 Sketch of the integrated constructed wetland (ICW) located at Glaslough in Ireland showing all sample points. Drawing showing the flow of the wastewater through a sedimentation tank, followed by five wetland cells in series and finally into a local river.

Table 3.12 Dimensions of the integrated constructed wetland (ICW) cells at Glaslough in County Monaghan, Ireland.

ICW section	Area (m^2)	Depth (m)	Total volume (m^3)	Effective volume (m^3)
Sedimentation pond 1	285	0.45	128.3	85.5
Sedimentation pond 2	365	0.45	164.3	109.5
Cell 1	4664	0.42	1958.9	1399.2
Cell 2	4500	0.38	1710.0	1350
Cell 3	12 660	0.32	4051.2	3798
Cell 4	9170	0.36	3301.2	2751
Cell 5	1460	0.29	423.4	423.4
Total wetland	33 104	—	11 737.3	9916.6

way, the build-up of sludge in the wetland cells, which could otherwise decrease the capacity of the cells, is prevented. From the sedimentation pond, the wastewater subsequently flows by gravity sequentially through the five earthen-lined cells. The effluent of the last cell discharges directly into the adjacent Mountain Water River.

The wetland cells were planted in a club pattern, and the main ones were *Carex riparia* Curtis, *Phragmites australis* (Cav.) Trin. ex Steud., *Typha latifolia* L., *Iris pseudacorus* L. and *Glyceria maxima* (Hartm.) Holmb. This currently includes a complex mixture of *Glyceria fluitans* (L.) R. Br., *Juncus effusus* L., *Sparganium erectum* L. emend Rchb, *Elisma natans* (L.) Raf. and *Scirpus pendulus* Muhl.

3.3.2.2 Monitoring Wells and Groundwater Sampling

Eight cassagrande (standpipe) piezometers were constructed around the ICW cells (labelled WS-1 to WS-F on Figure 3.19) in May 2011. The drilling of piezometers was done by using a hydraulic percussion drilling rig (IGSL Ltd., Business Park, Naas, Co. Kildare, Ireland). Depths of these wells ranged from 3 to 5 m below land surface (Table 3.13). PVC stand pipes of 50-mm diameter, typically

Table 3.13 Characteristics of piezometers within the integrated constructed wetland (ICW) system.

Well ID	Location	Well depth (m OD)	Casing depth (m OD)	Water bearing zone	Depth to water (m OD)
WS-A	Near Sedimentation pond 1 effluent	50.56	51.16	G/S/s/O	53.26
WS-B	Near cell 1 mid-point	50.26	50.26	G/S/O	51.76
WS-C	Near cell 5 effluent	49.43	50.63	s/C/G/O	52.43
WS-D	Near cell 2 effluent	49.82	50.72	s/G/O	52.82
WS-E	Near cell 3 effluent	51.46	51.46	s/O/S/G	52.96
WS-F	Off-site control well	51.08	51.08	S/s/O/	52.58
WS-1	Near cell 2 influent	49.04	49.04	S/O/s	51.24
WS-2	Near cell 4 effluent	48.88	48.88	s/S/O	51.78

C, Clay; G, Gravel; m OD, meters above ordnance datum; O, Organic soil; S, Sand; *s*, silt.

consisting of a solid casing down to the depth of the overburden, and a screened casing within the water bearing zone, were set into the ground. A filter medium consisting of washed pea gravel was constructed in the well annulus at the base and around the screened casing to prevent soil material from entering the wells. A seal of bentonite pellets was installed in the well annulus at the base of the solid casing to a thickness of at least 0.5 m, and the well annulus at the surface was sealed using cement grout and a steel cover.

Groundwater samples were collected weekly between May 2011 and April 2012 from the eight piezometers placed within the ICW system (Figure 3.19). Before a sample was taken each week, the water level in each piezometer was recorded using an electronic dip meter (Geotechnical Instruments Ltd., Leamington Spa, UK). The groundwater sampling involved purging the well to remove stagnant and contaminated water within the piezometer that may not be representative of in situ groundwater quality. Three times the volume of water contained in the well casing was purged using a WP 9012 submersible groundwater purging pump (Global Water, Gold River, CA, USA). Water samples were collected immediately after purging using 2-L polyethylene bottles and transported directly to the laboratory for immediate analysis. Samples for microbiological analyses were collected in 120-mL pre-sterilised polystyrene bottles and transported to the laboratory in cooled insulated boxes.

3.3.2.3 Monitoring and Analysis

A suite of automated sampling and monitoring instrumentation such as the ISCO 4700 Refrigerated Automatic Wastewater Sampler (Teledyne Isco, Inc., NE, USA) was used for weekly wetland water sampling (Figure 3.19). Additionally, all flows into and out of each ICW cell were measured and recorded with Siemens Electromagnetic Flow Meters FM MAGFLO and MAG5000 (Siemens Flow Instruments A/S, Nordborgrej, Nordborg, Demark) and their allied computer-linked data loggers. Mean flows were recorded at a one-minute interval frequency. A weather station is located beside the inlet pump sump to measure local temperature, precipitation and evapotranspiration.

The water samples were analysed weekly for several parameters including the five-days at 20°C N-allylthiourea biochemical oxygen demand (BOD_5), chemical oxygen demand (COD), total nitrogen (TN), ammonia-nitrogen (NH_3-N), nitrate-nitrogen (NO_3-N), total phosphorus (TP) and molybdate reactive phosphate (MRP; equivalent to soluble reactive phosphorus). All analyses were carried out at the Monaghan County Council wastewater laboratory, using kits supplied by HACH Lange (HACH Company, Loveland, CO, USA), and by following the standard operating procedures for the HACH DR/2010 portable data logging spectrophotometer (HACH Company 2000) or the standard methods for examination of water and wastewater (APHA 1998).

The BOD_5 was determined by APHA Method 5210 B (APHA 1998), using the respirometric (manometric) BOD OxiTop system (WTW GmbH, Weilheim, Germany), whereas COD was measured by the reactor digestion method, followed by colorimetric analysis (HACH Method 8000). The nitrogen compounds NH_3-N and NO_3-N were determined by the Nessler method (HACH Method 8038) and the cadmium reduction method (HACH Method 8171), respectively, whereas MRP was determined by the ascorbic acid method (HACH Method 8048). Low-range TN (HACH Method 10071) and high-range TN (HACH Method 10072) were determined following persulfate digestion. TP was determined by the acid persulfate digestion Test 'N Tube Method (HACH Method 8190).

Other ancillary water quality parameters such as water pH, temperature, oxidation–reduction potential (ORP), dissolved oxygen (DO) and electrical conductivity (EC) were measured on-site by using a WTW Multi 1970i portable multi-parameter meter and WTW ProfiLine Cond 197i portable conductivity meter (WTW GmbH, Weilheim, Germany).

Samples for microbiological and chloride analyses were transported to the Waterford County Council wastewater laboratory and analysed within 24 hours of collection. Total coliforms and *Escherichia coli* were analysed using the membrane filter method (APHA Method 9222).

3.3.2.4 Statistical Analyses

Data distributions were tested for normality by using the Shapiro–Wilk Normality Test (Shapiro and Wilk 1965). Statistically significant differences were all determined at $\alpha = 0.05$, unless stated otherwise. Comparisons of means were by paired samples t-tests and one-factor analysis of variance (ANOVA). Tukey's HSD (Honestly Significant Difference) tests were applied for post-ANOVA pairwise comparisons to identify significant differences among means. Exploratory factor analysis (FA) and principal component analysis (PCA) were employed to identify possible factors/sources of contamination in the groundwater and to estimate the contributions of the possible factors/sources identified on the concentrations of the measured physicochemical parameters. The raw analytical data sets were primarily standardised through a z-scale transformation in order to avoid misclassification due to the wide differences in data dimensionality. All statistical analyses were performed by using Minitab 16 (Minitab Inc., State College, PA, USA) and IBM SPSS Statistics 21 (IBM Corporation, Armonk, NY, USA).

PCA is concerned with establishing which linear components exist within the data set and how a particular variable might contribute to a component (Field 2005). This is achieved by transforming the original data variables into new, uncorrelated (orthogonal) variables (axes), called the principal components (PCs), which are linear combinations of the original data variables (Shrestha and Kazama 2007). The PC provides information on the most meaningful parameters, which describe whole data sets, affording data reduction with minimum loss of original information (Helena et al. 2000; Shrestha and Kazama 2007; Vega et al. 1998). These PCs are obtained by multiplying the original correlated variables with a list of coefficients (loadings or weightings) called eigenvectors. Thus, PCs are weighted linear combinations of the original data variables and can be expressed as shown in Eq. (3.1).

$$Z_{ij} = a_{i1}x_{1j} + a_{i2}x_{2j} + a_{i3}x_{3j} + \ldots + a_{im}x_{mj} \tag{3.6}$$

where Z is the component score; a is the component loading; x is the measured value of variable; j is the sample number; i is the component number and m is the total number of variables.

FA then extracts the eigenvalues and eigenvectors from a covariance/correlation matrix of the original data variables, which are subsequently rearranged in a manner that better explains the structure of the underlying system that produced the data set and further reduces the contributions of less significant variables obtained from PCA (Shrestha and Kazama 2007). The new group of variables, known as varifactors, are extracted through rotating the axis defined by PCA. The correlation coefficient matrix measures how well the variance of each constituent can be explained by relationships with each of the others. The correlation coefficients are computed as shown in Eq. (3.2).

$$r_{xy} = \frac{\sum(x - x_m)(y - y_m)}{\sqrt{[\sum(x - x_m)]^2 [y - y_m]^2}} \tag{3.7}$$

where the correlation coefficient (r_{xy}) is simply the sum (overall samples) of the products of the deviations of the x-measurements and the y-measurements on each sample, from the mean values of x and y, respectively, for the complete set of samples (Liu et al. 2003).

Overall, FA yields the general relationships among measured variables by showing multivariate patterns that may help to classify the original data (Liu et al. 2003). Equation (3.3) indicates FA.

$$Z_{ij} = a_{f1}f_{1i} + a_{f2}f_{2i} + a_{f3}f_{3i} + \dots + a_{fm}f_{mi} + e_{fi} \tag{3.8}$$

where Z is the measured variable; a is the factor loading; f is the factor score; e is the residual term accounting for errors or other source of variation; i is the sample number and m is the total number of factors.

In this study, PCA of the standardised groundwater quality data set was performed to extract significant PCs, which were further subjected to varimax rotation (a varimax rotation is a change of coordinates used in PCA and FA that maximises the sum of the variances of the squared loadings, i.e., squared correlations between variables and factors) to generate the varifactors. Exploratory FA, a statistical method used to uncover the underlying structure of a relatively large set of variables (Field 2009), was applied, which allowed for a better understanding of the complex set of variables by reducing them into a smaller number of factors.

Projections of the original variables on the subspace of the PCs, known as factor/component loadings, were used to determine the relative importance of a variable as compared to other variables in a particular PC and may not necessarily reflect the importance of the component itself (Ouyang et al. 2006). The terms 'strong', 'moderate' and 'weak', as applied to factor/component loadings, refer to absolute loading values of >0.75, 0.75–0.50 and 0.50–0.30, respectively (Liu et al. 2003).

3.3.3 Results and Discussion

3.3.3.1 Piezometer Hydrographs and Seasonal Fluctuations

The groundwater-level monitoring provided insight regarding the response of water levels in the aquifer underneath the ICW system to changing hydrologic conditions. All the piezometers displayed certain similarities, whereby the groundwater level in all piezometers showed variations in response to seasonal influences, with a clear recharge and natural discharge process evident in most instances. The relatively wet conditions that prevailed at the ICW site during the monitoring period saw a general upward trend in groundwater levels, with periodic minor recharge events associated with larger precipitation events.

Recharge events showed a rather quick response to larger precipitation events (Figure 3.20), with many of the piezometers generally exhibiting uniform increases in water levels following such precipitation events. Therefore, recharge to the aquifer seemed to originate from precipitation that infiltrated through the soil and percolated to the water table. Nevertheless, in the Blackwater catchment where the ICW system is located, Misstear and Brown (2010) noted that groundwater recharge mainly occurred indirectly through the fractured bedrock, which provide conduits for lateral flow from up-gradient areas and that direct recharge through the overlying till was only approximately 26 mm or 5% of the annual effective rainfall. This possibly explains the relatively quick increase of the groundwater table elevation to larger precipitation events that occurred during the monitoring period. At all other times during the monitoring period when precipitation was relatively low, fairly constant groundwater levels were recorded (Figure 3.20), with a relatively higher water table elevation being recorded during winter months.

The groundwater levels in most of the piezometers were always higher than the invert (base interior) levels of the wetland cells. The significance of the level of the groundwater table in this instance is, that water will flow from the wetland into the aquifer and thus degrade groundwater

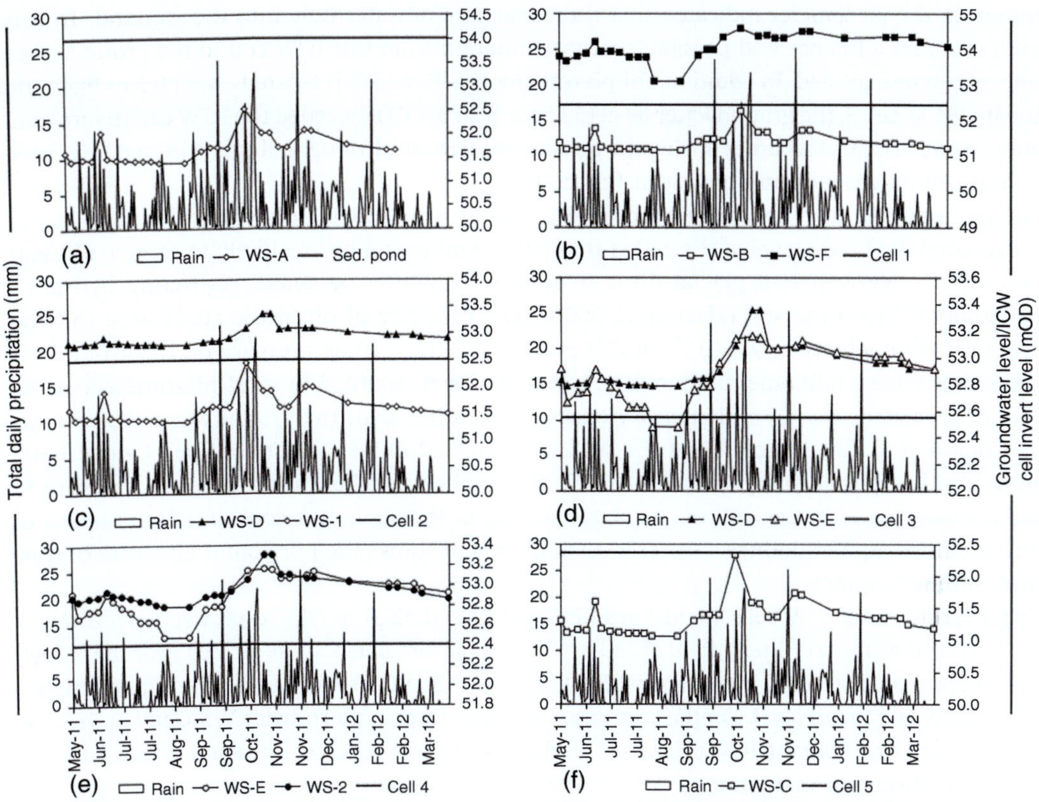

Figure 3.20 (a–f) Groundwater table elevation for monitoring wells within the integrated constructed wetland (ICW) system plotted in relation to bed invert level of individual ICW cells and total daily precipitation. Six graphs showing the relationship between precipitation and groundwater level in different parts of the ICW system over a period of about one year.

quality once the level of the groundwater table fell below the wetland cell invert level, albeit; the flow distance must be considered. In a review of the risk of groundwater pollution by on-site sanitation in developing countries, Lewis et al. (1980) noted that the groundwater quality tends to depend on the vertical separation allowed between the base of the pit latrines and the groundwater table elevation.

The mean ± SD water levels in piezometers WS-F and WS-B were 54.06 ± 0.37 m OD and 51.42 ± 0.25 m OD, respectively (Figure 3.20). Piezometer WS-F is located circa 100 m up-gradient from ICW cell 1, whereas WS-B is located about 0.5 m down-gradient (Figure 3.19). The invert level of the ICW cell is 52.46 m OD. The water level in piezometer WS-B was well below the invert level of the ICW cell during the monitoring period (Figure 3.20). The water level in piezometer WS-F (control) also exceeded the ICW cell invert level.

Piezometers WS-1 and WS-D are located approximately 0.5 m from the influent and effluent points of ICW cell 2, respectively. The invert level of the ICW cell is 52.50 m OD. The groundwater levels in the piezometers were 51.61 ± 0.26 m OD and 52.93 ± 0.16 m OD, respectively, for WS-1 and WS-D. The water level of piezometer WS-D was higher than the ICW cell invert level during the monitoring period. The fact that the invert level of this ICW cell sat below the groundwater elevation in the piezometer indicated that there was groundwater flow into the wetland. In this regard, minimal influence and possibly no contamination from this ICW cell to the groundwater quality might be expected. In addition, for piezometer WS-E, which is located about 0.5 m from the effluent of ICW cell 3, the groundwater level (52.88 ± 0.20 m OD) exceeded the ICW cell invert level slightly (52.56 m OD) at all times during the monitoring period, also suggesting a minimal influence from this cell on groundwater quality at this point.

The groundwater level in piezometer WS-D, located about 30 m from the influent of cell 3, also exceeded the invert level of the cell at that point, and would be less likely to receive effluent from that cell. Nevertheless, glacial till is frequently known to be dense, restricting hydraulic conductivity through the soil (Hart et al. 2006). The presence of till in the study area (having a hydraulic conductivity of 9×10^{-11} m s^{-1}) potentially exerts some pressure on the groundwater, creating some confinement. Therefore, the water level recorded in most piezometers at the ICW site represents the potentiometric surface level rather than the groundwater table elevation. Conlon et al. (2005) noted that if a well is completed in an aquifer confined by overlying materials of low permeability, groundwater in the aquifer may be under sufficient pressure to cause the water level in the well to rise above the top of the aquifer. A contour map of heads in such a confined aquifer defines a water level surface that shows the horizontal direction of flow in the confined aquifer.

The invert levels of ICW cells 4 and 5 are 52.41 m OD and 52.37 m OD, respectively. Piezometer WS-2 is located near the effluent point of cell 4, whereas piezometer WS-C is located near that of cell 5. The average water level in piezometer WS-2 (52.95 ± 0.13 m OD) was found to be slightly higher than the ICW cell invert level. On the contrary, the water level in piezometer WS-C fell at 51.35 ± 0.27 m OD far lower than the ICW cell invert level. This suggests a greater influence of that ICW cell on the groundwater at that point.

3.3.3.2 Contaminant Concentrations in Groundwater

Nitrogen in the groundwater underneath the ICW system was recorded mainly as ammonia. The overall average of ammonia (NH_3-N) for all sample points accounted for more than 80% of the total N recorded in the groundwater. The range for NH_3-N was 0.01–12.6 mg L^{-1} for all sample points, with over 60% of the total results falling below the WHO recommended threshold for odour and taste concentrations (WHO 2008). Total N and NH_3-N persisted quite strongly in piezometers WS-E, WS-D and WS-2 (Table 3.14), and tended to decrease significantly ($p < 0.05$) in concentration from piezometers up-gradient of the groundwater flow direction to the down-gradient piezometers. When compared with each other, there was also a significant ($p < 0.05$) variation in total N and NH_3-N concentrations, in all the down-gradient piezometers (i.e., WS-A, WS-B, WS-C and WS-1).

The concentrations in the control piezometer, which was constructed outside of the ICW system, were found to be generally similar to that in all the down-gradient piezometers. This indicated that concentrations in the down-gradient piezometers rarely exceeded the natural background levels.

The concentrations of NO_3-N recorded for all the piezometers were rather low and accounted for less than 10% of the total N recorded in the groundwater. The range for NO_3-N was 0–1.7 mg L^{-1} for all sample points, with all the total results falling far below the WHO recommended threshold for drinking water; nitrate could cause health problems at concentrations above 50 mg NO_3 L^{-1} or 10 mg NO_3-N L^{-1}(WHO 2008).

The presence of NH_3-N at higher than the natural (geogenic) levels of 0.2 mg L^{-1} (WHO 2008) is an important indicator of possible bacterial and faecal (human or animal) pollution (WHO 2003).

Table 3.14 Contaminant concentrations in groundwater recorded near the integrated constructed wetland (ICW) system between May 2011 and April 2012.

Parameter/well ID		WS-A	WS-B	WS-C	WS-D	WS-E	WS-F	WS-1	WS-2
Temperature (°C)	Mean	12.2	11.2	11.1	11.8	11.2	11.9	11.1	11.5
	SD[e]	1.30	1.56	1.75	2.23	1.64	2.03	1.87	1.84
pH	Mean	6.90	7.02	7.13	6.92	6.90	7.23	7.10	6.94
	SD[e]	0.12	0.17	0.17	0.16	0.16	0.13	0.23	0.16
EC[a] (μS/cm)	Mean	853	967	701	846	1317	1070	834	1202
	SD[e]	61	99	48	45	68	42	125	59
Dissolved oxygen (mg O_2 L^{-1})	Mean	3.3	3.7	3.1	3.0	3.0	7.7	4.6	3.5
	SD[e]	0.69	0.87	0.54	0.61	1.07	1.74	1.75	0.95
ORP[b] (mV)	Mean	10	4	−2	9	9	−7	−1	7
	SD[e]	3.7	7.7	7.1	7.1	5.3	5.3	9.9	5.7
BOD_5 (mg O_2 L^{-1})	Mean	7	6	3	5	17	2	6	12
	SD[e]	1.4	2.3	2.1	0.5	6.9	1.5	3.6	1.9
COD (mg O_2 L^{-1})	Mean	13	13	7	9	55	16	17	31
	SD[e]	2.8	4.5	3.6	5.1	10.9	6.6	8.3	8.2
N total (mg N L^{-1})	Mean	1.6	1.4	0.7	5.8	10.2	0.9	1.9	3.4
	SD[e]	0.60	0.65	0.43	1.41	1.14	0.44	0.88	0.84
Ammonia–N (mg N L^{-1})	Mean	0.7	0.4	0.2	5.2	8.1	0.4	1.1	3.4
	SD[e]	0.10	0.22	0.09	1.34	1.72	0.15	0.71	0.69
Nitrate–N (mg N L^{-1})	Mean	0.3	0.1	0.1	0.1	0.6	0.1	0.3	0.4
	SD[e]	0.58	0.08	0.16	0.10	0.41	0.08	0.14	0.28
P total (mg P L^{-1})	Mean	0.13	0.15	0.10	0.05	0.55	0.36	0.16	0.09
	SD[e]	0.08	0.17	0.13	0.08	0.25	0.48	0.17	0.07
MRP[c] (mg P L^{-1})	Mean	0.02	0.03	0.02	0.02	0.04	0.07	0.03	0.03
	SD[e]	0.02	0.02	0.01	0.01	0.03	0.11	0.02	0.02
Chloride (mg L^{-1})	Mean	29.8	27.0	38.6	61.2	19.9	14.1	36.8	13.2
	SD[e]	4.43	9.40	7.82	12.94	4.23	2.30	19.00	1.13
Total coliforms (MPN 100 mL^{-1})	Mean	26	145	46	99	46	214	996	596
	SD[e]	33	224	44	131	89	174	22	70
Escherichia coli (MPN[d] 100 mL^{-1})	Mean	<1	<1	<1	<1	<1	<1	<1	<1

[a] Electrical conductivity.
[b] Oxidation–reduction potential.
[c] Molybdate reactive phosphate.
[d] Most probable number.
[e] Standard deviation.

Therefore, the relatively high NH_3-N concentrations recorded in some up-gradient piezometers during the current study indicate the presence of an immediate source of organic or faecal pollution. However, it is also noted that anaerobic groundwater may contain up to 3 mg L^{-1} NH_3-N (WHO 2008) and that natural concentrations of up to 3 mg L^{-1} of NH_3-N could be found usually in strata rich in humic substances or iron or in forests (Dieter and Möller 1991; Tilaki and Kahe 2012).

Because the ICW site was originally a forestland, which was redeveloped, it is possible that the high NH_3-N concentrations recorded in some of the piezometers may have geogenic origin. In addition, the borehole logs from the affected piezometers indicated the presence of organic material in the stratigraphy of the site. Besides, Fay et al. (2007) estimated that the soils at the ICW site contain about 2.5% of Fe^{2+} and about 10% of organic carbon. These findings are consistent with the assertion that groundwater with naturally higher NH_3-N contents could occur in strata rich in humic substances or iron or in forests (Dieter and Möller 1991; Tilaki and Kahe 2012).

There was a slight reduction in N concentrations during recharge events, which generally occurred during high precipitation events. The relative decreases in N concentrations during such recharge events suggest possible dilution of N in the groundwater. This indicates that the source of N contamination can be found within the immediate surroundings of the sampling points. The recharge events were also associated with a rise in the groundwater table elevation, which exceeded the wetland cell invert levels, consequently discharging groundwater into the wetland cells. Thus, the wetlands were less likely to influence the groundwater quality. The differences in chloride concentrations, which were lower in the groundwater than that in the wetlands, further affirm this.

When compared using Figure 3.21, the quality of the wastewater contained in ICW cells 1 and 2, and the sedimentation pond(s) showed significantly higher ($p < 0.05$) concentrations of N than that in the groundwater at some of the sampling points; WS-A, WS-B, WS-1 and WS-D (Figure 3.19). At other sampling points (WS-E and WS-2), the concentrations of N in the groundwater were slightly higher than those contained in the nearby cells of the ICW system. For example, the concentration of N in the most contaminated piezometer (WS-E) was 10.2 ± 1.14 mg N total L^{-1}, 8.1 ± 1.72 mg NH_3-N L^{-1} and 0.6 ± 0.41 mg NO_3-N L^{-1}; BOD_5 was 17 ± 6.9 mg L^{-1}. This piezometer is located up-gradient of the effluent point of ICW cell 3, which contained 5.1 ± 4.26 mg N total L^{-1}, 4.0 ± 4.38 mg NH_3-N L^{-1}, 0.6 ± 0.30 mg NO_3-N $L^{-1}$and 10 ± 4.7 mg BOD_5 L^{-1}. It seems evident from here that the concentrations of N were generally higher in the groundwater at this point than that in the ICW cell, and therefore, N contamination was likely not originating from the ICW cell.

On the other hand, the ICW cell that contained the most contaminated wastewater is the sedimentation pond. The piezometer closest to this ICW cell is WS-A (down-gradient). The concentrations of N in the piezometer at that point were 1.6 ± 0.60 mg N total L^{-1}, 0.7 ± 0.10 mg NH_3-N L^{-1} and 0.3 ± 0.58 mg NO_3-N L^{-1}; BOD_5 was 7 ± 1.4 mg L^{-1}, whereas the concentrations of N in the ICW cell were 61 ± 24.11 mg N total L^{-1}, 42 ± 13.66 mg NH_3-N L^{-1} and 3.5 ± 2.90 mg NO_3-N L^{-1}; BOD_5 was 308 ± 16 mg L^{-1}. It seems evident that the concentrations of N were generally higher by quite a large margin in the ICW cell at this point than in the groundwater.

Findings were similar for all the other piezometers, which are down-gradient to the other proximal ICW cells. Overall, N contamination in the groundwater was higher in piezometers WS-E, WS-D and WS-2, which are located up-gradient and closer to the distal ICW cells containing the least contaminated wastewater compared to piezometers WS-A, WS-B and WS-1, which are down-gradient and closer to the proximal ICW cells containing the most contaminated wastewater. Thus, the proximal ICW cells pose a minimal risk to the groundwater quality.

Groundwater contamination with N includes several sources, which are frequently not evident, and sometimes occur with multiple sources. As chloride is ubiquitous to domestic wastewater and not subject to adsorption, ion exchange or oxidation–reduction reactions, it is commonly a useful and reliable chemical indicator of groundwater faecal contamination (Canter and Knox 1985;

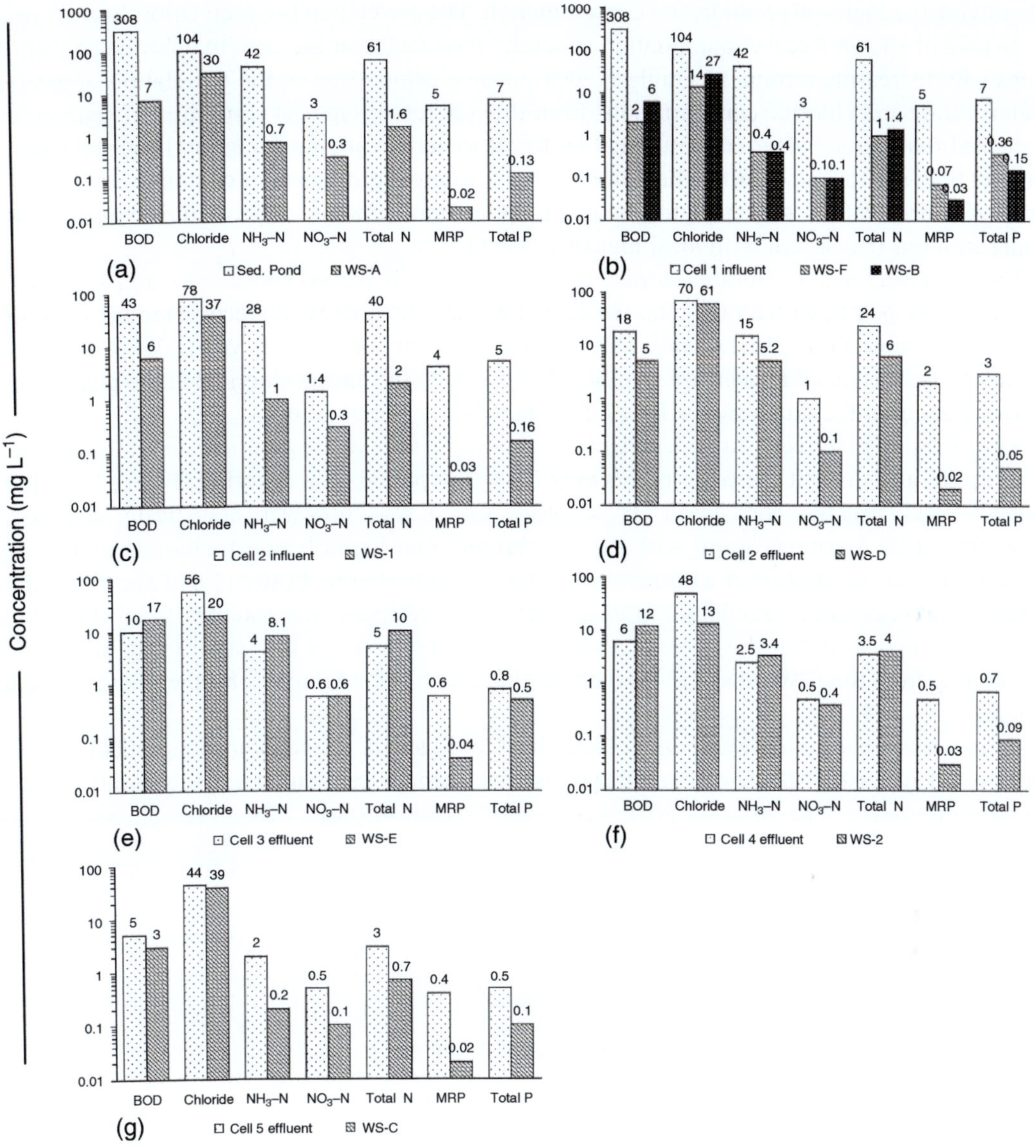

Figure 3.21 (a–g) Contaminant concentrations in groundwater in relation to contaminants contained in the integrated constructed wetland (ICW) cells. Seven graphs showing the relationship between various water quality parameters including BOD as well as nitrogen and phosphorus compounds in different parts of the ICW system

McQuillan 2004). Besides, nitrate is relatively mobile in soil and groundwater because, unlike ammonium, nitrate does not adsorb onto soil or aquifer geologic materials, and only precipitates as a mineral under dry conditions (Uhlman and Artiola 2011). Thus, nitrate tends to persist for long periods, sometimes migrating great distances down-gradient from the source (Taylor 2003).

The correlations of chloride with nitrate and dissolved organic carbon, which is otherwise distinctive for extremely low levels of major ions, consequently, often suggest contamination with wastewater (Foster 1990). Therefore, chloride and nitrate-N can be useful indicative parameters for identifying the source of groundwater contamination. The correlation between chloride and nitrate in an area of known faecal contamination generally shows a linear increase in chloride concentrations with increasing nitrate. McQuillan (2004) made a comparison of the correlation of groundwater nitrate and chloride concentrations from three different types of contaminant sources and indicated that groundwater contaminated by faecal material contained slightly lower concentrations of both nitrate and chloride than dairy sources, whereas groundwater contaminated by evaporative concentration of rainwater beneath ephemeral streams contained relatively lower concentrations (consistent with local meteoric water).

The concentrations of chloride recorded in the present study ranged between 8.0 and 75.6 mg L^{-1} for all sample points, with all the total results falling far below the WHO recommended threshold for taste concentrations (WHO 2008). Chloride was higher in piezometers WS-D, WS-C and WS-1 (Table 3.14) and tended to decrease significantly ($p < 0.05$) in concentration from the piezometers down-gradient of the groundwater flow direction to the up-gradient ones.

An evaluation of the correlation between chloride and nitrate in the groundwater at the ICW site recorded in all the eight sampling points over the entire period of monitoring (Figure 3.22) indicated a weak negative correlation ($R^2 = 0.098$), which was, however, statistically significant ($p < 0.05$). This is not consistent with the correlation often found between chloride and nitrate in groundwater contaminated with faecal materials as indicated by Foster (1990) and McQuillan (2004). The concentrations of both nitrate and chloride were rather consistent with local meteoric water. This indicated that the N contamination recorded in some of the piezometers may not be originating from the ICW cells, and that indeed the source of contamination might have geogenic origin.

Phosphorus in the groundwater underneath the ICW system occurred mainly in particulate form. The overall average of MRP for all sample points accounted for less than 15% of the total P recorded in the groundwater. The range for MRP was between 0 and 0.45 mg L^{-1} for all sample points, with

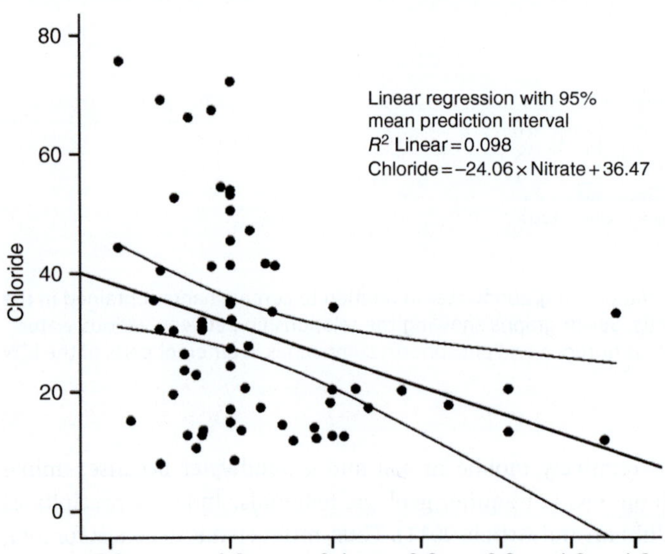

Figure 3.22 Correlation between chloride and nitrate-nitrogen in eight monitoring wells at the integrated constructed wetland (ICW) site. The graph shows a weak correlation between nitrate and chloride.

over 75% of the total results falling below the EQS for P in groundwater in Ireland, which is set at 0.03 mg L^{-1} (EPA 2003) or the threshold value for MRP in groundwater considered being at less than a good chemical status in Ireland, which is set at 0.035 mg L^{-1} (S.I. No. 9 of 2010).

In this study, MRP persisted quite strongly and somewhat variable in piezometer WS-F, which is the control piezometer constructed outside of the ICW system (Table 3.14). The highest concentrations (around 0.07 ± 0.11 mg L^{-1}) were recorded in this piezometer. The average concentrations of MRP were quite similar in all the other piezometers. The same was true for total P, but the individual values were quite variable.

Overall, the patterns of P concentrations recorded in the groundwater across the sampling sites provide no indication of P contamination originating from the ICW cells. When compared with Figure 3.21, the quality of wastewater contained in the ICW cells was associated with significantly higher ($p < 0.05$) concentrations of P than that in the groundwater at all sampling points. This suggested that either there was no P contamination originating from the ICW cells to the groundwater system underneath, or that any P contamination potentially originating from the ICW cells was being effectively attenuated in the sub-surface because of the sub-soil mineralogy. It is generally noted that P has the capacity to be adsorbed onto or form precipitates within mineral surfaces of soil particles (Gill et al. 2004), particularly clay, resulting in significantly lower P loads to the groundwater. Adsorbed P can further migrate into the interior of the minerals that adsorbed them and therefore becoming even less available (Troeh and Thompson 2005).

The overall average of BOD_5 for all sample points accounted for circa 35% of the COD recorded in the groundwater. The range for BOD_5 was 0–34 mg L^{-1} for all sample points, with over 75% of the total results falling below the recommended threshold values of greater than 10 mg BOD L^{-1} or 200 mg CODL^{-1} for water pollution due to wastewater discharges (Chapman and Kimstach 1996).

The concentrations of BOD_5 and COD persisted quite strongly in piezometer WS-E (Table 3.14), which is up-gradient to the ICW system, followed by WS-2. Furthermore, water samples from these two piezometers normally exhibited some colouration. Such colouration in water is usually primarily due to the presence of humic and fulvic acids associated with the humus fraction of soils (WHO 2008). Since the borehole logs from these piezometers indicated the presence of a considerable amount of organic soils in the stratigraphy at that point and also, because the ICW site was originally a forestland, the presence of relatively high BOD_5 and COD concentrations may be geogenically derived. Besides, the BOD_5/COD ratios in both piezometers were found to be relatively lower (approximately 0.3) than the range between 0.4 and 0.8 commonly found in domestic wastewaters (Srinivas 2008). Chapman and Kimstach (1996) noted that unpolluted waters typically have BOD concentrations of 2 mg L^{-1}or less and COD of 20 mg L^{-1}or less. In terms of COD, piezometers WS-1, WS-A, WS-B and WS-D actually fall into the class of unpolluted waters.

Additionally, the presence or absence of faecal indicator bacteria is a commonly used operational indicator of water quality. Coliform bacteria are the most prevalent genera used. In most waters, the predominant genus is *Escherichia*, with the detection of *E. coli* providing definite evidence of faecal pollution (APHA 1998; WHO 2008). Some members of the total coliform bacteria group, on the other hand, are commonly found in the environment (e.g., soil or vegetation) and are generally harmless. Consequently, if only total coliform bacteria are detected in water, the source may be environmental, and faecal contamination is not likely (Mañas et al. 2012).

In the current study, the total coliform count varied highly across sampling points (Table 3.14). Conversely, the *E. coli* count at all sample points was found to be <1 MPN 100 mL^{-1}. This provided an indication that there may not have been a source of faecal contamination originating from the ICW cells to the groundwater and that organic material found in the stratigraphy of the ICW site at some sample points may be contributing to the high levels of total coliforms recorded.

3.3.3.3 Factors Influencing the Variability in Groundwater Quality

The PCA undertaken on the groundwater quality dataset showed that 82.3% of the data variance extracted was contained within the first four PCs. These four components had eigenvalues greater than or close to unity and explained 48.7%, 15.1%, 10.3% and 8.2%, respectively, of the total variance extracted from the original dataset. The main correlations between the respective variables determined by the PCA are given in Table 3.15. Figure 3.23 provides a biplot of PC1 and PC2. Factor/component loadings of the four PCs are presented in Table 3.16.

PC1, which explained 48.7% of the total variance, exhibited strong positive loadings for inorganic nutrients and organic-related parameters, namely NH_3-N, NO_3-N, total N, total P, BOD and COD. It is evident from the component loadings of this PC that the contamination recorded in the groundwater at some sample points was predominantly due to an organic source of contamination. This is because most of the parameters, which recorded high loadings on the PC, are common to organic pollution and are also closely correlated (Figure 3.23), thereby confirming that they originated from the same source. Nevertheless, there was also a moderate negative loading of Cl^- on this PC. This suggested that the source of contamination is not likely to be of faecal origin. The moderate negative correlation between $Cl^-$and NH_3-N, NO_3-N, total N, total P, BOD and COD (Table 3.15) confirms this further. This provides evidence that the groundwater contamination recorded at some sample points during the monitoring period as stated early on, may not be originating from the ICW cells. With respect to the N species, a positive correlation of nearly unity was recorded between total N and NH_3-N, with each also having strong positive loading scores on PC1. This indicated that the majority of N in the groundwater was primarily in the form of NH_3-N.

PC2, which explained 15.1% of the total variance, exhibited strong positive loadings for ORP, weak positive loading for temperature and MRP and strong negative loading for pH. Therefore, this component related to environmental conditions. Commonly, pH has a significant influence on the reactivity of phosphate, with lower pH increasing the level of reactive phosphate (Carroll and Goonetilleke 2005). Similarly, as the pH of water increases, the level of reactivity is reduced.

The strong positive loading of ORP on this PC further confirms this. Generally, the solubility of many elements in water depends on whether they are oxidised or reduced. Iron, for example, is more soluble in the reduced state than it is in the oxidised state (Nelson 2002). Therefore, as ORP drops, the groundwater will have higher concentrations of Fe^{2+} available for precipitation reactions with P.

Furthermore, PC3, which explained 10.3% of the total variance in the dataset, exhibited strong positive loadings for groundwater table elevation and DO and a moderate positive loading for electrical conductivity. Therefore, this PC related to groundwater recharge events. Misstear and Brown (2010) noted that in the Blackwater catchment where the ICW system is located, groundwater recharge mainly occurred indirectly through the fractured bedrock, which provided conduits for lateral flow from up-gradient. Consequently, groundwater recharge at the ICW site was potentially associated with an influx of oxygenated water from the surface during periods of such rapid recharge. The influx of oxygenated water may have subsequently resulted in increased concentrations of DO, pH and ORP. These changes potentially led to decreases in concentrations of calcium and magnesium ions, heavy metals (particularly iron and manganese), hydrogen sulphide, ammonium and bulk organic parameters such as COD and BOD (EPA Victoria 2000). The weak negative correlations recorded between DO and both NH_3-N and total N (Table 3.15), and the moderate positive loading of electrical conductivity on this PC (Table 3.16) provide strong evidence to confirm this. In general, the oxidation of an organic molecule can result in, for example, Fe or Mn, in a solid oxide mineral, being dissolved into the groundwater in a reduced form as Fe^{2+} or Mn^{2+}

Table 3.15 Correlation matrix.

Variable	WL[a]	T[b]	pH	EC[c]	ORP[d]	DO[e]	N total	NH_3-N[f]	NO_3-N[g]	P total	MRP[h]	Cl^-	COD[i]	BOD[j]
WL[a]	1.000													
T[b]	0.128	1.000												
pH	0.018	−0.233	1.000											
EC[c]	0.551	0.103	−0.313	1.000										
ORP[d]	−0.055	0.226	−0.865	0.237	1.000									
DO[e]	0.508	−0.134	0.271	0.195	−0.342	1.000								
N total	0.272	0.045	−0.319	0.654	0.393	−0.358	1.000							
NH_3-N[f]	0.341	0.003	−0.366	0.672	0.401	−0.321	0.966	1.000						
NO_3-N[g]	0.226	0.078	−0.345	0.719	0.300	−0.160	0.753	0.719	1.000					
P total	0.374	0.011	−0.151	0.795	0.198	−0.064	0.815	0.753	0.795	1.000				
MRP[h]	0.140	−0.191	−0.303	0.462	0.214	−0.035	0.413	0.375	0.330	0.524	1.000			
Cl^-	−0.465	−0.176	0.298	−0.492	−0.319	0.077	−0.499	−0.435	−0.377	−0.517	−0.387	1.000		
COD[i]	0.327	−0.007	−0.243	0.862	0.289	−0.149	0.870	0.846	0.829	0.930	0.468	−0.512	1.000	
BOD[j]	0.182	0.112	−0.285	0.748	0.342	−0.220	0.797	0.795	0.795	0.780	0.170	−0.335	0.862	1.000

[a] Water level.
[b] Temperature.
[c] Electrical conductivity.
[d] Oxidation–reduction potential.
[e] Dissolved oxygen.
[f] Ammonia-nitrogen.
[g] Nitrate-nitrogen.
[h] Molybdate reactive phosphate.
[i] Chemical oxygen demand.
[j] Biochemical oxygen demand.

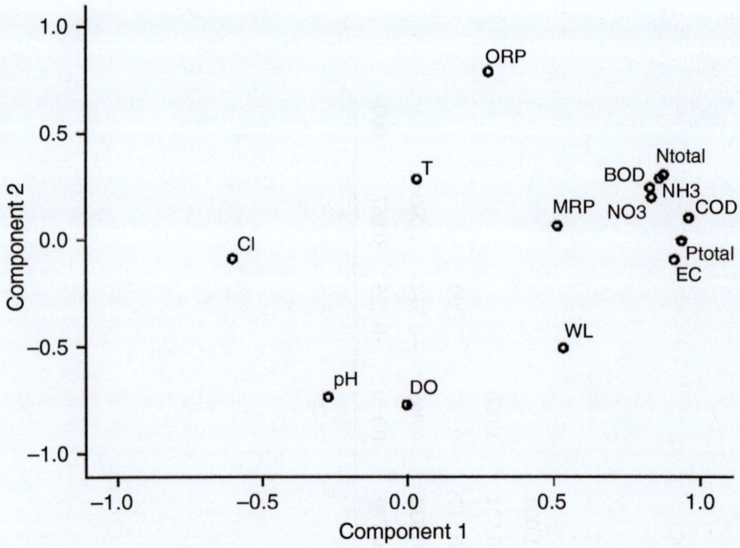

Figure 3.23 Component plot in rotated space. Graph showing the components 1 and 2, and indicating a clear cluster for the three nitrogen compounds together with the two indicators for wastewater strength.

Table 3.16 Rotated component matrix.

		Component			
Variable	**Communality**	**1**	**2**	**3**	**4**
COD[a]	0.957	0.950	0.128	0.132	0.138
BOD[b]	0.875	0.918	0.098	−0.052	−0.139
N total	0.894	0.911	0.236	−0.048	0.072
P total	0.892	0.899	0.070	0.216	0.177
NH_3-N[c]	0.849	0.886	0.243	−0.018	0.068
NO_3-N[d]	0.768	0.861	0.159	0.037	0.001
MRP[e]	0.776	0.313	0.381	0.206	0.700
EC[f]	0.861	0.770	0.181	0.481	0.066
pH	0.874	−0.153	−0.920	0.065	0.023
ORP[g]	0.858	0.201	0.888	−0.155	−0.076
Cl^-	0.551	−0.493	−0.441	−0.429	−0.040
WL[h]	0.822	0.281	−0.006	0.856	−0.100
DO[i]	0.779	−0.249	−0.296	0.788	0.093
T[j]	0.776	0.040	0.314	0.135	−0.811

[a] Chemical oxygen demand.
[b] Biochemical oxygen demand.
[c] Ammonia-nitrogen.
[d] Nitrate-nitrogen.
[e] Molybdate reactive phosphate.
[f] Electrical conductivity.
[g] Oxidation–reduction potential.
[h] Water level.
[i] Dissolved oxygen.
[j] Temperature.
Extraction method, Principal Component Analysis; Rotation method, Varimax with Kaiser Normalisation. Rotation converged in five iterations.

(Nelson 2002) and potentially causing a corresponding increase in the electrical conductivity of the groundwater.

The PC4 explained 8.2% of the total variance in the dataset. This PC exhibited a strong negative loading for temperature. Therefore, this component related well to seasonal temperature changes. There is a seeming negative effect of temperature on contaminant concentrations in the groundwater especially MRP, which also exhibited a moderate positive loading on this principal. It has been noted that the ability of soils to adsorb P was reduced at higher temperatures and that the rate of P desorption increases with increasing temperature (Mamo et al. 2005; Sallade and Sims 1997a, b).

3.3.4 Conclusions

Results from this study indicate some contamination with bulk organic matter and inorganic nutrients in the groundwater underlying the ICW system. Significantly higher contaminant concentrations were recorded in monitoring wells up-gradient of the groundwater flow direction and near to the distal wetland cells than in the down-gradient ones, which were near to the proximal cells.

For the down-gradient piezometers, concentrations rarely exceeded the natural background levels. Detailed analyses through the application of chemometrics models indicated that the source of contamination was largely of geogenic origin.

Four possible factors were identified as causing the variations in groundwater quality at the ICW site; namely, geogenically derived organic source of contamination, environmental conditions, groundwater recharge cycles and seasonal variations in water temperature. Findings suggest that ICW systems pose a minimal risk to the groundwater quality and that the risk was highest for the distal wetland cells.

References

Ahn, Y.H. (2006). Sustainable nitrogen elimination biotechnologies: a review. *Process Biochemistry* 41 (8): 1709–1721.

Aiken, G. (2002). Organic matter in ground water. In: *U.S. Geological Survey Artificial Recharge Workshop Proceedings* (ed. G.R. Aiken and E.L. Kuniansky), 21–23. Sacramento, CA: United States Geological Survey.

Akpor, O.B. and Muchie, M. (2011). Environmental and public health implications of wastewater quality. *African Journal of Biotechnology* 10 (13): 2379–2387.

APHA (1998). *Standard Methods for the Examination of Water and Wastewater*, 20e. Washington, DC: American Public Health Association/American Water Works Association/Water Environment Federation.

APHA (2005). *Standard Methods for the Examination of Water and Wastewater*, 21e. Washington, DC, USA: American Public Health Association/American Water Works Association/Water Environment Federation.

Barber, L.B., Leenheer, J.A., Noyes, T.I., and Stiles, E.A. (2001). Nature and transformation of dissolved organic matter in treatment wetlands. *Environmental Science & Technology* 35 (24): 4805–4816.

Boon, G.A. (1986). Report of a Visit by Members and Staff of WRc to Germany (GFR) to Investigate the Root Zone Method for Treatment of Waste Waters. WRc Report No. 376-S/1, Stevenage, UK.

Canter, L.W. and Knox, R.C. (1985). *Septic Tank System Effects on Groundwater Quality*. Chelsea, MI: CRC Press, Inc.

Carroll, S. and Goonetilleke, A. (2005). Assessment of high density of onsite wastewater treatment systems on a shallow groundwater coastal aquifer using PCA. *Environmetrics* 16 (3): 257–274.

Carty, A., Scholz, M., Heal, K. et al. (2008). The universal design, operation and maintenance guidelines for farm constructed wetlands (FCW) in temperate climates. *Bioresource Technology* 99 (15): 6780–6792.

Chapanova, G., Jank, M., Schlegel, S., and Koeser, H. (2007). Effect of temperature and salinity on the wastewater treatment performance of aerobic submerged fixed bed biofilm reactors. *Water Science and Technology* 55 (8–9): 159–164.

Chapman, D. and Kimstach, V. (1996). Selection of water quality variables. In: *Water Quality Assessments: a Guide to Use of Biota, Sediments and Water in Environmental Monitoring*, 2e (ed. D. Chapman), 59–126. London, UK: E&FN Spon.

Cloern, J.E. (2001). Our evolving conceptual model of the coastal eutrophication problem. *Marine Ecology Progress Series* 210: 223–253.

Conlon, T.D., Wozniak, K.C., Woodcock, D. et al. (2005). Ground-water hydrology of the Willamette basin, Oregon. U.S. Geological Survey Scientific Investigations Report 2005-5168. Reston, VA: U.S. Geological Survey.

Cooper, P. (2005). The performance of vertical flow constructed wetland systems with special reference to the significance of oxygen transfer and hydraulic loading rates. *Water Science and Technology* 51 (9): 81–90.

Dieter, H.H., and Möller. 1991. *Ammonia, in: Die Trinkwasserverordnung: Einführung und Erläuterungen für Wasserversorgungsunternehmen und Überwachungsbehörden (The Drinking Water Regulations: Introduction and Notes for Water Companies and Supervisory Authorities)*, eds. Aurand, K., Hässelbarth, V., Lange-Asschenfeldt, H., and W. Steuer, W. Berlin: Erich-Schmidt Verlag, 62–368.

Dinçer, A.R. and Kargi, F. (1999). Salt inhibition of nitrification and denitrification in saline wastewater. *Environmental Technology* 20 (11): 1147–1153.

Domingo, S., Dallas, S., Germain, M., and Ho, G. (2006). Heavy metals in a constructed wetlands treating industrial wastewater: distribution in the sediment and rhizome tissue. *Water Science and Technology* 60 (6): 1425–1432.

Dong, Y., Wiliński, P.R., Dzakpasu, M., and Scholz, M. (2011). Impact of hydraulic loading rate and season on water contaminant reductions within integrated constructed wetlands. *Wetlands* 31 (3): 499–509.

Dong, Y., Kayranli, B., Scholz, M., and Harrington, R. (2013). Nutrient release from integrated constructed wetlands sediment receiving farmyard runoff and domestic wastewater. *Water and Environment Journal* 27 (4): 439–452.

Dunne, E.J., Culleton, N., O'Donovan, G. et al. (2005a). Phosphorus retention and sorption by constructed wetland soils in Southeast Ireland. *Water Research* 39 (18): 4355–4362.

Dunne, E.J., Culleton, N., O'Donovan, G. et al. (2005b). An integrated constructed wetland to treat contaminants and nutrients from dairy farmyard dirty water. *Ecological Engineering* 24 (3): 221–234.

Dzakpasu, M., Hofmann, O., Scholz, M. et al. (2011). Nitrogen removal in an integrated constructed wetland treating domestic wastewater. *Journal of Environmental Science and Health, Part A: Toxic/Hazardous Substances and Environmental Engineering* 46 (7): 742–750.

Dzakpasu, M., Scholz, M., Harrington, R. et al. (2012). Characterising infiltration and contaminant migration beneath earthen-lined integrated constructed wetlands. *Ecological Engineering* 41: 41–51.

Dzakpasu, M., Scholz, M., Harrington, R. et al. (2014). Groundwater quality impacts from a full-scale integrated constructed wetland. *Groundwater Monitoring and Remediation* 34 (3): 51–64.

Dzakpasu, M., Scholz, M., McCarthy, V., and Jordan, S. (2015). Phosphorus retention and mass balance in an integrated constructed wetland treating domestic wastewater. *Water and Environment Journal* 29 (2): 298–306.

EPA (2003). Towards Setting Guideline Values for the Protection of Groundwater in Ireland (Interim Report). County Wexford, Ireland: Environmental Protection Agency.

EPA (2008). The Provision and Quality of Drinking Water in Ireland – a Report for the Years 2006–2007. County Wexford, Ireland: Environmental Protection Agency.

EU (European Union) (2007). Directive 2007/60/EC of the European Parliament and of the Council of 23 October 2007 on the assessment and management of flood risks. Official Journal of the European Union (Eur Lex). Ref L 288/27-34, dated 6 November 2007.

Fay, D., Kramers, G., Zhang, C. et al. (2007). *Soil Geochemical Atlas of Ireland.* Dublin: Teagasc and Environmental Protection Agency.

Field, A.P. (2009). *Discovering Statistics Using SPSS*, 3e. London: Sage Publications.

Foster, S.S.D. (1990). Impacts of urbanization on groundwater. *Proceedings of the Duisberg Symposium: Hydrological Processes and Water Management in Urban Areas.* IAHS publication No. 198, Duisberg: IAHS.

Gallardo, A. (2003). Spatial variability of soil properties in a floodplain forest in northwest Spain. *Ecosystems* 6: 564–576.

Gill, L., Johnston, P., Misstear, B., and O'Súilleabháin, C. (2004). *An Investigation into the Performance of Subsoils and Stratified Sand Filters for the Treatment of Wastewater from On-site Systems: Literature Review.* Dublin: Environmental Protection Agency.

Grad, H. (2006). Note on N-dimensional hermite polynomials. *Communications on Pure and Applied Mathematics* 2: 325–330.

HACH Company (2000). *Procedures Manual: Spectrophotometer DR 2010 49300-22*, 7e. Loveland, CO: HACH Company.

Hammer, D.A. and Knight, R.L. (1994). Designing constructed wetlands for nitrogen removal. *Water Science and Technology* 29 (4): 15–17.

Harrington, R. (2011) Applying the integrated constructed wetland (ICW) concept: a support strategy for sustainable water management. *Presented at Joint Meeting of Society of Wetland Scientists, Wetpol and Wetland Biogeochemistry Symposium*, Prague, Czech Republic.

Harrington, R., Dunne, D.J., Carroll, P. et al. (2005). The concept, design and performance of integrated constructed wetlands for the treatment of farmyard dirty water. In: *Nutrient Management in Agricultural Watersheds: A Wetlands Solution* (ed. E.J. Dunne, K.R. Reddy, and O. T. Carton), 179–188. Wageningen, The Netherlands: Wageningen Academic Publishers.

Harrington, R., Carroll, P., Carty, A.H. et al. (2007). Integrated constructed wetlands: concept, design, site evaluation and performance. *International Journal of Water* 3 (3): 243–256.

Hart, K., Lee, B., Franzmeier, D. et al. (2006). *Soil hydraulic conductivity and septic system performance. Purdue Extension Bulletins, RW-2-W, Rural Wastewater.* Purdue: Purdue University.

Hathaway, J.M., Cook, M.J., and Evans, R.O. (2010). Nutrient removal capability of a constructed wetland receiving groundwater contaminated by swine lagoon seepage. *Transactions of the ASABE* 53 (3): 741–749.

Helena, B., Pardo, R., Vega, M. et al. (2000). Temporal evolution of groundwater composition in an alluvial aquifer (Pisuerga River, Spain) by principal component analysis. *Water Research* 34 (3): 807–816.

Hunt, P.G. and Poach, M.E. (2001). State of the art for animal wastewater treatment in constructed wetlands. *Water Science and Technology* 44 (11–12): 19–25.

Jordan, W.R., Gilpin, M.E., and Aber, J.D. (1987). *Restoration Ecology: a Synthetic Approach to Ecological Research.* Cambridge: Cambridge University Press.

Jurado, G.B., Johnson, J., Feeley, H. et al. (2010). The potential of integrated constructed wetlands (ICWs) to enhance macroinvertebrate diversity in agricultural landscapes. *Wetlands* 30 (3): 393–404.

Kadlec, R.H. (2009). Wastewater treatment at the houghton lake wetland: soil and sediments. *Ecological Engineering* 35 (9): 1333–1348.

Kadlec, R.H. and Knight, R.L. (1996). *Treatment Wetlands*. Boca Raton, FL, USA: CRC Press.

Kang, H., Freeman, C., Lee, D., and Mitsch, W.J. (1998). Enzyme activities in constructed wetlands: implication for water quality amelioration. *Hydrobiologia* 368 (1–3): 231–235.

Karamouz, M., Kerachian, R., Akhbari, M., and Hafez, B. (2009). Design of river water quality monitoring networks: a case study. *Environmental Modeling and Assessment* 14: 705–714.

Kayranli, B., Scholz, M., Mustafa, A., and Hedmark, Å. (2010). Carbon storage and fluxes within freshwater wetlands: a critical review. *Wetlands* 30 (1): 111–124.

Kayranli, B., Scholz, M., Mustafa, A. et al. (2010). Performance evaluation of integrated constructed wetlands treating domestic wastewater. *Water, Air, and Soil Pollution* 210 (1–4): 435–451.

Khan, F.A. and Ansari, A.A. (2005). Eutrophication: an ecological vision. *Botanical Review* 71 (4): 449–482.

Knox, A.S., Paller, M.H., Nelson, E.A. et al. (2006). Metal distribution and stability in constructed wetland sediment. *Journal of Environmental Quality* 35 (5): 1948–1959.

Koh, D.C., Ko, K.S., Kim, Y. et al. (2007). Effect of agricultural land use on the chemistry of groundwater from basaltic aquifers, Jeju Island, South Korea. *Hydrogeology Journal* 15 (4): 727–743.

Langergraber, G. (2008). Modelling of processes in subsurface flow constructed wetlands: a review. *Vadose Zone Journal* 7 (2): 830–842.

Lee, C., Fletcher, T.D., and Sun, G. (2009). Nitrogen removal in constructed wetland systems. *Engineering in Life Sciences* 9 (1): 11–22.

Lesage, E., Rousseau, D.P., Meers, E. et al. (2007). Accumulation of metals in a horizontal subsurface flow constructed wetland treating domestic wastewater in Flanders, Belgium. *Science of the Total Environment* 380 (1–3): 102–115.

Lewis, J.W., Foster, S.D., and Draser, B.S. (1980). The risk of groundwater pollution by on-site sanitation in developing countries: a literature review. International Reference Centre for Wastes Disposal, Report No. 1–82, Dübendorf: International Reference Centre for Wastes Disposal.

Liu, C.W., Lin, K.H., and Kuo, Y.M. (2003). Application of factor analysis in the assessment of groundwater quality in a blackfoot disease area in Taiwan. *Science of the Total Environment* 313 (1–3): 77–89.

Liu, X., Wu, J., and Xu, J. (2006). Characterizing the risk assessment of heavy metals and sampling uncertainty analysis in paddy fields by geostatistics and GIS. *Environmental Pollution* 141: 257–264.

Majer Newman, J., Clausen, J.C., and Neafsey, J.A. (1999). Seasonal performance of a wetland constructed to process dairy milkhouse wastewater in connecticut. *Ecological Engineering* 14 (1–2): 181–198.

Mamo, M., Gupta, S.C., Rosen, C.J., and Singh, U.B. (2005). Phosphorus leaching at cold temperatures as affected by wastewater application and soil phosphorus levels. *Journal of Environmental Quality* 34 (4): 1243–1250.

Mañas, P., Castro, E., and de las Heras, J. (2012). Risks for human health of using wastewater for turf grass irrigation. In: *Irrigation – Water Management, Pollution and Alternative Strategies* (ed. I. García-Garizábal, R. Abrahao, and R. Rijeka), 185–208. In Tech.

Mann, R.A. and Bavor, H.J. (1993). Phosphorus removal in constructed wetlands using gravel and industrial waste substrata. *Water Science and Technology* 27 (1): 107–113.

McMinn, W.R., Yang, Q., and Scholz, M. (2010). Classification and assessment of water bodies as adaptive structural measures for flood risk management planning. *Journal of Environmental Management* 91: 1855–1863.

McQuillan, D. (2004). Ground-water quality impacts from on-site septic systems. *Proceedings of the National Onsite Wastewater Recycling Association, 13th Annual Conference* (7–10 November 2004). Albuquerque, NM: National Onsite Wastewater Recycling Association.

Misstear, B. and Brown, L. (2010). Groundwater recharge: influence of glacial tills on resource assessment and groundwater protection in Ireland. *Proceedings of the 30th Annual Groundwater Conference International Association of Hydrogeologists – Irish Group*, Tullamore, Co. Offaly, Ireland, pp. 1–12.

Mitsch, W.J. and Gosselink, J.G. (2007). *Wetlands*, 4e. New York, NY, USA: Wiley.

Moortel, V.D.A.M.K., Rousseau, D.P.L., Tack, F.M.G., and Pauw, N.D. (2009). A comparative study of surface and subsurface flow constructed wetlands for treatment of combined sewer overflows: a greenhouse experiment. *Ecological Engineering* 35 (2): 175–183.

Mulder, A., van de Graaf, A.A., Robertson, L.A., and Kuenen, J.G. (1995). Anaerobic ammonium oxidation discovered in a denitrifying fluidized bed reactor. *FEMS Microbiology Ecology* 16 (3): 177–184.

Mustafa, A., Scholz, M., Harrington, R., and Carroll, P. (2009). Long-term performance of a representative integrated constructed wetland treating farmyard runoff. *Ecological Engineering* 35 (5): 779–790.

Nelson, D. (2002). *Natural Variations in the Composition of Groundwater. Drinking Water Program.* Springfield, OR: Oregon Department of Human Services.

Nygaard, B. and Ejrnæs, R. (2009). The impact of hydrology and nutrients on species composition and richness: evidence from a microcosm experiment. *Wetlands* 29 (1): 187–195.

Oliver, M.A., Webster, R., and McGrath, S.P. (1996). Disjunctive kriging for environmental management. *Environmetrics* 7: 333–357.

Ouyang, Y., Nkedi-Kizza, P., Wu, Q.T. et al. (2006). Assessment of seasonal variations in surface water quality. *Water Research* 40 (20): 3800–3810.

Panswad, T. and Anan, C. (1999). Impact of high chloride wastewater on an anaerobic/anoxic/aerobic process with and without inoculation of chloride acclimated seeds. *Water Research* 33 (5): 1165–1172.

Power, J.F. and Schepers, J.S. (1989). Nitrate contamination of groundwater in North America. *Agriculture, Ecosystems & Environment* 26 (3–4): 165–187.

Reinhardt, M., Müller, B., Gächter, R., and Wehrli, B. (2006). Nitrogen removal in a small constructed wetland: an isotope mass balance approach. *Environmental Science & Technology* 40 (10): 3313–3319.

Rivoirard, J. (1994). *Introduction to Disjunctive Kriging and Non-linear Geostatistics.* Oxford, England, United Kingdom: Clarendon Press.

Sallade, Y.E. and Sims, J.T. (1997a). Phosphorus transformations in the sediments of Delaware's agricultural drainage ways: I. Phosphorus forms and sorption. *Journal of Environmental Quality* 26 (6): 1571–1579.

Sallade, Y.E. and Sims, J.T. (1997b). Phosphorus transformations in the sediments of Delaware's agricultural drainage ways: II. Effect of reducing conditions on phosphorus release. *Journal of Environmental Quality* 26 (6): 1579–1588.

Sauquet, E. (2006). Mapping mean annual river discharges: geostatistical developments for incorporating river network dependencies. *Journal of Hydrology* 331: 300–314.

Scholz, M. (2006). *Wetland Systems to Control Urban Runoff.* Amsterdam, The Netherlands: Elsevier.

Scholz, M. (2007). Classification methodology for sustainable flood retention basins. *Landscape and Urban Planning* 81: 246–256.

Scholz, M. and Sadowski, A.J. (2009). Conceptual classification model for sustainable flood retention basins. *Journal of Environmental Management* 90: 624–633.

Scholz, M. and Yang, Q. (2010). Guidance on variables characterizing water bodies including sustainable flood retention basins. *Landscape and Urban Planning* 98: 190–199.

Scholz, M., Harrington, R., Carroll, P., and Mustafa, A. (2007). The integrated constructed wetlands (ICW) concept. *Wetlands* 27 (2): 337–354.

Shapiro, S.S. and Wilk, M.B. (1965). An analysis of variance test for normality (complete samples). *Biometrika* 52 (3–4): 591–611.

Shepherd, D., Burgess, D., Jickells, T. et al. (2007). Modelling the effects and economics of managed realignment on the cycling and storage of nutrients, carbon and sediments in the blackwater estuary, UK. *Estuarine, Coastal and Shelf Science* 73 (3–4): 355–367.

Shih, S.C.W. and Nicholls, R.J. (2007). Urban managed realignment: application to the Thames estuary, London. *Journal of Coastal Research* 23: 1525–1534.

Shrestha, S. and Kazama, F. (2007). Assessment of surface water quality using multivariate statistical techniques: a case study of the Fuji river basin, Japan. *Environmental Modelling Software* 22 (4): 464–475.

Srinivas, T. (2008). *Environmental Biotechnology*. New Delhi: New Age International (P) Ltd.

Stottmeister, U., Wiessner, A., Kuschk, P. et al. (2003). Effects of plants and microorganisms in constructed wetlands for wastewater treatment. *Biotechnology Advances* 22 (1–2): 93–117.

Taylor, J.R. (2003). *Evaluating Groundwater Nitrates from On-lot Septic Systems, a Guidance Model for Land Planning in Pennsylvania*. Malvern, PA: Penn State Great Valley, School of Graduate Professional Studies.

Tilaki, R.A.D. and Kahe, D. (2012). Removal of ammonium ions from water by raw and alkali activated bentonite. In: *Advances in Environment, Biotechnology and Biomedicine* (ed. D. Pavelkova, J. Strouhal, and M. Pasekova), 169–174. Athens: WSEAS Press.

Tokida, T., Miyazaki, T., Mizoguchi, M., and Seki, K. (2005). In situ accumulation of methane bubbles in a natural wetland soil. *European Journal of Soil Science* 56 (3): 389–395.

Troeh, F.R. and Thompson, L.M. (2005). *Soils and Soil Fertility*, 6e. Ames, IA: Wiley-Blackwell.

Turcq, B., Albuquerque, A.L.S., Cordeiro, R.C. et al. (2002). Accumulation of organic carbon in five Brazilian lakes during the Holocene. *Sedimentary Geology* 148 (1–2): 319–342.

Uhlman, K. and Artiola, J.F. (2011). *Nitrate Contamination Potential in Arizona Groundwater: Implications for Drinking Water Wells*. Tucson, AZ: The University of Arizona Cooperative Extension, College of Agriculture and Life Sciences, The University of Arizona.

Vega, M., Pardo, R., Barrado, E., and Debán, L. (1998). Assessment of seasonal and polluting effects on the quality of river water by exploratory data analysis. *Water Research* 32 (12): 3581–3592.

Victoria, E.P.A. (2000). *Groundwater Sampling Guidelines*. Victoria, Australia: Environment Protection Authority.

Vohla, C., Alas, R., Nurk, K. et al. (2007). Dynamic of phosphorus, nitrogen and carbon removal in a horizontal subsurface flow constructed wetland. *Science of the Total Environment* 380 (1–3): 66–74.

Vymazal, J. (2007). Removal of nutrients in various types of constructed wetlands. *Science of the Total Environment* 380 (1–3): 48–65.

Wakida, F.T. and Lerner, D.N. (2005). Non-agricultural sources of groundwater nitrate: a review and case study. *Water Research* 39 (1): 3–16.

Wallace, S.D. and Knight, R.L. (2006). *Small-scale constructed wetland treatment systems: feasibility, design criteria, and o&m requirements*. London, UK: Water Environment Research Foundation.

Webster, R. and Oliver, M.A. (2001). *Geostatistics for Environmental Scientists*. Chichester, England, United Kingdom: Wiley.

WHO (2003). *Ammonia in Drinking-water. Background Document for Development of WHO Guidelines for Drinking-water Quality*. Geneva: World Health Organization.

WHO (2008). *Guidelines for Drinking-water Quality*, 3e, vol. 1. Recommendations. Geneva: World Health Organization.

Yang, Q., Scholz, M., and Shao, Y. (2012). Application of spatial statistics as a screening tool for sustainable flood retention basin management. *Water and Environment Journal* 26 (2): 155–164.

Yu, Z., Apps, M.J., and Bhatti, J.S. (2002). Implication of floristic and environmental variation for carbon cycle dynamics in boreal forest ecosystems of central Canada. *Journal of Vegetation Science* 13 (3): 327–340.

Zhang, L., Scholz, M., Mustafa, A., and Harrington, R. (2008). Assessment of the nutrient removal performance in integrated constructed wetlands with the self-organizing map. *Water Research* 42 (13): 3519–3527.

4

Water and Wastewater Treatment Technology and Modelling

4.1 Biological Activated Carbon Beds

4.1.1 Introduction

The treatment of river and reservoir water with biological activated carbon (BAC) is increasing because a greater understanding of the combined adsorption and biological regeneration process has been gained (Abed 1990; Jahangir 1994; Scholz and Martin 1997a). One of the major challenges concerning BAC filtration for water purification is the need to control biofilm development on the carbon granules without frequent backwashing that would destroy the natural biofilm equilibrium (Scholz and Martin 1997b). It is difficult to prevent biological activity because of the high adsorption potential of granular activated carbon (GAC) for organic matter and the ease of bacterial colonisation on the rough and porous GAC surface. Even a high dose of chlorine (2 mg dm^{-3} h^{-1}) may not be effective to sterilise the filter system (Abed 1990; Camper et al. 1986).

Frequent thermal regeneration of BAC is not necessary. Partial biological regeneration of GAC due to sorption and later metabolism of slowly biodegradable carbon can be observed in practice. This leads to an extended life of the carbon bed. BAC can further protect some bacteria against the negative influence of toxic substances. Surges of easily adsorbable matter in the feed can be damped. For small changes in concentration even virtual elimination of organic matter is possible (Jahangir 1994). The removal of organic and/or toxic substances is higher on BAC in comparison with GAC. This is particularly evident with reference to river water (Abed 1990).

Scholz and Martin (1997a, b) investigated the ecological equilibrium on BAC and possible parameters to control biomass growth. Good removal efficiencies of feed (mainly glycerol), nitrate, total organic carbon (TOC) and chemical oxygen demand (COD) were achieved. These researchers also showed that backwashing was reduced to a minimum for a steady-state biomass composition with limited growth of filamentous bacteria. Reductions of dissolved oxygen (DO), pH and redox potential in the bed were low during their study, but enough bacteria and protozoa were maintained to establish a strong protozoa–bacteria–fungi equilibrium by keeping pH values high and maintaining a balanced nutrition availability. Furthermore, high DO levels resulted in high numbers of crawling protozoa and rotifers (good indicators for removal efficiency), and in high bacteria counts leading to high numbers of ciliated protozoa. Finally, flagellates were present in significant numbers after the system set-up (exhaustion period dominated mainly by physical and chemical adsorption processes), toxic shocks and pH swings.

There are also disadvantages of BAC filtration without pH and DO control. The excessive growth of slime-producing bacteria on BAC may cause clogging due to mud-ball formation. This results in

Sustainable Water Systems, First Edition. Miklas Scholz.

high-pressure loss, so that backwashing must be undertaken more often. Furthermore, an increase in bacteria may be observed in the effluent from a BAC filter. This would require a higher chlorine demand for post chlorination (Jahangir 1994). In contrast, work undertaken by Donlan and Yohe (1983) shows that there is a low risk with respect to BAC filter operation; potentially pathogenic bacteria are reduced by higher micro-organisms including bacterivorous protozoa, if optimal growth conditions are present.

This study assesses the control of biofilm growth on BAC; parameters studied include empty bed contact time (EBCT), DO, pH and various nutrients. Correlations between changes in the population of micro-organisms in BAC beds and COD outflow concentrations were found and subjected to extensive statistical analysis. This assessment is the basis for two mathematical spreadsheet models estimating COD removal efficiencies achieved by BAC filters. Chapter 4.1 is based on original articles by Scholz and Martin (1998a, b).

4.1.2 Materials and Methods

A BAC pilot plant with four columns (BAC 1, BAC 2, BAC 3 and BAC 4) was established. Figure 4.1 shows a simplified flow chart of the BAC column system. The effects of different control and inflow water quality parameters on microorganisms were of interest. The following combinations were possible to investigate the influence of DO or EBCT on the composition of higher micro-organisms, including protozoa in BAC beds: (i) combination I: BAC 1 and BAC 2 (difference in DO); (ii) combination II: BAC 3 and BAC 4 (difference in DO); (iii) combination III: BAC 1 and BAC 3 (difference in EBCT); and (iv) combination IV: BAC 2 and BAC 4 (difference in EBCT).

All other parameters were kept constant. Table 4.1 indicates the parameters examined and initial conditions for the performance of the pressure upward flow system. Table 4.2 gives an overview concerning the different nutrient levels in the influent. High initial concentrations of glycerol as

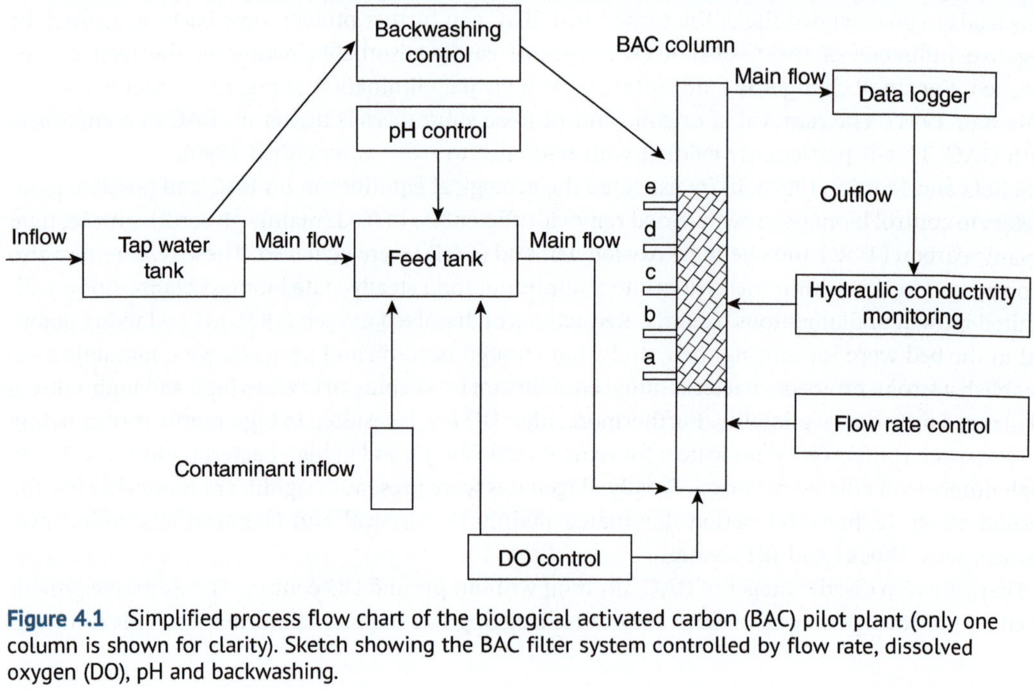

Figure 4.1 Simplified process flow chart of the biological activated carbon (BAC) pilot plant (only one column is shown for clarity). Sketch showing the BAC filter system controlled by flow rate, dissolved oxygen (DO), pH and backwashing.

Table 4.1 Purification of water by biological activated carbon (BAC): system parameters and performance of purification.

Parameter and performance	Explanations and approximate numerical values
Empty bed contact time	8 minutes for BAC 1 and BAC 2; 15 minutes for BAC 3 and BAC 4
Dissolved oxygen control	Artificial dissolved oxygen supply only for BAC 2 and BAC 4, if DO<4 $mg\ dm^{-3}$
Flow rate	0.54 $dm^3\ min^{-1}$ for BAC 1 and BAC 2; 0.29 $dm^3\ min^{-1}$ for BAC 3 and BAC 4
Flow mode	Pressure upward flow
Bed volume	4.34 dm^3 (initial)
Bed diameter : depth of granular activated carbon	12 cm : 38 cm (initial)
Granular activated carbon	Filtrasorb 400 (GAC between 0.85 and 1.2 mm in diameter)
Influent	Aerated (optional) tap water, glycerol, weak phosphate buffer, potassium nitrate, different trace minerals (optional) and natriumhydroxide (NaOH) (optional)

Table 4.2 Biological activated carbon (BAC) inflow water: own estimations of nutrients and trace minerals ($mg\ dm^{-3}$). Maximum possible biochemical oxygen demand (BOD) requirements and concentrations for tap water are also provided for comparison with BAC inflow water.

Nutrients and trace minerals	21 August 1996–7 October 1996	8 October 1996–6 November 1996	7 November 1996–17 November 1996	Maximum possible BOD ($mg\ dm^{-3}$) requirement for nutrients ($mg\ dm^{-3}$) after 7 November 1996[a]	Estimation for pure tap water[b]
Carbon (glycerol)	17.88	10.81	10.81	Unknown	0.00
Carbon (source unknown)	1.1	1.1	1.1	Unknown	1.1
Nitrogen	0.56	0.56	0.56	Unknown	0.36
Phosphorus	0.11	0.11	0.11	Unknown	0.00
Calcium	0.18	0.18	0.18	2903	0.18
Carbonate	29.38	29.38	29.38	10 881	29.38
Iron	0.08	0.08	0.67	56	0.08
Magnesia	1.10	5.04	5.04	1680	1.10
Manganese	0.01	0.07	0.07	700	0.01
Nickel	0.00	0.02	0.02	Unknown	0.00
Potassium	0.56	0.56	0.56	124	0.00
Sodium	4.76	4.76	4.76	95 200	4.60
Zinc	0.00	0.01	0.01	63	0.01

[a] Related to information provided by Eckenfelder (1980).

[b] According to Severn Trent Water Plc (water utility); mean of record period: 1 January 1995–29 November 1995; water supply source: Frankley Final Water.

a carbon source were used to shorten the exhaustion period of GAC and to simulate very weak wastewater or polluted river water.

With the exceptions of COD, total plate count tests and protozoa counting, all analytical procedures were performed according to the American Standard Methods for the Examination of Water and Wastewater (APHA, AWWA and WPCF 1985). The COD and total plate count tests with nutrient agar were based on British Standard Methods (Her Majesty's Stationery Office 1978). In the measurement of COD below concentrations of 100 mg dm^{-3}, it was necessary to concentrate the sample by a factor of 20 to make the analysis viable.

Protozoa and other micro-organisms were counted in the influent, effluent and at five or more positions in the carbon bed itself by using a counting chamber and a bright field/phase-contrast microscope (Olympus Vanox, Tokyo, Japan, 204899). Some protozoa were defined according to their movement in the liquid phase around carbon particles: free swimming, crawling, stalked and flagellated protozoa. This classification is partly subjective, easy to undertake on-site and like the subdivision of bacterivorous ciliates in sewage treatment processes.

A water sample (mean volume: 32 μL) was taken for analysis of free swimming, crawling, stalked and flagellated protozoa. Furthermore, rhizopoda, actinopoda, nematodes and rotifers were estimated by direct microscopic counting. The number of stalked protozoa attached to carbon granules could not be estimated with the present technology. However, glass slides exposed to the effluent showed the same order of magnitude with respect to the amount of stalked and crawling protozoa compared with values in the liquid phase of the column. Pictures were taken and observations were recorded on video tape for detailed determination of dominant species (magnitude: 40 × 40). Statistical analyses are based on examples by Fowler and Cohen (1996).

4.1.3 Results and Discussion

4.1.3.1 Biomass Growth Monitoring and Control

The purpose was to control and monitor biomass growth to optimise the COD removal efficiency. Glycerol was used as the main carbon source (Table 4.2). DO and pH control were applied to increase the abundance of ciliated protozoa and to decrease growth rates of filamentous bacteria and fungi. It had previously been found that numbers of ciliates correlate with adequately treated plant effluent, and that filamentous micro-organisms were responsible for filter-clogging phenomena (Scholz and Martin 1997b).

Average DO reductions (DOr) per week are expressed in Table 4.3 (for example: influent DO > effluent DO). BAC 2 and BAC 4 were subject to aeration, if the DO level in the effluent dropped below 4 mg dm^{-3} (the critical value of the aquatic species investigated in this study). pH change is expressed in terms of average reductions (pHr) per week (Table 4.3). The main decreases of pH took place in the first few centimetres of the filter length near the inlet of the BAC filter where the biofilm was particularly well established.

Turbidity and suspended solids (SS) are convenient parameters for monitoring biomass growth. Fluctuations of turbidity in terms of average increase rates (Tbi) per week are shown. Thus, effluent turbidity was greater than influent turbidity. Carbon fines were responsible for the presence of single peaks. Table 4.3 also shows the average increase of suspended solids (SSi) per week. The general level of SS was very low, if local maxima due to carbon losses (100 particles mm^{-3}, carbon fines <30 μm) during aeration periods are ignored (weeks 34–36).

The control of bacteria, fungi and protozoa growth rates by means of parameters including DO and pH was successful. It follows that the number of viable bacteria and fungi in the liquid phase of the carbon bed was relatively low (10^2–10^4 colonies cm^{-3}). Measurements with the respirometer

Table 4.3 Reduction of dissolved oxygen (DO) (mg dm^{-3}) and pH (influent level minus effluent level); increase of turbidity (NTU) and suspended solids (SS) (mg dm^{-3}) (effluent level minus influent level).

Calendar week	BAC 1				BAC 2				BAC 3				BAC 4			
	DOr	pHr	Tbi	SSi	DOr	pHr	Tbi	SSi	DOr	pHr	Tbi	SSi	DOr	pHr	Tbi	SSi
34	4.4	0.2	0.5	2	2.6	0.3	0.4	46	2.6	0.5	0.3	3	2.7	0.6	0.8	25
35	3.3	0.1	0.2	0	2.4	0.3	1.1	9	1.9	0.3	0.2	0	2.9	0.3	2.4	17
36	2.7	0.3	0.9	3	2.6	0.1	0.4	1	1.5	0.1	0.3	2	3.7	1.2	1.3	47
37	3.8	1.3	0.2	1	3.4	1.6	0.9	3	1.6	1.4	0.2	0	3.1	1.4	1.2	7
38	1.6	1.2	0.3	1	3.0	1.0	0.7	2	0.8	1.2	0.1	3	3.3	1.3	0.8	8
39	2.7	1.3	0.0	−1	3.6	1.3	0.5	1	2.1	1.5	0.1	1	4.4	1.4	0.3	6
40	4.5	1.3	0.9	1	4.1	1.3	0.7	9	1.6	1.3	1.6	12	3.7	1.3	0.5	5
41	2.9	1.3	0.4	1	4.3	1.3	0.4	13	1.4	1.5	1.4	3	3.5	1.6	0.7	5
42	3.4	1.1	0.7	2	3.6	1.1	0.8	10	1.5	0.9	2.3	2	4.8	1.1	0.7	9
43	3.2	0.4	1.2	−2	3.7	0.5	1.1	7	1.5	0.3	1.4	14	4.5	0.6	1.5	12
44	2.2	0.3	0.4	−1	2.2	0.6	0.3	3	1.3	0.6	0.8	8	2.1	0.8	0.6	5
45	1.7	1.2	0.1	0	1.8	1.3	0.0	0	1.0	1.2	0.4	5	2.4	1.4	0.5	6
46	1.0	0.9	0.0	−1	1.0	1.0	0.0	0	0.5	1.0	0.1	1	2.1	1.1	0.2	3

i, increase; NTU, nephelometric turbidity units; r, reduction; SS, suspended solids; Tb, turbidity.

Table 4.4 Correlation between chemical oxygen demand (COD) and microorganisms in %.

Location	FSP	CP	SP	FP	RZ	AC	NM	RT
Relative average for all columns	69	30	70	53	71	18	19	51
BAC 1	16	75	a	40	44	a	8	11
BAC 2	75	82	70	61	65	a	72	82
BAC 3	17	51	72	63	29	a	43	70
BAC 4	41	35	a	20	a	30	51	76

[a] Insufficient count data for a reliable analysis of correlation.
AC, actinopoda; CP, crawling protozoa; FP, flagellated protozoa; FSP, free swimming protozoa; NM, nematodes; RZ, rhizopoda (except AC); SP, stalked protozoa; RT, rotifers.

confirmed the results of total plate count tests. During the actual experiment, large protozoa were present in high numbers (10^2–10^3 counts cm^{-3}) throughout the experiment. Colourless flagellates like *Paraphysomonas vestita* showed maximum growth rates (5–10/day) directly after the system set-up and after great pH swings. Further analysis showed that high DO levels resulted in high numbers of crawling and stalked protozoa as well as rotifers (100–500 counts cm^{-3}) for each group. The numbers of these microorganisms correlated with good COD removal efficiencies (Table 4.4). Bacteria levels were high if ciliates were present in high numbers (400 bacteria colonies per ciliate).

Eckenfelder (1980) has defined the trace nutrients required for biological oxidation, and the fifth column of Table 4.2 summarises some of his findings. Of the trace minerals mentioned by Eckenfelder (1980), carbon trioxide (CO_3), magnesium (Mg), manganese (Mn) and sodium (Na) were present at high concentrations in the inflow water after the 8th of October. However, there was a lack of iron on the BAC granules. Therefore, iron was added as a trace mineral at the end of the experiment to investigate the biological response to that mineral (Table 4.2).

4.1.3.2 Biological Indicator Significance

Statistical analysis was necessary to compare the abundance of certain micro-organisms and COD effluent concentrations for different environmental conditions. Four BAC filters with slight differences in EBCT and DO concentration were in operation (Table 4.1). The combinations (mentioned in Chapter 4.1.2) showed statistically highly significant differences from each other with respect to a selection of eight micro-organism groups that are listed in Table 4.4. The analysis was based on the non-parametric G-test statistic for unmatched pairs of samples (Fowler and Cohen 1996).

The ecological diversity of micro-organisms in a BAC bed indicates how favourable environmental conditions are. It had previously been found that a high diversity is beneficial for an ecological equilibrium on BAC (Scholz and Martin 1997a). The Shannon–Weaver diversity index (H) considers how evenly the total number of individuals in a sample is apportioned between each species and was calculated regarding micro-organism diversity for all filters that are described in Table 4.1 (Fowler and Cohen 1996).

The diversity was low ($H = 0.4$) during the system set-up. The population diversity increased to $H = 1.0$ after the inflow pH value had been increased by 1.1. If the EBCT was low (eight minutes), the Shannon–Weaver diversity index was high ($H = 1.5$) for both simulation stages. The general development of H was different for all BAC filters after trace mineral levels were increased and glycerol as a carbon source was decreased by 7.07 mg dm^{-3} (Table 4.2). The Shannon–Weaver diversity index of BAC 1 decreased initially to $H = 0.3$ after about two weeks and then stepped up rapidly to 1.4 after having been constantly low for another week. It increased gradually to 1.3 for BAC 2 and fluctuated greatly between 0.1 and 1.4 for BAC 3. Small variations of the Shannon–Weaver index for BAC 4 at around 1.0 could be observed.

Another statistical tool can be selected to demonstrate that high species diversity is also an indicator for good COD removal efficiencies. The Spearman rank correlation coefficient was applied to estimate the correlation between the COD removal efficiency and the Shannon–Weaver diversity index regarding microorganism population diversity for all four BAC filters. Because observations, measurements and counts must be converted to ranks, the Spearman rank correlation coefficient is only used for ordinal scales. Ordinal numbers may not indicate absolute quantities. Counts as well as observations do not have to be distributed normally; they can be distribution free (Caswell 1995; Fowler and Cohen 1996).

The micro-organism diversity correlated positively with COD removal efficiency a few weeks after nutrient level control (Table 4.2) was introduced. A high diversity was an indicator for good COD removal efficiencies. This result is consistent with conservation theories that emphasise the importance of a rich population diversity as a necessity for an ecosystem that has reached an equilibrium state.

It remains to be investigated whether COD removal efficiencies of all filters are significantly different from each other, possibly because of significant differences in operational conditions. The G-test was used to show that COD removal efficiencies were not similar for filters with slight differences in EBCT and DO (Table 4.1). The G-test statistic was applied to a (time zones) × (average COD removal efficiencies) contingency table for different filter combinations (see above).

The level of probability for 'difference in proportions' concerning a combination can be estimated with the G-test, which is like the well-known Chi-test. In this case, the chi-test would probably be applied more often in practice. However, the G-test is much easier to conduct, more conservative and mathematically more advantageous.

There were large differences regarding all combinations of BAC filter COD removal efficiencies on a day-to-day basis. The probability that data sets concerning combination I are similar, was ≤5% and the probability that all other combinations are also similar was ≤1%. However, the absolute average COD removal efficiency difference between all combinations was low (≤5%). A comparison of the overall performance showed that a filter with a low EBCT and/or no DO control system was more successful. Therefore, BAC 1 (EBCT: eight minutes; no DO control) was the best option.

The G test statistic was further used for a (filter parameter; 'DO control on/off' and EBCT) × (average micro-organism distribution) contingency table for different filter combinations. This contingency table allows to investigate the association between the frequencies of operational parameters (EBCT and DO) and microorganism distribution. The level of 'difference in proportions' was statistically very highly significant for all combinations, with a probability of ≤1% that the proportions were similar. This reflected the fact that the differences of average numbers of microorganisms regarding the liquid phases of two BAC filters were high (approximately between 1000 and 6000). It was concluded that it is statistically very highly significant that there was a positive correlation between microorganisms and EBCT on one side and between microorganisms and DO concentration on the other side.

The G-test indicated that COD removal efficiencies were different on a day-to-day basis for all filter combinations. However, the Kruskal–Wallis test does not provide general support for this claim, as the two tests have different objectives. The Kruskal–Wallis test only compares medians of removal efficiencies for BAC filters and was carried out in Chapter 4.1 to compare its performance with the G-test. The simple non-parametric Kruskal–Wallis test statistic (K) for grouped average COD removal efficiencies of all four columns was applied to investigate possible differences regarding the medians as a measure of central tendency within the group of BAC filters as a whole (Fowler and Cohen 1996).

The Kruskal–Wallis test statistic compared medians of unmatched samples (different BAC filters) and a K value of 0.23 was considered to be very low. It follows that there were no statistically significant differences regarding the medians. Furthermore, the rank number distribution showed a similar optical pattern for all BAC filters but especially for the combinations I and II (see above). The influence of EBCT concerning the rank distribution was small, for example, between 22 September 1996 and 7 October 1996.

The Kruskal–Wallis test statistic was also applied for total and grouped average numbers of microorganisms of all four columns to investigate possible differences regarding the medians within the sum of BAC filters. A K-value of 5.82 for total average numbers of microorganisms showed that there were no great differences of the median variables with respect to microorganisms inhabiting BAC filters. The only statistically significant differences found within the rhizopoda group (except actinopoda) were between the medians for those populations ($K = 9.15$). This shows that the Kruskal–Wallis test is not appropriate to detect potential biological indicators (except for rhizopoda (without actinopoda)).

The statistics reported above provide evidence for a significant relationship between the control parameters studied and the abundance of microorganisms. It is important also for practical purposes to examine the strength of the correlation between COD removal efficiency and the presence of certain microorganisms (Tables 4.4 and 4.5). Overall TOC removal efficiencies were between 15% and 35%. The COD reductions expressed as efficiencies per week are given in Figure 4.2. Table 4.4 shows the strength of correlation between mean COD concentrations ($<18\ mg\ dm^{-3}$) and the means of microorganism groups in the BAC bed (counts cm^{-3} liquid phase of the carbon column).

Table 4.5 Distribution of points (total of 10 per organism) for genera of protozoa to estimate the chemical oxygen demand (COD) and total organic carbon (TOC) effluent of biological activated carbon (BAC) filters.

Genus	BAC No.	Indicator grade[a]	COD (TOC) effluent in mg dm^{-3} of the BAC filter					
			0–3	4–6	7–9	10–12	13–15	>15
Astasia	BAC 1	2 (2)	2 (2)	2 (1)	2 (1)	1 (2)	2 (2)	1 (2)
Chilodonella	BAC 1	2 (2)	2 (2)	1 (1)	1 (2)	2 (2)	2 (2)	2 (1)
Colpidium	BAC 1	2 (2)	2 (2)	1 (1)	1 (2)	2 (2)	2 (2)	2 (1)
Euplotes	BAC 1	2 (2)	2 (2)	1 (1)	1 (2)	2 (2)	2 (1)	2 (1)
Linotus	BAC 1	2 (2)	2 (2)	1 (1)	1 (2)	2 (2)	2 (2)	2 (1)
Oxytrichia	BAC 1	3 (1)	1 (3)	1 (2)	1 (2)	2 (1)	3 (1)	2 (1)
Polytoma	BAC 1	2 (2)	3 (1)	2 (1)	1 (2)	1 (2)	2 (2)	1 (2)
Anisonema	BAC 2	1 (2)	4 (2)	2 (2)	1 (2)	1 (2)	1 (1)	1 (1)
Chilodonella	BAC 2	1 (2)	3 (2)	2 (1)	2 (1)	1 (4)	1 (1)	1 (1)
Colpidium	BAC 2	2 (3)	2 (1)	2 (1)	2 (1)	2 (5)	1 (1)	1 (1)
Euplotes	BAC 1	1 (2)	3 (2)	2 (1)	2 (1)	1 (3)	1 (1)	1 (2)
Linotus	BAC 1	2 (2)	2 (2)	2 (1)	2 (1)	2 (3)	1 (1)	1 (2)
Loxodes	BAC 1	1 (3)	3 (1)	2 (1)	3 (1)	1 (5)	0 (1)	1 (1)
Oxytrichia	BAC 2	1 (3)	3 (1)	2 (1)	2 (1)	1 (3)	1 (2)	1 (2)
Polytoma	BAC 2	2 (2)	2 (1)	2 (3)	1 (2)	1 (2)	2 (1)	2 (1)
Urustyla	BAC 2	1 (2)	4 (2)	2 (1)	2 (1)	1 (4)	1 (1)	0 (1)
Bodo	BAC 3	2 (2)	1 (2)	2 (2)	2 (2)	2 (2)	2 (1)	1 (1)
Bryophyllum	BAC 3	2 (2)	1 (2)	2 (1)	2 (1)	2 (3)	2 (1)	1 (2)
Chilodonella	BAC 3	2 (2)	1 (1)	1 (1)	2 (2)	2 (3)	3 (1)	1 (2)
Colpidium	BAC 3	2 (2)	1 (1)	1 (1)	2 (2)	2 (3)	3 (1)	1 (2)
Linotus	BAC 3	2 (2)	1 (1)	1 (1)	2 (2)	2 (3)	3 (1)	1 (2)
Opisthonecta	BAC 3	1 (3)	3 (1)	2 (0)	2 (1)	2 (4)	1 (2)	0 (2)
Oxytrichia	BAC 3	2 (3)	1 (1)	2 (0)	2 (1)	2 (3)	2 (2)	1 (3)
Peranema	BAC 3	2 (2)	1 (1)	2 (1)	2 (2)	2 (3)	1 (1)	1 (2)
Polytoma	BAC 3	2 (2)	1 (1)	2 (1)	2 (2)	2 (3)	2 (2)	1 (1)
Vorticella	BAC 3	1 (3)	4 (1)	2 (0)	2 (1)	2 (4)	0 (2)	0 (2)
Chilodonella	BAC 4	3 (2)	1 (1)	1 (1)	1 (2)	2 (2)	2 (3)	3 (1)
Colpidium	BAC 4	3 (2)	1 (1)	1 (1)	1 (2)	2 (2)	2 (3)	3 (1)
Linotus	BAC 4	3 (2)	1 (1)	1 (1)	1 (2)	2 (2)	2 (3)	3 (1)
Loxodes	BAC 4	2 (2)	2 (1)	1 (1)	2 (2)	2 (2)	1 (2)	2 (2)
Oxytrichia	BAC 4	1 (3)	1 (0)	4 (1)	2 (1)	1 (1)	1 (4)	1 (3)
Vorticella	BAC 4	2 (np)	0 (np)	3 (np)	3 (np)	2 (np)	2 (np)	0 (np)

[a] 1, Indicator for good COD and/or TOC removal; 2, insufficient COD and/or TOC removal; 3, indicator for bad COD and/or TOC removal.

np, Not present in significant numbers.

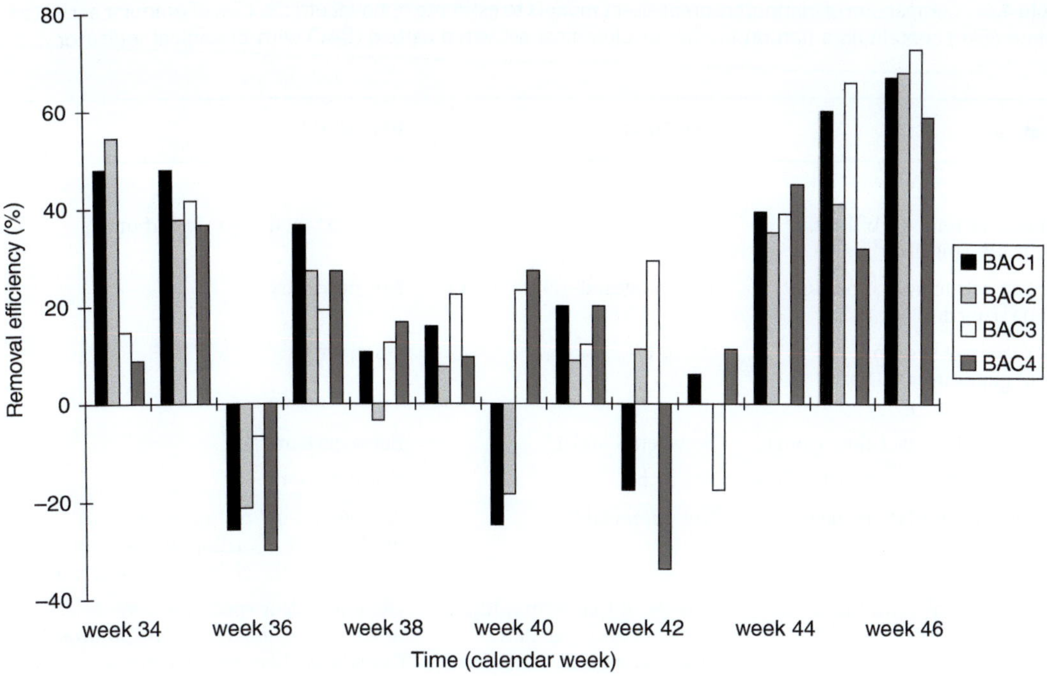

Figure 4.2 Average weekly chemical oxygen demand (COD) removal efficiency (empty bed contact time (EBCT) = 8 minutes for biological activated carbon (BAC) 1 and BAC 2, EBCT = 15 minutes for BAC 3 and BAC 4; dissolved oxygen (DO) manipulation for BAC 2 and BAC 4). Column graph showing an increase of the COD removal efficiency for all BAC filters towards the end of the experiment.

Furthermore, Table 4.5 is a guide for BAC filter operation managers to estimate COD and TOC removal efficiencies with indicator protozoa.

4.1.3.3 Spreadsheet Modelling of Filter Effluents

Two models were developed to give the BAC treatment plant manager a practical tool for estimating COD removal efficiencies. They are based on the statistical analysis discussed in the previous section. The first statistical spreadsheet model called BAC 2000 estimates COD removal efficiencies with the help of the following parameters: DO, EBCT, pH, free swimming protozoa, crawling protozoa, flagellates, rotifers and an estimated COD range. The model was verified with data obtained from BAC 2. There was a strong correlation between the predictions obtained with BAC 2000 and real measurements ($R^2 = 0.77$). The model is applicable for a wide range of GAC or BAC operations in the water industry. However, it does not refer to the level of confidence concerning the estimated values.

The second spreadsheet model entitled BAC 2001 estimates COD and TOC removal efficiencies. It is based on a few biological indicator genera and/or species and some physical and chemical control parameter (DO, EBCT, pH, nutrients and trace minerals). The operator can choose between four operation stages. The first stage 'system set-up' is applicable for fresh (not exhausted) BAC. The second stage 'increase of pH' is applicable for BAC containing a natural biofilm. All other stages are dominated by biological regeneration phenomena of the exhausted GAC. The BAC 2001 model is an empirical model applicable mainly for BAC columns. Furthermore, it calculates a confidence level where accuracy is subject to the limits of protozoa and/or rotifer count ranges defined in the model manual. Table 4.6 compares the main input and output features of both models with each other.

Table 4.6 Comparison of computer spreadsheet models to estimate removal efficiencies of granular activated carbon (GAC) containing a natural biofilm or biological activated carbon (BAC) with biological indicator organisms.

Features	BAC 2000	BAC 2001
Input data		
Choice of removal efficiency indicator (mg dm^{-3})	COD[a]	COD[a] or total organic carbon
Rough estimation of outflow COD[a] (mg dm^{-3})	Two ranges: 0–≤9 or >9–17	Not necessary
DO[b] control system ('on' guarantees a DO[b] of >4 mg dm^{-3})	On or off	On or off
Empty bed contact time (min)	Between 8 and 15	Between 8 and 15
Observation of recent pH swings	Yes or no	Not necessary
Choice of simulation stages	Not appreciable	'System set-up' or 'increase of pH' or 'decrease of carbon, increase of trace minerals' or 'increase of ferrous'
Available biological indicators	Groups of free swimming, crawling and flagellated protozoa as well as rotifers	Genera and/or species of protozoa or rotifers: *Chilodonella unicinatum*, *Colpidium*, *Lionotus* (*Litonotus*), *Ophryoglena*, *Oxytrichia fallax*, *Polytoma uvella*, *Cephalodella gracilis* and *Epiphanes senta*
Weighting of biological indicators	Possible	Possible
Graphical input representation	Yes	Yes
Output data		
Inflow COD[a], outflow COD[a] and estimated COD[a] within column centre (all in mg dm^{-3})	Yes	Not applicable
Removal efficiency in %	Yes	Yes
Confidence level in percentage (subject to biological indicator ranges)	None	Available as a guideline
Error messages	None	Service available
Graphical output presentation	Yes	Yes

[a] Chemical oxygen demand.
[b] Dissolved oxygen.

4.1.4 Conclusions

The aim of Chapter 4.1 to control micro-organism growth and to use protozoa as bio-indicators to predict COD removal efficiency rates of BAC filters was achieved. The analysis of experimental BAC filter data showed that physical and chemical parameters including DO, pH, turbidity and SS are in acceptable ranges for potable water production.

The EBCT, DO, pH and nutrient availability are suitable physical/chemical control parameters to manipulate growth of bacteria, fungi and protozoa. The micro-organism diversity is advantageous for good COD removal efficiencies. Median values of certain micro-organism groups are independent of EBCT and DO changes.

Certain groups of microorganisms including crawling and flagellated protozoa are indicators for the system performance. There is a strong correlation between the abundance of some microorganisms and acceptable COD and TOC removal efficiencies. The construction of mathematical spreadsheet models based on biological indicators and physical/chemical control parameters is an appropriate method of modelling biological regenerated granular activated carbon.

4.2 Constructed Wetlands Treating Sewage

4.2.1 Introduction

Constructed wetland systems are more effective and reliable alternatives for efficient treatment of various types of wastewaters in comparison to traditional wastewater treatment systems (Ansola et al. 2003; Scholz 2010; Tshihrintzis et al. 2007). Wetlands are particularly suitable for small communities due to their gravity-driven power independence and low construction, operation and maintenance costs (Kadlec and Knight 1996; Scholz 2010). Standard treatment wetlands are classified into two major types according to their hydraulic water flow characteristics: free water surface and sub-surface flow systems. Sub-surface flow systems can be further sub-divided into horizontal flow and vertical flow wetlands. General configurations of different types of constructed wetlands have been described by Kadlec and Knight (1996) and Scholz (2010).

In the UK, the application of constructed wetlands for the treatment of sewage (usually in a secondary or tertiary capacity) has been greatly expanded in recent years (Nuttall et al. 1997; Thomas et al. 2003). Current research on wetland treatment efficiency is mainly focused on the reduction of biochemical oxygen demand (BOD) (Brix et al. 2001), ammonia-nitrogen (NH_3-N), nitrate-nitrogen (NO_3-N), phosphorus (Scholz et al. 2010) and metals (Dunbabin and Bowner 1992).

The Wildfowl & Wetlands Trust (WWT) is a UK conservation organisation preserving and enhancing natural wetlands and ecosystems for wildlife and people. Sixteen wetland treatment systems have been constructed at nine wetland centres across the UK and are used to treat water coming into or leaving the reserves and wastewater generated from these centres (WWT 2008). Chapter 4.2 presents water quality monitoring data collected from five representative constructed wetland systems in WWT reserves (Figure 4.3) and assesses their treatment efficiencies in relation to key wastewater quality parameters. Any obvious trends in seasonal and long-term nutrient removal performance are highlighted.

This study is important because only relatively few constructed wetlands of the type have been studied for a sufficiently long period to reveal the long-term problems that may develop as the systems age. Chapter 4.2 particularly identifies long-term trends in nutrient removal within real constructed wetland systems operated in practice and is based on the original article by Dong et al. (2013).

4.2.2 Case Studies and Methods

4.2.2.1 Description of Wetland Systems

The locations of the studied constructed wetlands are shown in Figure 4.3. Each wetland system has a unique design configuration and operation period. Hydraulic loading rates and retention times vary considerably between the studied wetlands. A variety of plant species are planted in wetland beds and channels. More detailed information is provided in Table 4.7. The following paragraphs describe briefly the configurations of the five wetlands assessed in this study.

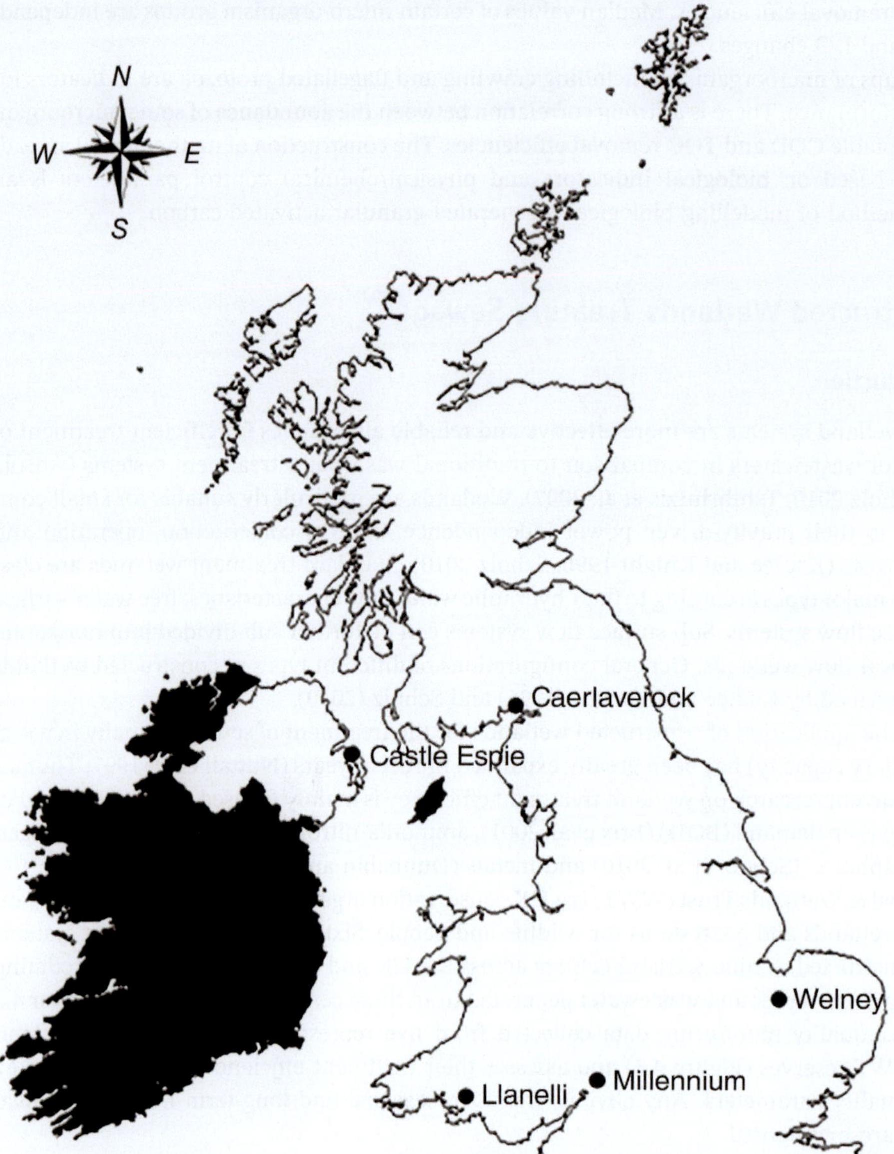

Figure 4.3 Location map of representative Wildfowl and Wetlands Trust constructed wetlands. Two wetlands in England and one each in Wales, Scotland and Northern Ireland are shown on a map.

Table 4.7 Summary information for Wildfowl and Wetlands Trust representative constructed wetlands.

Site	Construction	Influent	Plant species	System description	HRT (day)	HLR (cm day^{-1})
Caerlaverock	2001	Raw sewage	*Phragmites australis*	Stage 1: Septic tank Stage 2: Horizontal sub-surface flow bed (125 m^2) Stage 3: Free surface flow bed (125 m^2)	NA	NA
Castle Espie	1993	Raw sewage	*Phragmites australis*	Stage 1: Septic tank Stage 2: Two parallel wetland beds (horizontal sub-surface flow bed (150 m^2) and free surface flow bed (150 m^2))	30	2
Llanelli	2000	Tertiary treated sewage	*Phragmites australis* and *Iris pseudacorus*	Stage 1: Settling pond (3000 m^2) Stage 2: Two parallel free surface flow beds (2800 m^2 each) Stage 3: Two parallel free surface flow beds (5625 m^2 each)	0.5	1.4
Millennium	1999	Raw sewage	*Phragmites australis*, *Iris pseudacorus*, *Carex riparia* and others	Stage 1: Settlement tank Stage 2: Vertical sub-surface flow bed (400 m^2) Stage 3: Settlement pond Stage 4: Horizontal sub-surface flow bed (50 m^2) Stage 5: Two parallel free surface flow bed (75 m^2 each) Stage 6: Vertical flow phosphate removal bed (100 m^2)	5	11
Welney	2006	Raw sewage	*Phragmites australis*, *Typha latifolia*, *Sparganium etectum*, *Epilobium hirsutum*, *Glyceria maxima*, *Alisma plantago-aquatica*, *Lemna minor* and others	Stage 1: Septic tank Stage 2: Four horizontal sub-surface flow beds (105 m^2 each) Stage 3: Surface flow wetland ditch (100 m^2) Stage 4: Tertiary surface flow bed (100 m^2)	4	15

HLR, hydraulic loading rate; HRT, hydraulic retention time; NA, not available due to low reliability and infrequent measurements.

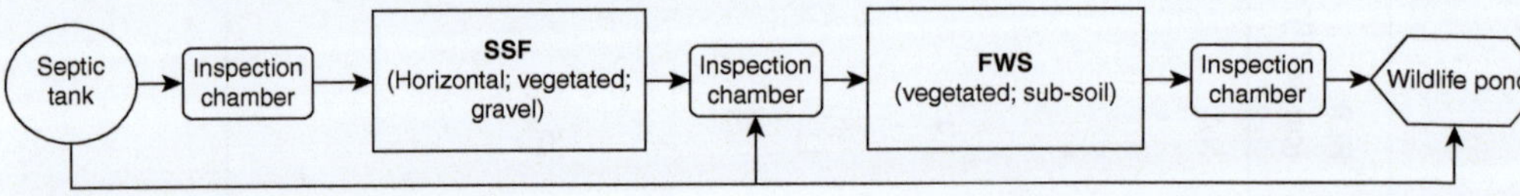

Figure 4.4 Schematic diagram of the Caerlaverock constructed wetland. FWS, free water surface flow wetland; SSF, sub-surface flow wetland. Diagram showing a septic tank, SSF, FWS and wildlife pond in series.

The Caerlaverock wetland is situated in the North Solway coast (Dumfriesshire, Scotland) and consists of two consecutive wetland beds (Figure 4.4). Following a septic tank for settling and pre-treatment, the sewage is piped to the first inspection chamber and subsequently to a gravel-filled horizontal sub-surface flow wetland bed planted with *Phragmites australis* (Common Reed). After that, the treated water flows into an intermediate inspection chamber and from there into a free water surface flow and planted (*P. australis*) wetland bed. The pre-treated water then flows through a third chamber and is finally piped into a wildlife pond.

The Castle Espie constructed wetland system, which comprises two parallel beds, is situated in County Down, Northern Ireland. Before entering the wetlands, sewage from various locations around the centre is collected in a septic tank. The wastewater then passes through a splitter chamber and distributes the sewage into two wetland beds. One of the two planted (*P. australis*) beds is a horizontal sub-surface flow wetland bed with a gravel substrate and the other is a free water surface flow wetland bed with a soil filled substrate. The treated sewage is discharged into a stream and eventually enters Strangford Lough (Figure 4.5).

Llanelli treatment wetland is located on the Burry Inlet. Tertiary treated sewage enters the settling pond first and is then distributed into two parallel primary free water surface flow constructed wetland beds. After the primary treatment, the effluent then flows into another two parallel secondary free water surface flow wetlands. Treated water is collected at the end of the secondary beds and discharged into a balancing pond, from where the water is distributed to the rest of the reserve and finally discharged into watercourses including the Burry Inlet Ramsar site (Figure 4.6). Vegetation cover at the primary and secondary treatment beds mainly consists of *Iris pseudacorus* L. (Yellow Iris or Yellow Flag) and *P. australis*, respectively.

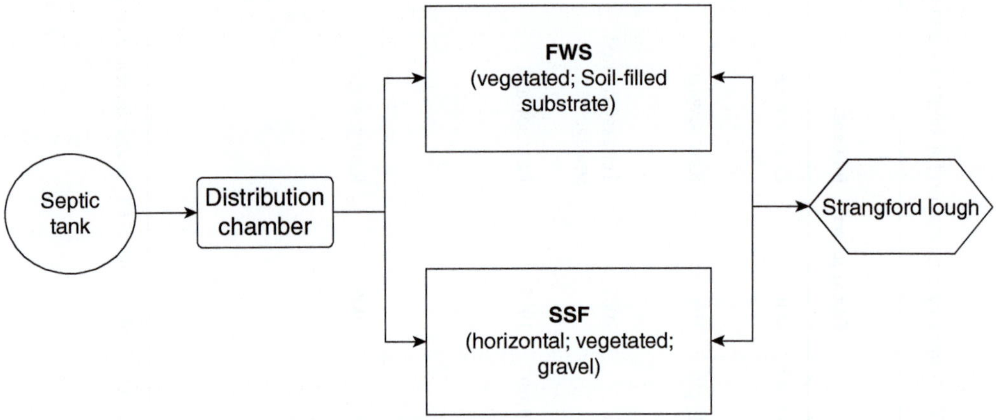

Figure 4.5 Schematic diagram of the Castle Espie constructed wetland. FWS, free water surface flow wetland; SSF, sub-surface flow wetland. Diagram showing a septic tank followed by a SSF and FWS in parallel.

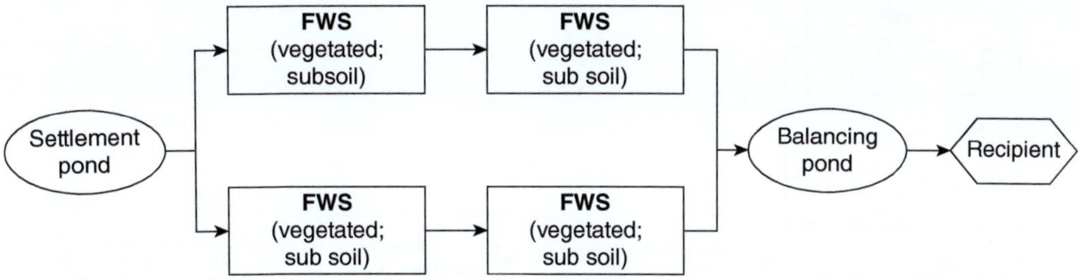

Figure 4.6 Schematic diagram of the Llanelli constructed wetland. FWS, free water surface flow wetland. Diagram showing a sediment pond, two FWS in series (two times; in parallel) and a balancing pond in series.

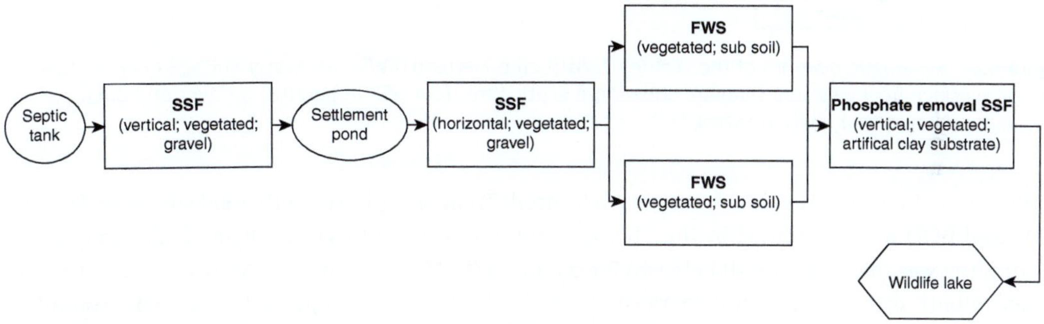

Figure 4.7 Schematic diagram of the Millennium constructed wetland. FWS, free water surface flow wetland; SSF, sub-surface flow wetland. Diagram showing a septic tank, SSF, settlement pond, SSF, FWS (two in parallel), phosphate removal SSF and a wildlife lake in series.

The Millennium Wetland was constructed at Slimbridge (England) in 1999. Treated sewage from the primary settlement tank is fed intermittently into a single vertical flow bed planted with *P. australis*. The effluent is further treated in a settlement pond, and subsequently introduced into a horizontal sub-surface flow wetland bed planted with *I. pseudacorus* and *Carex riparia* Curtis. From here, the effluent then flows into two parallel free surface flow wetland beds. Thereafter, the partially treated water enters a vertical flow bed that has been designed to remove phosphorus. The treated water ultimately flows into a wildlife lake (Figure 4.7).

Welney Wetland is located 26 miles north of Cambridge, England. Sewage from a visitor centre passes through a septic tank followed by four parallel horizontal sub-surface flow beds planted with *P. australis* before discharging into an overland flow ditch. From the ditch, effluent enters a tertiary free water surface flow wetland. After this, treated sewage is finally discharged into another ditch, where the water quality is monitored by the Environment Agency once per season (Figure 4.8).

4.2.2.2 Water Quality Monitoring and Limitations

The WWT initiated a chemical water quality monitoring programme in 2005 to evaluate the treatment performances of the selected constructed wetlands and to optimise their functioning. Water samples are taken from the inflow and outflow of wetland systems. Collected samples are analysed for a range of water quality parameters including NH_3-N, NO_3-N, nitrite-nitrogen (NO_2-N) and ortho-phosphate-phosphorus (PO_4-P) by using the Chemet system (Chemetrics 2010). Electrical

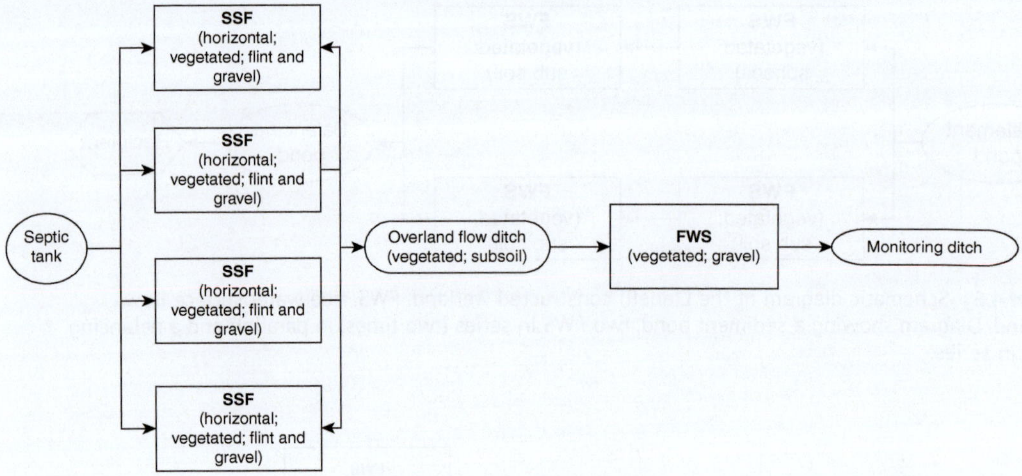

Figure 4.8 Schematic diagram of the Welney constructed wetland. FWS, free water surface flow wetland; SSF, sub-surface flow wetland. Diagram showing a septic tank, four SSF in parallel, an overland ditch, FSW and a (monitoring) ditch in series.

conductivity (EC) and pH values were also measured. Total phosphorus (TP), total suspended solids (TSS) and BOD were examined by the Phosyn laboratories based in York, England. The frequency of sampling was once per month between 2005 and 2008. However, samples were only taken every second month due to financial constraints from 2009 onwards. Except for Castle Espie, samples were sent to the Phosyn Laboratories for accuracy checks once per season.

The design inflow rates for the studied wetland systems are based on information provided by WWT. However, there is no reliable record of the inflow rates for the Caerlaverock wetland. Unfortunately, the inflow and outflow rates of some wetlands are not well monitored due to financial constraints. It follows that loading rate values should only be interpreted as rough estimates, particularly for the Caerlaverock wetland.

The electric conductivity values decreased between influent and effluent for most wetlands. This effect may be explained by dilution effects due to rainwater and groundwater inputs. However, a detailed mass and flow balance study was beyond the scope of Chapter 4.3.

4.2.3 Results and Discussion

4.2.3.1 Water Quality and Performance Efficiency

Water quality data and removal efficiencies for studied constructed wetlands during the monitoring period between 2005 and 2009 are presented in Table 4.8. Nitrogen has a complex cycle within wetland systems. The major nitrogen transformation processes are discussed in considerable detail by Kadlec and Knight (1996) and Vymazal (2007). In general, nitrogen removal in wetlands is accomplished primarily by physical settlement, nitrification, denitrification and plant and microbial uptake (Healy and Cawley 2002). Since the predominant forms of nitrogen within sewage and household wastewaters are ammonia and nitrate, nitrification followed by denitrification and assimilation to sediments or biomass are likely to be the principal nitrogen removal processes. Major factors that influence the magnitude of the reduction performances include temperature, pH, hydraulic loading rates (HLR), hydraulic retention time (HRT), dissolved oxygen, density and type of vegetation and biota (Lee et al. 2009; Reed et al. 1995).

Table 4.8 Summary performance data for Wildfowl and Wetlands Trust representative constructed wetlands.

		Caerlaverock	Castle Espie		Llanelli		Millennium	Welney
Number of samples		40	28		40		35	11
Monitoring period		06/2005–12/2009	06/2005–12/2008		06/2005–12/2009		06/2005–12/2009	03/2008–04/2009
NH_3-N	Unit		(G)	(S)	(R)	(L)		
Inflow	mg L^{-1}	27.23 ± 30.30	36.38 ± 38.93		0.79 ± 1.34		28.47 ± 18.66	10.63 ± 1.77
Outflow	mg L^{-1}	9.32 ± 9.48	24.79 ± 23.24	5.66 ± 2.82	0.28 ± 0.41	0.36 ± 0.62	0.90 ± 2.95	5.88 ± 5.33
Reduction	%	65.77	31.86	84.44	64.56	54.43	96.84	44.68
NO_3-N			(G)	(S)	(R)	(L)		
Inflow	mg L^{-1}	0.20 ± 0.52	0.66 ± 0.77		3.29 ± 1.76		0.27 ± 0.47	0
Outflow	mg L^{-1}	1.62 ± 1.38	0.41 ± 0.46	0.57 ± 0.72	1.08 ± 0.67	1.09 ± 0.93	10.65 ± 5.20	0.86 ± 1.86
Reduction	%	−710.00	37.88	13.64	67.17	66.87	−3844.44	NA
TON			(G)	(S)	(R)	(L)		
Inflow	mg L^{-1}	0.20 ± 0.53	0.67 ± 0.77		3.35 ± 1.76		0.29 ± 0.48	0
Outflow	mg L^{-1}	1.68 ± 1.37	0.42 ± 0.46	0.63 ± 0.79	1.08 ± 0.61	1.10 ± 0.94	11.21 ± 5.08	0.94 ± 1.98
Reduction	%	−740.00	37.31	5.97	67.76	67.16	−3765.52	NA
PO_4-P			(G)	(S)	(R)	(L)		
Inflow	mg L^{-1}	7.62 ± 2.67	8.31 ± 5.93		1.34 ± 1.30		4.31 ± 1.91	8.57 ± 1.90
Outflow	mg L^{-1}	6.28 ± 2.50	5.68 ± 4.26	6.26 ± 5.15	1.48 ± 0.77	1.12 ± 0.65	3.69 ± 0.87	1.06 ± 2.21
Reduction	%	17.59	31.65	24.67	−10.45	16.42	14.39	87.63
TP			(G)	(S)	(R)	(L)		
Inflow	mg L^{-1}	6.80 ± 3.31	8.44 ± 5.91		1.60 ± 1.48		5.13 ± 2.15	6.97 ± 1.61
Outflow	mg L^{-1}	4.18 ± 1.43	5.83 ± 4.29	6.84 ± 4.80	1.49 ± 0.56	1.30 ± 0.68	4.22 ± 1.45	0.52 ± 0.69
Reduction	%	38.53	30.92	18.96	6.88	18.75	17.74	92.54

(Continued)

Table 4.8 (Continued)

		Caerlaverock	Castle Espie		Llanelli		Millennium	Welney
TSS			(G)	(S)	(R)	(L)		
Inflow	mg L^{-1}	14.67 ± 19.52	32.17 ± 37.27		14.85 ± 18.03		15.91 ± 17.06	26.36 ± 29.21
Outflow	mg L^{-1}	9.74 ± 16.94	13.65 ± 21.11	112.33 ± 189.50	13.16 ± 12.93	15.98 ± 16.29	24.98 ± 32.22	28.45 ± 18.73
Reduction	%	33.61	57.57	−249.18	11.38	−7.61	−57.01	−7.93
BOD			(G)	(S)	(R)	(L)		
Inflow	mg L^{-1}	25.46 ± 15.08	34.00 ± 24.73		8.51 ± 6.82		32.67 ± 14.27	46.36 ± 9.80
Outflow	mg L^{-1}	11.94 ± 11.45	9.68 ± 9.50	13.18 ± 19.75	8.51 ± 6.82	8.01 ± 7.72	4.14 ± 5.50	16.96 ± 13.13
Reduction	%	53.10	71.53	61.24	0	5.88	87.33	63.42
pH			(G)	(S)	(R)	(L)		
Inflow	—	6.96 ± 0.75	6.80 ± 0.52		7.14 ± 0.26		7.88 ± 0.71	7.40 ± 0.43
Outflow	—	6.92 ± 0.66	6.92 ± 0.40	6.94 ± 0.35	7.29 ± 0.29	7.13 ± 0.26	7.95 ± 0.69	7.67 ± 0.39
EC			(G)	(S)	(R)	(L)		
Inflow	µs cm^{-1}	558.76 ± 328.46	605.74 ± 351.63		549.95 ± 455.17		702.13 ± 351.81	204.44 ± 10.19
Outflow	µs cm^{-1}	371.97 ± 224.36	541.65 ± 269.51	463.25 ± 228.61	483.82 ± 383.75	404.08 ± 260.82	749.13 ± 325.78	92.00 ± 35.75

BOD, biochemical oxygen demand; EC, electric conductivity; G, effluent from the horizontal sub-surface flow wetland bed with gravel substrate; L, effluent from the left free surface flow wetland bed; NA, not applicable; NH_3-N, ammonia-nitrogen; NO_3-N, nitrate-nitrogen; NO_3-N, nitrate-nitrogen; PO_4-P, ortho-phosphate-phosphorus; PO_4-P, ortho-phosphate-phosphorus; R, effluent from the right free surface flow wetland bed; S, effluent from the free surface flow wetland bed with soil substrate; TON, total oxidised nitrogen; TP, total phosphorus; NH_3-N, ammonia-nitrogen; TSS, total suspended solids.
Standard deviations are indicated by the symbol ±.

Reductions of NH_3-N concentrations ranged from 31.9% to 96.8% in the five monitored constructed wetlands, indicating that nitrification processes occur within all systems but that the removal performances are highly variable. At Castle Espie, the free water surface flow wetland bed achieved better NH_3-N treatment results than the horizontal sub-surface flow wetland bed, supporting the notions that the nitrification process is very oxygen demanding and that the free water surface flow wetland bed can produce more favourable aerobic conditions than the sub-surface flow wetland bed (Tanner and Kadlec 2003). In addition, almost complete nitrification (96.8%) occurred at the Millennium wetland where higher pH values were recorded in both influent and effluent. This result is consistent with the findings of Lee et al. (2009) who indicated that the pH value is important for the nitrification process since the nitrification rates swiftly decline when pH drops to values lower than seven.

In terms of total oxidised nitrogen (TON; $(NO_3 + NO_2)$-N), the most efficient removals were achieved in the Llanelli wetland. An increase in the HLR significantly deteriorated the TON reduction in other wetlands. This correlation between TON removal and HLR has also been observed during previous studies (Dong et al. 2011; Trang et al. 2010).

Principal mechanisms of phosphorus reduction within constructed wetlands include physical processes (e.g., sedimentation and precipitation), chemical processes (e.g., adsorption and complexation) and biological transformations (Greenway and Wooley 1999; Kadlec and Knight 1996; Yalcuk and Ugurlu 2009). Although uptake by plants and other biological processes can contribute to an initial rapid reduction of phosphorus, the storage of phosphorus in soils and sediments is commonly considered as the dominant long-term phosphorus removal mechanism. However, adsorption and storage are saturable processes and may become rapidly exhausted (Vymazal 2007).

Except for Welney wetlands, the average TP removal efficiencies ranged from 6.9% to 38.5%, indicating that these wetlands are not an effective method for reducing phosphorus in the long term. Wetland systems can usually effectively remove phosphorus during the early stages of operation. However, there tends to be an increasing proportion of phosphorus as the wetland ages; i.e., the sediments and subsoils become fully saturated (Scholz et al. 2007). This effect is well illustrated in Caerlaverock and Millennium wetlands (Figure 4.9), where there were initial mean reductions in PO_4-P concentrations of 43.8% and 14.3% in 2005, reducing to 3.2% and 4.4% in 2009, though some minor fluctuations were observed. In contrast, better phosphorus removal performance

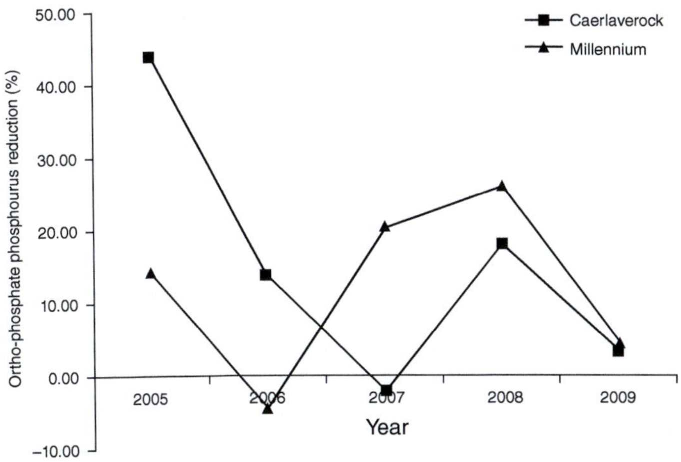

Figure 4.9 Annual mean ortho-phosphate-phosphorus removal efficiencies for the Caerlaverock and Millennium wetlands (2005–2009). Highly variable phosphorus reductions of around 15% for each wetland.

(87.6% in PO_4-P and 92.5% in TP) has been recorded for the Welney wetlands, presumably because substrate saturation has not yet occurred.

For Caerlaverock and Welney wetlands, the PO_4-P concentrations were greater than those of TP. This may be because TP values were measured at Phosyn Laboratories, potentially providing more accurate results than the field testing performed elsewhere. Nevertheless, records indicated a downward trend in terms of both PO_4-P and TP reductions at these wetlands over operation time.

In wetland systems, BOD and TSS are principally removed by processes such as photochemical oxidation, sedimentation, sorption and biodegradation (Kadlec 1992). Suspended and dissolved organic matter are broken down by microbiological activities including fermentation and/or respiration, and subsequently mineralised as a source of energy or assimilated into biomass (Kadlec and Knight 1996; Moshiri 1993).

Except for the wetland at Llanelli, the reduction in BOD concentration ranged from 53.1% to 87.3%, producing effluent with mean BOD concentrations in the range of 4–17 mg L^{-1}. At Llanelli, where the BOD level is relatively low in the influent, only 5.9% of the BOD in the effluent of the first wetland bed was removed, and no BOD reduction at all was typically recorded in the second wetland. This finding agrees with previous studies indicating low initial BOD loads are unlikely to be removed completely because of background levels produced within wetlands (Kadlec and Knight 1996).

Reductions in TSS concentrations ranged from −249.2% to 57.6% for the five wetlands. For most systems, solids in the effluent exceeded those in the influent, which suggests that a dispersion of soil and clay is occurring, presumably due to high flow rates and substantial wind action at this site. In addition, the lack of vegetation at the outlet of some wetlands might contribute to the initial increase in TSS (Greenway and Wooley 1999).

4.2.3.2 Seasonal Variations and Wetland Aging

Seasonal contaminant removals in four constructed wetlands are present in Table 4.9. Welney wetland data were not included due to the short monitoring period (2008–2009). Microbiological reactions, especially nitrification, can be significantly limited by cold temperature (Kadlec and Knight 1996). As shown in Figure 4.10, the reduction of NH_3-N was relatively high in summer but substantially lower in winter, which is consistent with findings presented by Vymazal (2007). The nitrification rates in wetland systems become increasingly inhibited at temperatures of about 10°C and drop rapidly at 6°C (Werker et al. 2002). In contrast, there is no obvious seasonal trend in NO_3-N removal, which is probably due to a relatively low denitrification rate for the studied wetlands. Since the main removal mechanism of phosphorus in a constructed wetland is adsorption, temperature effects on phosphorus removal efficiencies are unlikely to be substantial.

Few constructed wetlands have been studied for a long enough period to determine the long-term effects that will develop as the systems age. As can be observed in Table 4.10, the effectiveness of nitrogen removal has been consistent for the five representative wetlands regardless of the age of each system. However, phosphorus removal is highly variable and generally declines with time as the wetland sediments and soils become saturated (Figure 4.9). The nutrient removal also fluctuated due to variations in external factors such as local climate, water chemistry, hydraulic and contaminant loading rates and retention times. Some of the studied wetland systems might also temporarily release the accumulated nutrients due to flooding and change of water level.

4.2.4 Conclusions and Recommendations

Data collected from five representative constructed wetlands have been studied to demonstrate the long-term performance of wetland systems to improve the quality of sewage effluent by reducing nitrogen, phosphorus and BOD. However, the analysis shows that effective reductions in NO_3-N,

Table 4.9 Seasonal comparison of the nutrient removal efficiencies for representative constructed wetlands (2005–2009).

Season	Site	NH_3-N	NO_3-N	PO_4-P	TP
		%	%	%	%
Spring	Caerlaverock	40.90	NA	−3.88	22.59
	Castle Espie (G)	−171.32	NA	25.55	25.55
	Castle Espie (S)	−55.04	25.00	−37.27	−37.27
	Llanelli (R)	61.29	48.68	28.68	75.28
	Llanelli (L)	62.90	50.57	22.67	20.50
	Millennium	97.23	−39 066.67	−27.78	NA
Summer	Caerlaverock	79.78	−584.21	19.50	51.90
	Castle Espie (G)	47.17	41.48	32.69	32.69
	Castle Espie (S)	87.28	29.94	37.45	37.45
	Llanelli (R)	56.82	81.18	−73.61	−34.72
	Llanelli (L)	63.64	80.19	−8.87	−14.61
	Millennium	98.59	−3350.00	3.23	4.32
Autumn	Caerlaverock	67.08	−260.00	23.21	29.75
	Castle Espie (G)	17.87	19.79	39.26	39.26
	Castle Espie (S)	88.67	−58.00	30.64	30.64
	Llanelli (R)	80.23	69.01	10.48	18.83
	Llanelli (L)	50.14	65.79	24.22	31.27
	Millennium	98.79	−2461.54	30.03	32.53
Winter	Caerlaverock	21.55	−148.37	−5.79	29.51
	Castle Espie (G)	−166.67	NA	−11.11	−11.11
	Castle Espie (S)	−233.33	NA	−22.22	−22.22
	Llanelli (R)	52.17	34.22	25.76	36.70
	Llanelli (L)	50.36	40.54	46.83	45.25
	Millennium	82.31	−5735.00	4.26	NA

G, effluent from the horizontal sub-surface flow wetland bed with gravel substrate; L, effluent from the left free surface flow wetland bed; NA, not applicable; NH_3-N, ammonia-nitrogen; NO_3-N, nitrate-nitrogen; PO_4-P, ortho-phosphate-phosphorus; R, effluent from the right free surface flow wetland bed; S, effluent from the free surface flow wetland bed with soil substrate; TP, total phosphorus.

PO_4-P and TSS were not achieved in some wetland systems. Removal of NH_3-N has been generally effective in all studied wetlands and particularly effective in summer due to temperature-enhanced nitrification rate.

In general, the studies show that a wetland with low HLR and long HRT are the most effective and are suitable for effluent polishing. The NH_3-N removal was effective in all wetlands, although NO_3-N reduction was found to vary between summer and winter. For phosphorus, however, reduction is typically high during early stages, but efficiency decreases with time as the wetland substratum becomes saturated. There is a potential risk that some accumulated nutrients might be remobilised and released to the water phase as the wetland ages.

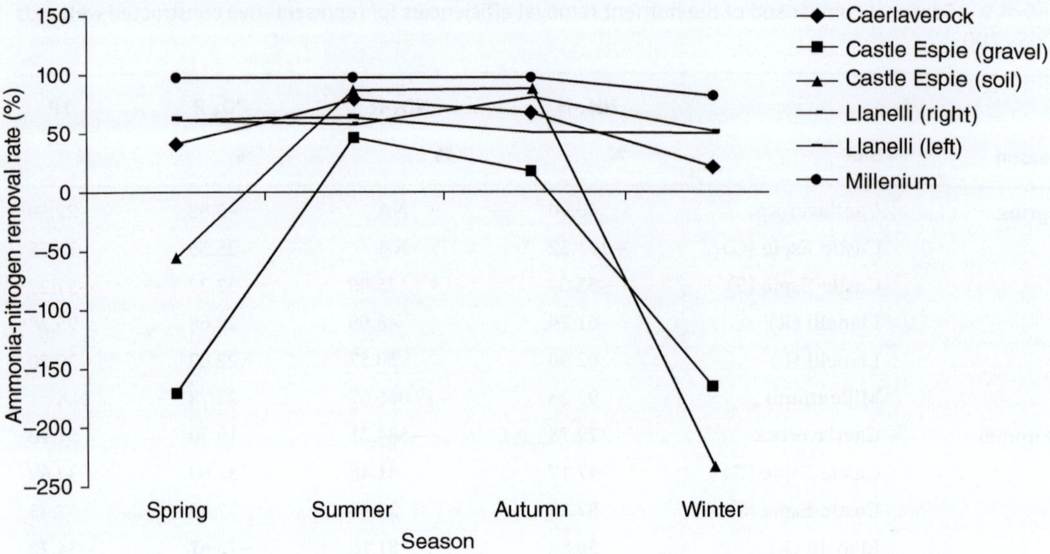

Figure 4.10 Seasonal mean ammonia-nitrogen removal efficiencies for the Caerlaverock, Castle Espie, Llanelli and Millennium constructed wetlands (2005–2009). Ammonia-nitrogen removal efficiencies are around 30–90%, except for Castle Espie, where the systems are failing with over 150% increases.

Future analysis of hydraulic data would permit more accurate quantification of the constructed wetland treatment performance and aid further wetland system design and management efforts. However, the determination of loads and fluxes as part of a complex nutrient balance model for each wetland would be rather costly.

The findings provide further evidence that constructed wetlands have a limited life span, particularly regarding phosphorus. Therefore, proper wetland management plans (e.g., sediment removal or separate phosphorus pre-treatment) need to be implemented within the wetland system design to allow for a long operation period.

4.3 Neural Network Simulation of the Chemical Oxygen Demand Reduction

4.3.1 Introduction

The history of neural networks as described by Bishop (1994) dates to 1943, when McCulloch and Pitts first introduced a simple mathematical model consisting of a single neuron which transforms a set of input variables into an output variable by a non-linear transfer function. Since then, much development has taken place, with major activities in the 1960s. However, in the 1970s, the interest in this area diminished as it was found that neural networks were not capable of solving problems of higher complexities mainly due to their linear structure. In the early 1980s with the discovery of the learning algorithm based on error back-propagation and easy availability of powerful computers, interest in neural computing increased and in the 1990s several successful consumer appliances such as intelligent washing machines and autoexposure video cameras became available.

Neural networks also have found application in solving complex process engineering problems, where it was difficult to develop models from the fundamental principles, particularly when dealing

Table 4.10 Annual nutrient concentrations (mean ± standard deviation) for Wildfowl & Wetland Trust representative constructed wetlands (2005–2009).

		NH_3-N (mg L^{-1})			NO_3-N (mg L^{-1})			PO_4-P (mg L^{-1})			TP (mg L^{-1})		
Site	**Year**	**Inflow**	**Outflow**		**Inflow**	**Outflow**		**Inflow**	**Outflow**		**Inflow**	**Outflow**	
Caerlaverock	2005	63.51 ± 37.28	17.66 ± 9.11		1.28 ± 0.65	0.67 ± 0.31		8.09 ± 2.79	4.55 ± 0.83		8.60 ± 2.71	5.04 ± 1.24	
	2006	43.50 ± 24.48	13.35 ± 9.34		0	0.81 ± 1.03		9.50 ± 1.51	8.19 ± 2.27		5.73 ± 1.91	3.94 ± 1.39	
	2007	21.10 ± 30.93	10.98 ± 11.69		0.01 ± 0.02	2.02 ± 1.22		6.57 ± 2.93	6.70 ± 3.27		6.36 ± 3.34	3.70 ± 1.02	
	2008	10.23 ± 6.75	3.65 ± 2.12		0	2.65 ± 1.55		7.23 ± 2.21	5.91 ± 1.43		5.09 ± 1.32	4.38 ± 1.47	
	2009	7.30 ± 3.31	2.04 ± 1.85		0	1.13 ± 0.85		7.75 ± 4.71	7.50 ± 4.75		9.34 ± 4.07	5.27 ± 2.05	
Castle Espie			(G)	(S)		(G)	(S)		(G)	(S)		(G)	(S)
	2005	75.45 ± 40.27	44.40 ± 23.29	6.48 ± 2.13	1.20 ± 0.77	0.75 ± 0.41	1.07 ± 0.75	8.06 ± 4.56	4.19 ±2.20	1.55 ± 0.86	8.55 ± 4.47	4.75 ± 2.74	3.77 ± 2.48
	2006	30.50 ± 18.04	24.17 ± 14.29	7.00 ± 3.58	0.17 ± 0.21	0.07 ± 0.12	0.13 ± 0.15	13.17 ± 8.47	10.25 ± 5.21	13.17 ± 4.02	13.17 ± 8.47	10.25 ± 5.21	13.17 ± 4.02
	2007	3.18 ± 2.76	2.00 ± 0.89	3.50 ± 0.96	0.05 ± 0.05	0.05 ± 0.05	0	5.40 ± 2.33	6.00 ± 2.45	6.70 ± 1.47	5.40 ± 2.33	6.00 ± 2.45	6.70 ± 1.47
	2008	NA	NA	NA	NA	NA	NA	6.14 ± 3.77	2.34 ± 0.07	3.69 ± 2.35	6.14 ± 3.77	2.34 ± 0.07	3.69 ± 2.35
Llanelli			(R)	(L)		(R)	(L)		(R)	(L)		(R)	(L)
	2005	1.11 ± 1.38	0.21 ± 0.11	0.27 ± 0.15	5.25 ± 2.50	1.12 ± 0.55	1.82 ± 1.47	1.30 ± 1.55	1.49 ± 0.36	1.32 ± 0.77	1.68 ± 1.94	1.67 ± 0.39	1.48 ± 0.89
	2006	0.16 ± 0.04	0.15 ± 0.07	0.17 ± 0.12	3.26 ± 1.53	1.49 ± 0.69	1.00 ± 0.95	NA	NA	NA	NA	NA	NA
	2007	0.72 ± 1.09	0.35 ± 0.34	0.38 ± 0.31	2.56 ± 0.79	0.83 ± 0.62	0.87 ± 0.63	2.05 ± 1.34	1.21 ± 0.52	1.21 ± 0.58	2.19 ± 1.29	1.33 ± 0.55	1.30 ± 0.55
	2008	1.73 ± 2.11	0.26 ± 0.31	0.48 ± 1.02	2.68 ± 2.06	0.57 ± 0.54	0.52 ± 0.39	0.68 ± 0.50	1.72 ± 1.38	0.72 ± 0.43	0.87 ± 0.56	1.38 ± 0.81	1.03 ± 0.46
	2009	0.03 ± 0.03	0.56 ± 1.10	0.63 ± 1.25	3.50 ± 0.91	1.45 ± 0.49	1.75 ± 0.65	NA	NA	NA	NA	NA	NA
Millennium	2005	27.72 ± 20.86	0.52 ± 0.24		0.87 ± 0.53	11.98 ± 6.82		4.31 ± 1.91	3.69 ± 0.87		5.13 ± 2.15	4.22 ± 1.45	
	2006	29.75 ± 18.18	0.490.89		0	9.65 ± 4.92		9.15 ± 3.60	9.55 ± 2.99		NA	NA	
	2007	28.56 ± 20.84	2.616.24		0.12 ± 0.19	11.00 ± 4.42		7.63 ± 2.00	6.06 ± 1.02		NA	NA	
	2008	34.50 ± 20.30	0.23 ± 0.24		NA	NA		10.43 ± 6.32	7.71 ± 0.76		NA	NA	
	2009	17.04 ± 11.08	0.54 ± 0.83		NA	NA		6.43 ± 2.54	6.15 ± 2.04		NA	NA	
Welney	2008	10.00 ± 0.00	4.57 ± 4.16		0	1.20 ± 2.17		8.80 ± 1.79	0.08 ± 0.11		6.63 ± 1.33	0.24 ± 0.10	
	2009	15.00 ± 0.00	15.00 ± 0.00		0	0		8.00 ± 2.83	1.00 ± 0.00		8.17 ± 2.51	1.52 ± 1.11	

G, effluent from the horizontal sub-surface flow wetland bed with gravel substrate; L, effluent from the left free surface flow wetland bed; NA, not applicable; NH_3-N, ammonia-nitrogen; NO_3-N, nitrate-nitrogen; PO_4-P, ortho-phosphate-phosphorus; R, effluent from the right free surface flow wetland bed; S, effluent from the free surface flow wetland bed with soil substrate; TP, total phosphorus.

with non-linear systems. They have attracted much interest mainly due to their ability to capture non-linear behaviour from the input–output data of a process. Several research papers have been published which show that neural networks are useful empirical tools for process simulation and control. Morris et al. (1994) and Hussain (1999) have presented excellent reviews of applications of neural networks to these topics.

A few papers have been published on the application of artificial neural networks (ANN) to advanced water treatment. Frith (1996) has suggested the use of neural network models (NNM) for simulating filter run length and estimation of water quality for a rapid gravity sand filter used for treatment of river water. These models can be used for minimising the cost of operation of such filters. Application of ANN to simulate the ammonia concentration in an industrial wastewater and purification of municipal water by coagulation has been presented by Boger (1992). Jose et al. (1999) have developed a neural-network-based inferential sensor for monitoring the phenol concentration in a biological wastewater treatment plant by measuring on-line biomass concentration by spectrophotometry. Syu and Chen (1998) have applied ANN to control the hydrogen peroxide and ferrous chloride dosage in treating synthetic wastewater (benzoic acid solution) to obtain the desired chemical oxygen demand (COD). Miller et al. (1997) have used a hybrid model which is a combination of NNM, principal component analysis (PCA) and semi-empirical correlations to control the interface level of a clarifier in a chemical wastewater treatment plant. Carriere et al. (1996) have applied ANN for optimising the parameters of a granular activated carbon (GAC) column adsorber system for removal of lead from wastewater. However, almost no work has been published on simulating the performance of a biological activated carbon (BAC) filter for treatment of potable water using ANN.

Preliminary studies by Scholz and Martin (1998b) show that conventional regression models do not give satisfactory estimations for such a non-linear system. In this chapter, a feed forward (also known as error back propagation) NNM has been used to simulate the (COD) reduction achieved by a BAC filter, based on the micro-organisms present in the water, dissolved oxygen of the inflow water and mean pH of the inflow and outflow water. Other parameters have been assumed to have negligible effect on the performance, but such neglect may be the source of imperfect simulation. Mohanty et al. (2002) provided the scientific basis for Chapter 4.3.

4.3.2 Materials and Methods

4.3.2.1 Source of Data

During the bio-regeneration mode of GAC filters with untreated river water, the filters naturally become BAC filters by formation of a viable biofilm. Partial biological regeneration of GAC due to sorption and later metabolism of slowly biodegradable organic matter can be observed in practice. This leads to an extended bed life of the carbon column. By maintaining a controlled ecological equilibrium between micro-organisms, a BAC filter can be more effective in improving the quality of the filtrate than a GAC filter (Scholz 1998). However, uncontrolled biological activity may result in an overproduction of biomass. Furthermore, large head losses due to clogging phenomena can occur. Curds and Cockburn (1970) and others have assessed the usefulness of ciliated protozoa as indicator organisms for the quality of effluents, in terms of biochemical oxygen demand (BOD), from activated sludge and trickling filter plants in England.

In the present study, raw data generated by Scholz (1998) were pre-processed and used for training and testing the network. The biofilm on BAC (based on GAC; Filtrasorb 400) beds was fed directly with river water. The bed was 0.12 m in diameter and 0.38 m deep. An empty bed contact time of eight minutes was applied for the upward flow system. The flow rate was 0.54 dm^3 min^{-1}.

All analytical procedures were performed according to the American Standard Methods for the Examination of Water and Wastewater (Water Environment Federation 1995) except for COD, total plate count tests and protozoa determination. Heterotrophic plate count tests and COD were based on British Standard Methods (Department of the Environment 1977–1988) for the examination of waters and associated materials.

Protozoa, diatoms and other micro-organisms were counted in the influent, effluent and at five or more positions (depending on the degree of bed expansion) in the carbon bed itself by using a counting chamber and a bright field/phase-contrast microscope (Olympus Vanox, Tokyo, Japan, 204899). A small but representative water sample was taken for analysis of protozoa, diatoms and rotifers. The simplified division of protozoa according to their movement is like the general subdivision of protozoa in UK sewage treatment processes (Scholz and Martin 1998a, b). The numbers of micro-organisms were estimated by direct microscopic counting. Pictures were taken and observations were recorded on video tape for detailed determination of dominant species at a magnitude of 40 × 40.

4.3.2.2 Neural Network Model

Although several neural network architectures have been proposed and used, the feed forward neural network (FFNN) is still a widely used architecture for solving process engineering problems. An FFNN can be viewed as a large regression model between input and output variables. The single network can estimate more than one output, unlike the conventional regression model where a single model can estimate only one output variable. The network learns the solution to a problem by being repeatedly presented with examples.

Figure 4.11 shows the topology of the NNM used in Chapter 4.3. A 3-layered feed-forward network consisting of 235 synaptic connections between an input layer (22 input variables and a bias), two hidden layers (8 and 5 processing elements, respectively, and a bias) and an output layer (single variable) has been used. Such a topology is referred to as a 22-8-5-1 network. Both the input and the hidden layers include a bias processing element, which has a constant output value of one. The bias is included to shift the space of non-linearity. The number of hidden layers and the number of

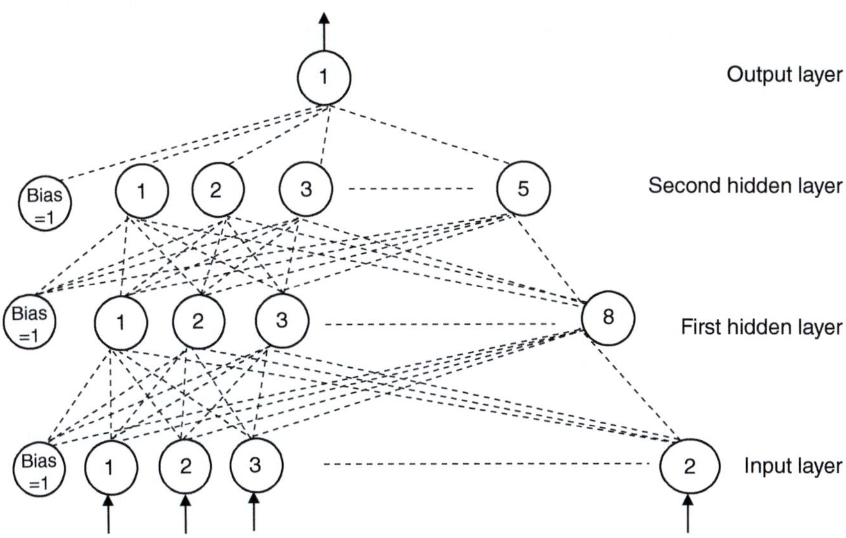

Figure 4.11 Topology of the feed forward neural network model. Diagram showing the input, first hidden, second hidden and output layers.

processing elements (PE) per layer were determined by trial and error to give a reasonable compromise between size and accuracy. The PE within the hidden layer sum up the weighted inputs to the PE and then perform a non-linear transformation of the resulting summation and pass it to the next layer. Equation (4.1) gives the summation and transformation performed by a PE.

$$x_j^s = f\left(\sum_i w_{ji}^s x_i^{s-1}\right) \tag{4.1}$$

where x_j^s is the current output state of the jth PE in layer s; w_{ji}^s represents the weight of the interconnection between the ith PE in layer s–1 to the jth PE in layer s; x_i^{s-1} indicates the current output state of the ith PE in the layer s–1 and f is any differentiable function.

Initially, training was performed with a single hidden layer and the number of processing elements increased from 8 to 25. However, satisfactory results could not be obtained with a single layer. The training was then repeated with two hidden layers. The number of processing elements in each hidden layer was increased, one at a time, until there was no significant change in the root mean square error. Too few PE hinder the learning process and too many degrade the generalising capability of the network.

The hidden layers and the output layer are connected to the previous layer by means of weights. The interconnections are such that every PE in a layer is connected to the PE in the adjacent layer (fully connected). The weights determine the strength of associations between the processing elements and are used to modify the input signals. All weights were constrained within the range −1 to +1.

The input consisted of 22 variables of which 20 were counts of micro-organisms (Table 4.11) and the other two were pH and dissolved oxygen concentration. It was assumed that other parameters such as turbidity and suspended solids have negligible effect on the performance of the filter. The output layer consisted of a single PE, which is the COD reduction extent. Daily data collected over a period of more than three months were used in the present investigation. The data were separated into training and testing sets. The purpose of separating the two sets is to check the generalising capability of the network. The training set consisted of 53 input–output pairs, taken alternately from the total data set so that the time spread of each set was approximately the same. Data for all the variables were scaled between −1.0 and +1.0 and introduced to the input layer of the network. The selection of the activation function (also called the transfer function) of the hidden and output layers is done based on minimisation of error. In the present model identical transfer functions (i.e., the hyperbolic tangent function given by Eq. (4.2) for the hidden and output layers) were found to best simulate the data.

$$f(a) = \tan h\, a = \frac{e^a - e^{-a}}{e^a + e^{-a}} \tag{4.2}$$

where the variable a is the summation of the weighted input to a PE from the PEs in the layer preceding it.

The information in the input layer is propagated in the forward direction of the network to the output layer. Initially all weights were set to random values between −1 and +1. At the output layer, the error between the actual output and the estimated output is calculated. The error is backpropagated to the network and the weights updated using the gradient descent rule; i.e., changing each weight according to the size and direction of negative gradient on the error surface. The input is then processed with these new weights to estimate the output.

During this iterative process, termed training or learning, the weights are adjusted so as to minimise the root mean square error between the actual and estimated output. To accelerate the

Table 4.11 Parameters used as input variables for the neural network model.

Input variable number	Parameter
1	Protozoa: free swimming protozoa (FSP) and crawling protozoa (CP)
2	Protozoa: flagellated protozoa (FP)
3	Rhizopoda and actinopoda
4	Worms
5	Rotifers
6	Bacteria: large, wet, white or yellow points or areas
7	Bacteria: wet and orange points
8	Bacteria: wet and red points
9	Bacteria: unclassified
10	Bacteria: unclassified
11	Fungi: large, unstructured, white and branched
12	Fungi: large and orange or red
13	Fungi: large, circular and brown
14	Fungi: circular and green
15	Diatoms: *Fragilaria, Asterionella, Diatoma, Meridion, Synedra and Tabellaria*
16	Diatoms: *Bacillariophyceae*
17	Algae: *Cyclotella*
18	Algae: *Chlorella*
19	Algae: *Scenedestius*
20	Algae: others
21	pH
22	Dissolved oxygen

convergence process, the weights were updated after all the data points of the training set were presented to the network. This is known as cumulative backpropagation where the differential changes in weights are accumulated until the complete data set is presented. The software packages used in this work were Neural Works Professional II/Plus (Neuralware Inc., USA) and Statistical Package for Social Sciences (SPSS).

4.3.3 Results and Discussion

The 22 input variables considered in this investigation include counts of different types of protozoa, algae (predominantly diatoms), bacteria and fungi as well as pH and dissolved oxygen concentration (Table 4.11). Input variables are protozoa (1–3), worms (4), rotifers (5), bacteria (6–10), fungi (11–14), algae (15–20), pH (21) and dissolved oxygen (22). Protozoa were classified according to Patterson (1996) and Scholz (1998) both by traditional taxonomy and life action via empirical observation.

Ciliated protozoa (free swimming protozoa (FSP) and crawling protozoa (CP)) represent relatively clean water and were therefore combined as a single-input variable. In contrast, flagellated protozoa (FP) as well as Sarcodina (rhizopoda in general and actinopoda in particular) indicate relatively polluted water. These micro-organisms were not combined because they belong to different sub-phyla and are different in size (Sarcodina normally larger than FP). Worms (predominantly nematodes) and rotifers are very large micro-organisms and very different from each other. Therefore, these were taken as separate variables. Worms were rarely counted and indicate very poor water quality.

Both bacteria and fungi were not defined on a genera level. However, an approximate classification according to their appearance on nutrient agar plates was undertaken (Scholz 1998). Bacteria were sub-grouped as follows: large, wet, white or yellow points or areas (6), wet and orange points (7) and wet and red points (8). Unclassified bacteria, but different from each other in terms of appearance and structure, were sub-grouped as (9) and (10). These were few.

Fungi were sub-grouped as follows: large, unstructured, white and branched (11), large and orange or red (12), large, circular and brown (13) and circular and green (14). Diatoms were the most abundant and important micro-organisms because of the spring algae bloom. They were grouped according to the classification key suggested by Cox (1996). The diatom variable (15) comprises *Fragilariophyceae* (predominantly Fragilariales (*Fragilaria*, *Asterionella*, *Diatoma*, *Meridion* and *Synedra*) and Tabellariales (only *Tabellaria*)) whereas the variable (16) comprises *Bacillariophyceae*.

In addition, single algae genera (17–19) were considered separately, because they belong to different taxonomic positions and were only present in small numbers. All other unclassified algae, although only a few, were combined as variable 20. pH and dissolved oxygen content were taken as separate variables.

To reduce dimensionality, attempts were made to delete some of the groups of micro-organisms as the input variables. The network performance was tested by deleting one input variable or a group of input variables at a time. However, it was found that the error in the network estimation of the COD reduction for the testing set was very large (except in the case of excluding pH and dissolved oxygen) although the fitting with the training set was good. Considering that the two variables, pH and dissolved oxygen, influence microbial growth and metabolism as well as the activity, these should be included as input variables. However, further investigations on this subject need to be done to draw more certain conclusions.

PCA for the 22 input variables was also carried out using SPSS and the results are presented in Table 4.12. Extracting only those components having eigenvalues greater than 1 showed that with retention of 7 components, 30% of the information was lost (7 components account for 70% of the total variance). Similarly, extracting only those components having eigenvalues greater than 0.5 showed that with 13 components, 9% of the information was lost. The analysis shows that all the 22 input variables are important. This is reasonable considering the complex nature of the microbial system. The NNM was simulated using Neural Works Professional II/Plus.

As the initial weights are set to random values, the final weights obtained for the same network vary when simulated again thus showing that the solution converges to a local minimum. Hence the network was simulated a few times and the weights, which gave the best results for the testing data set, were retained. Figure 4.12 shows the comparison between the estimated and the actual COD reduction of the BAC filter for the 53 data points used for training with 22 input variables. Figure 4.13 shows the parity plot for the COD reduction extent achieved by the BAC filter. It can be seen from Figures 4.12 and 4.13 that the training of the network was successful since the parity plot shows a coefficient of determination (R^2) of 0.99. The average absolute deviation (AAD), calculated according to Eq. (4.3), was 1.07%.

$$\text{AAD}\ (\%) = \left(\frac{100}{n}\right) \sum_{k=1}^{n} \left(\frac{|\,Y^{\text{exp}} - Y^{\text{est}}\,|}{Y^{\text{exp}}}\right)_k \tag{4.3}$$

Table 4.12 Principal component analysis for the input variables.

Component	Eigenvalue	% of variance	Cumulative %
1	4.89	22.21	22.21
2	2.96	13.44	35.65
3	2.18	9.90	45.55
4	1.77	8.03	53.58
5	1.45	6.61	60.19
6	1.26	5.73	65.92
7	1.07	4.85	70.78
8	0.95	4.32	75.10
9	0.88	4.00	79.10
10	0.81	3.68	82.78
11	0.68	3.09	85.87
12	0.62	2.80	88.67
13	0.52	2.36	91.02
14	0.45	2.05	93.07
15	0.40	1.80	94.87
16	0.30	1.35	96.22
17	0.24	1.10	97.31
18	0.23	1.03	98.35
19	0.15	0.67	99.02
20	0.11	0.50	99.52
21	0.06	0.28	99.80
22	0.04	0.20	100.00

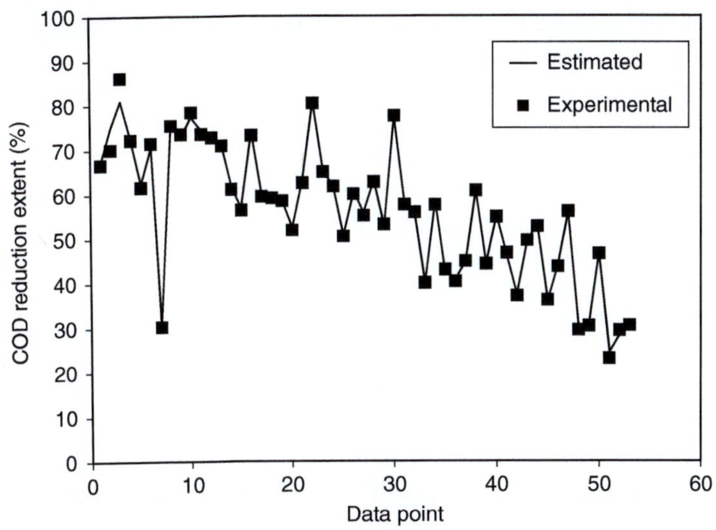

Figure 4.12 Estimated and experimental Chemical Oxygen Demand (COD) reduction extent for the training data set. Graph highlighting a near perfect fit between estimated and experimental data throughout training.

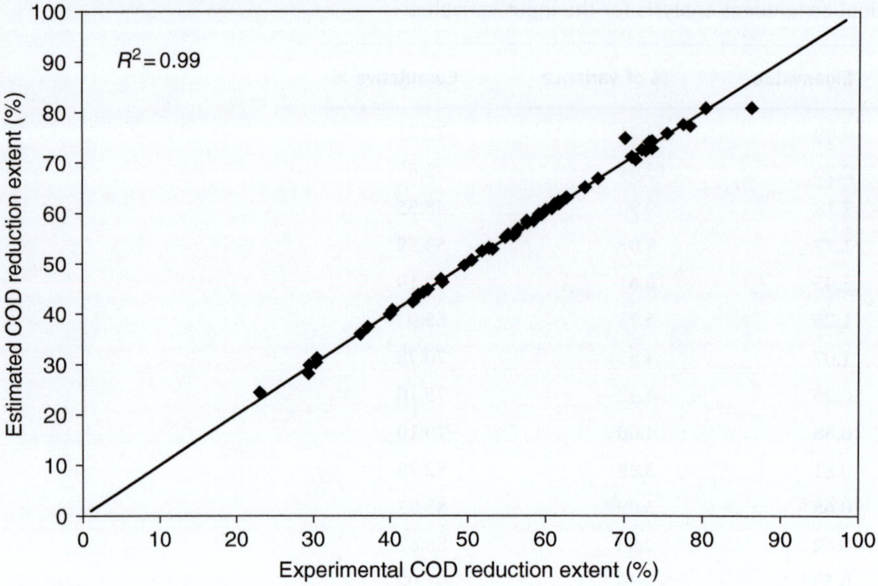

Figure 4.13 Parity plot for the estimated and experimental Chemical Oxygen Demand (COD) reduction extent for the training data set. Graph highlighting a perfect fit between estimated and experimental data during training.

where Y^{exp} is the experimental COD reduction extent and Y^{est} represents the NNM estimated COD reduction extent of the BAC filter of the kth data point and n indicates the number of data points presented to the network. This very good fit to the training data set is not surprising as it is already known that an FFNN with two hidden layers and with enough PE can approximate any continuous non-linear function to an arbitrary degree of accuracy (Cybenco 1989; Wang et al. 1992).

Of the 105 data points, 53 were used for training of the network and the remaining were retained for testing the network. However, 4 out of the 105 data points were found to be outliers and were excluded on the basis that extreme temporary divergence can be due to a detachment of solid biomass, which is not representative of the process. Therefore, 48 data points were used for testing the network. Figure 4.14 shows the plots of estimated and actual COD reduction achieved by the BAC filter for all the 48 data points used for the testing data set. The corresponding parity plot for the COD reduction extent is shown in Figure 4.15. The R^2 value was estimated to be 0.53 and an AAD was calculated according to Eq. (4.3) to be approximately 14.6%.

To compare the reliability of the NNM over a conventional regression model, a second-order polynomial was fitted to the 53 data points of the training data set. Figure 4.16 shows the plots of estimated and experimental COD reduction for the training data set and Figure 4.17 the corresponding parity plots. The average absolute error calculated according to Eq. (4.3) was 7.1% and the R^2 was 0.90. However, when the second-order polynomial was used to estimate the COD reduction extent for the testing data set, the errors were unacceptably high. The NNM was also trained using the 48 data points first applied for testing, and verified using 53 points first applied for training, but results obtained with the testing data set were less satisfactory. This is another indication that a larger data set is recommended. It also shows that one should be careful not to overtrain a neural network by having too little data compared to the number of PE, in which case the network also captures noise.

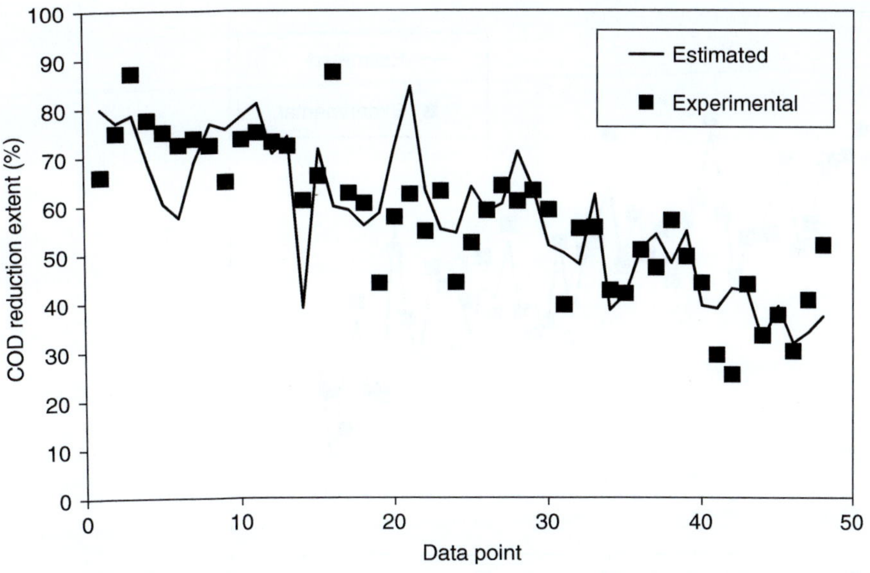

Figure 4.14 Estimated and experimental Chemical Oxygen Demand (COD) reduction extent for the testing data set. Graph indicating a rather poor fit (except for the general trend) between estimated and experimental data throughout testing.

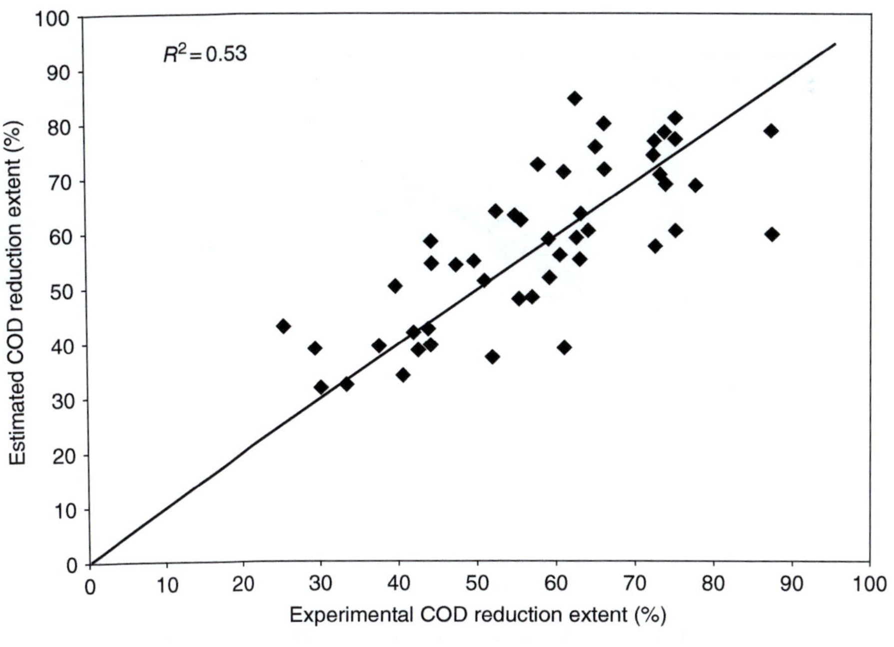

Figure 4.15 Parity plot for the estimated and experimental chemical oxygen demand (COD) reduction extent for the testing data set. Graph indicating a good fit between estimated and experimental data during testing.

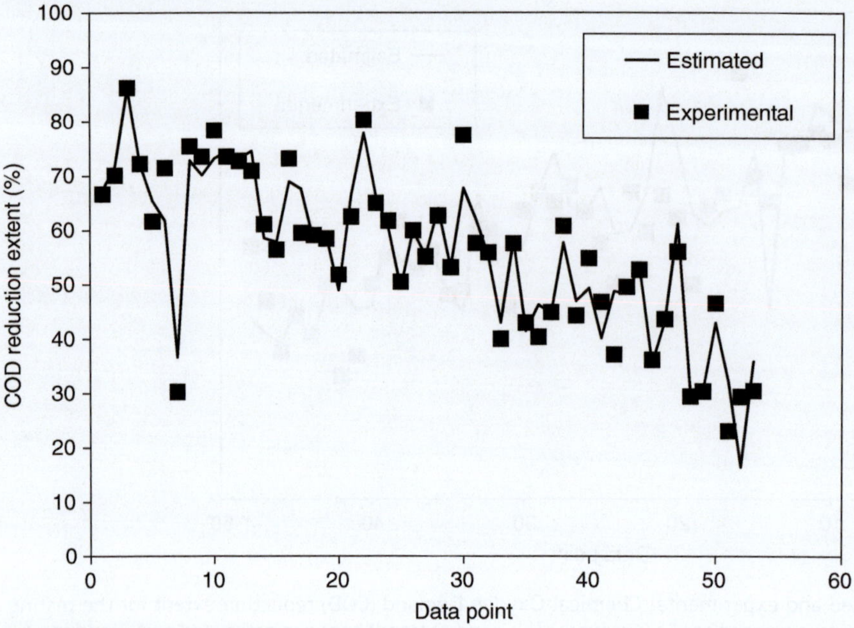

Figure 4.16 Estimated and experimental chemical oxygen demand (COD) reduction extent for the training data set (regression model). Graph indicating a good fit between estimated and experimental data throughout training.

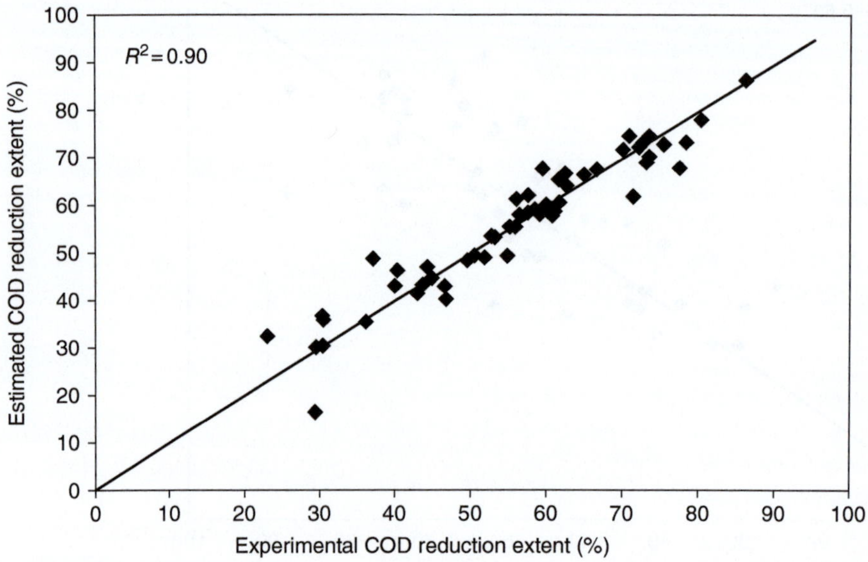

Figure 4.17 Parity plot for the estimated and experimental chemical oxygen demand (COD) reduction extent for the training data set (regression model). Graph indicating an almost perfect fit between estimated and experimental data during training.

4.3.4 Conclusions and Recommendations

Compared to conventional regression models, NNM have better capabilities of capturing the non-linear features of a process. Hence, practically any continuous non-linear function can be fitted accurately with neural networks. They are also an alternative for rapid development of models of complex processes where conventional techniques would take longer. Neural networks also have great potential as software sensors, where an on-line measurement of process variables is not possible, and as replacements for sensors having large time delay.

The major disadvantage of ANN is that they require a large quantity of good-quality data for training the network. The other area of concern is applying the network beyond the range for which it was trained. In this case, the neural network tries to extrapolate and may sometimes present erroneous results without giving any warning. The network should be subjected to regular retraining to account for any subtle change in the behaviour of the process with time.

In the case of many input variables, the regression model would require many adjustable coefficients to best fit the desired output. PCA indicates that all the 22 input variables are important. However, a smaller set of new variables (i.e., the principal components) can be used as input variables in place of the original 22 variables. Each component is then a linear function of these 22 input variables. The original data are to be pre-processed to obtain the new set of data, which are introduced to the network.

ANN can be used successfully for simulating the COD reduction of a BAC filter based on microbiological investigations of water samples. However, a larger data set is recommended, particularly for all seasons of the year, to validate the generalisation of the network. The accuracy of simulation (R^2) is much higher with NNM than with regression models.

References

Abed, A.F. (1990). The role of microbial activity in granular activated carbon beds. PhD thesis. The University of Birmingham.

APHA, AWWA & WPCF (1985). *Standard Methods for the Examination of Water and Wastewater*, 16e. Washington, DC.

Bishop, C.M. (1994). Neural networks and their applications. *The Review of Scientific Instruments* 65 (6): 1803–1832.

Boger, Z. (1992). Application of neural networks to water and wastewater treatment plant operation. *ISA Transactions* 31 (1): 25–33.

Brix, H., Arias, C.A., and del Bubba, M. (2001). Media selection for sustainable phosphorus removal in subsurface flow constructed wetlands. *Water Science and Technology* 44 (11–12): 47–54.

Camper, A.K., LeChevallier, M.W., Broadaway, S.C., and McFeters, G.A. (1986). Bacteria associated with granular activated carbon particles in drinking water. *American Society for Microbiology, Applied and Environmental Microbiology* 52: 434–438.

Carriere, P., Mohaghegh, S., Gaskari, R. et al. (1996). Performance of a virtual adsorber system for removal of lead. *Separation Science and Technology* 31 (7): 965–988.

Caswell, F. (1995). *Success in Statistics*, 3e. London: John Murray.

Chemetrics (2010). CHEMetrics By AquaPhoenix Scientific. http://www.chemetrics.com (accessed 12 April 2025).

Cox, E.J. (1996). *Identification of Freshwater Diatoms from Life Material*, 111–146. Oxford, UK: Chapman and Hall.

Curds, C.R. and Cockburn, A. (1970). Protozoa in biological sewage – treatment processes – I. Survey of the protozoan fauna of British percolating filters and activated sludge plants. *Water Research* 4: 225–236.

Cybenco, G. (1989). Approximation by superpositions of a sigmoidal function. *Mathematics of Control, Signals, and Systems* 2: 303–314.

Department of the Environment (1977–1988). *Standard Methods for the Examination of Waters and Associated Materials*. London, U.K.: HMSO.

Dong, Y., Wiliński, P.R., Dzakpasu, M., and Scholz, M. (2011). Impact of hydraulic loading rate and season on water contaminant reductions within integrated constructed wetlands. *Wetlands* 31 (3): 499–509.

Dong, Y., Scholz, M., and Mackenzie, S. (2013). Performance evaluation of representative wildfowl & wetlands trust constructed wetlands treating sewage. *Water and Environment Journal* 27 (3): 317–328.

Donlan, R.M. and Yohe, T.L. (1983). Microbial population dynamics on granular activated carbon used in treating surface impounded groundwater. In: *Treatment of Water by Granular Activated Carbon* (ed. M. J. McGuire and I.H. Suffet). American Chemical Society, Advances in Chemistry Series 202 MI, USA: Ann Arbor Science Publishers.

Dunbabin, J.S. and Bowmer, K.H. (1992). Potential use of constructed wetlands for treatment of industrial wastewater containing metals. *Science of the Total Environment* 111 (2–3): 151–168.

Eckenfelder, W.W. (1980). *Principles of Water Quality Management*. Boston, Massachusetts: CBI Publishing Company Inc.

Fowler, J. and Cohen, L. (1996). *Practical Statistics for Field Biology*. Chichester: Wiley.

Frith, A. (1996). Neural networks catch on in water treatment. *Process Engineer* s7–s8.

Greenway, M. and Woolley, A. (1999). Constructed wetlands in Queensland: performance efficiency and nutrient bioaccumulation. *Ecological Engineering* 12 (1–2): 39–55.

Healy, M. and Cawley, A.M. (2002). Nutrient processing capacity of a constructed wetland in western Ireland. *Journal of Environmental Quality* 31 (5): 1739–1747.

Her Majesty's Stationery Office (Department of the Environment, National Water Council) (1978). *Methods of the Examination of Waters and Associated Materials (Various Papers)*. London: HMSO.

Hussain, M.A. (1999). Review of the applications of neural networks in chemical process control – simulation and online implementation. *Artificial Intelligence in Engineering* 13: 55–68.

Jahangir, M.A.Q. (1994). Bioregeneration of granular activated carbon. PhD thesis. The University of Birmingham.

Jose, L., Sanchez, G., Robinson, W.C., and Budman, H. (1999). Developmental studies of an adaptive on-line softsensor for biological wastewater treatment. *Canadian Journal of Chemical Engineering* 77: 707–717.

Kadlec, R.H. (1992). Hydrological factors in wetland water treatment. In: *Constructed Wetlands for Wastewater Treatment: Municipal, Industrial and Agricultural* (ed. D.A. Hammer), 25–29. Chelsea: Lewis Publishers.

Kadlec, R. and Knight, R. (1996). *Treatment Wetlands*. Boca Raton: CRC Press.

Lee, C., Fletcher, T.D., and Sun, G. (2009). Nitrogen removal in constructed wetland systems. *Engineering in Life Sciences* 9 (1): 11–22.

Miller, R.M., Itoyama, K., Uda, A. et al. (1997). Modeling and control of a chemical wastewater treatment plant. *Computers and Chemical Engineering* 21 (supplement): s947–s952.

Mohanty, S., Scholz, M., and Slater, M.J. (2002). Neural network simulation of the chemical oxygen demand reduction in a biological activated carbon filter. *Journal of the Chartered Institution of Water and Environmental Management* 16 (1): 58–64.

Morris, A.J., Montague, G.A., and Willis, M.J. (1994). Artificial neural networks: studies in process modelling and control. *Chemical Engineering Research & Design* 72: 3–19.

Moshiri, G.A. (1993). *Constructed Wetlands for Water Quality Improvement*. New York: Lewis Publishers.

Nuttall, P.M., Boon, A.G., and Rowell, M.R. (1997). Review of the design and management of constructed wetlands. Construction Industry Research and Information Association (CIRIA), London.

Patterson (1996). *Free-living Freshwater Protozoa – A Colour Guide*. UK: Wolse Publishing Ltd.

Reed, S.C., Crites, R.W., and Middlebrooks, E.J. (1995). *Natural Systems for Waste Management and Treatment*, 2e. New York: McGraw-Hill.

Scholz M. (1998). Optimisation of biological activity in granular activated carbon beds. Ph.D. thesis. The University of Birmingham, UK.

Scholz, M. (2010). *Wetland Systems – Storm Water Management Control*. Berlin: Springer.

Scholz, M. and Martin, R.J. (1997a). Ecological equilibrium on biological activated carbon. *Water Research* 31: 2959–2968.

Scholz, M. and Martin, R.J. (1997b). Biological control in granular activated carbon beds. *Proceedings of the International Conference on Biofilms in Aquatic Systems (oral presentation, 13-16/04/97, Warwick, UK)*, Royal Society of Chemistry.

Scholz, M. and Martin, R.J. (1998a). Control of bio-regenerated granular activated carbon by spreadsheet modelling. *Journal of Chemical Technology and Biotechnology* 71: 253–261.

Scholz, M. and Martin, R.J. (1998b). Biological control in granular activated carbon beds. *International Review of Hydrobiology* 83: 657–664.

Scholz, M., Sadowski, A.J., Harrington, R., and Carroll, P. (2007). Integrated constructed wetlands assessment and design for phosphate removal. *Biosystems Engineering* 97 (3): 415–423.

Scholz, M., Harrington, R., Carroll, P., and Mustafa, A. (2010). Monitoring of nutrient removal within integrated constructed wetlands (ICW). *Desalination* 250 (1): 356–360.

Syu, M.J. and Chen, B.C. (1998). Back propagation neural network adaptive control of a continuous wastewater treatment process. *Industrial and Engineering Chemistry Research* 37 (9): 3625–3630.

Tanner, C.C. and Kadlec, R.H. (2003). Oxygen flux implications of observed nitrogen removal rates in subsurface-flow treatment wetlands. *Water Science and Technology* 48 (5): 191–198.

Thomas, R., Freeman, C., Rehman, N., and Fox, K. (2003). Removal of linear alkylbenzene sulphonate (LAS) in constructed wetlands. In: *Wetlands-nutrients, Metals and Mass Cycling* (ed. J. Vymazal), 35–47. Leiden: Backhuys Publishers.

Trang, N.T.D., Konnerup, D., Schierup, H.H. et al. (2010). Kinetics of pollutant removal from domestic wastewater in a tropical horizontal subsurface flow constructed wetland system: effects of hydraulic loading rate. *Ecological Engineering* 36: 527–535.

Tsihrintzis, V.A., Akratos, C.S., Gikas, G.D. et al. (2007). Performance and cost comparison of a FWs and a VSF constructed wetland system. *Environmental Technology* 28 (6): 621–628.

Vymazal, J. (2007). Removal of nutrients in various types of constructed wetlands. *Science of the Total Environment* 380 (1–3): 48–65.

Wang, Z., Tham, M.T., and Morris, A.J. (1992). Multilayer neural networks approximated canonical decomposition of non-linearity. *International Journal of Control* 56: 665–672.

Water Environment Federation (1995). *Standard Methods for the Examination of Water and Wastewater*, 16e. Washington DC, USA: American Public Health Association, American Water Works Association and Water Environment Federation.

Werker, A.G., Doughtery, J.M., Mchenry, J.L., and van Loon, V.A. (2002). Treatment variablility for wetland wastewater treatment design in cold climates. *Ecological Engineering* 34 (2): 137–146.

WWT (2008). Wildfowl & Wetlands Trust. http://www.wwt.org.uk (accessed 12 April 2025).

Yalcuk, A. and Ugurlu, A. (2009). Comparison of horizontal and vertical constructed wetland systems for landfill leachate treatment. *Bioresource Technology* 100 (9): 2521–2526.

5

Industrial Wastewater Treatment and Modelling

5.1 Membrane Bioreactors and Constructed Wetlands Treating Rendering Wastewater

5.1.1 Introduction

European legislation prohibits the on-site treatment of animal by-products such as offal, gut and blood (European Union 2002). These by-products are required to be transported for treatment to dedicated rendering plant sites. Primary rendering plant effluent treatment technologies include screening, settling and dissolved air flotation (DAF) as discussed elsewhere (Johns 1996; Manjunath et al. 2000).

Secondary or tertiary treatment of rendering plant wastewater is by some combination of aerobic and anaerobic wastewater ponds (Johns 1996), anaerobic digestion technologies (Banks and Wang 1999; Johns 1996; Masse and Masse 2000; Rivera et al. 1997; Wang and Banks 2003), activated sludge plants (Johns 1996), trickling filters (Johns 1996), constructed wetlands (Banks and Wang 1999; Johns 1996) and membrane bioreactors (MBRs) (Gander et al. 2000; Stephenson et al. 2000). Biological activated carbon filtration (Scholz and Martin 1997) could be a suitable treatment technology, but has not been tested, previously.

Considering activated sludge technologies, incomplete nitrification is common, despite low loading rates and high sludge retention times. The insufficient treatment performance can be explained by the limitation of phosphorus during ammonia oxidation processes (Nowak et al. 1999; Wehrle-Werk 2004).

Constructed wetlands could be effective for secondary and tertiary treatments of animal rendering plant effluent. They are usually associated with low capital and operational costs, and are a valuable resource for wildlife (Banks and Wang 1999; Cooper et al. 1996). Furthermore, the removal of chemical oxygen demand (COD), five-day @ 20°C biochemical oxygen demand (BOD), suspended solids (SS), organic nitrogen and faecal coliforms within wetlands is usually effective and stable (Banks and Wang 1999).

The author assessed two different novel treatment technologies used as a secondary treatment stage for pre-treated animal rendering plant wastewater following primary treatment with DAF: (i) a combined denitrification, nitrification and ultrafiltration system (operated essentially in sequence) called membrane bioreactor (MBR); and (ii) a flooded, actively aerated and passively ventilated vertical flow (downwards inflow followed by upwards outflow) constructed wetland.

The key objectives are to (i) determine if both treatment methods (based on a case study) are fit for purpose; (ii) highlight benefits and shortcomings associated with both treatment methods in

Sustainable Water Systems, First Edition. Miklas Scholz.

practice; (iii) assess the variability of MBR water quality data; and (iv) optimise sampling schemes to reduce effort and costs. O'Brien et al. (2005) provided the scientific basis for Chapter 5.1

5.1.2 Materials and Methods

5.1.2.1 Industrial Rendering Plant

Figure 5.1 shows the predominant process sequence of the rendering plant owned by Argent Group Europe Ltd at Newarthill near Glasgow in Scotland, UK. After preliminary treatment, the wastewater was treated by DAF with a retention time of 45 minutes. The manufacturer of the DAF unit is Robin Hawkins Engineering (Banks and Wang 1999).

DAF is followed by a sequence of denitrification, nitrification and ultrafiltration treatment steps. The denitrification and nitrification tank liquid volumes are 232 and 780 m^3, respectively. While the denitrification tank is kept anaerobic despite stirring activity, the nitrification tank is aerated. The total air requirement is 2200 $m^3\ h^{-1}$. An anti-foaming agent was supplied by Wehrle-Werke (Germany). The agent is called Drewplus L-674 and contains 85–100% oxygenated aliphatic hydrocarbons.

Phosphoric acid (H_3PO_4) is dosed directly into the nitrification unit to enhance biodegradation. Approximately 4.38 kg phosphoric acid (75% solution, $v\ v^{-1}$) is added per hour of plant running time.

The entire effluent of the DAF went to the MBR unit (except for 0.002% (approximately 0.8 $m^3\ d^{-1}$) that was used for two experimental pilot MBR; see below) for further treatment. The MBR was produced by Wehrle-Werk and developed in cooperation with the University of Stuttgart (Germany). This MBR is a hybrid of two commercial products called BIOMEMBRAT® (sequence of activated sludge and ultrafiltration) and BIOMEMBRAT®-plus (combination of denitrification, nitrification, ultrafiltration, adsorption and nanofiltration). The hybrid system (sequence of denitrification, nitrification and ultrafiltration as indicated in Figure 5.1) is usually applied for highly contaminated wastewater with complex high-molecular-weight organic compounds such as rendering plant effluent,

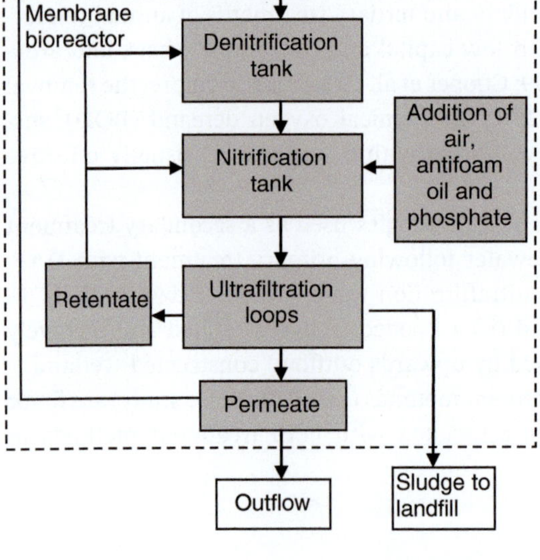

Figure 5.1 Overview of the sequence of the predominant at the industrial rendering plant near Glasgow. The elements of the membrane bioreactor are at the heart of the treatment train.

leachate from landfills and industrial effluents (e.g., chemical and textile industry) (Gander et al. 2000; Stephenson et al. 2000; Wherle-Werk 2004).

Ultrafiltration uses hydrostatic pressure differences as the driving force for the separation of water and small molecules from macromolecules, colloids and proteins via sieving. The ultrafiltration membrane with pores of 0.2 μm has an area of 265 m^2. The membrane is subject to fouling (deposition of degrading biomass and SS) and therefore routine membrane cleaning is required. The ultrafiltration unit (and subsequently the MBR) operates at a maximum flow rate of 427 $m^3\ d^{-1}$. The minimum flow rate during plant operation is 360 $m^3\ d^{-1}$. The approximate footprint of the overall system including the ultrafiltration unit is approximately 1000 m^2.

The final effluent is discharged into the public sewer when its quality complies with the standards set by Scottish Water (water utility) for the works and monitored by the Scottish Environmental Protection Agency; viz., pH between 6 and 10, SS of 500 mg L^{-1}, and ammonia of 200 mg L^{-1}. There was no consent level for nitrate in the effluent. If the discharge in terms of pH, ammonia and SS does not comply with the threshold concentrations, the insufficiently treated wastewater is recycled within the treatment process (Figure 5.1). However, the pH was only on three occasions (0.8% of the time) below 6.

5.1.2.2 Constructed Wetland

The constructed wetland (Figures 5.2 and 5.3) was designed in agreement with guidelines (Cooper et al. 1996; Scholz 2004; Scholz et al. 2002). The footprint of the wetland is approximately 0.63 m^2. The DAF unit effluent was distributed equally within the flooded and planted (see below) top layer of the wetland (Figure 5.2).

The top water layer of the flooded constructed wetland was aerated with an aquarium pump via two air stones at an average air rate of 3.5 L min^{-1}. The air stones and four vertical passive ventilation pipes (diameter 2 cm; buried 20 cm deep) were placed symmetrically within the wetland to provide equal air distribution (Figure 5.3).

The layer thicknesses (lengths within brackets) of the wetland from bottom to top were as follows (Figure 5.2): washed round gravel with diameters between 10 and 50 mm (20 cm); washed round pea-gravel with diameters between 2.5 and 10 mm (11 cm); Filtralite (light expanded clay aggregate provided by the Optiroc Group AB, Norway) with diameters between 1.5 and 2.5 mm (4 cm); washed sand with diameters between 0.2 and 2.5 mm (5 cm); barley straw (4 cm); wastewater and *Typha latifolia* L. (Reedmace or Broad-leaved Cattail) (5–10 cm). Twenty-four mature (at least four years old) individual *Typha latifolia* macrophytes were planted symmetrically within the sand and Filtralite layers in September 2002. The macrophytes were harvested (cut 10 cm above the minimum water level) on 19 December 2002 and 17 November 2003, and the associated dry plant mass was weighed. The purpose was to investigate the uptake of pollutants.

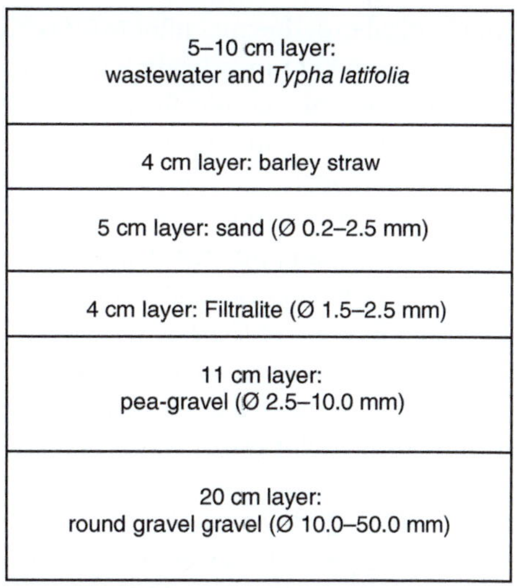

Figure 5.2 Overview of the layers of the flooded (top layer) constructed wetland. The planted filter is dominated by solid aggregates including Filtralite.

The wetland was flooded and a water level between 5 and 10 cm on top of the sand and barley straw was maintained by adding 5 L of wastewater (effluent of the DAF unit) approximately twice per week and counterbalancing the gain by abstracting treated

Figure 5.3 Photographs showing (a) the entire constructed wetland; and (b) the top surface layer during the growing period. A large square plastic tank was used as a wetland cell. The water surface area is still visible.

wastewater from the centrally located outflow pipe taking account of precipitation gains and evaporation losses. The mean hydraulic load was therefore approximately 0.060 mL h^{-1}.

5.1.2.3 Membrane Bioreactor

Two identical experimental pilot MBR (PP1 and PP2) were operated in the same way as the main MBR (Figure 5.1) to investigate temporal water quality data variation. The total volumes of the denitrification tank and nitrification tank were 0.232 and 0.780 m^3, respectively. Flow rates were chosen to give the same mean retention times as in full scale MBR to minimise scale effects. The air requirement was approximately 2.2 m^3 h^{-1}. However, the actual air input was frequently higher due to manual operation and therefore a lack of automatic control. The backflow rate from the nitrification tank into the denitrification tank was approximately 120 L h^{-1}.

The total volume of each pilot plant was approximately 1.1 m^3. Both plants were designed for a wastewater flow with a COD of 5125 mg L^{-1}. The daily inflow rate was approximately 17 L h^{-1}. The new membrane area was approximately 0.15 m^2. The relatively small membrane area was justified considering that the industrial-scale membrane was old and partly clogged. The trans-membrane pressure for the pilot plants was 0.24 MPa.

5.1.2.4 Water Quality Analysis

All analytical procedures were performed according to standard methods (Standard Methods for the Examination of Water and Wastewater 1998). However, unfiltered COD and all nutrient samples were determined with various Dr. Lange test kits and the associated Dr. Lange LASA 50 photometer.

The vegetation samples were analysed for various metals and heavy elements. Composite samples were collected from the vegetation at different sites within the filter and stored frozen at −18°C until analysis. The digestion method was adapted from the section 'Nitric Acid-Hydrochloric Acid Digestion' (Standard Methods for the Examination of Water and Wastewater 1998).

An Inductively Coupled Plasma Optical Emission Spectrometer (ICP-OES) was used for the analysis of metals and other heavy elements. Total concentrations of elements in filtered (Whatman 1.2 μm cellulose nitrate membrane filter) samples were determined by ICP-OES using a TJA IRIS instrument (ThermoElemental, USA). Multi-element calibration standards with a wide range of concentrations were used and the emission intensity measured at appropriate wavelengths.

Concerning the analysis of major nutrients, an extraction and digestion method was (Allen 1974). For analysis of total nitrogen (N_{total}), the following reactions occur: Ammonium (present in the digest) reacts with hypochlorite ions generated by alkaline hydrolysis of sodium dichloroisocyanurate. The reaction forms monochloroamine, which reacts with salicylate ions in the presence of sodium nitroprusside to form a blue indephenol complex. This complex is measured colorimetrically at 660 nm using a Bran & Luebbe autoanalyser (model AAIII).

For analysis of total phosphorus (P_{total}), the following reactions occur: Orthophosphate (present in the digest) reacts with ammonium molybdate in the presence of sulphuric acid to form a phosphomolybdenum complex. Potassium antimonyl tartrate and ascorbic acid are used to reduce the complex, forming a blue colour, which is proportional to the P_{total} concentration. Absorption was measured at 660 nm using a Bran & Luebbe autoanalyser (model AAIII).

For the analysis of total potassium (K_{total}), the digest was analysed by a flame atomic absorption spectrometer (Unicam 919, Cambridge, UK) at a wavelength of 766.5 nm and with a bandpass of 1.5 nm. The method was adapted from the section on 'Metals by Flame Atomic Absorption Spectrometry' (Standard Methods for the Examination of Water and Wastewater 1998). Descriptions for detailed chemical reaction equations for the above methods can be found elsewhere (Allen 1974; Standard Methods for the Examination of Water and Wastewater 1998).

5.1.3 Results and Discussion

5.1.3.1 Water Quality of the Dissolved Air Flotation Plant

Table 5.1 indicates the water quality associated with the primary treated DAF plant effluent that is used as influent for the industrial MBR, MBR PP1, MBR PP2 and constructed wetland. The inflow water contains high and very variable concentrations of COD, N_{total} and phosphate.

Table 5.1 Dissolved air flotation outflow water quality (21 August 2002–20 November 2003).

Variable	Unit	Number of samples	Minimum	Mean	Maximum	Standard deviation
Chemical oxygen demand	(mg L^{-1})	198	1450.0	5815.5	33 320.0	3005.03
Total nitrogen	(mg L^{-1})	43	0.8	907.8	3240.0	635.23
Ammonia	(mg L^{-1})	234	5.5	613.6	1500.0	268.72
Phosphate	(mg L^{-1})	69	0.2	5.2	75.0	10.35
pH	—	240	4.8	6.1	7.5	0.42
Temperature	(°C)	238	13.0	25.0	33.5	4.07

The wastewater qualities for slaughterhouse and rendering plants in countries such as Austria, France, Germany, Spain, The Netherlands and the United States have been reviewed elsewhere (Banks and Wang 1999; De Villiers and Pretorius 2001; Johns 1996; Manjunath et al. 2000; Nowak et al. 1999). The COD of the DAF plant effluent is similar to the values reported for France (4200–8500 mg L^{-1}), India (1100–7250 mg L^{-1}), Spain (6000 mg L^{-1}), The Netherlands (1925–11 118 mg L^{-1}) and the United States (4200–8500 mg L^{-1}). However, the concentrations for N_{total} and ammonia at the rendering plant operated by the Argent Group Europe are more variable and higher compared with the reference values reported in the literature. This is due to more scope for process optimisation (e.g., achieving a stable inflow concentration and continuous operation without shutdowns). In contrast, the concentrations of phosphate are lower but also more variable than the values reported elsewhere. This is due to manual phosphate addition (see below).

5.1.3.2 Water Quality of the Membrane Bioreactor

Table 5.2 summarises the quality of both the permeate (effluent) and sludge (to be transported to landfill sites) associated with the ultrafiltration unit. The permeate from the ultrafiltration unit is equal to the outflow of the rendering plant considering that this is the last treatment step. However, if the outflow water quality is not within the target values, it was recycled (see above and Figure 5.1).

In contrast to COD (see also Figure 5.4) and N_{total}, the concentrations of ammonia (Figure 5.5), nitrate, nitrite and phosphate are high and very variable. These findings compare well with recordings presented elsewhere (Fuchs et al. 2003; Johns 1996). The consent level for ammonia is exceeded 29 times between 16 September 2002 and 29 April 2003 (Figure 5.5).

The phosphoric acid dosing pump required flow adjustment from time to time (major operational adjustments on 17 December 2002 and 2 April 2003). However, phosphate concentrations in the

Table 5.2 Water quality (21 August 2002–20 November 2003) of the ultrafiltration unit permeate and the ultrafiltration sludge.

Variables for permeate of the ultrafiltration unit	Unit	Number of samples	Minimum	Mean	Maximum	Standard deviation
Chemical oxygen demand	(mg L^{-1})	206	1.5	37.2	111.0	21.96
Total nitrogen	(mg L^{-1})	29	11.6	111.3	298.0	70.62
Ammonia	(mg L^{-1})	258	0.0	85.9	1260.0	181.82
Nitrate	(mg L^{-1})	141	0.1	48.0	510.0	62.38
Nitrogen dioxide	(mg L^{-1})	185	0.0	14.5	192.0	21.06
Phosphate	(mg L^{-1})	124	0.0	7.3	85.2	16.97
pH	—	266	6.2	7.1	8.8	0.40
Temperature	(°C)	266	22.0	30.1	39.5	3.16
Variables for sludge of the ultrafiltration unit	**Unit**	**Number of samples**	**Minimum**	**Mean**	**Maximum**	**Standard deviation**
Mixed liquor suspended solids	(g L^{-1})	223	2.2	13.1	38.0	4.87
Mixed liquor volatile suspended solids	(g L^{-1})	189	2.1	11.0	31.9	4.15
pH	—	238	5.9	7.1	8.6	0.45
Temperature	(°C)	244	12.0	29.0	36.0	3.49

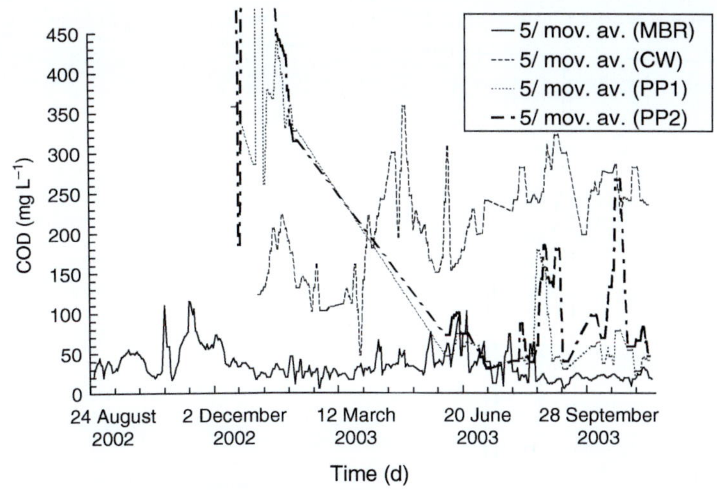

Figure 5.4 Comparison of the chemical oxygen demand distributions (four outliers are not shown). A very erratic data distribution can be seen.

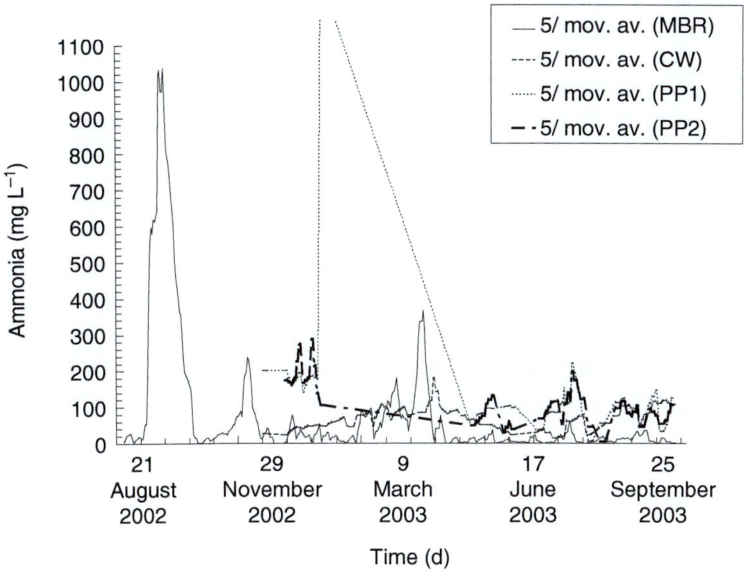

Figure 5.5 Comparison of the ammonia distributions (two outliers are not shown). A very erratic data distribution can be seen.

MBR effluent were elevated due to manual overdosing of phosphoric acid (see above, Table 5.2 and Figure 5.6). A reduction of the phosphoric acid dosing would also reduce the phosphate outflow concentrations at the same time. The dosing pump has been (at least temporarily) switched off since 1 October 2003 due to sufficient phosphate concentrations in the DAF effluent.

Furthermore, the COD, ammonia and phosphate fluctuations are high due to fluctuations of the organic strength in the inflow and maintenance processes such as membrane cleaning (Figures 5.4–5.6). Fluctuations of the organic strength meant also that hydraulic and sludge retention times varied resulting in varying MLSS values with a mean of 13.1 g L^{-1} and a standard deviation of 4.87 g L^{-1} (Table 5.2).

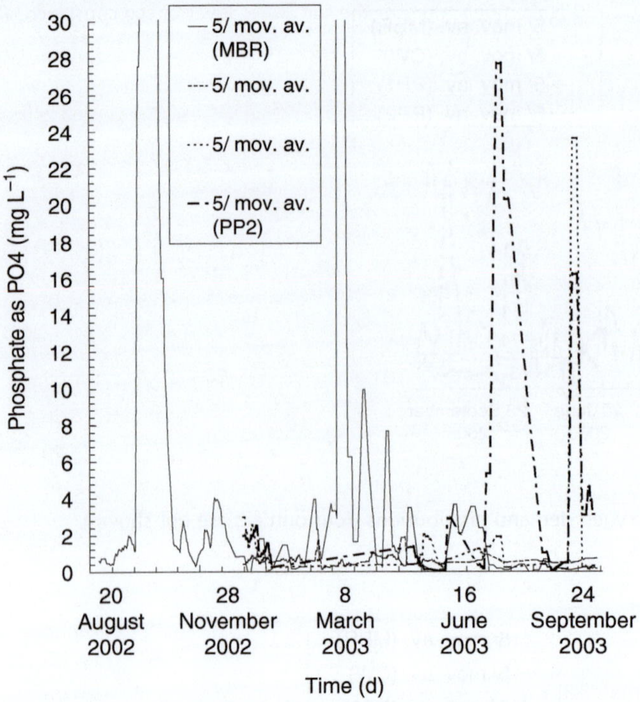

Figure 5.6 Comparison of the phosphate distributions (two outliers are not shown). A very erratic data distribution can be seen.

5.1.3.3 Water Quality of the Constructed Wetland

Table 5.3 summarises the water quality of both the top water layer (Figure 5.2) and the effluent of the constructed wetland (Figure 5.3). COD concentrations were relatively high (see also Figure 5.4). The concentrations of nitrate, nitrogen dioxide and SS were very variable.

The treatment performance of the aerobic top water layer (Figure 5.2) in terms of COD and ammonia removal is better if compared with the subsequent treatment by the anaerobic gravel and sand-based layers (Table 5.3 and Figure 5.2). However, the novel upward-flow outflow design helps to reduce the SS load (see above and Figure 5.3).

Ammonia toxicity to *T. latifolia* did not result in poor plant growth despite mean ammonia inflow concentrations of approximately 614 mg L^{-1} (Table 5.1). It must be noted that the effluent of the DAF is used as influent to the constructed wetland. The concentrations reported to be toxic to wetland plants were between 70 and 250 mg L^{-1}.

The high phosphate removal in the wetland (Tables 5.1 and 5.3) was surprising when compared with concentrations reported in the literature (Johns 1996). It is a possibility that Filtralite enhanced the removal of phosphate due to adsorption on the expanded clay. However, this has been previously shown to be only an insignificant process in terms of the overall treatment performance of constructed wetlands used to treat polluted urban surface water (Scholz 2004; Scholz et al. 2002). Furthermore, the introduction of air via air pumps into the system guaranteed the presence of enough oxygen particularly during winter and summer nights where algae do not contribute to the production of oxygen available for aerobic biodegradation.

The harvested biomass on 19 December 2002 and 27 November 2003 had dry weights of 136.3 and 84.7 g, respectively. The concentrations of most elements including aluminium, cobalt, chromium, nickel, lead and vanadium were below the detection limit of the ICP-OES: 0.22, 0.03, 0.05, 0.12, 0.23

Table 5.3 Water quality (16 December 2002–20 November 2003) of the top water layer of the constructed wetland and the effluent of the constructed wetland.

Variables of top water layer of constructed wetland	Unit	Number of samples	Minimum	Mean	Maximum	Standard deviation
Chemical oxygen demand	(mg L^{-1})	70	28.2	266.4	821.0	148.48
Ammonia	(mg L^{-1})	73	12.3	79.3	142.0	30.88
Nitrate	(mg L^{-1})	65	0.1	3.4	108.0	13.38
Nitrogen dioxide	(mg L^{-1})	54	0.0	1.7	40.2	6.58
Phosphate	(mg L^{-1})	62	0.0	0.5	5.1	0.70
Suspended solids	(mg L^{-1})	55	4.0	125.8	1242.0	192.57
Dissolved oxygen	(mg L^{-1})	52	0.1	5.1	16.1	3.26
pH	—	67	6.9	7.8	9.8	0.46
Temperature	(°C)	68	0.0	12.5	25.0	6.17
Variables of effluent of constructed wetland	**Unit**	**Number of samples**	**Minimum**	**Mean**	**Maximum**	**Standard deviation**
Chemical oxygen demand	(mg L^{-1})	72	48.0	205.0	358.0	64.02
Ammonia	(mg L^{-1})	74	15.0	67.1	179.0	30.45
Nitrate	(mg L^{-1})	66	0.3	5.7	85.3	12.59
Nitrogen dioxide	(mg L^{-1})	54	0.0	2.2	19.5	4.72
Phosphate	(mg L^{-1})	64	0.1	0.6	1.9	0.28
Suspended solids	(mg L^{-1})	53	1.0	57.3	619.0	82.77
Dissolved oxygen	(mg L^{-1})	52	0.0	3.5	11.6	2.49
pH	—	67	7.0	7.9	10.0	0.46
Temperature	(°C)	68	0.0	11.8	25.0	5.92

and 0.034 mg L^{-1} in that order. Furthermore, the estimated concentration of calcium (based on serial dilution) was approximately 4 g L^{-1}. Calcium and phosphorus concentrations are likely to be high in crushed animal bone and meat material. Therefore, plant litter removal from the litter zone of the constructed wetland receiving pre-treated rendering plant wastewater should lead to a reduction in total nutrient and mineral loadings of the wetland and subsequently also of the effluent.

Table 5.4 gives an overview of the major nutrient concentrations within aquatic plant matter of the constructed wetland. Harvesting of *T. latifolia* resulted in a relatively high overall reduction in N_{total} despite that most nitrogen is stored in the rhizomes, roots and organic sediment.

5.1.3.4 Comparison of Treatment Performances

Constructed wetlands are passive and industrial MBR are active treatment technologies (Gander et al. 2000; Scholz 2004; Scholz et al. 2002; Stephenson et al. 2000). A full-scale constructed wetland for the treatment of the same wastewater as the industrial MBR described in this case study would almost be as large as a football field.

In comparison with the constructed wetland, the industrial MBR was less effective in the removal of phosphate, because phosphoric acid was added frequently in potentially too high concentrations to the nitrification unit to enhance biodegradation (Tables 5.2 and 5.3). The mean concentrations for the outflow of the ultrafiltration unit and the constructed wetland were 7.3 and 0.6 mg L^{-1}, respectively.

Table 5.4 Mean nutrient concentrations for sediment and plant parts of *Typha latifolia* L.

Species	Plant part	Number of samples	N_{total}[a] (g kg^{-1})	P_{total}[b] (g kg^{-1})	K_{total}[c] (g kg^{-1})
T. latifolia	Flowers	3	10.465	0.302	0.907
T. latifolia	Leaves	3	13.601	0.393	2.475
P. australis	Leaves	3	20.839	0.427	5.916
N/A[d]	Sediment	3	23.646	1.425	1.780
T. latifolia	Roots	3	28.138	0.837	2.845
T. latifolia	Shoots	1	36.500	2.779	26.301
T. latifolia	Rhizomes	3	43.089	1.033	7.870

[a] total nitrogen.
[b] total phosphorus.
[c] total potassium.
[d] not applicable.
Average major nutrient concentrations for sediment (degraded aquatic plant matter and barley straw) and dried aquatic plants harvested on 17 November 2003. *Typha latifolia* L. (Reedmace or Broad-leaved Cattail) was deliberately planted and dominated the wetland (95% surface cover). *Phragmites australis* (Cav) Trin Ex Streud (Common Reed) was recorded as an alien invader (5% surface cover).

The phosphate removal efficiency of the industrial MBR plant was therefore variable. The standard deviation was 16.97 mg L^{-1} (Table 5.2). The phosphate distribution curve was subject to the implementation of cleaning routines (Figure 5.6). Optimising phosphate dosing would reduce the likelihood of polluting receiving waters (e.g., maximum phosphate concentration was 85.2 mg L^{-1}; Table 5.2).

However, the COD removal efficiency of the constructed wetland declined over time: outflow concentrations of 100 mg L^{-1} in March and 250 mg L^{-1} in September 2003 (Figure 5.4). It is likely that the constructed wetland started to suffer from overloading during the maturation process.

5.1.3.5 Water Quality Variability of the Membrane Bioreactor

Table 5.5 indicates the relatively high variability of COD and phosphate between the two similar experimental pilot MBR (PP1 and PP2). Moreover, the within-plant variabilities of COD, ammonia, nitrogen dioxide and phosphate were relatively high for PP1 and PP2 (except for ammonia). However, data variabilities for both pilot plants were less for ammonia and nitrate, and higher for COD, if compared with the full-scale industrial MBR (see also Table 5.2).

The consent level for ammonia at PP1 was exceeded five times between 19 December 2002 and 27 August 2003. The consent level for ammonia at PP2 was exceeded twice between 16 and 29 January 2003 (Figure 5.5). The results of small-scale experimental systems should always be assessed carefully, if compared or transferred to large-scale industrial systems due to potential artefacts in the data based on scaling problems.

5.1.3.6 Sampling Optimisation

Table 5.6 summarises the main findings from a correlation analysis for all systems and system units. For internal plant control, some expensive and time-consuming variables such as COD can be predicted with less expensive variables such as temperature and pH: e.g., corresponding pH and COD values of the ultrafiltration sludge are associated with a correlation coefficient of $R = 0.63$ (Table 5.6). This will lead to the optimisation of the sampling scheme. Moreover, regression analysis could be used for prediction purposes (Scholz et al. 2002).

Table 5.5 Water quality (16 December 2002–20 November 2003) of the first and second membrane bioreactor pilot plant (PP1 and PP2).

Variables of the first pilot plant PP1	Unit	Number of samples	Minimum	Mean	Maximum	Standard deviation
Chemical oxygen demand	($mg\,L^{-1}$)	42	22.4	185.6	2381.0	371.01
Ammonia	($mg\,L^{-1}$)	45	6.9	136.4	1260.0	180.56
Nitrate	($mg\,L^{-1}$)	31	0.8	79.1	167.4	50.63
Nitrogen dioxide	($mg\,L^{-1}$)	23	0.0	3.5	46.5	9.80
Phosphate	($mg\,L^{-1}$)	29	0.1	1.6	23.6	4.29
Sludge MLSS[a]	($g\,L^{-1}$)	24	2.3	9.2	23.1	6.29
Sludge MLVSS[b]	($g\,L^{-1}$)	12	3.8	10.5	18.9	5.37
Sludge dissolved oxygen	($mg\,L^{-1}$)	21	0.2	5.2	8.7	2.07
pH	—	42	3.4	6.5	8.3	1.20
Temperature	(°C)	43	6.0	16.5	29.0	5.82
Variables of the second pilot plant PP2	**Unit**	**Number of samples**	**Minimum**	**Mean**	**Maximum**	**Standard deviation**
Chemical oxygen demand	($mg\,L^{-1}$)	41	30.7	323.9	2347.0	566.47
Ammonia	($mg\,L^{-1}$)	42	5.0	106.5	286.0	60.44
Nitrate	($mg\,L^{-1}$)	28	0.5	75.7	194.0	53.75
Nitrogen dioxide	($mg\,L^{-1}$)	21	0.0	5.0	43.2	11.55
Phosphate	($mg\,L^{-1}$)	28	0.0	3.6	27.7	6.64
Sludge MLSS[a]	($g\,L^{-1}$)	23	2.3	8.2	16.8	3.94
Sludge MLVSS[b]	($g\,L^{-1}$)	12	5.2	7.8	11.0	1.71
Sludge dissolved oxygen	($mg\,L^{-1}$)	21	1.0	5.3	16.8	3.94
pH	—	39	3.5	6.8	8.3	1.13
Temperature	(°C)	41	6.0	17.0	37.0	6.38

[a] mixed liquor suspended solids.
[b] mixed liquor volatile suspended solids.

A very good correlation between ammonia and pH ($R = 0.72$) was recorded for the permeate of the ultrafiltration unit. It follows that there is less of a need to measure ammonia. Furthermore, there is absolutely no need to measure both the MLSS and the mixed liquor volatile suspended solids (MLVSS) for the ultrafiltration sludge, because they correlate very well ($R = 0.99$) with each other (Table 5.6).In comparison, COD correlates sufficiently with temperature and ammonia for the top water layer of the constructed wetland (Table 5.6). Moreover, nitrate correlates sufficiently with DO ($R = -0.41$).

5.1.4 Conclusions

The water quality of the DAF effluent was highly variable (COD and phosphate in particular), but concentrations for COD were comparable with the literature. Secondary effluent treatment with an industrial MBR plant or constructed wetland was effective to comply with Scottish Water standards.

Table 5.6 Correlation analysis.

Process unit	First variable	Second variable	Pairs	*R*
Dissolved air flotation	NH_4[a]	COD[b] (unfiltered)	187	0.44
	N_{total}[c]	NH_4[a]	48	0.45
Ultrafiltration permeate	NH_4[a]	pH	171	0.72
	PO_4[d]	T[e]	133	−0.43
	NO_2[f]	COD[b]	155	0.48
Ultrafiltration sludge	COD[b]	COD[b] (unfiltered)	37	0.62
	MLSS[g]	COD[b] (unfiltered)	31	0.87
	pH	COD[b] (unfiltered)	36	0.67
	MLSS[g]	COD[b]	32	0.64
	pH	COD[b]	38	0.63
	MLVSS[h]	MLSS[g]	190	0.99
Constructed wetland (top layer)	NO_3[I]	DO	48	−0.41
	COD[b]	T[e]	63	0.48
	COD[b]	NH_4[a]	67	0.46
Constructed wetland (effluent)	pH	DO[j]	48	0.44
	NO_3[I]	NO_2[6]	51	0.62
	NH_4[a]	NO_2[6]	52	−0.46
PP2	NH_4[a]	pH	33	0.52
	NH_4[a]	COD[b]	37	0.47

[a] ammonia (mg L^{-1}).
[b] chemical oxygen demand (mg L^{-1}).
[c] total nitrogen (mg L^{-1}).
[d] phosphate (mg L^{-1}).
[e] temperature (°C).
[f] nitrite (mg L^{-1}).
[g] mixed liquor suspended solids (mg L^{-1}).
[h] mixed liquor volatile suspended solids (mg L^{-1}).
[i] nitrate (mg L^{-1}).
[j] dissolved oxygen (mg L^{-1}).
Correlation analysis showing correlation coefficients (*R*) for dissolved air flotation, ultrafiltration permeate and ultrafiltration sludge data between 21 October 2002 and 20 November 2003 and constructed wetland (top layer and effluent) and the second membrane bioreactor pilot plant (PP2) data between 16 December 2002 and 20 November 2003. Only data for ≥30 corresponding pairs and where $R < -0.4$ or $R > 0.4$ are shown.

However, membrane flux rates were low due to fouling. The ammonia effluent qualities of both the industrial MBR (denitrification, nitrification and ultrafiltration in sequence) and novel experimental constructed wetland were similar. Ammonia concentrations were usually below the threshold of 200 mg L^{-1} permitted for discharge into the public sewer.

Chemical oxygen demand reduction efficiencies of the wetland were relatively low in comparison with the industrial MBR. The COD and ammonia removal within the aerated top water layer was relatively high.

All MBR required frequent cleaning and comprehensive maintenance. There is potential to optimise the addition of phosphoric acid to the nitrification unit to reduce the high phosphate outflow concentrations of the industrial MBR. In contrast, phosphate concentrations within the effluent of the constructed wetland were very low.

A constructed wetland has the advantage of low capital and running costs, virtually no waste production (e.g., sludge), low sensitivity to fluctuations in loading and little maintenance requirements. However, the high land footprint (due to the very low loading rates) of the treatment wetland could make its use uneconomical, and an industrial MBR might be a better solution.

A correlation analysis has shown that the sampling scheme can be optimised in this case study. For example, expensive variables such as COD can be predicted with inexpensive variables such as temperature and pH for internal process control purposes. It follows that COD does not require frequent sampling.

5.2 Benzene Removal with Constructed Treatment Wetlands

5.2.1 Introduction

5.2.1.1 Constructed Treatment Wetlands

Wetlands are complex and integrated systems in which water, animals, plants, micro-organisms and the environment interact to improve the water quality (Guirguis 2004). Man has started to mimic nature by building wetlands to treat a variety of waters, wastewaters, storm waters, gully pot liquor, acid-mine drainage waters, landfill leachate, irrigation waters, agricultural wastewater, runoff waters, industrial wastewater and produced waters (Moshiri 2000; Rew and Mulamoottil 1999; Scholz 2004; Vrhovsek et al. 1996; Yang et al. 2001). However, very little is known about the processes involved and the use of constructed treatment wetlands in the removal of petroleum hydrocarbons from processed water. This has limited the effective application of this technology in the oil and gas industries.

Vertical-flow and horizontal-flow constructed reed beds based on soil, sand and/or gravel are used frequently to treat domestic and industrial wastewater (Kadlec and Knight 1996). Constructed wetlands for wastewater treatment have several advantages such as low operational costs and habitat enhancement over conventional treatment methods (Cooper et al. 1996; Rew and Mulamoottil 1999). The wetland treatment performance depends on the interactions between many different physical and biochemical components. It follows that some functions such as the role of aggregates (also called substrate or wetland media) within constructed wetlands are not completely understood (Scholz 2006).

Vertical-flow wetland systems have a structural make-up of several layers of aggregates. Water is forced to flow perpendicular to the length of the wetland. Wetlands consist of different sizes of gravel and sand, and are often planted with macrophytes, which provide oxygen to the rhizosphere, thus creating an aerobic environment (Scholz 2006). This in turn supports microbial communities, which can either directly biodegrade organics or catalyse chemical reactions, and subsequently support biotransformation processes. Bacteria capable of degrading volatile organics such as benzene, toluene, ethylbenzene and p-xylene have been found in the rhizosphere (Hiegel 2004; Sugai et al. 1997).

Numerous benzene-degrading aerobic microorganisms have been identified; the most notable are the *Pseudomonas* species, which may account for up to 87% of the gasoline-degrading microorganisms in contaminated aquifers (Ridgeway et al. 1990). Petroleum wastes degrade in natural wetland environments (Wallace and Knight 2006; Wemple and Hendricks 2000). Benzene is biodegradable, particularly in the presence of oxygen (Alexander 1999). Benzene degradation has also been demonstrated in the presence of nitrate-nitrogen (Burland and Edwards 1999). Many studies have shown that microbial degradation of petroleum hydrocarbons in the environment is strongly influenced by physical and chemical factors such as temperature, oxygen, nutrients, salinity, pressure, water activity, pH and the chemical composition, physical state and concentration of the contaminant and by biological factors such as the composition and adaptability of the microbial population (Zhou and Crawford 1995).

Kadlec (2001) states that aeration is an important component of subsurface-flow wetland design, because an active aeration system enhances both volatilisation and aerobic degradation of hydrocarbons. The transfer of oxygen to contaminated aquifers to stimulate aerobic degradation is a common bioremediation practice (Lovley and Lloyd 2000). The biodegradability of the most water-soluble components of gasoline such as benzene (Paje et al. 1997; Solano-Serena et al. 1999), toluene (Leahy and Olsen 1997), ethylbenzene (Di Lecce et al. 1997) and xylene (Di Lecce et al. 1997) compounds has been established.

Because biodegradation and evaporation processes compete in removing petroleum hydrocarbons, biodegradative losses cannot be differentiated clearly from volatility losses (Zhou and Crawford 1995). Mitsch and Gosselink (1993) reveal that freshwater wetlands are typically considered to be nutrient-limited due to the heavy demand for nutrients by the plants, and they could also be nutrient traps, as a substantial amount of nutrients may be bound in below-ground biomass. Hence the addition of nutrients is necessary to enhance the biodegradation of oil pollutants (Choi et al. 2000; Kim et al. 2005). However, studies in the past (Chaillan et al. 2006) have shown that excessive nutrient concentrations can inhibit the biodegradation activity, and several authors have reported the negative effect of high nitrogen, phosphorus and potassium levels on the biodegradation of hydrocarbons (Chaıneau et al. 2005; Oudot et al. 1998), and particularly on the aromatics (Carmichael and Pfaender 1997).

5.2.1.2 Benzene Removal

There is a gap in knowledge concerning the optimisation of those process conditions that would foster a more efficient application of the constructed wetland technology particularly for the removal of hydrocarbon under different climatic and other variable environmental conditions. However, the number of research projects concerning constructed wetlands for hydrocarbon treatment has increased (Baris et al. 2001; Eke and Scholz 2006; Myers and Jackson 2001; Omari 2003; Xia 2003).

Since 1995, journal articles and symposia proceedings indicate the petroleum industry's interest in using constructed wetlands to manage process wastewater and storm water at a variety of installations including refineries, oil and gas wells and pumping stations (Knight 1999). The area of emphasis in this research is the use of constructed wetlands for the treatment of dissolved petroleum hydrocarbon compounds. Hence, there is a need to have a better understanding of what constitutes petroleum and its degradation products. There are four basic petroleum compounds (Harayama et al. 2004): (i) saturated hydrocarbons, which are the primary component; (ii) aromatic hydrocarbons that have at least one aromatic ring within the molecule; (iii) resins, which are mostly unknown; and (iv) asphaltenes.

These compounds are relatively large molecules and tend to possess small amounts of nitrogen, sulphur and/or oxygen (Harayama et al. 2004). Untreated petroleum industry wastewaters contain phenolics (Knight 1999), sulphides and various trace metals (Knight 1999). Hydrocarbons are highly soluble, neurotoxic and cause cancer (Hiegel 2004). Monoaromatic BTEX hydrocarbons (i.e., benzene, toluene, ethylbenzene and xylene) are commonly found in gasoline and are highly volatile substances (Coates et al. 2002). Due to their relatively high solubility and toxicity, they represent a significant health risk in contaminated environments.

Of all the BTEX compounds, benzene is of most concern, because it is the most toxic and a well-known human carcinogen. The benzene ring is a chemical structure that is common in nature. Moreover, the thermodynamic stability of the benzene ring increases its persistence in the environment; therefore, many aromatic compounds are major environmental pollutants (Dagley 1986; Díaz 2004).

The major industrial source of aromatic compounds is petroleum and natural gas, formed geochemically from biomass under high pressure and temperature (Heider et al. 1998). Aromatic hydrocarbons are part of one of the most abundant classes of organic compounds and constituents of petroleum and its refined products. Monocyclic aromatic hydrocarbons are of major concern, because of their toxicity, high solubility and ability to migrate within groundwater. These BTEX compounds are of primary discharge concern for the water quality of receiving waters (Caswell et al. 1992). The BTEX fraction of total volatile hydrocarbons is primarily responsible for most of the total toxicity in gasoline-contaminated groundwater. Hence, an attempt to reduce toxicity requires targeting these compounds for destruction.

Many components of hydrocarbon mixtures are toxic and relatively soluble in water. In natural gas, benzene concentrations typically range from about 0 to 1000 mg L^{-1}; in crude oils from virtually zero to10 000 mg L^{-1} (Janks and Cadena 1991). Benzene has a relatively high-water solubility (1780 mg L^{-1}). Water contamination by oil exploration and production operations, tank farms, below-ground storage tank leakage and refineries has become a concern to the oil and gas industry.

Hydrocarbon degradation is less dependent on the actual reactions taking place than it is on the processes occurring in the surrounding ecosystem (Sugai et al. 1997). Aerobic biodegradation and volatilisation constitute a coupled pathway that contributes significantly to the natural attenuation of hydrocarbon (Lahvis et al. 1999). Achieving high treatment performances within a short time is critical, so the design can be amended to allow manipulations of environmental conditions to enhance dissolved hydrocarbon treatment. Environmental conditions to be taken into consideration include dissolved oxygen (DO), pH, temperature and nutrient requirements (i.e., nitrogen and phosphorous) of the wetland plants and microbes. Associated contamination is therefore a major environmental problem due to the manufacture, transportation and distribution of petroleum (Atlas and Cerniglia 1995).

5.2.1.3 Aim and Objectives

The research uses data gathered from experimental small-scale wetlands to assess the efficiency of benzene removal in each wetland and to compare different operational conditions. The research covers the assessment of environmental, physical and microbial processes. This enhances operational knowledge and understanding of treatment wetlands and provides data that could be used to design full-scale wetland systems for efficient hydrocarbon treatment, and to model biodegradation and operational processes. Improved system control should include knowledge and understanding concerning environmental requirements such as oxygen availability, water inundation duration and temperature variability, fertiliser requirements for wetland microbes and characterisation of microbes capable of degrading petroleum hydrocarbons.

The overall aim is therefore to advance the understanding of the application of constructed treatment wetlands for benzene removal. The objectives are to assess: (i) the current literature on benzene removal with constructed wetland systems; (ii) variables and boundary conditions impacting on the operation and treatment performance (e.g., temperature level and variability, macrophytes and aggregates); (iii) the efficiency of different wetland set-ups in removing benzene, chemical oxygen demand (COD), five-day @ 20°C N-Allylthiourea biochemical oxygen demand (BOD) and nutrients; (iv) the effect of nutrient concentration increases on benzene removal within wetlands; (v) the impact of seasonal change and environmental control on the treatment efficiency of benzene and other water quality variables such as COD, BOD, DO, redox potential, turbidity and nutrients; and (vi) the main benzene removal pathways. Eke and Scholz (2008) provided the scientific foundation for Chapter 5.2.

5.2.2 Materials and Methods

5.2.2.1 System Design and Operation

Two experimental small-scale constructed wetland rigs with six wetlands each are located on The King's Buildings campus in Edinburgh, Scotland, UK (Table 5.7). The systems are in operation since April 2005. The wetlands were designed to simulate physical, chemical and microbiological processes occurring in full-scale semi-natural wetlands. One experimental rig was situated outdoors to assess seasonal changes, while the other system was placed indoors to allow for a better control over the environmental changes. The word 'control' indicates in this case that the data variability of most environmental variables was relatively low due to the absence of seasons. Since summer 2006, a temperature and humidity control unit was in operation to reduce the variability of temperature even further. The indoor rig was located below three plant grow lights (Sylvania 15 000 Hour, 36 W, 1200 mm, T8 Grolux Fluorescent Tube; supplied by Lyco Direct Limited (Bletchely, Milton Keynes, England), product code number: EV1768) simulating day and night conditions.

The wetland rigs were designed to optimise the chemical, physical and microbiological processes naturally occurring within wetlands. Phenomena studied include biomass manipulations, loading rate variations and changes to the cycle of filling and emptying of the wetlands. Round grey polyvinyl chloride drainage pipes (height: 75 cm; diameter: 10 cm), which are resistant to hydrocarbons, were used to construct vertical-flow wetlands. The outlet valves are located at the centre of the bottom plate of each wetland and are used for the regulation of flow and sampling. Passive aeration is encouraged with 1.3 cm diameter ventilation pipes reaching down to 10 cm above the bottom of each wetland. The wetlands are very small in comparison to large-scale systems used in industry, but previous findings based on similar column experiments have been accepted by the scientific community (Eke and Scholz 2006; Hiegel 2004; Myers and Jackson 2001; Scholz 2004, 2006; Wallace and Knight 2006).

The wetland systems have different wetland bed volumes depending on the composition of layers of aggregates such as stones, gravel and sand. Selected wetlands were planted with *Phragmites australis* (Cav.) Trin. ex Steud. (Table 5.7). Nine plants of roughly equal biomass and strength obtained from a local supplier (Alba Trees Public, Lower Winton, Gladsmuir, East Lothian, Scotland) were planted in each 'planted wetland'. Wetlands 5 and 6 are strictly speaking rather pond systems (extended storage) considering that they do not contain any aggregates. Moreover, wetland 6 containing only water and fertiliser can be considered as a 'blank'.

Wetlands 1, 3 and 5 of both rigs received tap water mixed with a concentration of $1\ g\ L^{-1}$ benzene two times per week. Benzene was used as an example volatile hydrocarbon to assess the removal of low-molecular-weight petroleum compounds. Benzene (BDH analytical reagent, C_6H_6 (99.7%)), supplied by VWR International Ltd (Hunter Boulevard, Lutterworth, England), was used. Benzene was chosen for various reasons: (i) it is a common constituent of liquid fuels; (ii) it can be used as a surrogate for a mixture of hydrocarbons to allow for easy interpretation of the data and subsequent modelling; (iii) the traditional treatment technologies used by the oil industry such as hydrocyclones and separators predominantly remove heavy hydrocarbons but not aromatics components in the dissolved water phase; and (iv) the thermodynamic stability of the benzene ring increases its persistence in the environment; therefore, many aromatic compounds are major environmental pollutants.

While hydrocarbons dissolved in water are a suitable source of carbon and energy for microbes, they are an 'incomplete food source' since they do not contain nutrients such as nitrogen, phosphorus and potassium required for microbial growth. To stimulate the growth of microbiological biomass and

Table 5.7 Packing order of the experimental constructed wetland set-up for inside and outside wetlands. All wetlands were alternatingly inundated and subsequently fully drained two times per week.

Height (cm)	Wetland 1	Wetland 2	Wetland 3	Wetland 4	Wetland 5	Wetland 6
61–75 (top)	W + B + F	W + F	W + B + F	W + F	W + B + F	W + F
56–60	5 + P + W + B + F	5 + P + W + F	5 + W + B + F	5 + W + F	W + B + F	W + F
51–55	5 + P + W + B + F	5 + P + W + F	5 + W + B + F	5 + W + F	W + B + F	W + F
36–50	4 + P + W + B + F	4 + P + W + F	4 + W + B + F	4 + W + F	W + B + F	W + F
26–35	3 + W + B + F	3 + W + F	3 + W + B + F	3 + W + F	W + B + F	W + F
11–25	2 + W + B + F	2 + W + F	2 + W + B + F	2 + W + F	W + B + F	W + F
0–10 (bottom)	1 + W + B + F	1 + W + F	1 + W + B + F	1 + W + F	W + B + F	W + F

1, stones (37.5–75 mm); 2, large gravel (10–20 mm); 3, medium gravel (5–10 mm); 4, small gravel (1.2–5 mm); 5, sand (0.6–1.2 mm); B, benzene; F, fertiliser (8 g of N-P-K Miracle-Gro fertiliser were added to all wetlands every two weeks until 29 May 2006 when the concentration was increased to 30 g. From 26 June 2006 onwards, the concentration was lowered to 15 g every two weeks.); P, *Phragmites australis* (Cav.) Trin. ex Steud. (nine plants of roughly equal biomass and strength per wetland); W, water.

P. australis, 8 g of the well-balanced N-P-K Miracle-Gro (formerly Osmocote, produced by Scot Europe B. V., The Netherlands) fertiliser were added to all wetlands every two weeks until 29 May 2006 when the concentration was increased to 30 g to assess the effect of nutrient concentration increases on benzene removal. From 26 June 2006 onwards, the concentration was lowered to 15 g every two weeks, which was seen as a more realistic fertiliser concentration to reduce the fertiliser budget and to avoid too high outflow concentrations, which would have been detrimental to the environment (Table 5.7). The system was designed to operate in batch-flow mode to reduce pumping costs.

All wetlands were alternatingly inundated and subsequently fully drained two times per week to encourage rapid air penetration through the soil (Table 5.7) according to Cooper et al. (1996). During draining, the water flowed rapidly out of the system, and air was subsequently drawn deeply into the lower part of the wetland. This aerated the system and sped up the rate of hydrocarbon degradation during the two or three days when the wetlands were not inundated.

Water samples were tested twice per week for COD, pH, DO, turbidity, redox, conductivity, ammonia-nitrogen, nitrate-nitrogen, ortho-phosphate-phosphorus and temperature. The BOD was analysed in all water samples using a respirometric method with the help of the OxiTOP IS 12-6 system, supplied by the Wissenschaftlich-Technische Werkstatten (WTW), Weilheim, Germany. This equipment uses the principle of piezo-resistive measure of pressure differences. Samples were analysed for benzene removal once per month until January 2007. Samples were testing twice per month afterwards. Benzene was determined with the Perkin Elmer GC-FID and headspace sampler (models 9700 and HS-101, respectively) instrument.

All water quality variables were determined according to the Standard Methods for the Examination of Water and Wastewater (1998), unless stated otherwise. Statistical differences were assessed by analysis of variance and Tukey's Honestly Significantly Different (HSD) test ($p < 0.05$), which is based on the studentised range distribution (Rice Virtual Lab, Lane D [http://davidmlane.com/hyperstat/B95118.html and http://onlinestatbook.com/glossary/tukey_hsd.html]).

5.2.2.2 Biodegradation and Volatilisation

Biodegradation and volatilisation were also tested in separate experiments. Two extra wetlands (height: 24 cm; diameter: 5 cm) were set up under controlled environmental conditions; one wetland comprised aggregates and detritus containing mature microbial biomass (284 g detritus was taken from the upper layer of the contaminated parent wetland 3 located indoors) and another wetland was left empty. The small wetlands were constructed in the same way as the large wetlands except for the absence of the ventilation pipes (see above). The purpose of this auxiliary experiment was to assess the main removal pathways of benzene (combined biodegradation and adsorption *versus* volatilisation) in constructed treatment wetland. Samples were taken after 1, 2, 3, 6 and 9 d, and benzene was subsequently determined using headspace and gas chromatography (see above).

5.2.3 Results and Discussion

5.2.3.1 Treatment Performance

Table 5.8 shows the benzene outflow concentrations for both the indoor and outdoor rigs. Considering the benzene inflow concentration of approximately 1.3 g L^{-1}, the findings indicate high percentage mean removal efficiencies (85–95%) for wetlands containing *P. australis* and aggregates (Wetlands 1 and 3 of the indoor and outdoor rig; Table 5.7) and are comparable to data published by Myers and Jackson (2001). The findings indicate a high variability of the benzene outflow concentrations. In contrast, wetlands without aggregates and *P. australis* (i.e., Wetland 5 of the indoor and outdoor rig; Table 5.7) had slightly lower mean treatment performances (86–93%). The removal rate achieved in Wetland 5 has been predominantly attributed to the presence of fertiliser (Table 5.7) enhancing the biodegradation rate, because some microbial communities are able to utilise the nitrogen component (i.e., nitrate-nitrogen) of the fertiliser.

The relatively low removal rate for Wetland 5 is an indication that aggregates play an indirect role in the treatment of benzene by providing habitat for nitrifying micro-organisms. In comparison, the treatment efficiencies for wetlands planted with *P. australis* were similar to the efficiency of the corresponding unplanted wetlands. The treatment performances reduced during the winter of the second year (2006/2007) with increasing hydrocarbon accumulation within the corresponding wetlands (Tables 5.8 and 5.9). This finding is in accordance with data presented, previously (Cooper et al. 1996; Scholz 2006).

Table 5.8 Mean benzene effluent concentrations (mg L^{-1}) for the indoor (i) and outdoor (o) wetlands (W1, W3 and W5).

Season	W1i	W1o	W3i	W3o	W5i	W5o
Overall mean	68	200	97	194	96	184
Spring 2005	14.7	36.1	94.1	38.3	103.2	122.4
Summer 2005	0.0	95.4	79.8	61.3	160.1	121.8
Autumn 2005	0.0	0.0	10.6	0.0	0.2	10.3
Winter 2005/2006	0.0	23.3	33.3	19.2	0.0	121.0
Spring 2006	0.0	0.0	0.0	0.1	0.0	28.7
Summer 2006	55.2	297.4	22.0	86.7	0.1	114.6
Autumn 2006	89.9	126.4	98.1	149.9	96.8	79.4
Winter 2006/7	168.5	478.4	223.4	477.1	174.1	322.0
Spring 2007	235.6	250.7	188.6	1,013.9	357.2	1,016.9

Table 5.9 Mean effluent concentrations for the (a) indoor rig; and (b) outdoor rig (4 April 2005–19 March 2007).

(a)

		Wetland 1		Wetland 2		Wetland 3		Wetland 4		Wetland 5		Wetland 6	
Var.	***n***	**ov**	**sp5**	**ov**	**sp5**	**ov**	**sp5**	**ov**	**sp5**	**ov**	**sp5**	**ov**	**sp5**
COD	135	143.2	97.0	21.6	47.8	120.9	18.0	18.7	38.8	156.3	15.5	18.7	22.3
BOD	114	38.1	31.5	13.1	7.1	47.1	22.5	15.0	5.5	44.7	7.5	14.1	8.5
NH_4	127	26.58	8.7	28.1	7.5	39.5	5.4	46.4	1.8	28.5	5.5	31.6	1.5
NO_3	127	18.3	0.2	31.5	1.4	35.3	0.8	41.0	1.7	20.5	1.7	20.0	0.3
PO_4	127	15.8	7.9	19.3	11.8	18.3	3.9	17.9	3.5	17.8	12.0	19.9	3.0
Tem	131	17.8	23.0	17.8	23.0	17.8	23.0	17.8	23.0	17.8	23.0	17.8	23.0
Var.	***n***	**su5**	**au5**	**su5**	**au5**	**su5**	**au5**	**su5**	**au5**	**su5**	**au5**	**su5**	**au5**
COD	135	59.6	30.3	27.9	41.0	56.6	85.5	22.3	32.5	76.8	49.3	24.4	28.3
BOD	114	19.8	22.1	7.8	14.4	23.5	39.6	8.5	17.1	15.5	17.2	9.5	9.7
NH_4	127	6.4	5.4	5.1	4.6	5.7	4.9	4.7	3.5	5.2	4.7	4.1	4.5
NO_3	127	0.8	2.1	1.8	2.0	1.8	2.1	2.1	2.2	1.9	2.6	2.0	2.7
PO_4	127	5.4	8.4	9.9	8.0	5.8	7.5	6.3	6.3	6.5	7.6	7.1	8.3
Tem	131	24.0	19.7	24.0	19.7	24.0	19.7	24.0	19.7	24.0	19.7	24.0	19.7
Var.	***n***	**wi5/6**	**sp6**	**wi5/6**	**sp6**	**wi5/6**	**sp6**	**wi5/6**	**sp6**	**wi5/6**	**sp6**	**wi5/6**	**sp6**
COD	135	59.6	30.3	27.9	41.0	56.6	85.5	22.3	32.5	76.8	49.3	24.4	28.3
BOD	114	95.0	37.1	76.7	25.6	127.2	43.5	86.7	30.4	83.3	37.9	93.9	23.2
NH_4	127	5.5	6.6	7.1	5.6	5.9	4.2	6.6	3.6	4.5	4.9	5.1	3.3

(Continued)

Table 5.9 (Continued)

(a)

		Wetland 1		Wetland 2		Wetland 3		Wetland 4		Wetland 5		Wetland 6	
Var.	***n***	**ov**	**sp5**	**ov**	**sp5**	**ov**	**sp5**	**ov**	**sp5**	**ov**	**sp5**	**ov**	**sp5**
NO_3	127	1.8	3.1	1.7	3.7	2.3	4.0	1.6	3.5	1.6	3.4	2.4	3.7
PO_4	127	8.5	9.3	7.9	8.1	9.8	6.3	7.6	7.5	9.1	8.1	8.1	7.6
Tem	131	18.8	20.6	18.8	20.6	18.8	20.6	18.8	20.6	18.8	20.6	18.8	20.6
Var.	***n***	**su6**	**au6**	**su6**	**au6**	**su6**	**au6**	**su6**	**au6**	**su6**	**au6**	**su6**	**au6**
COD	135	30.9	89.8	2.0	7.4	38.2	114.6	0.2	4.9	97.5	106.9	0.7	7.1
BOD	114	22.7	37.9	3.6	1.6	29.3	46.1	4.0	3.3	37.6	47.5	3.0	1.9
NH_4	127	16.8	81.8	14.6	95.5	13.7	88.4	12.7	104.6	13.0	87.2	12.5	94.3
NO_3	127	8.3	17.3	8.6	42.1	8.4	34.0	8.6	51.8	7.7	35.1	8.2	42.5
PO_4	127	17.1	47.3	15.7	66.1	14.6	63.6	14.4	62.7	14.9	61.0	15.0	74.5
Tem	131	15.0	15.0	15.0	15.0	15.0	15.0	15.0	15.0	15.0	15.0	15.0	15.0
Var.	***n***	**wi6/7**		**wi6/7**		**wi6/7**		**wi6/7**		**wi6/7**		**wi6/7**	
COD	135	602.4		5.2		411.7		5.5		581.8		7.5	
BOD	114	54.5		0.9		61.0		0.6		80.0		2.9	
NH_4	127	40.8		41.6		122.9		161.3		49.0		73.8	
NO_3	127	75.7		131.5		165.8		191.4		53.5		62.6	
PO_4	127	6.3		8.1		10.7		10.2		5.2		8.1	
Tem	131	15.0		15.0		15.0		15.0		15.0		15.0	

(b)

		Wetland 1		Wetland 2		Wetland 3		Wetland 4		Wetland 5		Wetland 6	
Var.	***n***	**ov**	**sp5**	**ov**	**sp5**	**ov**	**sp5**	**ov**	**sp5**	**ov**	**sp5**	**ov**	**sp5**
COD	135	225.5	31.3	29.1	72.5	207.5	10.5	24.4	33.0	216.2	55.5	22.2	30.0
BOD	114	48.5	13.5	17.6	21.1	49.5	12.5	16.2	7.5	45.5	10.0	13.9	5.0
NH_4	127	35.5	7.4	45.5	9.0	32.0	6.2	40.3	2.3	45.8	5.6	31.3	1.9
NO_3	127	29.2	0.3	42.0	1.2	24.5	1.2	29.4	1.6	42.2	1.6	27.9	0.3
PO_4	127	19.7	7.7	28.7	11.8	17.1	3.8	16.3	3.8	16.2	11.8	20.1	2.8
Tem	130	12.6	17.0	12.6	17.0	12.6	17.0	12.6	17.0	12.6	17.0	12.6	17.0
Var.	***n***	**su5**	**au5**	**su5**	**au5**	**su5**	**au5**	**su5**	**au5**	**su5**	**au5**	**su5**	**au5**
COD	135	87.8	45.0	37.1	65.3	38.7	86.1	31.9	37.2	42.5	104.0	21.9	27.6
BOD	114	30.8	46.8	17.5	20.3	17.8	59.1	10.5	24.4	17.0	25.8	7.8	11.6
NH_4	127	5.3	3.9	4.6	3.6	7.2	3.6	5.5	4.2	4.3	3.5	6.1	3.8
NO_3	127	2.0	2.3	2.1	1.9	1.8	2.3	2.0	2.4	1.4	2.1	1.3	1.8
PO_4	127	9.6	8.7	10.3	9.4	5.9	6.9	6.2	6.2	4.2	7.3	5.7	6.5
Tem°	130	19.6	11.1	19.6	11.1	19.6	11.1	19.6	11.1	19.6	11.1	19.6	11.1
Var.	***n***	**wi5/6**	**sp6**	**wi5/6**	**sp6**	**wi5/6**	**sp6**	**wi5/6**	**sp6**	**wi5/6**	**sp6**	**wi5/6**	**sp6**
COD	135	45.0	49.2	56.5	45.8	88.9	56.7	69.7	44.1	35.1	44.9	66.5	61.7
BOD	114	98.9	35.	86.1	36.2	125.6	41.2	71.7	40.5	141.1	41.4	80.0	27.1
NH_4	127	3.8	6.4	3.9	6.0	5.3	4.7	5.4	3.9	5.1	4.1	5.5	4.1
NO_3	127	3.3	3.7	1.3	4.1	1.6	4.5	1.7	3.9	1.9	3.7	1.6	4.1
PO_4	127	9.0	9.9	8.3	8.6	9.7	6.8	7.3	7.8	6.4	8.3	6.4	7.6
Tem	130	11.1	13.8	11.1	13.8	11.1	13.8	11.1	13.8	11.1	13.8	11.1	13.8

(*Continued*)

Table 5.9 (Continued)

(b)

		Wetland 1		Wetland 2		Wetland 3		Wetland 4		Wetland 5		Wetland 6	
Var.	***n***	**ov**	**sp5**	**ov**	**sp5**	**ov**	**sp5**	**ov**	**sp5**	**ov**	**sp5**	**ov**	**sp5**
Var.	***n***	**su6**	**au6**	**su6**	**au6**	**su6**	**au6**	**su6**	**au6**	**su6**	**au6**	**su6**	**au6**
COD	135	22.1	126.3	3.2	8.3	34.4	104.0	3.4	2.9	82.0	64.6	1.5	4.5
BOD	114	27.6	61.0	5.5	2.0	24.2	51.9	4.4	3.6	33.6	42.9	5.1	4.6
NH_4	127	17.6	122.0	15.8	121.8	14.2	121.1	12.6	103.2	13.1	82.0	12.6	92.7
NO_3	127	8.2	24.6	8.6	55.6	8.5	26.1	8.3	36.7	8.0	21.5	8.0	29.1
PO_4	127	18.2	60.0	16.1	119.0	15.0	56.6	14.9	51.3	14.9	37.2	14.9	67.8
Tem	130	16.7	9.4	16.7	9.4	16.7	9.4	16.7	9.4	16.7	9.4	16.7	9.4
Var.	***n***	**wi6/7**		**wi6/7**		**wi6/7**		**wi6/7**		**wi6/7**		**wi6/7**	
COD	135	1035.1		7.5		934.5		14.0		959.8		10.0	
BOD	114	72.0		0.4		73.7		0.9		56.5		0.7	
NH_4	127	57.4		121.6		46.1		112.4		76.0		30.5	
NO_3	127	133.0		160.8		87.5		102.7		81.8		60.2	
PO_4	127	11.8		10.9		9.0		10.7		6.8		9.0	
Tem	130	7.0		7.0		7.0		7.0		7.0		7.0	

n, sample number; COD, chemical oxygen demand (mg L^{-1}); BOD, five-day @ 20°C N-Allythiourea biochemical oxygen demand (mg L^{-1}); NH_4, ammonia-nitrogen (mg L^{-1}); NO_3, nitrate-nitrogen (mg L^{-1}); PO_4, ortho-phosphate-phosphorus (mg L^{-1}); tem, temperature (°C); sp5, mean for 20 March 2005–20 June 2005; su5, mean for 21 June 2005–21 September 2005; au5, mean for 22 September 2005–20 December 2005; wi5/6, mean for 21 December 2005–19 March 2006; sp6, mean for 20 March 2006–20 June 2006; su6, mean for 21 June 2006–22 September 2006; au6, mean for 23 September 2006–21 December 2006; ov, 4 April 2005–19 March 2007; wi6/7, mean for winter 22 December 2006–20 March 2007).

Concentrations of DO were variable (0–12 mg L^{-1}). The low DO concentrations (0–1 mg L^{-1}) in the system during inundation could be due to the lasting effects of rapid aerobic degradation processes observed after drawing air into the system, thus causing a sudden change of the microbiology and subsequent imbalance promoting predominantly aerobic microorganisms in the wetlands. High DO variability was expected considering the relatively high influent concentrations of benzene, low retention time, small size of the experimental wetlands, and sudden environmental changes due to rapid operational changes (see above and Table 5.7).

Control of the variability of environmental variables such as temperature, pressure and light resulted in improved overall treatment performances of the wetland wetlands. The temperature data (standard deviation: 3.7°C) of the indoor rig were relatively stable, particularly after the temperature was fully controlled (Tables 5.8 and 5.9, and Figure 5.7). Benzene treatment performances were better for the indoor rig in comparison to the outdoor rig. The overall removal efficiencies were lower for the outdoor experimental rig (e.g. benzene, 85–86%) in comparison to the experimental rig placed indoors (e.g. benzene, 93–95%).

Furthermore, the mean removal efficiencies for the water quality variables were lowest for the experimental rig placed outside (benzene, 85%; COD, 70%; ammonia-nitrogen, 83%; nitrate-nitrogen, 88%; ortho-phosphate-phosphorus, 58%). This is likely due to relatively low and relatively variable temperatures in Scotland (annual mean of approximately 8°C). The best overall mean treatment performances were obtained for the experimental rig placed indoors (benzene, 95%; COD, 80%; ammonia-nitrogen, 90%; nitrate-nitrogen, 94%; ortho-phosphate-phosphorus, 66%). The relatively high treatment performance observed indoors is influenced by stable and usually relatively high temperatures, particularly after the temperature was fully controlled as described above (Figure 5.7). The BOD removal is greater in the uncontaminated wetlands (2, 4 and 6) compared to the contaminated wetlands (1, 3 and 5) as indicated in Table 5.9.

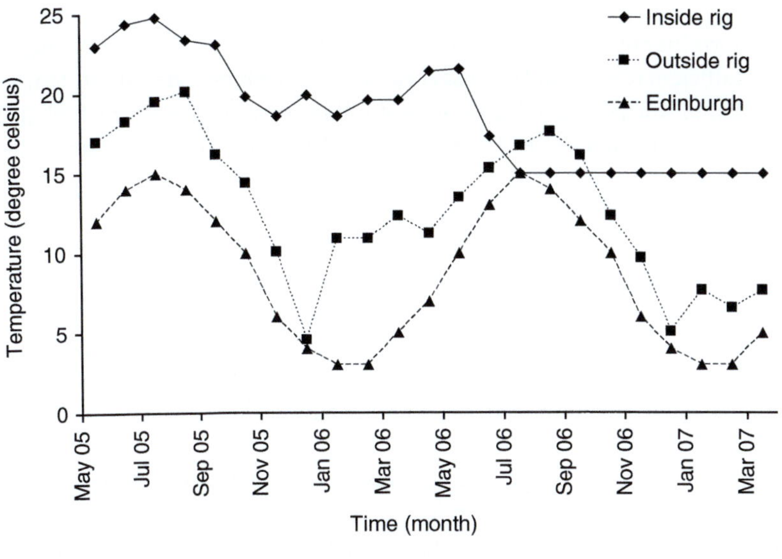

Figure 5.7 Comparison of the monthly mean temperatures for the inside and outside rigs, and Edinburgh (official Met Office data). The outside temperature is below inside temperature and seasonal temperature fluctuations are clearly visible.

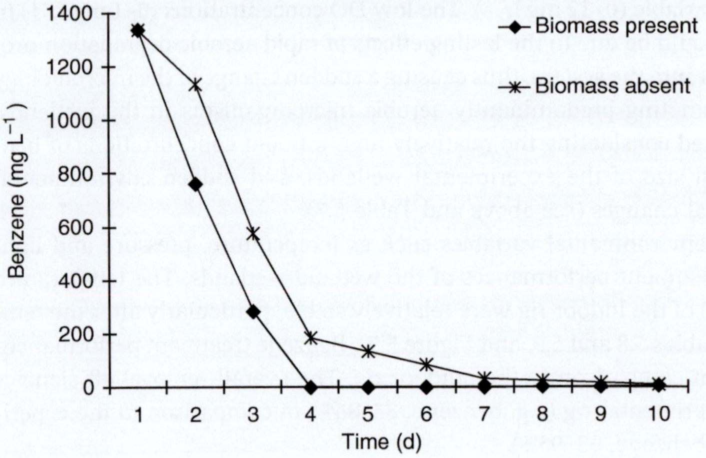

Figure 5.8 Comparison of benzene removal for wetlands with and without biomass. The presence of biomass leads to lower benzene concentrations.

5.2.3.2 Impact of Volatilisation

Figure 5.8 shows a comparison of benzene removal for wetlands with and without biomass. The impacts of volatilisation, biodegradation and adsorption on the benzene removal efficiency are often difficult to separate quantitatively from each other. Preliminary findings indicate that biodegradation and adsorption dominate for a retention time of one day. However, volatilisation becomes the major removal mechanism afterwards. Water and oil are likely to separate if the inflow is not in motion.

5.2.4 Conclusions

The findings suggest that intermittently flooded vertical-flow constructed wetlands treat benzene effectively in the presence of sufficient oxygen and fertiliser (Table 5.7), which provides nitrate as used as an alternative electron acceptor during anaerobic periods of full inundation. Relatively high and stable temperatures (i.e., no seasonal variations) lead to improved hydrocarbon treatment efficiencies.

As benzene and its degradation products started to accumulate in the wetlands, removal efficiencies subsequently reduced. Findings show also that benzene removal was highest in wetlands with aggregates and biomass (Table 5.7) providing habitat for hydrocarbon-degrading microbes. However, further studies on estimating the microbial biomass are encouraged.

Metabolic processes of microorganisms are likely to play an important role in removing hydrocarbon compounds in both controlled and semi-natural wetlands. The results show also that *Phragmites australis* (Cav.) Trin. ex Steud. does not play a significant role (despite providing additional oxygen via its rhizomes) in removing benzene, unless sufficient nutrients (including fertiliser; Table 5.7) are available.

Findings indicate also that both biodegradation and volatilisation support the treatment. Volatilisation is the dominant mechanism for benzene removal after one day of retention time. However, optimising environmental conditions such as locating wetlands in areas with relatively high temperatures enhances the biodegradation rate. Further research is required to quantify volatilisation, aerobic and anaerobic biodegradation, adsorption, absorption, mineralisation and other removal mechanisms in large-scale constructed treatment wetlands.

5.3 Diesel Oil Spillage Removal Using Agricultural Waste Products

5.3.1 Introduction

Liquid petroleum in the form of crude oil and its refined products is one of the most important energy and raw material source in modern society, but its widespread use also inevitably results in accidental and intentional releases. Oil spill accidents may occur during its exploration, transportation and storage (Amini et al. 2008; Burgherr 2006; Nakata et al. 1997).

Whenever and wherever spillages occur, they cause serious long-term environmental and socio-economic impacts on terrestrial and marine ecosystems and the subsistence of, for example, coastal communities. The spilled oil has an undesirable taste and odour, and causes severe environmental damage to soil, water and wildlife such as waterfowl, and negatively affects tourism and the economy (Annunciado et al. 2005; Hofer 2003; Kingston 2002; Peterson et al. 2003; Miraaglia 2002). Unfortunately, experience shows that even the best safety efforts cannot prevent occasional oil spill accidents on sea. Therefore, finding new cost-effective and sustainable methods on how to treat the oil spill accidents rapidly and efficiently has become a global concern.

The methods commonly used to combat accidental spills of oil and oil products can be classified as mechanical, chemical, physicochemical and biological (Choi 1992; Deschamps et al. 2003). Although many advanced technologies have been developed for oily water treatment, the expensive initial and operating costs prohibited the wide application of some of these techniques (Banerjee et al. 2006).

Among all the existing techniques used for oil treatment, sorption by sorbents may be one of the most effective techniques for complete removal of spilled oil under ambient conditions. Oil sorbents are classified into inorganic mineral materials (Bastani et al. 2006; Karakasi and Moutsatsou 2010; Rajakovic et al. 2007), synthetic organic products (Bayat et al. 2005; Ceylan et al. 2009; Zahid et al. 1972) and natural organic materials (Choi 1992). The poor buoyancy and low oil sorption capacity properties limited the wide application of inorganic mineral materials. By contrast, synthetic organic products such as polypropylene and polyurethane have emerged as polymeric materials with significantly high sorption capacities. However, their non-biodegradability and high cost is a major disadvantage (Teas et al. 2001). Among the various sorbents used, natural organic sorbents appear to be the most attractive materials in terms of their low cost, potentially good oil sorption performance, eco-friendly characteristics and abundant resource supply in nature.

Some agricultural products that might be suitable to remove oil have been widely assessed by different research groups. Corresponding materials include rice hull (Angelova et al. 2011; Bayat et al. 2005), walnut shell media (Asha and Thiruvenkatachari 2008), bark (Haussard et al. 2003), bagasse (Bayat et al. 2005; Chung et al. 2011; El-Aziz et al. 2009; Husssein et al. 2008), cotton (Hussein et al. 2011; Johnson and Manjrekar 1973), kapok (Lim and Huang 2007), sawdust (Annunciado et al. 2005) and straw (Hussein et al. 2009b). Sugarcane bagasse treated by acetylation reaction to enhance its abilities for both wicking oil and promoting biodegradation in saturated environments has also been studied (Chung et al. 2011). Most of these agricultural products and residues are inexpensive and available locally. Previous studies mainly focused on the oil sorption capacity, but detailed scientific studies with a practical focus about the floating performance properties of these agricultural wastes within combined oil and water environments and the optimum conditions in disposing oil from spillages have not been reported in high-impact scientific journals.

The aim of this work is to assess the oil sorption capacity and floating performance of three common wastes from the agriculture and forestry industries. The levels of the experimental variables affecting oil removal rates were assessed after conducting an optimised orthogonal experiment. The findings presented in this chapter are based on those by Li et al. (2013).

5.3.2 Materials and Methods

5.3.2.1 Materials

The chemicals (analytical reagents) used in the experimental work were sulphuric acid (H_2SO_4), *n*-hexane and sodium chloride (NaCl). Diesel oil was purchased from the Sinopec gas station located in the Binhai New Area of Tianjin near Beijing, China. The oil was placed under ventilation conditions for 20 days before use to avoid that diesel oil volatilisation may affect experimental results.

Sawdust produced from hemlock spruce (*Tsuga canadensis*) was purchased from a local carpentry workshop located in the Binhai New Area of Tianjin, China. Corn stalk was obtained from a farm in the suburb of the Binhai New Area. Wheat straw was collected from a farm situated in the suburb of Ji County (Tianjin, China). The materials were pre-treated before using them in the experiments. All materials (except sawdust) were crushed by a disintegrator and washed with distilled water to remove impurities. The washed materials were then oven-dried at 105°C for 24 hours and subsequently cooled. The dried materials were classified by using a set of sieves (10, 20, 32, 40 and 60 meshes) by the American Society for Testing and Materials, and subsequently stored in a drying apparatus before use.

The following key experimental instruments were used: Ultraviolet visible spectrophotometer (TU-1901; Beijing Purkinje General Instrument Company, Beijing, China), constant temperature bath oscillator (SHZ-88; Wenhua Instruments Company, Jintan, China), electronic analytical balance (ESJ200-4; Shenyang Longting Electronic Company, Shenyang, China), electro-thermostatic blast oven (CFO30; Tianjin Huabei Experimenting Company, Tianjin, China), stainless steel net, separation funnel (250 mL), straw disintegrator and a sampling sieve.

5.3.2.2 Methods

5.3.2.2.1 Sorption Capacity Tests All methods used were performed according to the American Public Health Association (2005) and standard methods accepted in the specific research field (Amini et al. 2008; Annunciado et al. 2005; Burgherr 2006; Nakata et al. 1997) unless specified otherwise. For the sorption experiments, 75 mL of diesel oil was poured into a beaker (250 mL). An excess amount of oil was chosen so that there was still plenty of oil remaining in the beaker after completion of the sorption test. Half of a gram of the sorbent material (particle size between 500 and 830 μm) was gently and evenly placed into the beaker containing the oil with the help of a stainless-steel net. After 5, 15, 30, 45, 60 and 90 minutes, the material covered in oil was taken out of the beaker. Excess oil was allowed to drain for five minutes before weighing. All tests were carried out at 20 ± 1°C.

All tests were undertaken in duplicates and the mean data are represented in this chapter. The sorption was calculated as the ratio of sorbed material to dry sorbent mass according to Eq. (5.1).

$$\mathrm{OSC} = (S_t - S_1 - S_2 - S_0)/S_0 \tag{5.1}$$

where OSC is the oil sorption capacity (–); S_t is the total weight (g) of the oil sorbed by the sorbent, the stainless steel net, the oil sorbed by the stainless steel net and the dry sorbent; S_1 is the weight (g) of the stainless steel net; S_2 is the weight of the oil sorbed by the stainless steel net (g); and S_0 is the dry weight of the sorbent (g).

After placing a stainless-steel net into the beaker (250 mL), exactly 6 mL (4.92 g) of diesel oil was poured into the beaker containing already 100 mL of distilled water to obtain an oil layer of about 1.65 mm above the water. Thereafter, the sorbent material (particle size: 500–830 μm) weighing 0.5 g was gently and evenly placed on the oil surface. After 5, 15, 30, 45, 60 and 90 minutes, the stainless-steel net was taken out of the beaker and placed on the top of the beaker for 5 minutes. The remaining diesel oil in the beaker was extracted using n-hexane and subsequently determined by ultraviolet visible spectrophotometer (Amini et al. 2008). Absorbance measurements were performed on a TU-1901 UV–VIS Spectrometer using a quartz corvette of 10 mm and at a wavelength of 223 nm. N-hexane was used as a reference measurement absorbent. The corresponding mathematical relationship can be expressed in Eq. (5.2).

$$Y = 0.0185 \times X + 0.0141, R^2 = 0.9998 \tag{5.2}$$

where Y is the absorbance (–) and X is the diesel oil concentration (mg L^{-1}). Note that R^2 is the coefficient of determination (–).

The amounts of oil and sorbent materials were chosen to allow plenty of oil to remain in the beaker after completion of the sorption experiment. Moreover, the right amounts were also chosen to simulate a realistic remaining oil film thickness after mechanical oil removal when an oil spill occurs. All tests were performed in duplicates and the average of the two runs was used for subsequent calculations.

5.3.2.2.2 Floating Performance Experiments For the buoyancy tests, 6 mL of diesel oil was poured into a beaker (250 mL) containing 100 mL of distilled water. Subsequently, 0.5 g of the sorbent material (particle size: 500–830 μm) was gently and evenly placed onto the oil surface. After 0.2, 1, 3, 10, 24, 48 and 72 hours, both the floating and submerged sorbent materials were withdrawn separately, and subsequently drained with the help of a stainless-steel net for five minutes. All weights were determined using an analytical balance. The tests were conducted separately under stagnant and shaking (oscillation frequency of 150 r min^{-1}) conditions. Shaking experiments simulated wave movements. All tests were performed in duplicates and the averages of the two runs were taken for subsequent calculations. The floating rate was calculated using Eq. (5.3).

$$\text{FR} = S_F/(S_F + S_L) \tag{5.3}$$

where FR is the floating rate (–), S_F is the weight of the floating sorbent materials (g) and S_L is the weight of the submerged materials (g).

5.3.2.2.3 Orthogonal Design Experiments and Limitations An orthogonal experimental design of the form L16 (4 × 5) was selected from the design module of Statistica 8.0. The main feature of an orthogonal design is that the total variation in the response can be decomposed into individual components due to each factor and interaction. Such a performed decomposition allows for the ranking of the importance of factors regarding their contribution to the total performance variance. The form with a total of 16 experiments was used to investigate the effect of five variables (particle size, adsorbent dosage, oscillation frequency, oil film thickness and salinity) on the removal rate of the three agricultural wastes. Each of the variables was selected at four levels (Table 5.10). Sixteen experiments were arranged according to the design. The experiments were repeated two times. The selection of levels for all variables was influenced by the above literature review and the known material characteristics. An adsorption time of 30 minutes was selected to determine the three oil removal rates associated with the materials to ensure that all materials reached a point of full saturation in terms of adsorption.

This is a practical chapter with methodologies performed that are rather simple due to the straightforward nature of the study under controlled conditions. Nevertheless, the study and its presentation have its limitations.

Table 5.10 List of levels per variable.

Factors	Particle size (μm)	Oscillation frequency ($r\ min^{-1}$)	Dosage (g)	Oil film thickness (mm)	Salinity ($mol\ L^{-1}$)
Level 1	830–1700	0	0.3	0.55	0.0
Level 2	500–830	80	0.5	1.10	0.3
Level 3	380–500	120	0.7	1.65	0.6
Level 4	250–380	150	0.9	2.20	0.9

In the orthogonal design, the oscillation frequency was assessed as a key variable to estimate the impact of shaking due to simulated wave action on oil adsorption. Initial oscillation experiments showed that oil sticks to the beaker walls, but that the corresponding overall amount was relatively low in comparison to the oil associated with the agricultural waste and the free water surface.

Adsorption is usually described through isotherms, if the studied environments are complex. Isotherms describe the amount of adsorbate on the adsorbent as a function of its concentration at constant temperature. The quantity adsorbed can be normalised by the mass of the adsorbent to allow for a comparison of various materials. Isotherms are not presented in this chapter, because of the rather simple and straightforward methodology and low complexity of the studied controlled environment.

5.3.3 Results and Discussion

5.3.3.1 Sorption Capacity of the Agricultural Wastes

The kinetics of sorption for the three agricultural wastes within an oil medium is rather trivial. The sorption capacities are approximately 7.03, 8.54 and 8.20 g of oil for 1.00 g of corn stalk, wheat straw and sawdust, respectively, regardless of time for the first 90 minutes. The agricultural wastes absorbed diesel oil quickly and a point of saturation was reached within only five minutes, which is consistent with those times reported by Banerjee et al. (2006). Hussein et al. (2009b) reported that the saturated adsorption amounts of corn stalk and modified corn stalk on gasoline were 4.4 and 8.6 $g\ g^{-1}$, respectively. These values are relatively small, because oil sorption is considerably affected by viscosity: the higher the viscosity, the higher the adsorption.

The saturated adsorption amounts of spill-sorb as well as both carbonised rap and bamboo on diesel oil are 6.69, 5.71 and 0.61 $g\ g^{-1}$, respectively (Annunciado et al. 2005; Chen et al. 2011). Compared to these materials, the agricultural wastes used in Chapter 5.3 have larger oil adsorption amounts and therefore good prospects of application. Wheat straw showed the highest sorption capacity followed by sawdust. Corn stalk had the lowest sorption capacity. The results of this study conducted with pure oil show that the oil sorption capacities of the three agricultural wastes have certain differences. This may be due to the different pore structure of the three materials. Pore size and its distribution have impact on the oil sorption capacity of the sorbent material. Interconnected pore structures can promote the oil drops formed by capillary action to connect rapidly, forming sucking oil layers, and therefore improve the properties of oil absorption.

The kinetics of sorption for the three agricultural wastes within a combined oil and water medium are also trivial, if compared to the kinetics for the oil medium only. The sorption capacities are approximately 7.02, 8.54 and 8.08 g of oil for 1 g of corn stalk, wheat straw and sawdust, respectively, regardless of time for the first 90 minutes. The mean data are similar compared to the oil medium only. All agricultural wastes absorbed diesel oil rapidly and could reach saturation within five minutes as well. It was observed that the oil absorption rate and the oil sorption capacity of the three agricultural wastes have no significant differences within both the oil and combined oil and water medium.

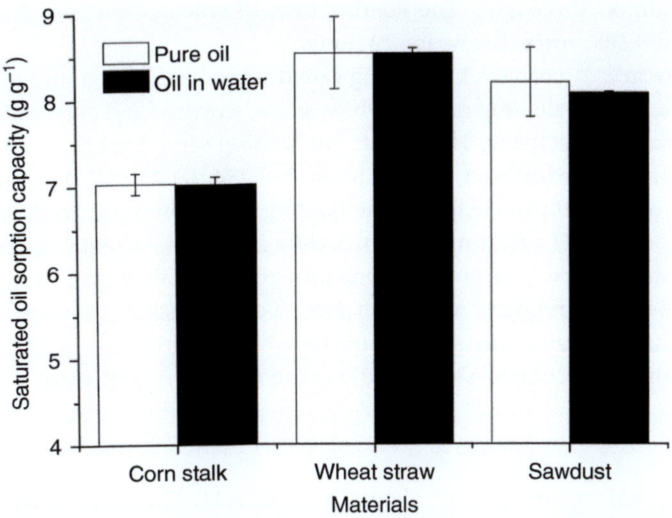

Figure 5.9 Comparison of the saturated sorption capacities for pure oil and combined oil and aqueous medium. The sorption capacities of pure oil and oil in water are almost the same for all three materials.

Figure 5.9 shows a comparison of the sorption capacities of the three wastes for diesel oil. The rapid oil sorption rates and good oil sorption capacities (Vlaev et al. 2011) indicate that wheat straw, corn stalk and sawdust can be used as sorbent materials for removing accidentally spilled oil.

5.3.3.2 Floating Performance of the Wastes

Figure 5.10 shows the effect of time on floating rate. There are great differences between wheat straw, corn stalk and sawdust concerning floating rates. When the oscillation frequency is 0 $r\,\text{min}^{-1}$ (i.e., no shaking), the floating rate of all materials declined quickly within three hours, Corn stalk's floating rate dropped the fastest among the three materials within three hours and then the

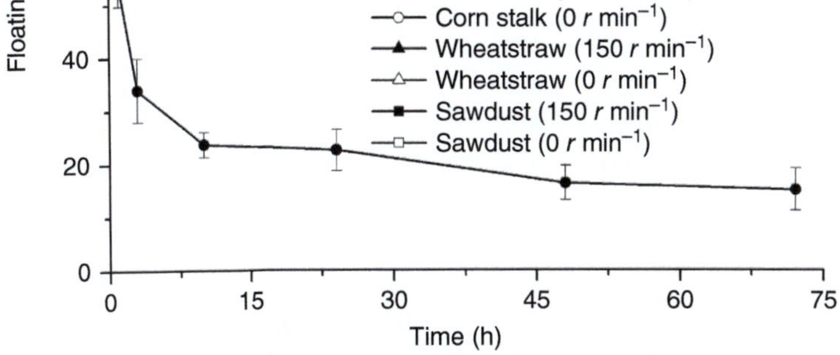

Figure 5.10 Floating rate as a function of time for the three agricultural wastes at 0 and 150 $r\,\text{min}^{-1}$. The floating rates are highest for wheat straw and sawdust at both 0 $r\,\text{min}^{-1}$.

floating rate remained constant for almost three days. The floating rates of wheat straw, sawdust and corn stalk were 0.921, 0.917 and 0.584, respectively, for 72 hours.

At an oscillation frequency of 150 r min^{-1}, corn stalk's floating rate dropped the fastest and the one for wheat straw dropped the slowest. The floating rates of wheat straw, sawdust and corn stalk were 0.773, 0.762 and 0.237 at 10 hours, respectively. Thereafter, the floating rates remained constant. The floating rates of wheat straw, sawdust and corn stalk were 0.754, 0.675 and 0.152 at 72 hours in that order. Shaking led to a significant decline of the floating rates compared to those obtained during the static state experiments. The floating capacities were as follows: wheat straw > sawdust > corn stalk. Wheat straw and sawdust have great floating properties on oily waters at relatively high wave action. Corn stalk's floating property was not so good; it can be used to remove oil from calm water surfaces. Moreover, the saturated corn stalk should be salvaged within a short time to avoid it to sink to the bottom of the watercourse, which would cause secondary pollution.

5.3.3.3 Orthogonal Design

The results obtained from the orthogonal experiments are shown in Table 5.11. The range of particle sizes for sawdust and corn stalk were between 380 and 4000 μm, and those for wheat straw were between 380 and 4000 μm. The results of the orthogonal experiments using corn stalk show that salinity had little effect on the oil removal rate and the oil sorption capacity, which is like findings from previous reports (Chen et al. 2011; Zhou et al. 2010). Therefore, salinity was removed as a variable from subsequent orthogonal experiments using wheat straw and sawdust. Results of the orthogonal experiments were assessed by range analysis.

Range analysis findings for the oil sorption capacities and oil removal rates of corn stalk, wheat straw and sawdust are shown in Tables 5.12–5.14. Table 5.12 indicates that the effect of the variables

Table 5.11 Results from the orthogonal experiments.

						Corn stalk		Wheat straw		Sawdust	
No.	Particle size	Oscillation frequency	Dosage	Oil film thickness	Salinity	Oil removal rate	Sorption capacity	Oil removal rate	Sorption capacity	Oil removal rate	Sorption capacity
1	1	1	1	1	1	0.642	3.51	0.797	4.36	0.943	5.31
2	1	2	2	2	2	0.440	2.88	0.497	3.26	0.950	6.36
3	1	3	3	3	3	0.989	6.95	0.508	3.57	0.998	7.02
4	1	4	4	4	4	0.965	7.03	0.479	3.49	0.943	7.01
5	2	1	2	3	4	0.699	6.88	0.415	4.09	0.777	7.78
6	2	2	1	4	3	0.315	6.88	0.463	8.47	0.377	8.21
7	2	3	4	1	2	0.995	1.81	0.993	1.81	0.993	1.81
8	2	4	3	2	1	0.734	3.44	0.839	3.93	0.929	4.38
9	3	1	3	4	2	0.723	6.77	0.909	8.52	0.802	7.43
10	3	2	4	3	1	0.524	2.86	0.991	5.42	0.993	5.42
11	3	3	1	2	4	0.413	4.52	0.507	5.55	0.577	6.40
12	3	4	2	1	3	0.819	2.69	0.965	3.16	0.980	3.22
13	4	1	4	2	3	0.885	3.23	0.999	3.64	0.999	3.64
14	4	2	3	1	4	0.603	1.41	0.998	2.34	0.995	2.33
15	4	3	2	4	1	0.378	4.96	0.651	8.54	0.551	7.24
16	4	4	1	3	2	0.353	5.79	0.460	7.55	0.522	8.42

on corn stalk's oil removal rates and oil sorption capacities are as follows: dosage > oscillation frequency > particle size > salinity > oil film thickness, and oil film thickness > oscillation frequency > dosage > salinity > particle size, correspondingly. As can be seen from Table 5.13, the four variables impacted on the wheat straw's oil removal rates and oil sorption capacities are like this: Oil film thickness > dosage > particle size > oscillation frequency. Finally, Table 5.14 indicates the significance of the impacts of the variables on the sawdust's oil removal rate: dosage > oil film thickness > particle size > oscillation frequency. In comparison, the effect on oil sorption capacity of sawdust was oil film thickness > dosage > particle size > oscillation frequency.

In terms of oil removal rate, the optimum conditions for absorbing diesel onto corn stalk and sawdust were summed up in this manner: particle size between 830 and 1700 μm, oscillation frequency of 0 *r* min^{-1}, adsorbent dosage of 0.9 g and oil film thickness of 0.55 mm. For wheat straw, the best particle size was between 500 and 830 μm. In this case, the three agriculture wastes' removal rates could reach above 99%. While in terms of oil sorption capacity, the optimum conditions for absorbing diesel onto corn stalk and sawdust were like so: particle size between 830 and 1700 μm, oscillation frequency of 0 *r* min^{-1}, adsorbent dosage of 0.3 g and oil film thickness of 2.2 mm. For wheat straw, the best particle size was between 500 and 830 μm. At these conditions, the agriculture wastes could reach their maximum oil sorption capacities.

Table 5.12 Range analysis of oil sorption capacity and oil removal rate for corn stalk.

	Particle size		Oscillation frequency		Dosage		Oil film thickness		Salinity	
K^a	Removal rate	Sorption capacity	Removal rate	Sorption capacity	Removal rate	Sorption capacity	Removal rate	Sorption capacity	Removal rate	Sorption capacity
K_1	0.759	5.095	0.737	5.097	0.431	5.173	0.765	2.356	0.569	3.693
K_2	0.686	4.753	0.470	3.510	0.584	4.352	0.618	3.516	0.627	4.314
K_3	0.620	4.210	0.694	4.559	0.762	4.645	0.641	5.621	0.752	4.936
K_4	0.555	3.846	0.718	4.737	0.842	3.734	0.595	6.411	0.670	4.961
R^b	0.204	1.249	0.267	1.587	0.411	1.439	0.170	4.055	0.183	1.268

[a] K_1, K_2, K_3 and K_4, mean values for levels of different variables.
[b] range of the mean values for all variables.

Table 5.13 Range analysis of oil sorption capacity and oil removal rate for wheat straw.

	Particle size		Oscillation frequency		Dosage		Oil film thickness		Salinity	
K^a	Removal rate	Sorption capacity	Removal rate	Sorption capacity	Removal rate	Sorption capacity	Removal rate	Sorption capacity	Removal rate	Sorption capacity
K_1	0.570	3.670	0.780	5.152	0.538	6.483	0.938	2.917	0.820	5.563
K_2	0.658	4.575	0.718	4.872	0.632	4.762	0.711	4.095	0.715	5.285
K_3	0.843	5.662	0.665	4.867	0.814	4.590	0.594	5.157	0.715	4.710
K_4	0.777	5.518	0.686	4.532	0.866	3.590	0.607	7.255	0.600	3.867
R^b	0.273	1.992	0.115	0.620	0.328	2.893	0.344	4.338	0.220	1.696

[a] K_1, K_2, K_3 and K_4, mean values for levels of different variables.
[b] range of the mean values for all variables.

Table 5.14 Range analysis of oil sorption capacity and oil removal rate for sawdust.

	Particle size		Oscillation frequency		Dosage		Oil film thickness		Salinity	
K^a	Removal rate	Sorption capacity	Removal rate	Sorption capacity	Removal rate	Sorption capacity	Removal rate	Sorption capacity	Removal rate	Sorption capacity
K_1	0.959	6.425	0.880	6.040	0.605	7.085	0.978	3.167	0.854	5.588
K_2	0.769	5.545	0.829	5.580	0.815	6.150	0.864	5.195	0.817	6.005
K_3	0.838	5.617	0.780	5.617	0.931	5.290	0.823	7.160	0.839	5.522
K_4	0.767	5.408	0.843	5.758	0.982	4.470	0.668	7.473	0.823	5.880
R^b	0.192	1.017	0.100	0.460	0.377	2.615	0.310	4.306	0.037	0.483

[a] K_1, K_2, K_3 and K_4, mean values for levels of different variables.
[b] range of the mean values for all variables.

When the oil removal rate reached its highest value, however, the oil sorption amount was the lowest. Previous studies have shown that both the oil removal rate and the oil sorption capacity could reach the highest values when spilled the oil amount reached 80% of the maximum oil sorption capacity of the sorbent materials (Lu and Zhou 2002).

The optimum conditions could be set as follows: corn stalk particle size between 830 and 1700 μm, oscillation frequency of 0 r min^{-1}, oil film thickness of 0.55 mm and adsorbent dosage of 0.29 g; sawdust particle size between 830 and 1700 μm, oscillation frequency of 0 r min^{-1}, oil film thickness of 0.55 mm and adsorbent dosage of 0.25 g; wheat straw particle size between 500 and 830 μm, oscillation frequency of 0 r min^{-1}, oil film thickness of 0.55 mm and adsorbent dosage of 0.24 g. At these conditions, the wastes showed a larger adsorption capacity to diesel oil, which is advantageous for removing oil from water at a degreasing rate of 99%.

The key variables impacting on the oil removal rate and sorption amount of the three agriculture and forestry wastes are summarised in Figures 5.11 and 5.12, respectively. Under experimental conditions, the oil removal rate (Figure 5.11) and sorption amount (Figure 5.12) generally increased with an increase of the particle sizes for corn stalk and sawdust. However, the oil removal rate was relatively erratic for sawdust due to experimental data variability (Figure 5.11). Both materials reached their highest values at particle sizes between 830 and 1700 μm.

In terms of wheat straw, the oil removal rate and the sorption amount initially increased and subsequently decreased with an increase of particle size. At particle sizes between 500 and 830 μ m, oil removal reached its highest value. This observation is due to the inner pore structure of the different materials. Pore size and structure tightness impact on the capillary action during sorption, affecting the different materials' optimal adsorption particle sizes.

The oscillating frequency had no significant effect on the oil removal rate and sorption capacity, but the values were slightly lower than those obtained from the static state (Figures 5.11 and 5.12). This is because shaking reduced the sorption amount of the oil absorption materials (Annunciado et al. 2005). This observation is important for 'real-world' applications where water surfaces are rarely permanently calm.

With an increase in dosage, the total diesel amount absorbed on the materials increased, so that the oil removal rate rose significantly. As the dosage was increased even further, the oil removal rate continued to increase, but more slowly (Figure 5.11). In comparison, the oil adsorption amount decreased considerable (Figure 5.12).

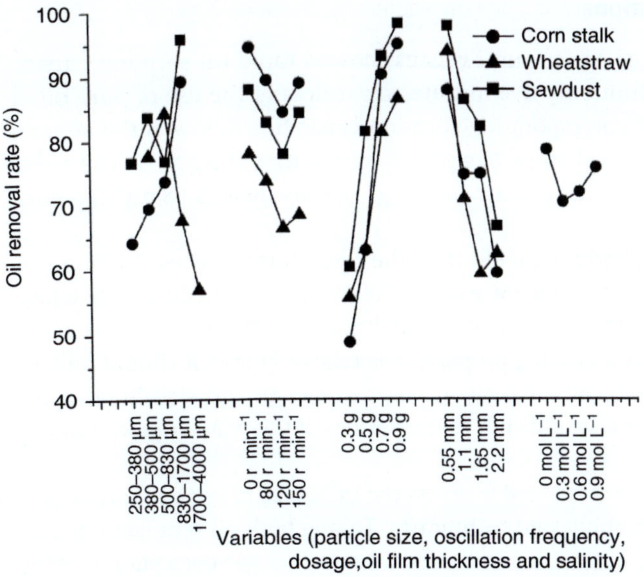

Figure 5.11 Effect of different variables and levels on oil removal rate. The oil removal rate is highest for sawdust with a high particle size and high oil dosage, but with a low film thickness.

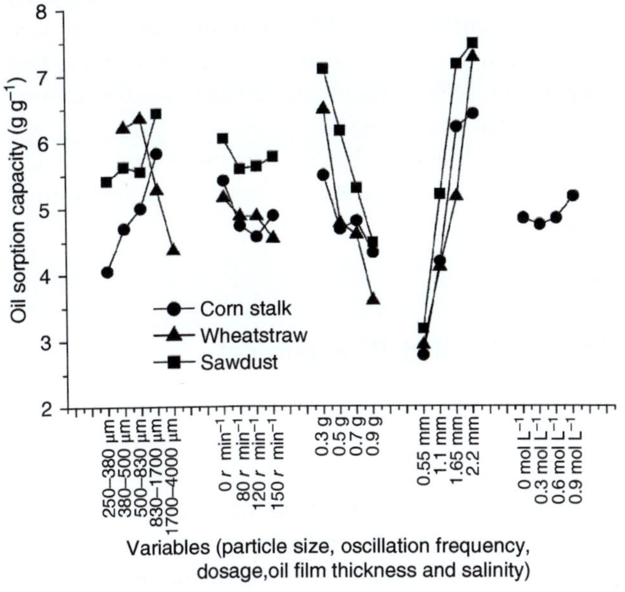

Figure 5.12 Effect of different variables and levels on oil sorption capacity. The oil sorption capacity is highest for sawdust with a high particle size and low oil dosage, but with a high film thickness.

In general, as the oil film thickness increased, the oil removal rate decreased. For example, as the thickness of the oil film enhanced, corn stalk gradually saturates, and the remaining diesel cannot be adsorbed effectively anymore (Figures 5.11 and 5.12). Furthermore, the salinity had little influence on the oil removal rate and oil sorption capacity of corn stalk. This result is important, because it indicates that the findings are generally relevant for both freshwaters and salty waters (including brackish waters).

5.3.4 Conclusions and Recommendations

The sorption tests revealed that the three agricultural wastes showed rapid oil sorption capacities. A saturation stage was reached within only five minutes regardless of the use of pure oil or a combined oil and water medium. The corresponding sorption capacities followed the general trend for oil: wheat straw ($8.54\ g\ g^{-1}$) > sawdust ($8.20\ g\ g^{-1}$) > corn stalk ($7.03\ g\ g^{-1}$). The trend for combined oil and water was similar: wheat straw ($8.54\ g\ g^{-1}$) > sawdust ($8.08\ g\ g^{-1}$) > corn stalk ($7.02\ g\ g^{-1}$).

Buoyancy tests indicated that shacking had a great effect on the agricultural wastes' floating performances. The floating rates of 72 hours at 0 and 150 $r\ min^{-1}$ followed the general trends wheat straw (0.921) > sawdust (0.917) > corn stalk (0.584), and wheat straw (0.754) > sawdust (0.675) > corn stalk (0.152), respectively. Corn stalk's floating property was relatively low; it should only be used at calm water surface scenarios for oil spill removal purposes, and engineers should salvage it within a relatively short period of time, or it would sink to the bottom of the watercourse and may cause secondary pollution.

An experimental orthogonal design was introduced to assess the influence of various variables to optimise oil spill accident remediation methods and techniques. Taking both oil removal rate and sorption capacity into consideration, the optimum conditions were as follows: corn stalk particle size between 830 and 1700 μm, oscillation frequency of 0 $r\ min^{-1}$ (i.e., no wave movement induced by shaking), oil film thickness of 0.55 mm and adsorbent dosage of 0.29 g; sawdust particle size between 830 and 1700 μm, oscillation frequency of 0 $r\ min^{-1}$, oil film thickness of 0.55 mm and adsorbent dosage of 0.25 g; wheat straw particle size between 500 and 830 μm, oscillation frequency of 0 $r\ min^{-1}$, oil film thickness of 0.55 mm and adsorbent dosage of 0.24 g. For these cases, the agriculture wastes were associated with a larger adsorption capacity to diesel oil, which is very good for removing oil from water at a degreasing rate of 99%. Wheat straw and sawdust should be more widely used for cases of water and oil spills compared to corn stalk.

Further experiments with other agricultural waste products that may be more common elsewhere should be undertaken. Sustainable and cost-effective disposal solutions of the agricultural waste contaminated with oil need to be found.

5.4 Kohonen Self-Organising Map to Predict Biochemical Oxygen Demand

5.4.1 Introduction

The dissolved oxygen used during the bacterial oxidation of organic pollutants, typically determined under standard conditions (Standard Methods for the Examination of Water and Wastewater 1998), is known as the five days at 20°C biochemical oxygen demand (BOD_5). This method is applied for assessing water quality, biological treatability of wastewater, performance of wastewater treatment operations and organic loadings to treatment plants for the purpose of sizing aeration facilities. The BOD_5 is also important for discharge consents for wastewater treatment plant facilities in many countries (European Economic Community 1991).

However, the usefulness of the BOD_5 for the effective monitoring and control of water pollution is hampered by the long time it takes to obtain an estimate, which precludes its use in real-time control. For example, considering a sustained low dissolved oxygen content in a river resulting from the discharge of large concentrations of organic pollutants, it will take at least five days before the problem can be effectively diagnosed, and any remedial actions to be taken are likely to be too late. This makes it very desirable to have a rapid and accurate inferential model for BOD_5 prediction, thus removing the time delay associated with the laboratory-based bioassay method.

Some researchers have argued that BOD_5 data are not useful for the purpose of process control and have therefore advocated the use of chemical oxygen demand (COD) data instead (Mohanty et al. 2002; Scholz 2006). However, in contrast to COD, the BOD_5 only characterises the biodegradable component of organic matter in wastewater; it is, therefore, the most important water quality parameter for assessing biological treatability. Furthermore, previous attempts to calibrate BOD_5 against COD had only limited success because of the wide variability often observed in the relationship between the two water-quality variables (Mohanty et al. 2002; Scholz 2006). Thus, although the COD is faster to determine than the BOD_5, it is certainly not a direct substitute.

An alternative to measure BOD_5 is the development of biosensors, which are devices for the detection of an analyte that combines a biological component with a physicochemical detector component. Biosensors indirectly measure the BOD_5 via a short-term (i.e., <0.5 hour) BOD (BOD_{st}) and a calibration curve, which relates this BOD_{st} to the BOD_5 (see Karube et al. (1977) for the pioneering work in this area). Consequently, biosensors are now commercially available, but they do have several limitations, such as their high maintenance costs, limited run lengths and the need for frequent reactivation and the inability to respond to changing water quality characteristics as they would typically occur in wastewater treatment streams (Praet et al. 1995). Another important limitation with the use of biosensors is the uncertainty associated with the calibration function for translating the BOD_{st} to the BOD_5 or BOD_7 (after 7 days) or even BOD_{25} (after 25 days).

Chapter 5.4 is proposing to use the Kohonen self-organising map (KSOM) to develop intelligent models for making rapid inferences about BOD_5 using other easy-to-measure water quality parameters, which, unlike BOD_5, can be obtained directly and reliably using online hardware sensors. This will make the use of BOD_5 for online process monitoring and control a more plausible proposition. Compared with other data-driven modelling paradigms, such as multilayer perceptron artificial neural networks (MLP ANN) and classical multivariate regression analysis, the KSOM is not inhibited by missing data. Moreover, time sequencing of data is not a problem when compared with classical time series analysis (Lee and Scholz 2006a, b). Chapter 5.4 is based on an article by Rustum et al. (2008).

5.4.2 Literature Review

Attempts to improve the rapidity with which the BOD_5 estimates are obtained have led to the development of BOD_5 biosensors. A microbial BOD_5 biosensor typically consists of microbial cells immobilised on an oxygen electrode. The first generation of such sensors was developed by Karube et al. (1977) for BOD_5 determination within minutes. However, these biosensors do not measure the BOD_5 directly, but they record the enhanced respiration rate of the immobilised microbes caused by the influx of biodegradable organics in the presence of sufficient oxygen (Kim and Kwon 1999). The graphical representation of the signals of this activity, which is picked up by the dissolved oxygen probe, is known as the respirogram, which can be correlated with the BOD_5 after calibration. Therefore, it is to be expected that the outcome of any BOD_5 determination exercise with the biological electrodes will depend on the species composition of the immobilised microorganism community, in addition to other boundary conditions and variables, such as temperature, pH, quantity of biodegradable organics and presence of toxins and other growth-inhibiting substances (Scholz 2006).

The original biosensor by Karube et al. (1977) used *Trichosporon cutaneum*, but later researchers applied other microbial species such as *Bacillus subtilis* (Riedel et al. 1988; Tan et al. 1992), *Klebsiella oxytoca* (Ohki et al. 1994) and activated sludge (Sakai et al. 1995). Praet et al. (1995) provided a detailed and critical review of available biological electrodes. They concluded that current BOD_5 biosensors are not a panacea for the limitations caused by the five-day time delay associated with conventional bioassay BOD_5 determination.

Considering that hardware sensors do not provide a satisfactory solution, work has also been carried out on the development of model-based software sensors for the rapid online estimation of BOD_5 and other water quality variables such as COD, heavy metals and nutrients (Lee and Scholz 2006a, b; Mohanty et al. 2002). In contrast to hardware biosensors, software sensors attempt to obtain the BOD_5 directly and require little or no maintenance. Furthermore, software sensors are readily updated, if more data become available, unlike hardware sensors, which generally require great expense and time to regenerate when the biological cells within the probe are becoming ineffective (Lee and Scholz 2006a, b).

The majority of the available software sensors are based on data-driven modelling approaches such as artificial neural networks. The areas of application in water management predominantly include water and wastewater treatment plants, but also sewer systems and stormwater runoff management control systems (Lee and Scholz 2006a; Lee and Stenstrom 2005; Mohanty et al. 2002; Shen et al. 2006). For example, Brydon and Frodsham (2001) developed an MLP ANN to predict the settled wastewater BOD_5 for a wastewater treatment plant in England. Two models with different numbers of input variables were successfully developed to predict the overall trends in BOD_5 time series data. However, the measured peaks were often not well predicted.

Other applications of artificial neural networks to predict water quality parameters, including BOD_5, were discussed by Ellis et al. (1990), Hiraoka et al. (1990), Lee and Scholz (2006a, b) and Scholz (2006). The main objective of these studies was to provide a BOD_5 estimate in a fraction of the time required for conventional bioassay determination, thus making the BOD_5 a more realistic water quality parameter for real-time process control.

In Chapter 5.4, the author has applied the unsupervised KSOM instead of the MLP ANN. The advantages of the KSOM over the MLP ANN are that the former can handle missing values and gaps in the data set without the need for a priori data infilling exercise. Moreover, KSOM can be used to visualise features in the data, thus providing easy-to-assess evidence of correlations between important water quality parameters. This chapter is based on an original article by Rustum et al. (2008).

5.4.3 Methodology

The KSOM (also referred to as feature map or Kohonen map) is one of the most widely used artificial neural network algorithms (Kohonen et al. 1996). It is typically presented as a dimensional grid or map whose units (i.e., nodes or neurons) become tuned to different input data patterns. The principal goal of the KSOM is to transform an incoming signal pattern of arbitrary dimension into a two-dimensional discrete map. This mapping preserves the most important topological and metric relationships of the original data, implying that no essential information is lost during mapping.

To train the KSOM, the multidimensional input data are first standardised by deducting the mean and subsequently dividing the results by the corresponding standard deviation. Then, a standardised input vector is chosen at random and presented to each of the individual neurons for comparison with their code vectors to identify the code vector most similar to the presented input vector. The identification uses the Euclidian distance, which is defined in Eq. (5.1).

$$D_i = \sqrt{\sum_{j=1}^{n} (x_{ij} - m_{ij})^2}; i = 1, 2, ..., M \quad (5.4)$$

where D_i = Euclidian distance between the input vector and the weight vector i; n = dimensionality of the input vector; x_{ij} = j-th element of the current input vector; m_{ij} = j-th element of the weight vector i; and M = number of neurons in the KSOM.

The neuron whose vector most closely matches the input data vector (i.e., for which the D_i is a minimum) is chosen as a winning node or the best matching unit (BMU) as indicated in Figure 5.13. The vector weights of this winning neuron and those of its adjacent neurons are then adjusted to match the input vector using an appropriate algorithm, thus bringing the two vectors further into agreement. It follows that each neuron in the map internally develops the ability to recognise input vectors similar to itself. This characteristic is referred to as self-organising, because no external information is supplied to lead to a classification (Penn 2005).

The process of comparison and adjustment continues until some specified error criteria are attained. The training stops at this moment in time. The following two error criteria are normally used: the quantisation error and the topographical error, as defined by Eqs. (5.2) and (5.3), respectively (Garcia and Gonzalez 2004). Garcia and Gonzalez (2004) and Hong et al. (2003) provided further details regarding the step-by-step procedure of the KSOM algorithms, including particular descriptions of the weight-updating and neighbourhood functions.

$$q_e = \frac{1}{N}\sum_{i=1}^{N} \|x_i - m_c\| \quad (5.5)$$

$$t_e = \frac{1}{N}\sum_{i=1}^{N} u(x_i) \quad (5.6)$$

where q_e = quantisation error; N = number of samples; x_i = i-th data sample or vector; m_c = prototype vector of the best matching unit for x_i; $\|...\|$ denotes distance; t_e = topographical error and u = binary integer, which is equal to 1 if the first and second BMUs for the argument of u are not adjacent units on the map; otherwise, $u = 0$.

The application of the KSOM for prediction purposes is illustrated in Figure 5.14 (Obu-Cann et al. 2001). First, the model is trained using the training data set. Then, to predict a set of variables as part

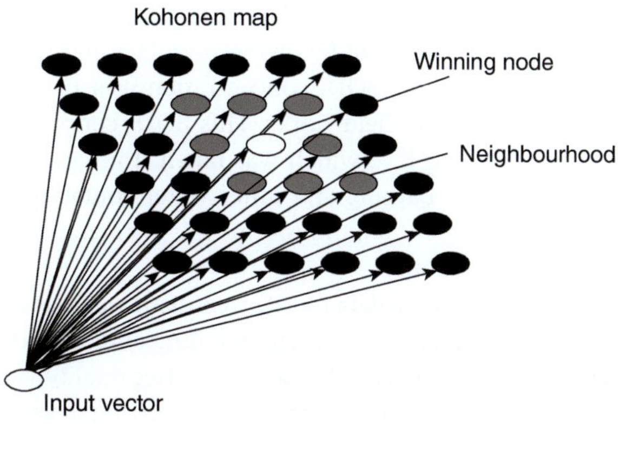

Figure 5.13 Sketch of the Kohonen self-organising map model. Illustration of the winning node and its neighbourhood in the Kohonen self-organising map.

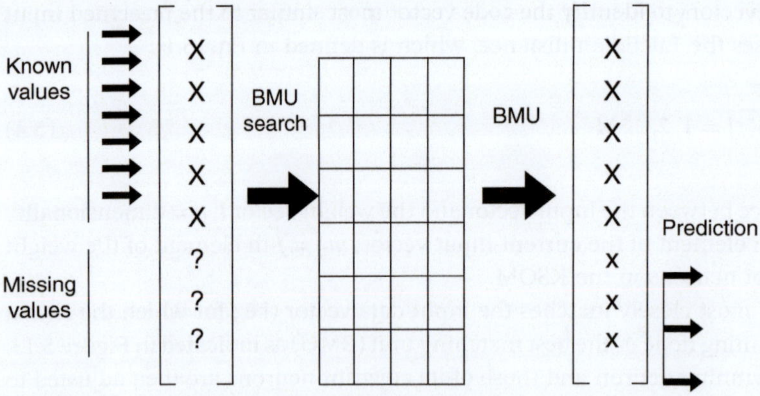

Figure 5.14 Prediction of missing components of the input vector using the Kohonen self-organising map. Outline showing the use of the best matching unit (BMU) in the prediction process using known values.

of an input vector, they are first removed from the vector, and the depleted vector is subsequently presented to the KSOM to identify its BMU. The values for the missing variables are then obtained by their corresponding values in the BMU.

5.4.4 Case Study

The KSOM was applied to data from three wastewater treatment plants in Scotland. Data from the relatively large Seafield treatment works in Edinburgh were used for model development (calibration and verification), while data from the relatively small works at East Calder and New Bridge, both outside Edinburgh, were used for independently validating the model. All works use the activated sludge treatment process for secondary biological oxidation. However, given that the emphasis of this study was on the influent (i.e., raw wastewater) stream, the type of secondary biological process used has no bearing on the validity of the modelling results. Consequently, the model would also apply to other secondary treatment processes, such as trickling filters or rotating biological contactors.

The Seafield works comprises six circular sedimentation tanks, six rectangular non-nitrifying aeration lanes and eight circular final settlement tanks. The main treatment is preceded by six screens (spacings of 6 mm) and four detritor grit removal units. The works is located in the eastern part of Edinburgh and discharges its effluent directly to the sea. The maximum flow is 674 000 $m^3\ d^{-1}$, which equates to a BOD_5 population equivalent of 1 136 825. All incoming flow is treated by primary clarification and secondary biological aeration. However, only a fraction of the throughput is given tertiary treatment by disinfection (UV light) before discharge.

The final effluent BOD_5 concentration was 10 mg L^{-1} with respect to the example data record provided by the Seafield works. This BOD_5 concentration was considerably lower than the design consent of 25 mg L^{-1}, which was taken from an internal report commissioned by the treatment plant operator. Bathing and other water contact activities occurring at nearby beaches close to the discharge point require routine UV disinfection of the effluent during summer to comply with bathing regulations.

A summary of the data for the treatment works is shown in Table 5.15. All data were provided by Thames Water Utilities plc (London, England, UK), the plant operators. The Seafield data comprise daily measurements of 16 raw wastewater quality variables, including flow, BOD_5, COD and suspended solids for a period of approximately three years, giving a total of 1066. An important feature

Table 5.15 Water quality of the inflow. These variables were used for the Kohonen self-organising map development.

Plant	Symbol	Description	Unit	Min.	Max.	Mean	Missing values
Seafield (1066 data points)	Inflow	Flow rate to the treatment plant	$m^3\ d^{-1}$	100 000	674 000	300 261	2
	COD	Chemical oxygen demand	$mg\ L^{-1}$	74	880	350.58	143
	SS	Suspended solids	$mg\ L^{-1}$	15	580	164.27	41
	NH4	Ammonia-nitrogen	$mg\ L^{-1}$	0.50	35.84	13.83	144
	pH	pH	—	7.09	9.20	6.10	162
	T	Temperature	°C	9	19	14	187
	BOD	Five days @ 20°C biological oxygen demand	$mg\ L^{-1}$	19	244	117	45
East Calder (1091 data points)	Inflow	Flow rate to the treatment plant	$m^3\ d^{-1}$	18 956	56 052	30 921	117
	COD	Chemical oxygen demand	$mg\ L^{-1}$	186	730	410	384
	SS	Suspended solids	$mg\ L^{-1}$	34	354	169	124
	pH	pH		6.8	7.8	7.33	173
	NH4	Ammonia-nitrogen	$mg\ L^{-1}$	3.85	30.14	18.23	124
	BOD	Five days @ 20°C biological oxygen demand	$mg\ L^{-1}$	30.5	152.5	96.4	161
New Bridge (1375 data points)	Inflow	Flow rate to the treatment plant	$m^3\ d^{-1}$	7235	28 389	13 550	29
	COD	Chemical oxygen demand	$mg\ L^{-1}$	113	778	395	44
	SS	Suspended solids	$mg\ L^{-1}$	38	361	166	49
	NH4	Ammonia-nitrogen	$mg\ L^{-1}$	3.18	38.33	17.51	48
	pH	pH		6.1	7.7	7.2	60
	BOD	Five days @ 20°C biological oxygen demand	$mg\ L^{-1}$	34.93	189.1	103.8	58

Max., maximum; Min., minimum.

of the three data sets is the large number of missing values, which would have made modelling by other methods, such as regression analysis and MLP ANN, impractical.

Of the 1006 Seafield vectors, 800 were used for model development, and the remaining 266 were used for verification of the model. Moreover, 1091 data vectors (East Calder works) and 1375 data vectors (New Bridge works) were used for independently validating the model. As shown in Table 5.15, the main characteristics at both East Calder and New Bridge are within the range of the Seafield works values.

5.4.5 Numerical Analysis and Modelling

The self-organising map (SOM) toolbox for Matlab 5 was used for this case study. The toolbox was developed by the SOM team at the Helsinki University of Technology, Finland (http://www.cis.hut.fi). An initial assessment of the data provided by Thames Water for input to Matlab 5 was carried out using Microsoft Excel (Microsoft Corporation, Redmond, Washington).

Considering that one of the objectives of this study was to develop a model for the rapid estimation of BOD_5, a decision had to be made on which of the available water quality variables to include with BOD_5 in the KSOM. This decision was based on two considerations: the rapidity with which the variables could be measured or evaluated using hardware sensors and the degree of association between the variable and the BOD_5. The former consideration is important for the rapid prediction of the BOD_5. Therefore, the variables inflow, COD, suspended solids, ammonia-nitrogen, pH and temperature (Table 5.15) were chosen, because they can be measured within three hours or by online hardware sensors, to allow BOD_5 prediction in virtually real time.

Apart from pH, the variables are also strongly related to the influent BOD_5 based on the visual inspection of their KSOM component planes (see below). Both temperature and pH have been known to affect reaction rates and kinetics in water and wastewater (Scholz 2006), so the lack of a strong and distinct connection between pH and BOD_5 was a surprise. Nevertheless, it was still decided to include pH as an input variable to further investigate the potential lack of correlation between pH and BOD_5.

For modelling purposes, the complete data set was subdivided into two sets: a first set of 800 observations to train the model (training set) and a second set of 266 observations to test the model (validation set). Three different KSOM models were trained with the training data set, with each KSOM model having different input variables with BOD_5. The corresponding list of variables for each KSOM is given in Table 5.16.

The estimated BOD_5 concentrations from the BMU of the training set were compared with the actual concentrations to evaluate the performance during training. Following training, the models were then validated with the validation data set, which was not used during the training phase. To establish the ability of the KSOM to generalise, it was verified using 266 data vectors. In the verification phase, the BOD_5 was omitted from the input vectors, essentially implying that BOD_5 values were missing. The BMU for each verification input vector were then determined to derive the missing BOD_5 values. Finally, BOD_5 concentrations obtained with the BMU were then compared with the actual observations to evaluate the performance during verification. The performance of the models during training and verification was evaluated using the following criteria: The average absolute error (AAE) measures the mean error of the predictions (Eq. 5.7).

$$\mathrm{AAE} = \frac{1}{N}\sum_{i=1}^{N}\left(\left|x_i - x_i'\right|\right) \tag{5.7}$$

Table 5.16 The structure of the trained Kohonen self-organising map models with a map size of 18 × 8 and a hexagonal lattice (also called triangular lattice).

Map no.	Input variables	Quantisation error	Topographic error
1	Influent flow, COD, SS, NH4, pH, Temperature and BOD	1.176	0.059
2	Influent flow, COD, SS, NH4, pH and BOD	0.975	0.055
3	Influent Flow, COD, SS, NH4 and BOD	0.723	0.022

BOD, five days @ 20°C biochemical oxygen demand; COD, chemical oxygen demand; NH4, ammonia-nitrogen; SS, suspended solids.

where AAE = average absolute error, N = number of samples, x = actual value and x' = value predicted by the model.

The relative average absolute error (RAAE) is the AAE scaled by the range (maximum subtracted by the minimum) of the data. For a well-performing model, the RAAE should be as small as possible (Eq. 5.8). The normalised root mean squared error (NRMSE) measures the mean root squared error scaled by the standard deviation of the values (Eq. 5.9). The correlation coefficient (R) measures the similarity of the shapes of the original and predicted time series and ranges between −1 and 1; the absolute value of the correlation coefficient for perfect predictions is unity (i.e., 1) (Eq. 5.10).

$$\text{RAAE} = \frac{\text{AAE}}{\text{Max-Min}} \tag{5.8}$$

where RAAE = relative average absolute error, AAE = average absolute error, and Max and Min = maximum and minimum data points of the range.

$$\text{NRMSE} = \frac{\sqrt{\frac{1}{N}\sum_{i=1}^{N}(x_i - x_i')^2}}{S} \tag{5.9}$$

where NRMSE = normalised root mean squared error, N = number of samples, x = actual value, x' = value predicted by the model and S = standard deviation.

$$R = \frac{N\sum x \times x' - \sum x \sum x'}{\sqrt{\left[N\sum x^2 - \left(\sum x\right)^2\right]\left[N\sum x'^2 - \left(\sum x'\right)^2\right]}} \tag{5.10}$$

where R = correlation coefficient, N = number of samples, x = actual value and x' = value predicted by the model.

The classification error measures the fraction of over- and under-predictions for a variable $z = x - x'$, where x and x' are the measured and predicted values, respectively (Eqs. 5.11 and 5.12). For purely random residuals, one would expect a 50 : 50 split between the positive rate (PR) and negative rate (NR). Otherwise, the model could be either upward- or downward-biased. The PR is the fraction of overprediction in the set of predicted values (Eq. 5.11). In contrast, the NR is the fraction of underprediction in the set of predicted values (Eq. 5.12).

$$\text{PR} = \frac{\text{count } z \text{ (if } z < 0)}{N} \times 100\% \tag{5.11}$$

where PR = positive rate; $z = x - x'$, where x and x' are the measured and predicted values, respectively; and N = number of samples.

$$\text{NR} = \frac{\text{count } z \text{ (if } z > 0)}{N} \times 100\% \tag{5.12}$$

where NR = negative rate; $z = x - x'$, where x and x' are the measured and predicted values, respectively; and N = number of samples.

5.4.6 Results

The component planes for each variable of the three KSOM are shown in Figure 5.15 to illustrate the associations between variables. Each component plane can be viewed as a sliced version of the KSOM, because it consists of the values of single vector variables in all map units. In other words, the component planes show the values of the variables in each map unit (Vesanto et al. 2000). These planes are filled using grey shades to show the individual values of a given feature of each SOM unit

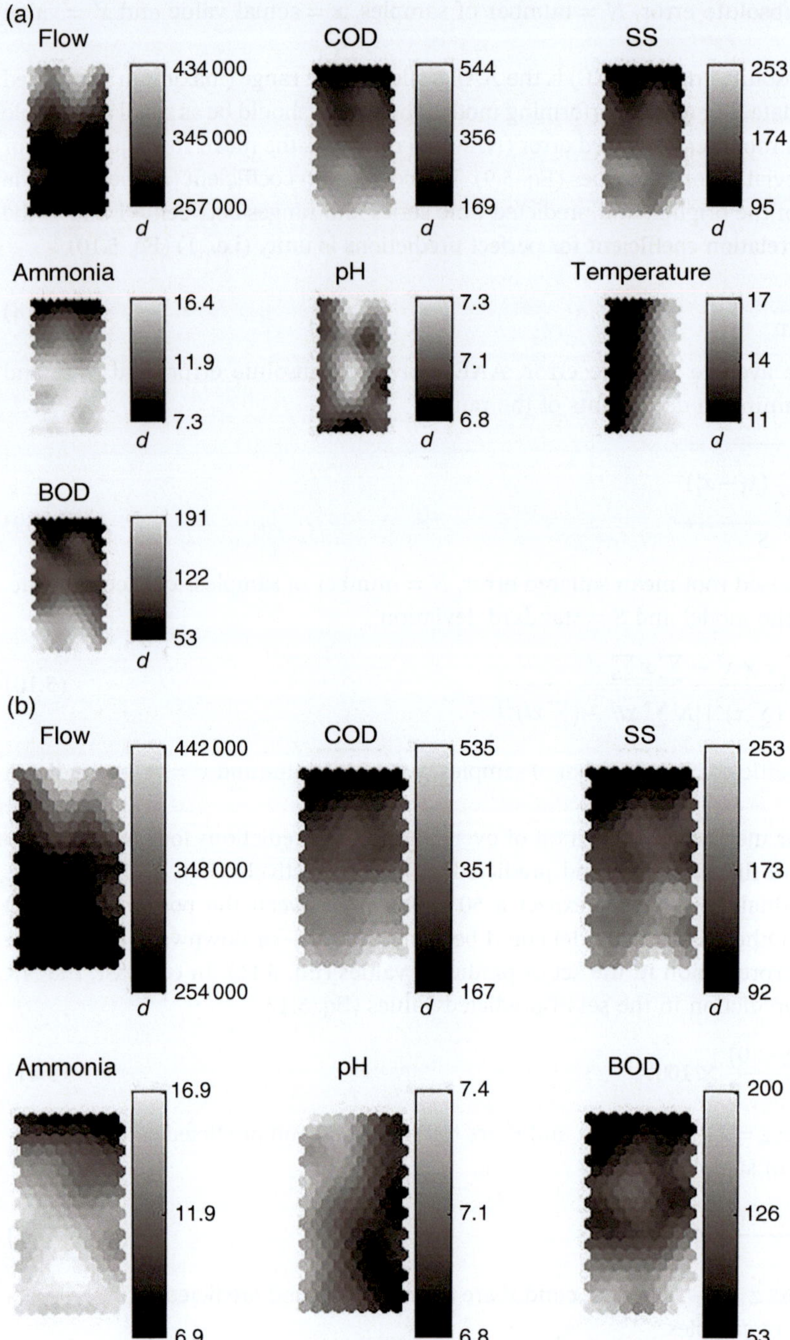

Figure 5.15 Component planes for the (a) Kohonen self-organising map (KSOM) model 1, (b) KSOM model 2 and (c) KSOM model 3. The unit for flow is $m^3\ d^{-1}$. COD = chemical oxygen demand ($mg\ L^{-1}$), SS = suspended solids ($mg\ L^{-1}$), NH4 = ammonia-nitrogen ($mg\ L^{-1}$), temperature (°C) and BOD5 = five days @ 20°C biochemical oxygen demand. Three sets of two-dimensional maps highlighting the influence of various water quality parameters on the prediction process for each of the three models used are shown.

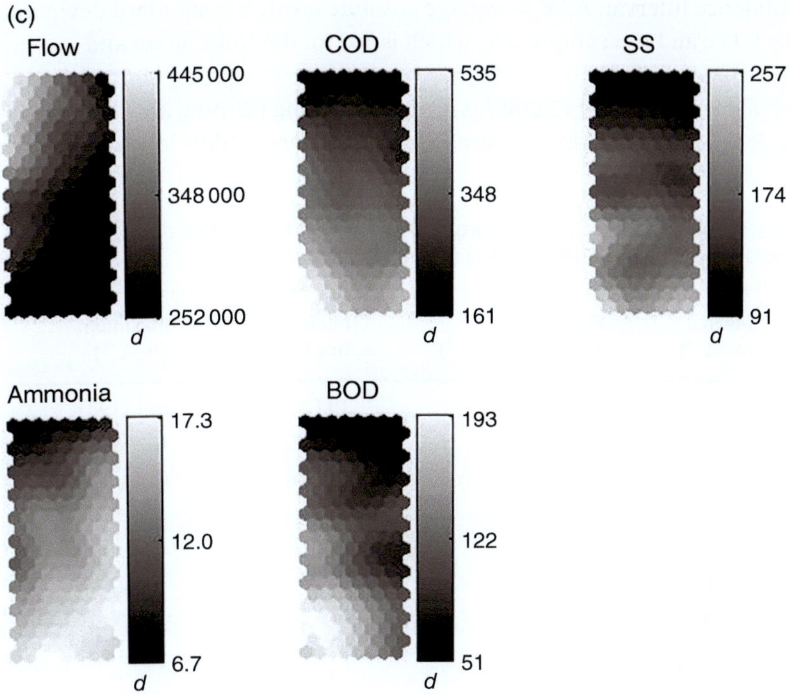

Figure 5.15 (Continued)

in the two-dimensional lattice, such that the lighter the colour grey, the higher the relative component value of the corresponding weight vector.

The component planes help to illustrate the relationship between BOD_5 and other variables considered in this analysis. For example, by looking at the upper left hand of the component planes, it can be seen that low BOD_5 concentrations are associated with a high influent flowrate, which is expected, given the dilution effects of high inflow rates. Furthermore, low BOD_5 concentrations in the influent are associated with low COD, suspended solids and ammonia-nitrogen concentrations. While BOD_5 is positively correlated with suspended solids, ammonia-nitrogen and COD, it is negatively correlated with the inflow rate. Furthermore, visual inspections of the component planes of pH and temperature do not reveal any obvious association with BOD_5. In particular, the relationship between BOD_5 and pH is weak (Figure 5.15).

The structures of the three trained KSOMs are summarised in Table 5.16. All KSOMs have the same map size. However, the third map has the smallest quantisation and topographic errors, which indicate that this is the best model for prediction purposes.

Table 5.17 summarises the errors during training and verification, which reinforce the superiority of the third KSOM model. For example, the correlation coefficient between the measured and KSOM-predicted BOD_5 during training was 0.93 for map 3, compared with 0.86 and 0.91 for maps 1 and 2, respectively. Map 3 also performs much better than maps 1 and 2 based on the AAE (4.34 as opposed to 18.86 and 14.78 for maps 1 and 2, respectively, during training). Assuming that the errors are normally distributed, the approximate 95% confidence interval is shown by Eq. (5.13).

$$95\%\,\text{Conf} = \text{AAE} \pm 1.96\frac{S}{\sqrt{n}} \tag{5.13}$$

where 95% Conf = 95% confidence interval; AAE = average absolute error; S = standard deviation of the absolute errors (Table 5.17); and n = sample size, which is 800 for the training set and 266 for the validation set.

The 95% confidence interval for the AAE of KSOM 3 is 4.34 ± 0.4 during training and 4.92 ± 0.55 during verification (Table 5.18). In comparison, these errors are considerably lower than the

Table 5.17 Summary statistics of the three Kohonen self-organising map models for five days @ 20°C biochemical oxygen demand (BOD) prediction during training (and verification).

Map no.	Statistics	Mean (mg L^{-1})	Standard deviation (mg L^{-1})	Minimum (mg L^{-1})	Maximum (mg L^{-1})
1	Measured	126 (92)	45.83 (32.32)	19 (22)	244 (192)
	Estimated	124 (99)	36.99 (26.24)	52 (52)	192 (186)
	Absolute error	18.86 (14.00)	14.560 (16.000)	0.03 (0.02)	89.92 (61.00)
	Correlation	0.86 (0.87)			
	AAE	18.86 (14.00)			
	RMAE	0.084 (0.082)			
	NRMSE	0.520 (0.547)			
	Fraction positive	45% (64%)			
	Fraction negative	55% (36%)			
2	Measured	126 (92)	45.83 (32.32)	19 (22)	244 (192)
	Estimated	124 (98)	38.92 (27.14)	52 (52)	201 (193)
	Absolute error	14.78 (12.16)	12.320 (9.590)	0.01 (0.02)	93.91 (43.32)
	Correlation	0.91 (0.90)			
	AAE	14.78 (12.16)			
	RMAE	0.066 (0.072)			
	NRMSE	0.420 (0.479)			
	Fraction positive	47% (66%)			
	Fraction negative	53% (34%)			
3	Measured	126 (92)	45.83 (32.32)	19 (22)	244 (192)
	Estimated	125 (96)	40.67 (28.33)	50 (50)	195 (187)
	Absolute error	4.34 (4.92)	11.530 (8.890)	0.00 (0.00)	68.16 (47.06)
	Correlation	0.93 (0.91)			
	AAE	4.34 (4.92)			
	RMAE	0.055 (0.050)			
	NRMSE	0.360 (0.414)			
	Fraction positive	52 (53%)			
	Fraction negative	48 (47%)			

AAE, average absolute error; NRMSE, normalised root mean squared error; RMAE, relative mean absolute error.

Table 5.18 Summary statistics of the validation for the Kohonen self-organising map model number 3 for five days @ 20°C biochemical oxygen demand predictions with data from Seafield, East Calder and New Bridge.

	Seafield	East Calder	New Bridge
MSE	79	98	112
AAE	4.92	7.99	8.82
Correlation	0.91	0.88	0.87
Fraction positive (%)	53	44	53
Fraction negative (%)	47	56	47
Minimum measured	22	31	35
Minimum predicted	52	50	50
Maximum measured	192	152	189
Maximum predicted	186	154	182
Mean measured	92	96	104
Mean predicted	96	96	103

AAE, average absolute error; MSE, mean squared error.

errors obtained by Brydon and Frosham (2001) for their two MLP ANN models also predicting BOD_5.

The superiority of map 3 is also highlighted by the classification error (Table 5.17). The fraction of the false PR is nearly equal to the fraction of the false NR. This implies little bias in predictions obtained by map 3 compared with predictions obtained by maps 1 and 2. The classification error for maps 1 and 2 during verification was worse than during training, which is clear evidence of the significant upward bias in the BOD_5 prediction for these two models.

Both temperature and pH correlated weakly with the influent BOD_5 (see above) but were included because of the theoretical importance of these two variables for the reaction kinetics in water and wastewater treatment (Mohanty et al. 2002; Scholz 2006). The fact that the third KSOM, which contains neither pH nor temperature, is performing better than the first two models, which included these variables (Table 5.16), confirms that both temperature and pH are not directly linked to BOD_5.

Because of the relative superiority of the third map, further research was only undertaken with this map. Figure 5.16a, b show the time series of the observed and predicted BOD_5 during testing and verification, respectively. These show that the performance of the KSOM is very good and in agreement with corresponding evidence presented in Table 5.18. The plot of residuals is shown in Figure 5.16c, indicating that the residuals are random, as expected, from the approximate 50 : 50 split of the positive and negative prediction fractions of map 3.

The observed and modelled values have been presented in Figure 5.17a (training) and 5.17b (verification), to better demonstrate how well KSOM 3 is predicting the BOD_5. Most data points are contained within the 95% prediction limits, particularly during verification. These findings support the use of the KSOM model for predicting mean BOD_5 concentrations. Moreover, the 95% prediction limits during verification completely enclosed all the plotted points, implying that individual BOD_5 concentrations can be predicted with sufficient accuracy using KSOM 3.

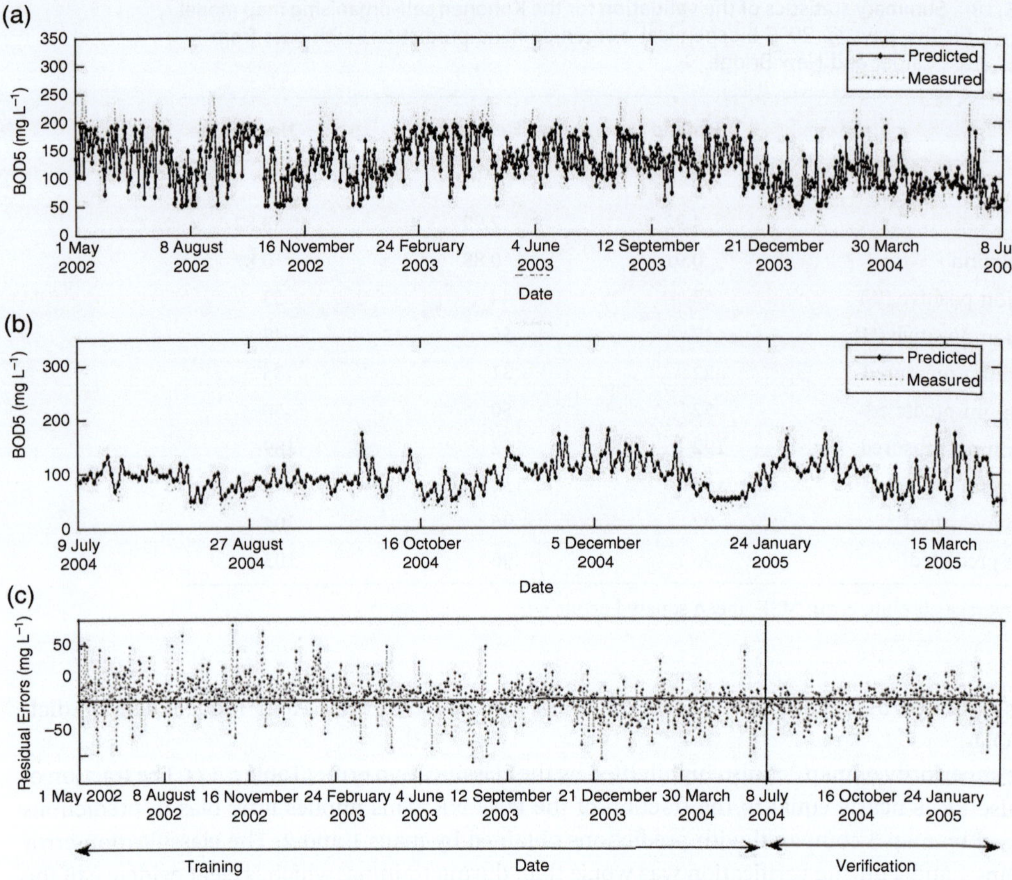

Figure 5.16 Comparison of the observed and predicted five days @ 20°C biochemical oxygen demand (BOD_5) concentrations of the Kohonen self-organising map (KSOM) 3 during (a) training and (b) verification; the corresponding residuals are shown in part (c). The verification process shows good predictions, but relatively high residuals are apparent and need to be controlled.

5.4.7 Discussion

The KSOM performed reasonably well during calibration and verification with the Seafield data. However, this cannot be taken as evidence that the model is sufficiently general as to perform well when applied to an independent data set. In other words, to ascertain the true ability of the proposed model to generalise, it must be applied to at least one independent data set unrelated to that used for the model development.

Consequently, the KSOM model 3 was applied to the East Calder and New Bridge data sets for the purpose of validation. The validation exercise followed the same procedure used for verifying the model earlier, in which the BOD_5 measurement in each measured vector was removed, assumed missing. The validation results are summarised in Table 5.18. In general, the performance of the model is good. In particular, the correlation coefficient between the observed and predicted BOD_5 during validation was very close to that obtained for the verification of the Seafield data. However, much more satisfying is that the model was able to better match the mean, minimum and maximum concentrations concerning the data of the validation sites in comparison with

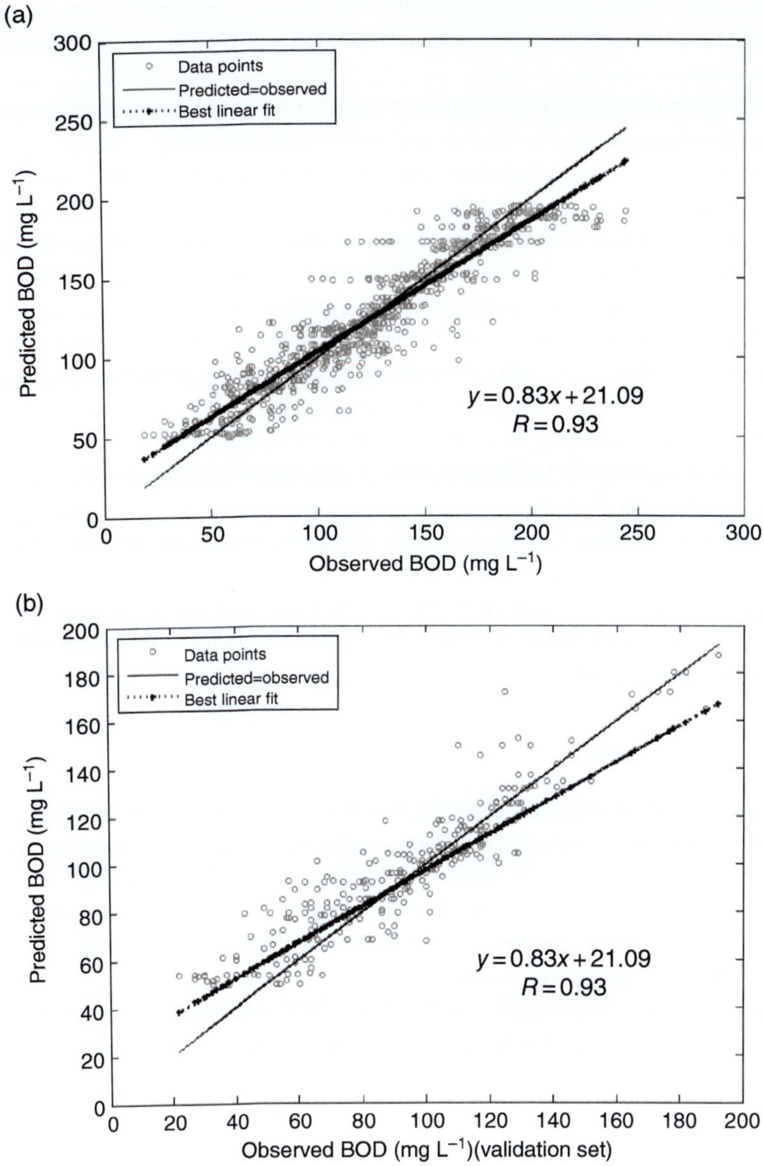

Figure 5.17 The performance of the Kohonen self-organising map (KSOM) model 3 in predicting the five days @ 20°C biochemical oxygen demand (BOD_5) during (a) training and (b) verification. Good predictions (correlation coefficient $R = 0.93$) are shown for both the training and validation processes.

the data of the Seafield calibration site. Finally, the distribution of positive and negative predictions at the validation sites was relatively even, again suggesting that the residuals are random during the validation.

All the above evidence is proof that the KSOM model developed in Chapter 5.4 is an adequate predictor of the BOD_5, thus meeting the objective to develop a rapid software sensor for the estimation of BOD_5. However, this model has its limitations. The first limitation is that the model has been developed using historical daily data only. The number of samples taken for each variable differs from one variable to another, which could have affected the accuracy and precision of the measurements. This is a result of inconsistencies in the real industrial sampling regime, which was not designed for research purposes.

Furthermore, the number of data points used to develop the model limits its overall performance. As a general rule, if more data points are available for the model development, the performance of the model is likely to increase, because it can extract more patterns out of a larger data set. Finally, the model has been developed using raw wastewater quality variables. However, its application can be widened if more data from other sources, such as river water or treated wastewater, are available for analysis. More data from different sources would allow the model to capture the patterns of data from a wider range of scenarios. This may also lead to an increase of the model performance.

The proposed software sensor is fundamentally different from any biosensor. Therefore, only an empirical comparison, rather than an objective statistical comparison, is possible. Biosensors do not measure BOD_5 directly or estimate actual BOD_5 concentrations based on historical BOD_5 data, such as the proposed model, but they rely on a good correlation between the respiration of sensor organisms and historical BOD_5 data. Considering that a large variety of organisms could be used as biosensors, a comparison with the proposed model is meaningless. Moreover, large numbers of biosensor data are rare and were not available for the case study site or the sites used for validation purposes.

The only journal paper so far that refers to biosensors and neural networks for BOD_5 determination in the same context was written by Roche et al. (2006). However, neural networks were only used to model direct biosensor performances.

5.4.8 Conclusions and Recommendations

The current work presents a novel methodology based on the use of KSOM models to predict BOD_5 concentrations in wastewater, using raw wastewater data obtained at three wastewater treatment plants in Scotland. Extensive testing and validation show that the model is sufficiently general to predict the BOD_5, applying variables that can be measured within three hours or in real time using online hardware sensors, thus making it possible to estimate BOD_5 very rapidly. This allows for a timely intervention and cost reduction during problem diagnosis.

The proposed BOD_5 software sensor methodology is preferred over BOD_5 biosensors, because the BOD_5 can be estimated directly, and no costly maintenance is required. The software sensor does not require calibration and cannot be negatively affected by toxins and other inhibitors. Moreover, the software sensor is very dynamic and can be readily updated when additional data become available, thus enhancing its accuracy.

The KSOM tool used for the development of the software sensor can readily deal with missing values in one or more of the input variables, without significantly affecting the accuracy of the model. The proposed methodology is applicable for other water and environmental engineering problems, and Chapter 5.4 could therefore also be regarded as a teaching aid.

References

Alexander, M. (1999). *Biodegradation and Bioremediation*, 2e. New York: Academic Press.

Allen, S.E. (1974). *Chemical Analysis of Ecological Materials*, 1e, 89. Oxford, UK: Blackwell Scientific Publications.

American Public Health Association (2005). *Standard Methods for the Examination of Water and Wastewater*, 21e. Washington, DC: American Public Health Association/American Water Works Association/Water Environment Federation.

Amini, A., Bollaert, E., Boillat, J.-L., and Schleiss, A.-J. (2008). Dynamics of low-viscosity oils retained by rigid and flexible barriers. *Ocean Engineering* 35 (14–15): 1479.

Angelova, D., Uzunov, I., Uzunova, S. et al. (2011). Kinetics of oil and oil products adsorption by carbonized rice husks. *Chemical Engineering* 172 (1): 306.

Annunciado, T.R., Sydenstricher, T.H.D., and Amico, S.C. (2005). Experimental investigation of various vegetable fibers adsorbent materials for oil spills. *Marine Pollution Bulletin* 50 (11): 1340.

Asha, S. and Thiruvenkatachari, V. (2008). Removal of oil by walnut shell media. *Bioresource Technology* 99 (17): 8217.

Atlas, R.M. and Cerniglia, C.E. (1995). Bioremediation of petroleum pollutants. *BioScience* 45 (5): 332–338.

Banerjee, S.S., Joshi, M.V., and Jayaram, R.V. (2006). Treatment of oil spill by sorption technique using fatty acid grafted sawdust. *Chemosphere* 64 (6): 1026.

Banks, C.J. and Wang, Z. (1999). Development of two phase anaerobic digester for the treatment of mixed abattoir wastes. *Water Science and Technology* 40 (1): 69.

Baris, A.J., Eifert, W.H., Klotzer, K., and McGuckin, C.J. (2001). *Use of a Sub-surface Flow Constructed Wetland for Collection and Treatment of Water Containing BTEX*. Inslandia, New York: Roux Associates Inc.

Bastani, D., Safekordi, A.A., Alihosseini, A., and Taghikhani, V. (2006). Study of oil sorption by expanded perlite at 298.15K. *Separation and Purification Technology* 52 (2): 295.

Bayat, A., Aghamiri, S.F., Moheb, A., and Vakili-Nezhaad, G.-R. (2005). Oil spill clean from sea water by sorbent materials. *Chemical Engineering and Technology* 28 (12): 1525.

Brydon, D.A. and Frosham, D.A. (2001). A model-based approach to predicting BOD5 in settled sewage. *Water Science and Technology* 44 (2–3): 9–16.

Burgherr, P. (2006). In-depth analysis of accidental oil spills from tankers in the context of global spill trends from all sources. *Journal of Hazardous Materials* 140 (1–2): 245.

Burland, S.B. and Edwards, E.A. (1999). Anaerobic benzene biodegradation linked to nitrate reduction. *Applied and Environmental Microbiology* 65 (2): 529–533.

Carmichael, L.M. and Pfaender, F.K. (1997). The effect of inorganic and organic supplements on the microbial degradation of phenanthrene and pyrene in soils. *Biodegradation* 8 (1): 1–13.

Caswell, P.C., Gelb, D., Marinello, S.A. et al. (1992). Evaluation of constructed surface-flow wetlands systems for the treatment of discharged waters from oil and gas operations in Wyoming. *SPE Rocky Mountain Regional Conference. Paper SPE 24331*, Casper, Wyoming.

Ceylan, D., Dogu, S., Karacik, B. et al. (2009). Evaluation of butyl rubber as sorbent material for the removal of oil and polycyclic aromatic hydrocarbons from seawater. *Environmental Science and Technology* 43 (10): 3846.

Chaillan, F., Chaıneau, C.H., Point, V. et al. (2006). Factors inhibiting bioremediation of soil contaminated with weathered oils and drill cuttings. *Environmental Pollution* 144 (1): 255–265.

Chaıneau, C.H., Rougeux, G., Yepremian, C., and Oudot, J. (2005). Effects of nutrient concentration on the biodegradation of crude oil and associated microbial populations in the soil. *Soil Biology and Biochemistry* 37 (8): 1490–1497.

Chen, L., Xia, W., Li, J. et al. (2011). Adsorption of diesel in seawater on spill-sorb and their influencing factors. *Journal of Environmental Engineering* 5 (3): 584.

Choi, H.M. (1992). Natural sorbents in oil spill cleanup. *Environmental Science and Technology* 26 (4): 772.

Choi, S.-C., Kwon, K.K., Sohn, J.H., and Kim, S.-J. (2000). Evaluation of fertilizer additions to stimulate oil biodegradation in sand seashore mescocosms. *Journal of Microbiology and Biotechnology* 12 (3): 431–436.

Chung, S., Suidan, M.T., and Venosa, A.D. (2011). Partially acetylated sugarcane bagasse for wicking oil from contaminated wetlands. *Chemical Engineering and Technology* 34 (12): 1989.

Coates, J.D., Chakraborty, R., and McInerney, M.J. (2002). Anaerobic benzene biodegradation – a new era. *Research in Microbiology* 153 (10): 621–628.

Cooper, P.F., Job, G.D., Green, M.B., and Shutes, R.B.E. (1996). *Reed Beds and Constructed Wetlands for Wastewater Treatment*. Swindon: Water Research Centre.

Dagley, S. (1986). Biochemistry of aromatic hydrocarbon degradation in Pseudomonas. In: *The Bacteria*, vol. 10 (ed. J. Sokatch and L.N. Ornston), 527–555. Orlando: Academic Press.

Deschamps, G., Caruel, H., Borredon, M.-E. et al. (2003). Oil removal from water by selective sorption on hydrophobic cotton fibers. 1. Study of sorption properties and comparison with other cotton fiber-based sorbents. *Environmental Science and Technology* 37 (5): 1013.

Di Lecce, C., Accarino, M., Bolognese, F. et al. (1997). Isolation and metabolic characterization of a *Pseudomonas stutzeri* mutant able to grow on the three isomers of xylene. *Applied and Environmental Microbiology* 63 (8): 3279–3281.

Díaz, E. (2004). Bacterial degradation of aromatic pollutants: a paradigm of metabolic versatility. *International Microbiology* 7 (3): 173–180. ISSN 1139-6709.

Eke, P.E. and Scholz, M. (2006). Hydrocarbon removal with constructed treatment wetlands for the benefit of the petroleum industry. *10th International Conference on Wetland Systems for Water Pollution Control* (23–29 September 2006) (ed. V. Dias and J. Vymazal). International Water Association, Lisbon, Portugal, Volume 3:1707–1714, ISBN: 989-20-0361-6.

Eke, P.E. and Scholz, M. (2008). Benzene removal with vertical-flow constructed treatment wetlands. *Journal of Chemical Technology and Biotechnology* 83 (1): 55–63.

El-Aziz, A., Said, A., Ludwick, A.G., and Aglan, H.A. (2009). Usefulness of raw bagasse for oil absorption: a comparison of raw and acylated bagasse and their components. *Bioresource Technology* 100 (7): 2219.

Ellis, G.W., Ge, X., Grasso, D. (1990). Time series analysis of wastewater quality. *Proceedings of the 5th International Association on Water Pollution Research and Control Workshop on Instrumentation, Control, and Automation of Water and Wastewater Treatment and Transport Systems*, Tokyo, Japan (July 26–August 3). Oxford, United Kingdom: International Association on Water Pollution Research and Control, 595–600.

European Economic Community (1991). Directive concerning urban wastewater treatment, 91/271/EEC. European Economic Community Official Journal L135/40, http://eur-lex.europa.eu/LexUriServ/LexUriServ.do?uri=CELEX:31991L0271:EN:HTML (accessed 1 December 2006).

European Union (2002). Regulation (EC) No 1774/2002 of the European Parliament and of the Council of 3 October 2002 Laying Down Health Rules Concerning Animal By-products not Intended for Human Consumption. Official Journal, L 273, 10 October 2002, 1–95.

Fuchs, W., Binder, H., Mavrias, G., and Braun, R. (2003). Anaerobic treatment of wastewater with high organic content using a stirred tank reactor coupled with a membrane filtration unit. *Water Research* 37 (4): 902.

Gander, M., Jefferson, B., and Judd, S. (2000). Aerobic MBRs for domestic wastewater treatment: a review with cost considerations. *Separation and Purification Technology* 18 (2): 119.

Garcia, H. and Gonzalez, L. (2004). Self-organizing map and clustering for wastewater treatment monitoring. *Engineering Applications of Artificial Intelligence* 17 (3): 215–225.

Guirguis, M. (2004). Treatment of waste water: a reed bed environmental case history. *Paper SPE86673 presented at the 2004 SPE International Conference on Health, Safety, and Environment in Oil and Gas Exploration and Production* (29–31 March 2004). Calgary, Alberta Canada.

Harayama, S., Kasai, Y., and Hara, A.A. (2004). Microbial communities in oil contaminated seawater. *Current Opinion in Biotechnology* 15 (3): 205–214.

Haussard, M., Gaballah, I., Kanari, N. et al. (2003). Separation of hydrocarbons and lipid from water using treated bark. *Water Research* 37 (2): 362.

Heider, J., Spormann, A.M., Beller, H.R., and Widdel, F. (1998). Anaerobic bacterial metabolism of hydrocarbons. *FEMS Microbiology Reviews* 22 (5): 459–473.

Hiegel, T. (2004). Analysis of pilot scale constructed wetland treatment of petroleum contaminated groundwater. MSc thesis. Department of Civil Engineering, University of Wyoming.

Hiraoka, M., Tsumura, K., Fujita, I., and Kanaya, T. (1990). System Identification and Control of Activated Sludge Process by Use of Autoregressive Model, Instrumentation, Control and Automation of Water and Wastewater Treatment and Transport Systems. Advanced Water Pollution Control; Report No. 10; pp. 121–128; Briggs, R. (ed.); Pergamon; IAWPRC: London, UK.

Hofer, T. (2003). Tanker safety and coastal environment: prestige, Erika, and what else. *Environmental Science and Pollution Research* 10 (1): 1.

Hong, Y., Rosen, M., and Bhamidimarri, R. (2003). Analysis of a municipal wastewater treatment plant using a neural network-based pattern analysis. *Water Research* 37 (7): 1608–1618.

Hussein, M., Amer, A.A., and Sawsan, I.I. (2008). Oil spill sorption using carbonized pith bagasse: 1. Preparation and characterization of carbonized pith bagasse. *Analytical and Applied Pyrolysis* 82 (2): 205.

Hussein, M., Amer, A.A., El-Maghraby, A., and Taha, N.A. (2009). Availability of barley straw application on oil spill clean up. *Journal of Environmental Science and Technology* 6 (1): 123.

Hussein, M., Amer, A.A., and Sawsan, I.I. (2011). A comprehensive characterization of corn stalk and study of carbonized corn stalk in dye and gas oil sorption. *Journal of Analytical and Applied Pyrolysis* 2 (8): 132.

Janks, J.S. and Cadena, F. (1991). Identification and properties of modified zeolites for the removal of benzene, toluene and xylene from aqueous solutions. *Paper SPE 22833, in 1991 Society of Petroleum Enginers (SPE) Annual Technical Conference and Exhibition* (6–9 October 1991). Dallas, Texas, USA.

Johns, M.R. (1996). Developments in wastewater treatment in the meat processing industry: a review. *Bioresource Technology* 54 (3): 203.

Johnson, R.F. and Manjrekar, T.G. (1973). Removal of oil from water surfaces by sorption on unstructured fibers. *Environmental Science and Technology* 7 (5): 439.

Kadlec, R.H. (2001). Thermal environments of subsurface treatment wetlands. *Water Science and Technology* 44 (11): 251–258.

Kadlec, R.H. and Knight, R.L. (1996). *Treatment Wetlands*. Boca Raton, Florida: Lewis Publishers.

Karakasi, O.K. and Moutsatsou, A. (2010). Surface modification of high calcium fly ash for its application in oil spill clean up. *Fuel* 89 (12): 3966.

Karube, I., Matsunaga, T., Mitsuda, S., and Suzuki, S. (1977). Biochemical electrode BOD sensors. *Biotechnology and Bioengineering* 19 (10): 1535–1547.

Kim, M.N. and Kwon, H.S. (1999). Biochemical oxygen demand sensor using serratia marcescens LSY 4. *Biosensors and Bioelectronics* 14 (1): 1–7.

Kim, S., Choi, D.H., Sim, D.S., and Oh, Y. (2005). Evaluation of bioremediation effectiveness on crude oil-contaminated sand. *Chemosphere* 59 (6): 845–852.

Kingston, P.F. (2002). Long-term environmental impact of oil spills. *Spill Science and Technology Bulletin* 7 (1–2): 53.

Knight, R.L. (1999). The use of treatment wetlands for petroleum industry effluents. *Environmental Science & Technology* 33 (7): 973–980.

Kohonen, T., Oja, E., Simula, O. et al. (1996). Engineering applications of the self organizing map. *Proceedings IEEE (Institute of Electrical and Electronics Engineers)* 84 (10): 1358–1384.

Lahvis, M.A., Baehr, A.L., and Baker, R.J. (1999). Quantification of aerobic biodegradation and volatilization rates of gasoline hydrocarbons near the water table under natural attenuation conditions. *Water Resources Research* 35 (3): 753–765.

Leahy, J.G. and Olsen, R.H. (1997). Kinetics of toluene degradation by toluene-oxidizing bacteria as a function of oxygen concentration, and the effect of nitrate. *FEMS Microbiology Ecology* 23 (1): 23–30.

Lee, B.-H. and Scholz, M. (2006a). A Comparative study: prediction of constructed treatment wetland performance with K-nearest neighbors and neural networks. *Water, Air, and Soil Pollution* 174 (1–4): 279–301.

Lee, B.-H. and Scholz, M. (2006b). Application of the self-organizing map (SOM) to assess the heavy metal removal performance in experimental constructed wetlands. *Water Research* 40 (18): 3367–3374.

Lee, H. and Stenstrom, M.K. (2005). Utility of stormwater monitoring. *Water Environment Research* 77 (3): 219–228.

Li, M.M., Pan, H.-C., Huang, S.-L., and Scholz, M. (2013). Controlled experimental study on removing diesel oil spillages using agricultural waste products. *Chemical Engineering and Technology* 36 (4): 673–680.

Lim, T.T. and Huang, X.F. (2007). Evaluation of kapok (Ceiba pentandra (L.) Gaertn.) as a natural hollow hydrophobic–oleophilic fibrous sorbent for oil spill cleanup. *Chemosphere* 66 (5): 955.

Lovley, D.R. and Lloyd, J.R. (2000). Microbes with a mettle for bioremediation. *Nature Biotechnology* 18 (6): 600–601.

Lu, J. and Zhou, M. (2002). The development of oil sorbent materials. *Journal of Dong Hua University* 28 (1): 126. (in Mandarin).

Manjunath, N.T., Mehrotra, I., and Mathur, R.P. (2000). Treatment of wastewater from slaughterhouse by DAF-UASB System. *Water Research* 34 (6): 1930.

Masse, D. and Masse, L. (2000). Treatment of slaughterhouse wastewater in anaerobic sequencing batch reactors. *Canadian Agricultural Engineering* 42 (3): 131.

Miraglia, R.A. (2002). The cultural and behavioral impact of the Exxon Valdez oil spill on the native peoples of Prince William Sound, Alaska. *Spill Science and Technology Bulletin* 7 (1-2): 75.

Mitsch, W.J. and Gosselink, J.G. (1993). *Wetlands*, 2e. New York: Wiley.

Mohanty, S., Scholz, M., and Slater, M.J. (2002). Neural network simulation of the chemical oxygen demand reduction in a biological activated carbon filter. *Water and Environment Journal* 16 (1): 58–64.

Moshiri, G.A. (2000). *Constructed Wetland for Water Quality Improvement*. Florida, USA: CRC Press.

Myers, J.E. and Jackson, L.M. (2001). An evaluation of the Department of Energy Naval Petroleum Reserve No. 3 - Produced water bio-treatment facility. *Proceedings of SPE/EPA/DOE Exploration and Production Environmental Conference, SPE Paper 66522*, San Antonio, Texas.

Nakata, K., Sugioka, S.-I., and Hosaka, T. (1997). Hindcast of a Japan sea oil sill. *Spill Science and Technology Bulletin* 4 (4): 219.

Nowak, O., Wandl, G., and Svardal, K. (1999). Long-term experience with biological pre-treatment of rendering plant effluent. *Water Science and Technology* 40 (1): 37.

O'Brien, C.A., Scholz, M., and McConnachie, G.L. (2005). Membrane bioreactors and constructed wetlands for treatment of rendering plant wastewater. *Water and Environment Journal* 19 (3): 189–198.

Obu-Cann, K., Morita, Y., Fujimura, K. et al. (2001). Data mining of power transformer database using self-organizing maps. *Proceedings of the 6th International Conference on Soft Computing, IZUKA 2000*, Iizuka, Fukuoka, Japan (1–4 October). New York, NY, USA: Institute of Electrical and Electronics Engineers, 201–206; http://ieeexplore.ieee.org/iel5/7719/21162/00983717.pdf (accessed 1 December 2006).

Ohki, A., Shinohara, K., Ito, O. et al. (1994). A BOD sensor using Klebsiella oxytoca AS1. *International Journal of Environmental Analytical Chemistry* 56: 261–269.

Omari, K. (2003). Hydrocarbon removal in an experimental gravel bed constructed wetland. *Water Science and Technology* 48 (5): 275–281.

Oudot, J., Merlin, F.X., and Pinvidic, P. (1998). Weathering rates of oil components in a bioremediation experiment in estuarine sediments. *Marine Environmental Research* 45 (2): 113–125.

Paje, M.L.F., Neilan, B.A., and Couperwhite, I.A. (1997). Rhodococcus species that thrives on medium saturated with liquid benzene. *Microbiology* 143 (9): 2975–2981.

Penn, B.S. (2005). Using self-organizing maps to visualize high dimensional data. *Computational Geosciences* 31 (5): 531–544.

Peterson, C.H., Rice, S.D., Short, J.W. et al. (2003). Long-term ecosystem response to the Exxon Valdez oil spill. *Science* 302 (5653): 2082.

Praet, E., Reuter, V., Gaillard, T., and Vasel, J.L. (1995). Bioreactors and biomembrane for biochemical oxygen demand estimation. *TrAC Trends in Analytical Chemistry* 14 (7): 371–378.

Rajakovic, V., Aleksic, G., Radetic, M., and Rajakovic, L.J. (2007). Efficiency of oil removal from real wastewater with different sorbent materials. *Journal of Hazardous Materials* 143 (1–2): 494.

Rew, S. and Mulamoottil, G. (1999). A cost comparison of leachate treatment alternatives. In: *Constructed Wetlands for the Treatment of Landfill Leachates* (ed. G. Mulamoottil, E.A. McBean, and F. Rovers). Boca Raton: Lewis.

Ridgeway, H.F., Safarik, J., Phipps, D. et al. (1990). Identification and catabolic activity of well-derived gasoline-degrading bacteria and a contaminated aquifer. *Applied and Environmental Microbiology* 56 (11): 3565–3575.

Riedel, K., Renneberg, R., Kuhn, M., and Scheller, F. (1988). A fast estimation of biochemical oxygen demand using microbial sensors. *Applied Microbiology and Biotechnology* 28: 316–318.

Rivera, F., Warren, A., Curds, C.R. et al. (1997). The application of the root zone method for the treatment and reuse of high-strength abattoir waste in Mexico. *Water Science and Technology* 35 (5): 271–278.

Roche, P., Al-Jowder, R., Narayanaswamy, R. et al. (2006). A novel luminescent lifetime-based optrode for the detection of gaseous and dissolved oxygen utilising a mixed ormosil matrix containing ruthenium (4,7-diphenyl-1,10-phenanthroline)(3)Cl-2 (Ru.dpp). *Analytical and Bioanalytical Chemistry* 386: 1245–1257.

Rustum, R., Adeloye, A.J., and Scholz, M. (2008). Applying kohonen self-organizing map as a software sensor to predict biochemical oxygen demand. *Water Environment Research* 80 (1): 32–40.

Sakai, Y., Abe, N., Takeuchi, S., and Fermen, J. (1995). BOD sensor using magnetic activated sludge. *Journal of Fermentation and Bioengineering* 80 (3): 300–303.

Scholz, M. (2004). Treatment of gully pot effluent containing nickel and copper with constructed wetlands in a cold climate. *Journal of Chemical Technology and Biotechnology* 79 (2): 153–162.

Scholz, M. (2006). *Wetland Systems to Control Urban Runoff*. Amsterdam: Elsevier.

Scholz, M. and Martin, R.J. (1997). Ecological equilibrium on biological activated carbon. *Water Research* 31 (12): 2959.

Scholz, M., Höhn, P., and Minall, R. (2002). Mature experimental constructed wetlands treating urban water receiving high metal loads. *Biotechnology Progress* 18 (6): 1257.

Shen, J., Shen, W., Chang, J., and Gong, N. (2006). Fuzzy neural network for flow estimation in sewer systems during wet weather. *Water Environment Research* 78 (2): 100–109.

Solano-Serena, F., Marchal, R., Ropars, M. et al. (1999). Biodegradation of gasoline: kinetics, mass balance and fate of individual hydrocarbons. *Journal of Applied Microbiology* 86 (6): 1008–1016.

(1998). *Standard Methods for the Examination of Water and Wastewater*, 20e. Washington DC: American Public Health Association/American Water Works Association/Water and Environment Federation.

Stephenson, T., Judd, S., Jefferson, B., and Brindle, K. (2000). *Membrane Bioreactors for Wastewater Treatment*. London: International Water Association Publishing.

Sugai, S.F., Lindstrom, J.E., and Braddock, J.F. (1997). Environmental influences on the microbial degradation of Exxon Valdez oil on the shorelines of Prince William Sound, Alaska. *Environmental Science & Technology* 31 (5): 1564–1572.

Tan, T.C., Li, F., Neoh, K.G., and Lee, Y.K. (1992). Microbial membrane-modified dissolved oxygen probe for rapid biochemical oxygen demand measurement. *Sensors and Actuators B: Chemical* 8: 167–172.

Teas, C., Kalligeros, S., Stournas, S. et al. (2001). Investigation of the effectiveness of absorbent materials in oil spill cleanup. *Desalination* 140 (3): 259.

Vesanto, J., Himberge, J., Alhoniemi, E., and Parhankangas, J. (2000). Self-Organizing Map (SOM) Toolbox for Matlab 5, Report No. A57. Helsinki University of Technology, Laboratory of Computer and Information Science: Helsinki, Finland.

de Villiers, G.H. and Pretorius, W.A. (2001). Abattoir effluent treatment and protein production. *Water Science and Technology* 43 (4): 343.

Vlaev, L., Petkov, P., Dimitrov, A., and Genieva, S. (2011). Cleanup of water polluted with crude oil or diesel fuel using rice husks ash. *Taiwan Institute of Chemical Engineers* 42 (6): 957.

Vrhovsek, D., Kukanja, V., and Bulc, T. (1996). Constructed wetland (CW) for industrial wastewater treatment. *Water Research* 30 (10): 2287–2292.

Wallace, S.D. and Knight, R.L. (2006). Small-scale constructed treatment systems: feasibility, design criteria, and O&M requirements. Final report, Project 01-CTS-5. Water Environment Research Foundation, Alexandria, Virginia.

Wang, Z. and Banks, C.J. (2003). Evaluation of a two stage anaerobic digester for the treatment of mixed abattoir wastes. *Process Biochemistry* 38 (9): 1267.

Wemple, C. and Hendricks, L. (2000). Documenting the recovery of hydrocarbon-impacted wetlands: a multi-disciplinary approach. *Wetlands and Remediation: An International Conference* (ed. J.L Means and R.E. Hinchee). Battelle Press, Columbus, Ohio, USA, 73–78.

Xia, H. (2003). Ecological effectiveness of vetiver constructed wetlands in treating oil-refined wastewater. *Proceedings of the Third International Vetiver Conference in Guangzhou* (6–9 October 2003). China.

Yang, L., Chang, H., and Huang, M. (2001). Nutrient removal in gravel- and soil-based wetland microcosms with and without vegetation. *Ecological Engineering* 18 (1): 91–105.

Zahid, M.A., Halligan, J.E., and Johnson, R.F. (1972). Oil slick removal using matrices of polypropylene filaments. *Industrial & Engineering Chemistry Process Design and Development* 11 (4): 550.

Zhou, E. and Crawford, R.L. (1995). Effects of oxygen, nitrogen, and temperature on gasoline biodegradation in soil. *Biodegradation* 6 (2): 127–140.

Zhou, J., Tang, T., Zhong, M. et al. (2010). Adsorption performance of diesel by Kapok fiber. *Environmental Science and Technology* 33 (7): 17. (in Mandarin).

6

Sludge Dewatering Tests

6.1 Dewaterability Assessment Including the Capillary Suction Time Test

6.1.1 Introduction

Wherever sludges must be disposed of in restricted land areas or transported over long distances for ultimate disposal, some form of volume reduction is usually necessary. Sludge dewatering is an effective method of volume reduction in such cases. Pre-treatment is essential when incineration is required. Dewatering processes in common use, such as pressure filters, vacuum filters and centrifuges, require for their design some measure of the sludge dewatering characteristics. Most dewaterability tests are simple but empirical and shed little light on dewaterability mechanisms that could be described mathematically.

The capillary suction time (CST) apparatus has been studied both theoretically and experimentally (Chen et al. 1996; Guan et al. 2004; Herwijn et al. 1995; Singh et al. 2004). A review paper on CST summarised research trends (Scholz 2005). Various measurements of dewaterability such as the CST, specific resistance to filtration (SRF), the conditioned filtrate (liquid stream) streaming current and filtrate total solids were assessed. A significant correlation between filtrate viscosity and CST was established, with both related to dewaterability. There was also a clear relationship between the CST, the filtrate streaming current and solids recovery. A minimum CST, filtrate streaming current values close to zero for both filtrate and conditioning sludge, and minimum filtrate viscosity correlated well with maximum solids recovery (Dentel and Abuorf 1995; Scholz 2005).

The CST equipment has long been established as a practical but empirical method for the determination of sludge dewaterability. The CST was shown as a good index for sludge filterability, if only the product of solid concentration and average SRF is of interest. On the other hand, the bound water content cannot be directly evaluated from the CST data (Chen et al. 1996).

The CST values usually correlate well with SRF measurements. However, the CST values do not allow for the solids content, and therefore it is necessary to specify the solids content of the sludge experimentally. It is possible for sludge with a low CST value and high solids content to be easily dewatered (Scholz 2005).

The radial CST apparatus, and the paper used by it, were investigated to obtain the capillary suction pressure, the porosity of the paper and the water saturation within the paper as a function of the radial distance from the suspension to be filtered. Air was shown to coexist with water in the wetted Whatman No. 17 chromatographic (chr) paper, but the capillary suction pressure and the

Sustainable Water Systems, First Edition. Miklas Scholz.

effective water porosity were shown to be practically constant in the paper as used in the CST method (Meeten and Smeulders 1995).

The aim of Chapter 6.1 is to revise the CST dewaterability test to reduce consumable costs and to improve its interpretation. The objectives are to assess (i) current and relevant literature on CST; (ii) CST data variability for typical light and heavy sludges, and to compare the findings with a cost-effective reference synthetic sludge; (iii) the application of various papers to reduce consumable costs; (iv) the modified CST test using a rectangular funnel; and (v) the influence of stirring the sludge to reduce the impact of sedimentation, and subsequently to improve the interpretation of using the CST as an indicator for dewaterability. Note that Chapter 6.1 is based on material provided by Scholz (2006).

6.1.2 Materials and Methodology

6.1.2.1 Capillary Suction Time Devices and Measurements

The CST was measured with two standard CST devices; single and five radii test heads incorporating circular funnels. The Model 304B CST (Figure 6.1) can only be used with one single radius test head. In comparison, the Model 319 Multi-purpose CST (Figure 6.2) can be used with either five single radius test heads or with one multi-radii test head comprising five radii for the rapid assessment of SRF. The SRF can be calculated from the slope of the filtration curve. The latter application option allows for better mathematical treatment of the corresponding CST values.

Two 10- and 18-mm diameter funnels were applied for fast and slow filtering light and heavy sludges, respectively. The 10-mm diameter funnel had a cone-shaped inlet at the top. The funnel volumes were 5.62 and 6.36 cm^2 for the 10- and 18-mm funnel, respectively.

Concerning the CST apparatus application procedure, sludge is poured into a small open funnel (circular tube) resting on a piece of paper. Filtrate is extracted by capillary suction, and a cake is formed at the bottom of the funnel. However, the cake formation is enhanced for heavy sludges by sedimentation that leads to artificially high CST values (Scholz 2005; Smiles 1998).

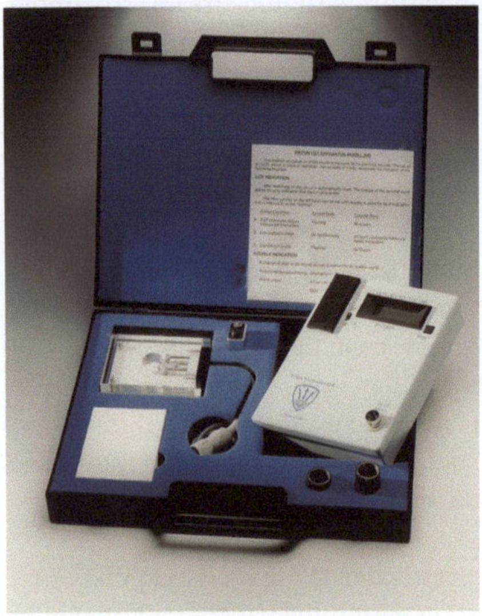

Figure 6.1 Standard apparatus (Model 304B CST) to measure the capillary suction time (CST) at one fixed position. Picture showing a box containing the CST device together with accessories and papers.

The CST is obtained from two electrodes placed at a standard interval from the funnel (Figure 6.1). The time taken for the waterfront to pass between these two electrodes constitutes the CST. The pressure generated by capillary suction is by an order of magnitude greater than the hydrostatic head within the funnel, so that the test is relatively independent of the amount of sludge, as long as there is sufficient to generate a CST measurement (Guan et al. 2004). This is therefore usually the case except for very heavy sludges.

The rate at which filtrate is extracted depends on the resistance of the cake. By measuring the distance, which the filtrate requires to travel along the paper as a function of time, the cake resistance can be determined. The method relies upon the varying pressure related to the movement of the advancing waterfront through the paper, so that a theoretically complete mathematical treatment of the CST data is difficult (Scholz 2005).

A rectangular instead of a circular funnel allows for easy mathematical treatment of CST data. Therefore, a further set of experiments was performed with a rectangular funnel (not supplied as part of the standard kit) of 18 mm × 18 mm × 20 mm (width, length and height, respectively) replacing the circular funnel. The rectangular funnel volume of 6.48 cm^3 was like the 18 mm circular funnel. A rectangular funnel should principally be applicable for light and heavy sludges (Scholz 2005).

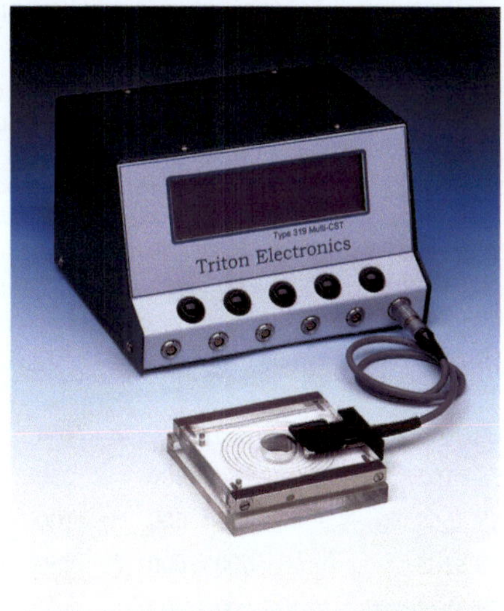

Figure 6.2 Revised apparatus (Model 319 Multi-purpose CST) to measure multiple capillary suction times (CST) or CST measurements at five fixed positions. Picture showing the CST device together with accessories.

6.1.2.2 Testing of Different Papers

Filter, chromatographic and other papers (all called papers for the purpose of Chapter 6) from Whatman in Brentford (UK) (distributed by Camlab in Over [UK]), Hollingsworth and Vose in Kentmore (UK) (HOVO), Carlson Filtration in Barnoldswick (UK) (Carlson), Schleicher and Schuell Microscience in Dassel (Germany) (SS; part of the Whatman group since 2004) and Fisher Scientific in Loughborough (UK) (Fisher; paper manufacturer was not disclosed) were tested. Table 6.1 shows a summary of various paper characteristics.

Whatman No. 17 chr is a smooth cellulose paper with a relatively high flow rate. However, the wet front moves faster along the grains than across, resulting in an elliptical shape of the wet front, because of its anisotropic properties (Scholz 2005). The papers SS 2668 chr, SS3324 chr and Fisher 200 chr are also anisotropic. In contrast, papers HOVO TO w/s, Carlson EE1.0H, SS 1107 and SS 3205 are isotropic.

Table 6.1 Summary of paper properties (data volunteered by the manufacturer and/or distributor).

Name	A	B	C	D	E	F	G	H	I
Whatman No. 17 chr	Cellu-lose	190	413	920	4.13	3.36		8	
HOVO TO w/s			280			3.47		6.7	3.0
Carlson EE1.0H			425–475	800–1200		1.53	0.354–0.593		
SS 1107	Cellu-lose	121	140	280	25.64	>11.00	0.50	5–6	5–6
SS 2668 chr	Linters	>180	320	900	5.45	>11.00	0.36	6–12	6–12
SS 3205	Cellu-lose	136	95	200	19.23	>11.00	0.48	4–7	4–7
SS 3324 chr	Linters	>180	280	730	6.92	>12.00	0.38	6–12	6–12
Fisher 200 chr	Cellu-lose	180–220	440	900					

A, paper type; B, flow rate (mm/30 minutes); C, basis weight ($g\,m^{-2}$); D, thickness (μm); E, Gurley Air Porosity ($s/100\ mL\,cm^{-2}$); F, dry burst ($kg\,cm^{-2}$); G, density ($g\,cm^{-3}$); H, pore diameter (μm); I, retention size (μm).

Table 6.2 Summary of paper costs (data volunteered by the manufacturer and/or distributor for 2005 and 2006).

Name	Sheet size (mm × mm)	Minimum sheets (number)	Cost (£)	Maximum CST papers (number)	Cost per CST paper (£)	Cost per CST (bulk supply) (£)
Whatman No. 17 chr (via Camlab)	460 × 570	25	141.00	1000	0.1410	0.1034 (bulk of 26k papers)
HOVO TO w/s	420 × 540	400	116.08	14 400	0.0081	0.0073 (bulk of 100.8M papers)
Carlson EE1.0H	420 × 540	100	79.90	3600	0.0222	(no discount)
SS 1107	580 × 580	250	219.78	12 000	0.0183	
SS 2668 chr	580 × 600	100	221.86	4800	0.0462	
SS 3205	600 × 600	250	140.05	12 000	0.0117	
SS 3324 chr	700 × 900		1733.25			
Fisher 200 chr	580 × 600	25	78.72	1200	0.0656	

A summary of a paper cost analysis can be found in Table 6.2. Whatman provides the standard CST paper that is by far the most expensive one in this comparison but sets the benchmark criteria. The alternative papers are cheaper but have different properties (Table 6.1) that may influence their performance. The following sections explore the influence of these properties on CST testing.

6.1.2.3 Testing of Different Sludges

Experiments were undertaken with heavy primary sludge and light surplus activated sludge (sludge blanket of a CYCLAZUR sequencing batch reactor) delivered by Scottish Water (Bo'Ness wastewater treatment works, West Lothian, UK). The sludges were constantly aerated and stored at 4°C in the fridge for not more than 35 days before analysis. The CST experiments were conducted with various papers (Table 6.1) and three different funnels (see above).

Synthetic activated sludge (Sanin and Vesilind 1999) can be prepared by flocculating negatively charged polystyrene latex particles (bacteria replacement) using sodium alginate (a polysaccharide produced from brown seaweed (*Phaeophyceae*) representing the extracellular polymeric substances) and calcium chloride (act as general bridging cations). However, the polystyrene latex particles are expensive. At least £1270 (for 100 mL based on 2006 prices) would need to be spent to enable one series of simple experiments requiring a few litres of synthetic sludge (e.g., catalogue of Interfacial Dynamics Corporation in Tualatin, Oregon, USA). These economic constraints severely limit the use of the proposed synthetic sludge for reference purposes. However, the polystyrene latex particles are not essential for alginate floc formation. Therefore, 180 mg L^{-1} alginate was mixed with 1000 mg L^{-1} calcium chloride only to simulate a cheap and stable synthetic sludge for reference purposes. Quantities are based on optimisation calculations published by Sanin and Vesilind (1999).

Mean total solids (mg L^{-1}), suspended solids (mg L^{-1}) and turbidity (NTU) figures for primary sludge, surplus activated sludge and synthetic sludge were 30 733, 24 681 and >99 999, and 3733, 1999 and 1960, and 1282, 296 and 9, respectively.

Different papers (Table 6.1) were tested with all sludges as indicated in Tables 6.3–6.7. Sludges were collected on different days and subsequently stored for different durations to yield high variabilities of the most essential water quality variables.

Table 6.3 Summary of capillary suction time test results for primary sludge.

Name	Sample no.	Minimum	Mean	Median	Maximum	SD[a]	SD[a]/mean
10 mm diameter funnel (cone) for fast filtering light sludges							
Whatman No. 17 chr	10	248.6	513.6	490.6	744.1	194.95	0.38
HOVO TO w/s	11	57.9	401.3	387.0	780.3	191.56	0.48
Carlson EE1.0H	10	323.0	1544.7	1596.8	3280.0	892.54	0.58
SS 1107	11	66.2	180.3	185.2	478.4	117.05	0.65
SS 2668 chr	10	173.6	300.2	292.4	496.6	120.14	0.40
SS 3205	10	65.8	140.9	125.4	378.1	86.99	0.62
SS 3324 chr	10	123.6	317.2	340.8	639.8	175.03	0.55
Fisher 200 chr	10	287.5	316.0	301.4	383.2	33.25	0.11
18 mm diameter funnel for slow filtering heavy sludges such as primary sludge							
Whatman No. 17 chr	11	40.7	82.8	84.8	118	28.12	0.34
HOVO TO w/s	10	13.7	48.7	36.8	93.2	25.85	0.53
Carlson EE1.0H	10	52.7	142.8	148.1	234.3	64.78	0.45
SS 1107	11	14.9	37.4	32.3	68.0	14.96	0.40
SS 2668 chr	11	25.2	51.0	51.6	72.5	17.61	0.35
SS 3205	10	8.4	23.2	20.8	44.5	12.47	0.54
SS 3324 chr	10	22.4	66.1	65.8	100.7	29.84	0.45
Fisher 200 chr	10	40.9	61.0	63.1	72.1	9.60	0.16
Rectangular funnel							
Whatman No. 17 chr	10	33.1	38.3	38.5	45.6	3.74	0.10
HOVO TO w/s	10	11.7	13.0	13.1	14.9	1.04	0.08
Carlson EE1.0H	10	77.6	93.0	93.3	106.9	9.92	0.11
SS 1107	10	14.0	17.9	18.2	22.8	2.84	0.16
SS 2668 chr	10	25.4	28.4	27.2	36.4	3.46	0.12
SS 3205	10	9.8	11.5	11.8	13.4	1.15	0.10
SS 3324 chr	10	25.2	29.8	30.1	32.0	1.91	0.06
Fisher 200 chr	10	29.8	34.9	35.3	37.3	2.41	0.07

[a] standard deviation.

6.1.2.4 Stirred Capillary Suction Time Test

A modified Typhoon Europe Frother Product Code 35712 with an L-shaped shaft of 1.6 mm diameter was applied as a micro-mixer to stir the sludges during an additional experiment to assess the influence of hindered sedimentation. The length of the horizontal part of the shaft was only 4 mm to reduce damage of sludge flocs during rotation. The maximum rotations per minute possible with the modified device were 5000. The deviation from the centre during rotation was ≤1 mm. The stirrer was inserted at a right-hand angle to a depth of approximately 2 mm above the bottom of the funnel. Moreover, the individual price of a modified frother is approximately £2 (based on a price in 2005), and does therefore add only an insignificant cost to any CST apparatus.

Table 6.4 Summary of capillary suction time test results for surplus activated sludge.

Name	Sample no.	Minimum	Mean	Median	Maximum	SD[a]	SD[a]/mean
10 mm diameter funnel (cone) for fast filtering light sludges such as surplus activated sludge							
Whatman No. 17 chr	10	17.8	21.6	20.2	28.2	3.55	0.16
HOVO TO w/s	7	6.3	14.6	18.2	22.7	7.54	0.52
Carlson EE1.0H	7	22.5	28.5	27.1	36.4	4.38	0.15
SS 1107	8	14.2	21.4	20.6	33.0	6.96	0.33
SS 2668 chr	8	7.8	12.1	11.5	19.1	4.12	0.34
SS 3205	5	0.6	17.2	18.6	34.0	12.36	0.72
SS 3324 chr	5	7.8	12.8	13.5	16.7	3.54	0.28
Fisher 200 chr	7	20.6	23.3	23.9	25.9	1.84	0.08
18 mm diameter funnel for slow filtering heavy sludges							
Whatman No. 17 chr	8	1.3	7.5	8.4	12.8	3.99	0.53
HOVO TO w/s	6	1.3	7.1	8.4	12.8	4.63	0.65
Carlson EE1.0H	6	0.7	6.6	5.8	14.6	5.05	0.77
SS 1107	8	2.3	9.3	10.2	14.6	3.76	0.40
SS 2668 chr	8	0.1	3.2	4.0	5.9	2.52	0.79
SS 3205	7	0.1	2.8	3.6	5.6	2.45	0.88
SS 3324 chr	4	0.7	3.6	3.6	6.6	3.32	0.92
Fisher 200 chr	4	3.8	4.9	4.2	7.3	1.66	0.34
Rectangular funnel							
Whatman No. 17 chr	7	5.6	7.0	7.2	7.7	0.69	0.10
HOVO TO w/s	7	4.0	4.3	4.2	4.8	0.26	0.06
Carlson EE1.0H	7	9.2	12.0	11.3	15.4	2.38	0.20
SS 1107	7	14.5	15.3	15.0	16.2	0.68	0.04
SS 2668 chr	7	3.6	4.0	4.0	4.4	0.26	0.07
SS 3205	7	3.2	4.2	4.6	5.1	0.80	0.19
SS 3324 chr	7	3.2	3.5	3.5	4.0	0.27	0.08
Fisher 200 chr	7	4.2	4.9	5.0	5.4	0.39	0.08

[a] standard deviation.

6.1.3 Results and Discussion

6.1.3.1 Single Radii Test and Paper Application

Since the invention of the CST test in 1967, the same paper (Whatman No. 17 chr; Table 6.1) has been used to allow for an easy comparison of test results worldwide. However, the paper type was virtually selected at random (Gale and Baskerville 1968; Scholz 2005). The use of a CST apparatus with a rectangular funnel can overcome the problem of anisotropic chromatographic paper (Table 6.1) by making use of only one direction of the paper.

Paper water flow can physically be described by the theory of water flow in porous media. This theory is based on the material balance for water and Darcy's law. The matric potential or suction

Table 6.5 Summary of capillary suction time test results for synthetic sludge.

Name	Sample no.	Minimum	Mean	Median	Maximum	SD[a]	SD[a]/mean
10 mm diameter funnel (cone) for fast filtering light sludges such as synthetic sludge							
Whatman No. 17 chr	10	10.3	11.2	11.3	12.4	0.59	0.05
HOVO TO w/s	10	4.1	6.0	6.0	7.9	1.38	0.23
Carlson EE1.0H	10	10.3	17.0	18.3	19.7	3.01	0.18
SS 1107	10	8.4	16.5	16.2	27.6	5.69	0.34
SS 2668 chr	10	5.0	6.6	6.8	8.0	1.03	0.16
SS 3205	10	4.3	6.0	6.1	7.2	1.00	0.17
SS 3324 chr	10	6.2	7.8	8.1	9.0	1.01	0.13
Fisher 200 chr	10	7.4	8.9	8.9	10.2	0.77	0.09
18 mm diameter funnel for slow filtering heavy sludges							
Whatman No. 17 chr	10	5.8	6.5	6.5	7.1	0.47	0.07
HOVO TO w/s	10	1.5	2.4	2.4	3.5	0.63	0.26
Carlson EE1.0H	10	8.7	9.5	9.5	10.2	0.47	0.05
SS 1107	10	1.2	3.3	3.5	5.7	1.34	0.41
SS 2668 chr	10	2.5	3.6	3.6	4.8	0.72	0.20
SS 3205	10	1.2	3.0	2.9	4.3	0.92	0.31
SS 3324 chr	10	2.0	4.1	4.5	5.2	1.02	0.25
Fisher 200 chr	10	3.2	4.3	4.5	4.8	0.56	0.13
Rectangular funnel							
Whatman No. 17 chr	10	4.8	5.2	5.2	5.7	0.25	0.05
HOVO TO w/s	10	2.3	3.7	3.9	4.4	0.71	0.20
Carlson EE1.0H	10	5.8	7.0	6.8	9.7	1.29	0.18
SS 1107	10	3.6	5.1	5.2	6.5	0.96	0.19
SS 2668 chr	10	3.4	4.9	4.4	7.2	1.28	0.26
SS 3205	10	4.2	6.5	6.9	8.4	1.33	0.20
SS 3324 chr	10	2.6	3.4	3.3	4.4	0.59	0.17
Fisher 200 chr	10	3.5	4.2	4.2	4.9	0.44	0.10

[a] standard deviation.

and the hydraulic conductivity of both the paper and the suspension are functions of the water content (Chen et al. 1996; Lee and Hsu 1992; Scholz 2005; Smiles 1998).

The advance of the wetting front in the paper is a function of the interaction of the sorptivity and matric potential functions of both the paper and the suspension (Smiles 1998). Extension of this approach to radial flow in the paper is complicated and does not permit a simple mathematical interpretation of the wetting front advance.

The experiments with a rectangular funnel (see above) summarised in Tables 6.3 and 6.4 show that the ratio of standard deviation over mean is consistently lower in comparison to the application of a circular funnel. However, findings for synthetic sludge are less consistent (Table 6.5). The reasons for this observation are the relatively low overall CST mean and maximum times of 5.2 seconds

Table 6.6 Summary of capillary suction time test results for stirred primary sludges.

Name	Sample no.	Minimum	Mean	Median	Maximum	SD[a]	SD[a]/mean
10 mm diameter funnel (cone) for fast filtering light sludges							
Whatman No. 17 chr	10	78.4	434.4	381.2	735.0	192.19	0.44
HOVO TO w/s	10	54.8	341.9	371.8	521.0	122.11	0.36
Carlson EE1.0H	3	68.8	1882.9	2300.0	3280.0	1645.73	0.87
SS 1107	10	51.0	163.3	172.1	218.4	49.69	0.26
SS 2668 chr	10	116.4	344.0	329.4	658.0	131.06	0.38
SS 3205	10	11.7	133.9	146.2	179.3	47.99	0.36
SS 3324 chr	10	175.4	419.8	429.4	639.8	148.36	0.35
Fisher 200 chr	10	254.9	271.9	271.4	284.4	9.98	0.04
18 mm diameter funnel for slow filtering heavy sludges such as primary sludge							
Whatman No. 17 chr	10	46.8	64.7	60.6	102.4	15.41	0.25
HOVO TO w/s	10	6.1	29.4	32.4	60.0	16.33	0.56
Carlson EE1.0H	10	32.0	114.3	111.0	184.9	50.43	0.44
SS 1107	10	17.8	38.0	34.7	68.0	16.56	0.44
SS 2668 chr	10	30.9	51.6	48.0	71.7	13.54	0.26
SS 3205	10	8.4	33.9	34.8	52.9	14.04	0.41
SS3324 chr	10	28.0	61.0	60.0	95.8	20.54	0.34
Fisher 200 chr	10	50.4	60.0	59.2	74.0	7.46	0.12

[a] standard deviation.

Table 6.7 Decision support matrix for selecting an alternative paper for Whatman No. 17 chr (see also Tables 6.1 and 6.2 as well as 6.4–6.6).

Name	A	B	C	D	E	F	G	H	I
HOVO TO w/s	X	√√	X	√	X	√√	X	X	X
Carlson EE1.0H	X	√	X	X	√	X	√	√	X
SS 1107	X	√	√	√	√	√	X	√√	X
SS 2668 chr	√	√	√	√	X	X	X	X	√
SS 3205	X	√√	X	X	X	X	X	√	X
SS 3324 chr	X	X	√	√	X	X	√	X	√
Fisher 200 chr	√√	√	√√	√√	√√	√√	√√	√	√√

A, paper properties most closely related to Whatman; B, costs compared to Whatman; C, testing of heavy sludge (circular funnel); D, testing of heavy sludge (rectangular funnel); E, testing of light sludge (circular funnel); F, testing of light sludge (rectangular funnel); G, testing of synthetic sludge for reference purposes (circular funnel); H, testing of synthetic sludge for reference purposes (rectangular funnel); I, testing of stirred heavy sludge (circular funnel); X, not related, more expensive, or should not be used; √, related, cheaper, or may be used; √√, very related, considerably cheaper, or should be used.

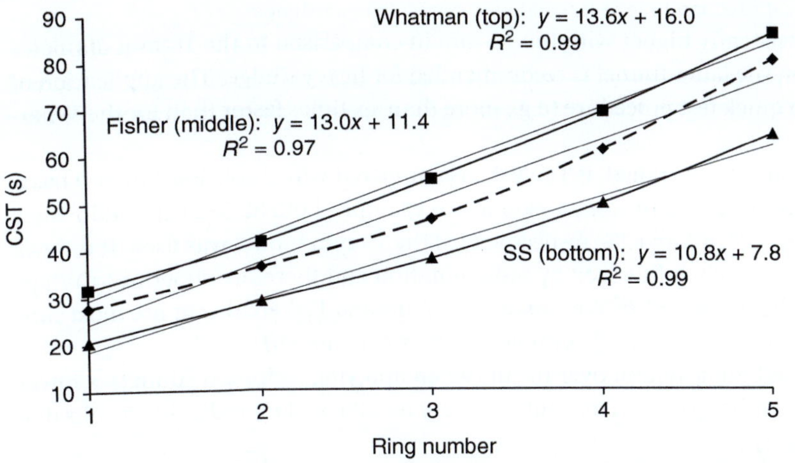

Figure 6.3 Use of the Model 319 Multi-purpose CST for primary sludge (28 days old) applying the 18-mm diameter funnel to measure multiple capillary suction times (CST): comparison of different papers (Whatman No. 17 chr (Whatman), Fisher 200 chr (Fisher) and Scheicher and Schuell 2668 chr (SS)). Graph showing the CST as a function of the ring number for different paper applications with coefficient of determination (R^2) values of close to one.

and 5.7 seconds, respectively, for Whatman No. 17 chr, for example. It follows that the errors due to the CST apparatus and procedure have a great relative impact.

6.1.3.2 Multi-radii Test and Paper Application

Figure 6.3 shows the findings for primary sludge. The slopes of the linear curves relate to the SRF. The slopes for heavy primary sludge are all of the same order of magnitude (therefore, similar SRF), and that the standard Whatman No. 17 chr paper has a similar curve if compared to the alternative Fisher 200 chr paper.

6.1.3.3 Rectangular Testing Device

Fluid flow and cake formation in a rectangular CST apparatus were investigated experimentally and theoretically (Lee and Hsu 1994a, b; Lee et al. 1994). The wet front shape is taken as an ellipse in a CST test using a cylindrical funnel and as a straight line using a rectangular funnel (Lee et al. 1994; Scholz 2005). The Darcy fluid flow field in the paper (applying a rectangular funnel) is one-dimensional (straight line). Therefore, the standard deviations of the CST measurements are therefore lower (less variable mean) in comparison to the CST data, if circular funnels are used (Tables 6.3–6.5). Particle sedimentation affects the CST in a cylindrical capillary suction apparatus. In contrast, sedimentation effects should be insignificant in rectangular CST tests (Lee et al. 1994; Scholz 2005).

6.1.3.4 Stirring of Sludge to Avoid Sedimentation

One problem with the conventional CST test is the effect of sedimentation, which cannot be neglected for a majority of flocculated slurries and sludges (Scholz 2005). Sedimentation may lead to an overestimation of the cake resistance. The application of a rectangular chamber and/or a stirrer might solve this problem. However, recent results presented in Table 6.6 show that stirring had only a relatively low impact on CST measurements in comparison to unstirred (allowing for sedimentation) primary sludge.

6.1.3.5 Primary Sludge

CST measurements are consistently higher with the 10-mm in comparison to the 18-mm diameter funnel (Table 6.3). However, the latter funnel is recommended for heavy sludge. The application of this large funnel results in a quick test procedure (e.g., more than six times faster than for the Whatman No. 17 chr paper).

The effect of sedimentation can be studied, if Table 6.3 is compared with Table 6.6. Stirring has a greater impact on Whatman No. 17 chr in comparison to, for example, both SS 3324 chr and Fisher 200 chr. The reductions were 22%, 8% and 2%, respectively, if the 18-mm funnel was used. It follows that Fisher 200 chr is not negatively influenced by sedimentation and therefore superior to Whatman No. 17 chr. However, the mean CST of Whatman No. 17 chr and Fisher 200 chr are most similar for comparable unstirred (but not stirred) samples (Tables 6.3 and 6.6).

Moreover, the ratio of standard deviation over mean (when applying a circular funnel) is lower for Fisher 200 chr than for all other papers. This indicates very reliable and reproducible results due to the high filter paper quality.

Using a rectangular funnel, CST mean measurements for Fisher 200 chr are only 9% lower in comparison to those for Whatman No. 17 chr. The data variability for Fisher 200 chr is also less than for Whatman No. 17 chr indicating a greater reliability of the findings due to a reduced impact of sedimentation (see above). Both papers are, therefore, for all intents and purposes, similar. Nevertheless, Fisher 200 chr would be an adequate and cheap replacement for the standard CST paper even for light sludges.

6.1.3.6 Surplus Activated Sludge

The SS 1107 and Fisher 200 chr papers were most like Whatman No. 17 chr in terms of their mean time using the 10-mm funnel. However, the ratio of standard deviation over mean was the lowest for Fisher 200 chr, if compared to all other papers (Table 6.4).

Using a rectangular funnel, Fisher 200 chr was the most similar paper to Whatman No. 17 chr. The ratio of standard deviation over mean for Fisher 200 chr was lower in comparison to Whatman NO. 17 chr, but higher compared to HOVO TO w/s, SS 1107 and SS2668 chr, and the same as for SS 3324 chr (Table 6.4).

6.1.3.7 Synthetic Sludge

Conducting physical and chemical tests with any natural sludge is difficult because of its constantly changing properties. To overcome this difficulty, a synthetic sludge that physically and chemically resembles natural sludge and is stable over time was created.

A simplified but affordable synthetic sludge (see above) was used as a reliable reference material to further develop the CST test. The synthetic sludge proposed by Sanin and Vesilind (1999) has a CST of 12.6 ± 0.24 (based on eight measurements). In comparison, the simplified synthetic sludge used for the purpose of this paper has a CST of only 5.2 ± 0.25 seconds (based on 10 measurements; Table 6.5). The difference is due to the absence of polystyrene latex particles that tend to clog large pores.

Experiments with synthetic sludge frequently resulted in more consistent results, but synthetic sludge (large and heavy flocs) had the tendency to settle faster than natural sludges (small and light flocs) and required therefore vigorous stirring before the start of the experiment. Nevertheless, Fisher 200 chr was most like Whatman No. 17 chr, if a 10 mm funnel was used. However, the ratio of standard deviation over mean was higher for the former, if compared for the latter paper regardless of funnel type (Table 6.5).

As expected for CST tests using a rectangular funnel, relative synthetic sludge variability (ratio of standard deviation over mean) was frequently lower than the variabilities for natural sludges

(27 out of 32 comparisons; Tables 6.3–6.5). Synthetic sludge could therefore be used as a benchmark for light sludges such as surplus activated sludge but not for heavy sludges such as primary sludge (Tables 6.3–6.5). Considering the greater need to test for dewaterability of heavy sludges, the need for synthetic sludge in practice (except for predictive modelling purposes) is questionable.

However, synthetic sludge was not a reliable reference sludge for CST tests using the rectangular funnel. Only five out of sixteen relative standard deviations improved (Tables 6.3–6.5). Consistent improvements for both natural sludges were noted only for Whatman No. 17 chr indicating the advantage of large pore size for synthetic sludge lacking fine particles (see above).

6.1.3.8 Selection of an Alternative Test Apparatus, Procedure and Paper

An alternative improved CST apparatus in comparison to the standard Model 304B CST using a circular funnel should have the following attributes: more accurate (i.e., fewer variable results (Tables 6.3–6.5); results should relate more closely to the dewaterability problem (advantage of using a rectangular funnel according to Scholz (2005)); and cheaper to construct (standard and modified CST apparatus have similar construction costs).

The application of an alternative improved CST measurement procedure should result in the following benefits: the negative effect of sedimentation (e.g., by using a stirrer) should be reduced; the procedure should be easy and time-efficient (e.g., use of a stirrer adds only approximately 30 seconds to the test); and the associated costs should be low (see above).

An alternative cheaper paper that has the potential to replace the expensive standard Whatman No. 17 chr paper should have the following attributes: similar in terms of mean CST results for various sludges and procedures, so that past and future CST values are comparable with each other (Tables 6.3–6.6); and the standard deviation for a set of measurements of the same sludge should be smaller (or at least similar) indicating higher reliability (Tables 6.3–6.6); the individual and bulk costs for the paper should be lower so that the consumable costs for the user can be reduced (Table 6.2).

Table 6.7 shows a practical overview of all results indicating which alternative paper could be used for the CST test in the future. It is obvious from the simplified graphical representation that the Fisher 200 chr paper is the best replacement for Whatman No. 17 chr even though other papers are relatively cheaper (Table 6.2). However, the Fisher 200 chr paper has very stable CST values that are like the Whatman No. 17 chr paper.

6.1.4 Conclusions and Further Research

Current key literature on the CST test was critically assessed. The CST test is easy to conduct, cost effective and accurate, if the product of solid concentration and average SRF is of interest. On the other hand, the bound water content cannot be directly evaluated from the CST.

A study of different papers showed that Fisher 200 chr can replace Whatman No. 17 chr. If the sludge is stirred to avoid sedimentation, Fisher 200 chr, and possibly also SS 2668 chr and SS3324 chr can replace the standard Whatman paper. All alternative papers are cheaper than the currently used standard Whatman No. 17 chr.

Modified CST tests were assessed. The use of a rectangular instead of a circular funnel was the most promising innovation, because CST measurements were more stable. Moreover, stirring of the sludges during the CST test resulted in a small (however mostly insignificant) reduction of the sedimentation effect for the majority of sludges.

The potential for modelling processes within the conventional CST apparatus that uses a circular funnel was assessed. However, the conventional test kit is empirical and requires therefore modification to enable dynamic modelling of filter wetting processes, which is possible, if a rectangular funnel is used. Moreover, the water content and pressure relationships can be clearly described for the modified apparatus (prototype).

Concerning further research needs, the influence of sedimentation on filtration for cake-forming sludges should be studied. It should be assessed if a constant current within the sludge chamber to avoid sedimentation might be beneficial for sludges heavier than primary sludge. Moreover, the effect of stirring (at varying speeds) on sludge floc breakage, turbulent flow impact on flocs, temperature changes and flow patterns within the paper should be assessed for both circular and rectangular funnels.

The use of a rectangular chamber and applying a constant pressure head during filtration should be researched. This would be the basis for spreadsheet (Scholz and Martin 1998) and subsequently dynamic modelling in the future.

6.2 Improved Design and Precision of the Capillary Suction Time Testing Device

6.2.1 Introduction

Dewaterability tests are used to quantify the ease of removing liquid from sludge with high moisture content (Scholz 2005). Gale and Baskerville developed the standard capillary suction time (CST) test and its corresponding apparatus in 1967, which has been widely used in various applications and disciplines. Worldwide, the CST test is one of the most used tests to determine sludge dewaterability (Lee and Hsu 1994a). Other tests, such as the specific resistance to filtration (SRF), are still being used in the dewaterability industry. The SRF test, however, is associated with several problems; it is hard to execute and time consuming and there are no special devices for measurement.

One of the major drawbacks of the standard CST test is the high variability of its test results, which undermines its value as a standard dewaterability test (Sawalha and Scholz 2007). The CST apparatus and testing methodology is, therefore, being revised, and research is in progress to address several potential problems with the test and its device such as perceived low testing precision (Sawalha and Scholz 2007; Scholz 2005). This is being done by assessing the applicability of several alternative papers (less expensive, isotropic and smaller pore diameter), altering the funnel geometries and incorporating stirring activity.

Chapter 6.2 assesses the use of adding a sealant at the bottom of the funnel to eliminate or at least minimize unwanted filtrate leakage between the edge of the funnel and the filter paper to improve testing repeatability. Accuracy of the CST test to predict SRF also was studied by introducing a funnel sealant. Synthetic sludge with consistent CST values and rheological properties was used to test alternative equipment components and methodologies. Sawalha and Scholz (2009) provided the original scientific basis of Chapter 6.2.

6.2.2 Methodology

6.2.2.1 Preparation of Synthetic Sludge

Real sludge commonly used for CST testing is difficult to keep stable because it contains live microorganisms, which can result in inconsistent properties of the test depending on the age and condition of the sample (Örmeci and Vesilind 2000). Therefore, proper testing requires use of synthetic sludge with similar CST and physical characteristics as natural sludge but with consistent and stable properties. Synthetic sludge also allows experiments to be run over a long period of time with samples that have similar properties. Synthetic sludge with different properties to simulate different sludge types can be produced by changing ingredients and corresponding proportions (Dursun et al. 2004). Different recipes for synthetic sludge with varying properties have been investigated and are summarized and assessed elsewhere (Sawalha and Scholz 2007).

The author used the method outlined in previous research to prepare synthetic sludge (Dursun et al. 2004) but using kaolin and bentonite instead of fresh yeast. The synthetic sludge used contained sodium alginate to simulate microbial extracellular polymers. Calcium ions were used as bridging cations. Calcium ions have a high affinity for extra-cellular polymer structures, and calcium is involved in bioflocculation (Sanin and Vesilind 1999). The calcium and alginate concentrations were chosen to simulate common ranges of microorganisms that normally exist in wastewater treatment sludge (Sanin and Vesilind 1996). Bentonite and kaolin were applied as surrogates to simulate the individual bacteria (Chu and Lee 2005). Cellulose fibres were used to simulate filamentous microorganisms, which are considered the backbone of sludge flocs.

Physical and chemical properties such as rheology, pH, suspended solids and supernatant turbidity were measured for the developed synthetic sludge and compared to real sludge. Viscosity is one of the most important characteristics because it can be used to describe deformation of sludge components under the influences of stresses (Barnes et al. 1993). Because of the non-Newtonian nature of sludge flow, viscosity depends on the shear rate gradient under certain pressure and temperature conditions (Chen et al. 2005; Seyssiecq et al. 2003). This means that shear stress is not linearly related to shear rate.

Rheological properties such as viscosity are important in sludge management operations such as estimating the design parameters of transportation and pumping schemes. Rheological parameters have been used to determine sludge network strength to evaluate sludge conditioning and improve dewatering (Abu-Orf and Örmeci 2005; Abu-Orf et al. 2003). In addition, rheological properties can be used to estimate the quality of the flocculated sludge (Tixier et al. 2003).

Synthetic sludge was developed using 5 g of sodium alginate dissolved in 1000 mL of distilled water to which was cellulose fibres of 6 g were added. This was then mixed to create a suspension. Next, 23.3 g of kaolin and 11.7 g of bentonite were mixed and added slowly to the suspension by gently stirring the mixture for 120 minutes at 10 rpm to allow alginate to incorporate into particle surfaces (Dursun et al. 2004; Sawalha and Scholz 2007). Then 250 mL of 146 $g\,L^{-1}$ potassium chloride solution was added and stirred to suspension at 10 rpm for two minutes. Thereafter, 250 mL of 5 $g\,L^{-1}$ calcium chloride was added and stirred at 10 rpm for three minutes. The synthetic sludge was diluted with distilled water at a ratio of 3 : 1 to reduce salt concentration. After 24 hours of settling, the synthetic sludge was returned to its original volume to reduce salinity, which is known to affect SRF and, consequently, CST (Lai et al. 2004). The synthetic sludge was diluted with distilled water at a ratio of 1–9 because it was deemed to be too concentrated.

6.2.2.2 Physical and Chemical Properties of Synthetic Sludge

Physical and chemical properties were determined for the synthetic and real sludge samples. These properties included total suspended solids, pH measured by HANNA HI 991300 and turbidity of the supernatant, which was measured using HACH 2100N after 30 minutes of settling. The rheological properties were measured using an AR 2000 rheometer with concentric cylinder geometry. The concentric cylinders had a 500-μm double gap rotor with 20.00, 20.38 and 21.96 mm stator outer, rotor inner and rotor outer radii, respectively.

A multi-radii CST apparatus (model 319) provided by Triton Electronics Ltd. (Great Dunmow, England, United Kingdom) was used for Chapter 6.2. This device can produce five CST values (CST1, CST2, CST3, CST4 and CST5) instead of one, as with the standard single radius CST device. The CST value from the standard device is equivalent to the second CST value of the

Table 6.8 Distances between electrodes and the funnel centre.

Ring number	Distance (cm)
1	1.86
2[a]	2.23
3	2.54
4	2.82
5	3.09

[a] Identical to the standard single radius capillary suction time test device.

multi-radii device, which was used throughout this study. Findings can be used to predict the SRF. The device consists of a funnel resting on filter paper (Whatman No. 17) fitted between two plates with electrode sensors placed at standard intervals across the top plate and connected to a timer (Lee and Hsu 1993; Sawalha and Scholz 2007). The electrodes are used to measure the travel times of the waterfront. Table 6.8 summarizes the distances between the electrodes and the centre of the funnel.

The CST test is executed by pouring a suspension of a homogenous sample into a funnel until the liquid is level with its top rim. The capillary suction pressure is generated within the porous filter paper, forcing the filtrate to be sucked from the suspension at the point of contact with the paper. The filtrate moves across the filter paper and timing starts when the wet front of the filtrate reaches the starting electrode placed at 1.59 cm from the funnel centre. The time required for the wet front to reach each of the five electrodes is automatically recorded by the device producing five CST recordings. The tests typically were repeated ten times for each test combination (i.e., two funnel types; two types of sludge; five filter papers; with or without sealant), resulting in 402 tests being carried out in total.

6.2.2.3 Funnel Sealant

Several types of elastic materials were tested as funnel sealants to improve product design and precision of the CST test. The ideal sealant should have several properties: (i) high elasticity to close the gap between the metallic funnel edge and the filter paper without having to apply extra pressure to avoid blocking the pores of the filter paper at the point of impact; (ii) insignificant overall thickness and material to avoid interference with standard testing conditions and to maintain funnel volume and geometry; (iii) high ease of fitting (and removing) material to the funnel and filter paper to avoid changing the funnel properties; (iv) low price and ease of availability of material to avoid increasing the consumable costs of the test and (v) high effectiveness in sealing to improve test precision.

Several standard and unconventional types of sealing materials were tested in a small pilot study. For example, rubber material which is made of natural rubber latex was wrapped around the bottom edge of the funnel. The expectation was that the elastic rubber would prevent or at least reduce the leakage of water between the funnel and the filter paper. The natural rubber latex satisfied the above criteria and produced significant improvements in the test.

O-rings (ELE International, Bedfordshire, England, United Kingdom) were tested unsuccessfully, because it was not possible to fit them to the bottom of the funnel, which produced inconsistent results. Elastic silicon (Homebase, Stafford, England, United Kingdom) was also tested, but was not able to be applied uniformly to the bottom of the funnel. Nevertheless, silicon showed some promising preliminary results. Finally, adhesive bonding material such as blue tack (Bostik, Stafford, England, United Kingdom) was tried, resulting in some improvements in CST test repeatability.

6.2.2.4 Experimental Design and Testing

The material used in this sub-chapter was natural rubber latex, because it showed the most promising results in preliminary studies, was practical to fit to the funnel, and could be used repeatedly over the course of the research. Rubber material from different large supermarket (e.g., Tesco Store [Cheshunt, England, United Kingdom] and Sainsbury's Supermarket [London, England, United Kingdom]) was tested to assess, if there were any differences in applicability (i.e., simple party balloon). Supplementary experiments showed that the rubber materials performed similarly in the context of the CST test. The material used in this research was obtained from Sainsbury's Supermarket (own product brand; product reference numbers 747/405 and #81808; internal supplier code: UO295; bar code 0107-9852).

The rubber latex material was attached to the bottom edge of both a circular (18 mm diameter) and a rectangular funnel (18 mm long, 18 mm wide and 20 mm deep). Five different filter papers were tested: Whatman No. 17 (Whatman Plc., Brentford, England, United Kingdom), Fisher 200 chr (Fisher Scientific Ltd., Loughborough, England, United Kingdom), FN30, FN8 and BF4 (Munktell and Filtrak GmbH, Bärenstein, Germany). The tests were carried out using both synthetic sludge and primary sludge obtained from the Seafield wastewater treatment plant located in Edinburgh, Scotland, United Kingdom.

A similar rectangular funnel was tested in a previous study alongside 12 alternative filter papers in which the rectangular funnel showed a considerable improvement in the test repeatability, as shown previously by Lee (1994) and Lee and Hsu (1994a) according to Sawalha and Scholz (2007). The sensor positions were not altered, and the wetting front moved to a right angle to the paper-grain lines. Moreover, several alternative filter papers were identified as potentially superior substitutes for the currently used Whatman No. 17 paper and were included in Chapter 6.2.

A graduated pipette fixed at the top of the circular funnel placed over the filter paper was used to quantify the amount of leakage (Figure 6.4). Approximately 5 mL of synthetic sludge was poured

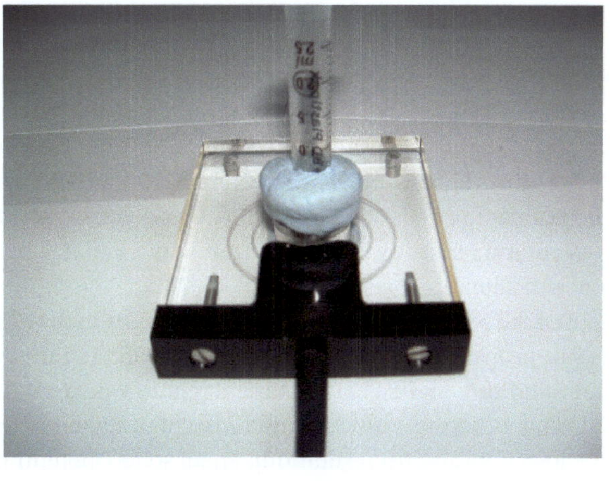

Figure 6.4 Graduated pipette, which was fixed to the top of the circular funnel (18 mm) of the capillary suction time test equipment. Picture showing a pipette that is fixed on the funnel with a lot of sealant to avoid any leakage.

into the funnel, and the volume that disappeared after one minute was recorded. This process was repeated ten times for each of the five filter papers with and without using a sealant. The difference in the volumes between the funnel with and without sealant is considered to roughly represent the filtrate escaping from the gap between the funnel and the filter paper.

The SRF can be predicted using CST values generated by the multi-radii CST device. This is done by undertaking a regression analysis based on the five CST values. The slope of the regression line represents the predicted SRF value. The CST data with and without the funnel sealant were studied in relation to SRF using linear regression.

The variability of the CST test with and without the sealant was assessed based on the coefficients of variation. The means of the CST data were compared for the presence or absence of sealant using an analysis of variance. All analyses were undertaken using the Statistical Analysis System (SAS) Program 9.1 (SAS Institute Inc., Cary, New York, USA).

6.2.3 Results and Discussion

6.2.3.1 Synthetic Sludge Properties

Physical and chemical properties such as the total suspended solids content, pH, supernatant turbidity, rheological properties and dewaterability were measured for both the synthetic and activated sludges. Viscosity has been shown to have a main influence on sludge dewaterability according to Forster (2002) and Abu-Orf and Örmeci (2005). The primary sludge contained 8.2 g L^{-1} total suspended solids and had a pH of 5.56 and a turbidity of 201 NTU. The characteristics of synthetic and activated sludges were compared. The main physical and chemical properties of both sludge types were similar.

Values after 30 minutes of settling were 2.3 g L^{-1} total suspended solids, pH 6.1 and 16.5 NTU supernatant turbidity for synthetic sludge, and 2.24 g L^{-1}, 6.34 and 10.8 NTU, respectively, for activated sludge. The supernatant turbidity of the synthetic sludge was slightly higher than that for the activated sludge. The rheological properties at different shear rates at 10ºC for both synthetic and activated sludge types are shown in Figure 6.5.

Viscosity decreased as shear rates increased for both types of sludge. This is because of the expected break-up of the flocs, which usually occurs at higher shear rates (Örmeci and Vesilind 2000). Synthetic and real sludge are known to be thixotropic, which means that their rheological properties are time-dependent and can change during a rheological testing period (Seyssiecq et al. 2003). The change in viscosity becomes less. Viscosity is expected to become constant at higher shear rates. When approaching linearity at higher shear rates, it is called limit viscosity, which is the viscosity of sludge that corresponds to the maximum dispersion of flocs under influence of the high shear rate (Tixier et al. 2003).

Figure 6.5 shows that the viscosity values of the activated sludge were 15.37000, 1.56500, 0.11840 and 0.01618 pa × s at shear rates of 0.01, 0.10, 1.00 and 10.00 L s^{-1}, respectively. For synthetic sludge, the values were 13.38000, 1.62100, 0.17230 and 0.02775 pa × s at the same shear rates. The temperature for all experiments was constant at 10°C. Findings indicate that viscosity values of synthetic and activated sludge are very similar when they have similar total suspended solids concentrations.

Total suspended solids content has been recognized as a major factor influencing sludge rheological properties (Dentel and Abu-Orf 1995; Tixier et al. 2003). Viscosity is positively correlated with solid content in a mathematical relationship that can be described as predominantly exponential. This is because solid content affects particle-to-particle interaction, which influences sludge rheological properties. Additionally, as solid concentration increases, sludge suspension tends to

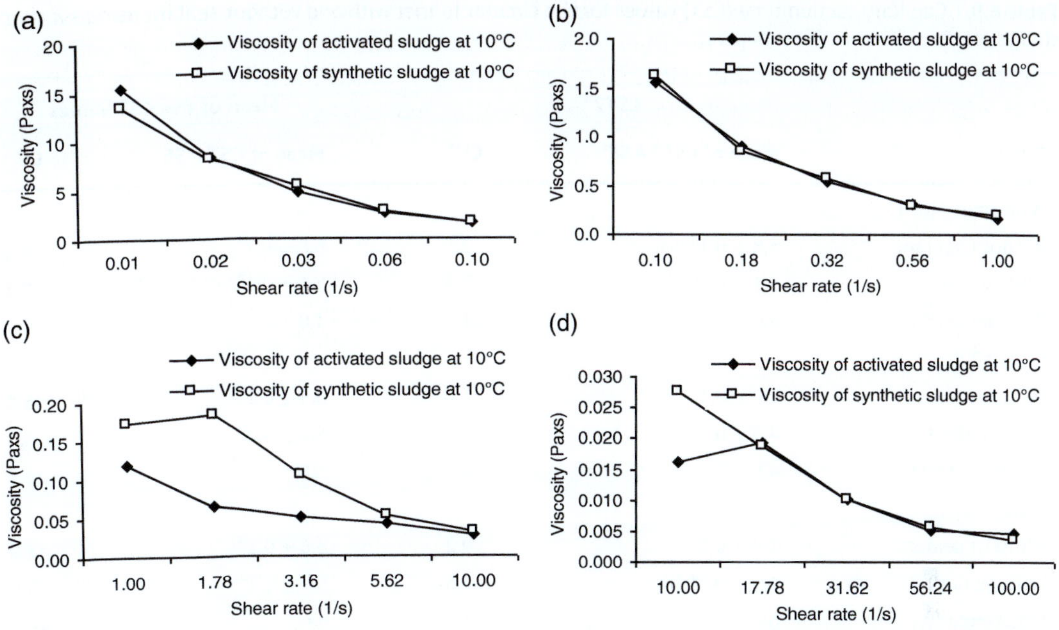

Figure 6.5 Rheograms for different shear rates (1/s) between (a) 0.01–0.1, (b) 0.1–1.0, (c) 1.0–10 and (d) 10–100 for activated and synthetic sludges, respectively, at 10°C. For figure parts (a), (b) and (d), both sludges have very similar behaviours. In contrast, the viscosity of the synthetic sludge is higher than that for activated sludge regarding figure part (c).

develop non-Newtonian characteristics (Forster 2002; Hasar et al. 2004; Tixier et al. 2003). As a result, synthetic activated sludge was developed to have similar solid content in comparison to real activated sludge.

Other parameters such as flocculation state, particle size, sludge composition and bound-water content also can affect sludge viscosity. Viscosity is also sensitive to filamentous bacteria, which is associated with a large hysteresis area parameter even at low suspended solid concentrations. The presence of filamentous bacteria can increase viscosity. This can occur even at low solid contents (Tixier et al. 2003). Therefore, cellulose fibres were used in the synthetic sludge to simulate filamentous bacteria common in real sludge.

6.2.3.2 Effect of Sealant on Test Accuracy

Tables 6.9–6.12 compare the means and corresponding variations (expressed as standard errors and coefficients of variation) in CST results based on tests with and without applying a funnel sealant. The tests were conducted using primary and synthetic sludge types with circular and rectangular funnels for five different filter papers (Whatman No. 17, Fisher 200 chr, FN30, FN8 and BF3). Results showed that there was a substantial reduction in test variability, as determined by the coefficients of variation values, when the funnel sealant was used for most of the filter papers tested for both types of funnels (circular and rectangular). These findings were more apparent for the synthetic sludge.

The test showed an increase in mean CST values (for most cases, statistically significant differences were noted) when using the sealant compared to tests without using the sealant. Results shown in Tables 6.9–6.12 are for CST2 (which represents the standard CST test) and the mean

Table 6.9 Capillary suction time (CST) values for the circular funnel with and without sealant using synthetic sludge for five different filter papers.

	CST2[b]		Mean of five CST values	
Paper	**Mean of CST2 ± SE[c]**	**CV[d]**	**Mean of CST ± SE**	**CV**
Whatman No. 17				
Without sealant	5.8 ± 0.1^a	3.5	6.8 ± 0.1^a	3.4
With sealant	6.1 ± 0.1^a	2.4	7.1 ± 0.1^a	2.4
Difference (%)	5.0	31	4.0	29
Fisher 200 chr				
Without sealant	3.0 ± 0.1^a	5.4	3.4 ± 0.1^a	5.4
With sealant	4.3 ± 0.1^a	4.6	5.1 ± 0.1^a	3.0
Difference (%)	43	16	50	44
FN30				
Without sealant	2.9 ± 0.1^a	9.2	3.6 ± 0.1^a	5.9
With sealant	4.9 ± 0.1^a	4.6	5.9 ± 0.1^a	3.0
Difference (%)	69	51	64	49
FN8				
Without sealant	5.0 ± 0.2^a	10.8	5.9 ± 0.2^a	8.7
With sealant	7.4 ± 0.2^a	9.8	8.4 ± 0.2^a	5.6
Difference (%)	48	9	42	36
BF3				
Without sealant	7.3 ± 0.1	6.8	8.3 ± 0.1	5.4
With sealant	7.0 ± 0.1	4.5	8.1 ± 0.1	4.7
Difference (%)	−4.0	35	−2.0	13

[a] Different superscripts for the same filter paper and funnel are significantly different ($p < 0.05$).
[b] CST2 is the second CST value measured by the second electrode sensor of the multi-radii CST device.
[c] SE, standard error
[d] CV, coefficient of variability (100 × standard deviation/mean).

of five CST values. The difference between means as a percentage was calculated as follows: 100 × [(CST with sealant) – (CST without sealant)]/(CST without sealant). Coefficients of variation (CV) differences (%) were calculated as follows: 100 × [(CV with sealant) – (CV without sealant)]/(CV without sealant).

Concerning the synthetic sludge types tested with the circular funnel, the coefficients of variation of CST2 decreased between 9% and 51% for different filter papers (Table 6.9). Similarly, the coefficients of variation of the mean of all five CST values decreased between 13% and 49% for the same type of sludge and using the same funnel type for the five different filter papers. Furthermore, a statistically significant increase between 5% and 69%, and between 4% and 64% for the means of CST2 and the five CST values, respectively, were recorded, when using a sealant to test synthetic sludge dewaterability using a circular funnel for four out of five tested filter papers.

Similar but even more consistent results were obtained when the rectangular funnel was used for CST tests using synthetic sludge (Table 6.10). When the sealant was applied, the reductions in the coefficients of variation values ranged between 12% and 57% for CST2 and between 16% and 47% for

Table 6.10 Capillary suction time (CST) values for the rectangular funnel with and without sealant using synthetic sludge for five different filter papers.

	CST2[b]		Mean of five CST values	
Paper	**Mean of CST2 ± SE[c]**	**CV[d]**	**Mean of CST ± SE**	**CV**
Whatman No. 17				
Without sealant	4.8 ± 0.1^a	8.1	5.3 ± 0.2^a	12.8
With sealant	5.7 ± 0.1^a	6.5	6.6 ± 0.1^a	7.1
Difference (%)	19	20	25	45
Fisher 200 chr				
Without sealant	3.5 ± 0.1	7.9	4.0 ± 0.1	6.8
With sealant	3.5 ± 0.1	6.7	4.1 ± 0.1	5.7
Difference (%)	0.0	15	3.0	16
FN30				
Without sealant	3.3 ± 0.1	7.6	3.7 ± 0.1	5.9
With sealant	3.2 ± 0.1	6.6	3.7 ± 0.1	4.8
Difference (%)	−3.0	12	0.0	19
FN8				
Without sealant	4.7 ± 0.1^a	9.1	5.3 ± 0.1^a	6.4
With sealant	6.1 ± 0.1^a	3.9	7.0 ± 0.1^a	3.4
Difference (%)	30	57	32	47
BF3				
Without sealant	6.7 ± 0.1	6.4	7.4 ± 0.1^a	4.7
With sealant	6.8 ± 0.1	5.1	7.9 ± 0.1^a	3.8
Difference (%)	1.5	20	6.8	19

[a] Different superscripts for the same filter paper and funnel are significantly different ($p < 0.05$).
[b] CST2 is the second CST value measured by the second electrode sensor of the multi-radii CST device.
[c] SE, standard error.
[d] CV, coefficient of variability (100 × standard deviation/mean).

the means of the five CST values for the five tested filter papers. The changes in the means for synthetic sludge tested with the rectangular funnel for the five filter papers were between −3% and 30% and between 0% and 32% when the sealant was used for CST2 and the mean of the five CST values, respectively. Negative values indicate that there was a decrease in the CST value when the sealant was used. However, this happened only once when the FN30 filter paper was tested.

When primary sludge was used for testing with the circular funnel, there were reductions in coefficients of variation for CST2 for all filter papers except for BF3 when the sealant was used (Table 6.11). Reductions in the coefficients of variation for CST2 regarding the papers Whatman No. 17, Fisher 200 chr, FN30 and FN8 were 13%, 34%, 52% and 11%, respectively. In comparison to tests undertaken without sealant, there were increases and decreases in the means for CST2 when the sealant was used for the five filter papers. These differences were 28%, −27%, −12%, 11% and −76 for Whatman No. 17, Fisher 200 chr, FN30, FN8 and BF3, respectively. These differences were significant ($p < 0.05$) for the papers Whatman No. 17, Fisher 200 chr and BF3. The same trend was also observed for the mean of all five CST values. These CST values were carried out using

Table 6.11 Capillary suction time (CST) values for the circular funnel with and without sealant using primary sludge for five different filter papers.

	CST2[b]		Mean of five CST values	
Paper	**Mean of CST2 ± SE[c]**	**CV[d]**	**Mean of CST ± SE**	**CV**
Whatman No. 17				
Without sealant	74.7 ± 4.5[a]	14.9	89.8 ± 4.8[a]	15.5
With sealant	95.7 ± 3.6[a]	12.9	114.1 ± 3.8[a]	10.5
Difference (%)	28	13	27	32
Fisher 200 chr				
Without sealant	17.4 ± 0.5[a]	16.1	20.8 ± 0.7[a]	18.6
With sealant	12.7 ± 0.7[a]	10.7	15.4 ± 1.0[a]	13.3
Difference (%)	−27	34	−26	28
FN30				
Without sealant	13.9 ± 0.8	23.1	17.0 ± 0.9[a]	23.5
With sealant	12.3 ± 0.7	10.9	14.2 ± 0.9[a]	11.1
Difference (%)	−12	52	−16	53
FN8				
Without sealant	40.1 ± 2.2	15.9	51.2 ± 2.1	9.0
With sealant	44.4 ± 2.4	14.1	55.8 ± 2.3	12.7
Difference (%)	11	11	9.0	−41
BF3				
Without sealant	97.9 ± 7.0[a]	24.9	112.6 ± 8.3[a]	25.1
With sealant	23.1 ± 5.8[a]	26.4	28.7 ± 6.7[a]	30.1
Difference (%)	−76	−6.0	−74	−20

[a] Different superscripts for the same filter paper and funnel are significantly different ($p < 0.05$).
[b] CST2 is the second CST value measured by the second electrode sensor of the multi-radii CST device.
[c] SE, standard error.
[d] CV, coefficient of variability (100 × standard deviation/mean).

real primary sludge, which can have inconsistent properties over the period of testing extending to several days.

Consistent reductions in the CV for CST2 values with respect to the five filter papers were observed when the rectangular funnel with the sealant was used for primary sludge (Table 6.12). The reductions were 51%, 63%, 50%, 28% and 26% for Whatman No. 17, Fisher 200 chr, FN30, FN8 and BF3, respectively. There were also differences in the mean CST2 values when the sealant was used. These differences were significant ($p < 0.05$) for all filter papers except for BF3. The mean CST2 values increased when the sealant was applied for Whatman No.17 (55%), Fisher 200 chr (36%) and FN8 (55%), and decreased for FN30 (33%) and BF3 (5%). The results were similar for the means of the five CST values. These reductions were 69%, 59%, 46%, 12% and 30% for coefficients of variation values for Whatman No. 17, Fisher 200 chr, FN30, FN8 and BF3, respectively.

There were also differences in the means of the average of the five CST values with and without using the sealant. These differences indicated significant increases in the CST values for Whatman No. 17, Fisher 200 chr and FN8. In comparison, decreases were observed for FN30.

Table 6.12 Capillary suction time (CST) values for the rectangular funnel with and without sealant using primary sludge for five different filter papers.

Paper	CST2[b]		Mean of five CST values	
	Mean of CST2 ± SE[c]	CV[d]	Mean of CST ± SE	CV
Whatman No. 17				
Without sealant	38.4 ± 2.6^a	22.0	46.1 ± 3.1^a	25.1
With sealant	59.6 ± 2.3^a	10.8	75.9 ± 2.8^a	7.8
Difference (%)	55	51	65	69
Fisher 200 chr				
Without sealant	9.2 ± 0.6^a	26.1	11.1 ± 0.7^a	21.5
With sealant	12.5 ± 0.5^a	9.7	15.8 ± 0.6^a	8.9
Difference (%)	36	63	42	59
FN30				
Without sealant	7.4 ± 0.4^a	25.2	8.5 ± 0.5^a	25.7
With sealant	4.9 ± 0.5^a	12.5	5.7 ± 0.5^a	14.0
Difference (%)	−33	50	−33	46
FN8				
Without sealant	31.2 ± 4.7^a	40.8	36.8 ± 5.7^a	37.7
With sealant	48.3 ± 5.5^a	29.6	56.9 ± 6.6^a	33
Difference (%)	55	28	55	12
BF3				
Without sealant	5.9 ± 0.4	24.4	7.0 ± 0.5	25.8
With sealant	5.6 ± 0.4	18.0	6.5 ± 0.5	18
Difference (%)	−5	26	−7	30

[a] Different superscripts for the same filter paper and funnel are significantly different ($p < 0.05$).
[b] CST2 is the second CST value measured by the second electrode sensor of the multi-radii CST device.
[c] SE, standard error.
[d] CV, coefficient of variability (100 × standard deviation/mean).

Findings indicate that the CST test precision can be improved significantly ($p < 0.05$) when incorporating the funnel sealant into the new product design. The improvement seems to be because of the prevention or reduction of unwanted sample liquid escaping between the edge of the funnel and the filter paper. Evidence can be found by both observing the increase in the mean CST values and the reduction in the amount of sample liquid disappearance per unit time when the funnel sealant was used.

It is important to improve CST test precision, because this is one of the most widely used tests for determining the ease of dewatering sludge. The CST test can be used to calculate the dose of costly polymer conditioners needed to achieve optimal sludge dewaterability. Some polymer costs may compose as much as 50% of the overall operational costs (Abu-Orf and Dentel 1999). Improving CST test precision can, therefore, improve test credibility and its use and adoption in the wastewater treatment industry.

Several researchers have studied the CST test to reduce its variability. They indicated that the relatively high variability is mostly associated with filter paper, because it is not uniform, or anisotropic, which means there is greater capillary suction pressure along the grains than across them (Lee 1994; Sawalha and Scholz 2007; Tiller and Wenping 2001; Tiller et al. 1990). The use of a rectangular funnel instead of the recommended circular funnel was suggested to achieve unidirectional flow of the filtrate within the filter paper (Lee and Hsu 1994a; Leu 1981).

In a previous study, researchers tested various filter papers and rectangular funnel geometry and found little improvement in CST test precision for most sludges tested (Sawalha and Scholz 2007). In this study, however, improvements in test repeatability and the reduction in test variability were clear. The author, therefore, expects that commercial adoption of the funnel sealant will greatly enhance CST results precision under field conditions.

6.2.3.3 Quantifying the Leakage

The volumes of the synthetic sludge that disappeared within one minute were recorded for the five different filter papers with and without using a sealant (Table 6.13). For all filter papers, the volume of sludge samples that disappeared was higher for the synthetic sludge without than with using the sealant. The reductions in the amount of liquid that disappeared when the sealant was used ranged from 0.1 to 0.4 mL. These reductions of 25–67% were significant ($p < 0.05$) and prove that there is significant leakage occurring between the funnel and the filter paper, leading to unwanted increases in the variability of the CST test data. Using the sealant prevented, or at least reduced leakages, which improved and minimized variability of the CST tests (see Tables 6.9–6.12).

Because viscosity affects the movement of filtrate through porous material, the amount of leakage of filtrate between paper and funnel is expected to be affected by the properties of the tested suspension in this study. Results concerning the quantification of leakage are only relevant for sludge with similar properties such as viscosity.

The CST test assumes that all filtrate is constrained within the paper. Preventing leakage between the funnel and the paper is, therefore, essential to reduce data variability. The leakage that occurs during testing without a sealant is random in its nature. Besides other variables, it also depends on how well the funnel is attached to the device (e.g., flatness and uniformity of the surface of the paper). The hypothesis is that when leakage is prevented, the CST becomes more consistent because of control over random leakage. Preventing leakage is expected to increase the mean CST value, which was observed in the majority of tests in this study.

Table 6.13 The mean volume of the leakage after one minute with and without using the funnel sealant for different filter papers.

	Sealant		
Filter paper	**With**	**Without**	**Volume of the leakage (mL)**
Whatman No. 17	0.2^a	0.4^a	0.2
Fisher 200 chr	0.3^a	0.4^a	0.1
FN8	0.3^a	0.4^a	0.1
FN30	0.6^a	0.9^a	0.3
BF3	0.2^a	0.6^a	0.4

[a] Different superscripts for the same filter paper and the volume with and without the sealant are significantly different ($p < 0.05$).

6.2.3.4 Specific Resistance to Filtration Accuracy

When analysing the CST data with linear regression, the corresponding slope of the predicted line of CST values regressed over the five ring numbers can be used to represent the SRF (Sawalha and Scholz 2007; Scholz 2006). Meeten and Lebreton (1993) previously have proven this. Lee and Hsu (1994b) used the mean SRF of a filter cake to describe filterability. The desorptivity can be related to the mean SRF (Meeten and Smeulders 1995).

The CST test and SRF are highly correlated and show similar behaviour (Buyukkamaci 2004; Pollice et al. 2007). The SRF typically is measured using the Buchner funnel but can be predicted using regression analyses based on the CST values generated by the multi-radii CST test device. The accuracy of the prediction of SRF from the linear regression equation of the CST values can be evaluated using the coefficient of determination (R^2) and the standard error of prediction. The accuracy of SRF prediction, if a funnel sealant is used (Figure 6.6a), improved compared to the CST test without the sealant (Figure 6.6b). The R^2 value was higher (0.73 versus 0.56), and the standard error was smaller (2.3 versus 3.6) when using the funnel sealant compared to experiments undertaken without the sealant. Data are only shown for the Fisher filter paper using the circular funnel and primary sludge. Similar results also were obtained for the other four filter papers and for the rectangular funnel using synthetic sludge.

With the standard CST test, filtrate leakage was noted because colouration of the wetted filter paper was observed (Sawalha and Scholz 2007). The leakage of the filtrate across the top of the filter paper is uncontrolled and can vary subject to individual test conditions. Therefore, data variability randomly affects the test results.

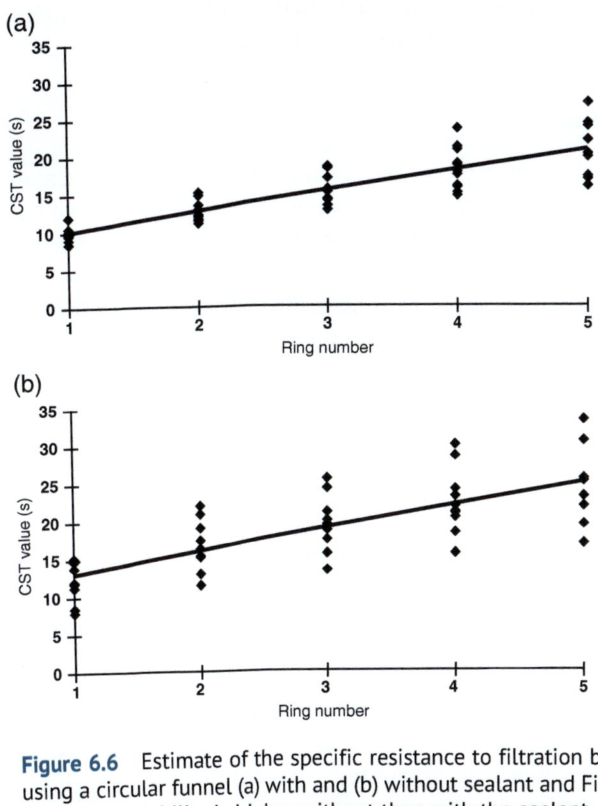

Figure 6.6 Estimate of the specific resistance to filtration based on the capillary suction time (CST) test using a circular funnel (a) with and (b) without sealant and Fisher 200 chr filter paper for primary sludge. The CST data variability is higher without than with the sealant.

Findings suggest that the funnel sealant was effective in significantly reducing or even preventing unwanted filtrate leakage. This resulted in an improvement of test repeatability. If the sealant prevents leakage, the whole volume of fluid is constrained to that within the pores of the filter paper. This is the cause of the increase of the mean CST values observed for most tests.

Several other types of sealants (i.e., latex membranes, silicon sealants and O-rings) were tested, but the results were unsuccessful. Large amounts of filtrate leakage between the funnel and the sealants were observed when these materials were tested. Moreover, it is important that the latex membrane is sealed to the funnel itself. The same holds true for silicon.

The CST test measures the time for a specific volume of permeable filter paper to be filled with a specific volume of the filtrate (water moving through the filter paper). The amount of the filtrate produced during the test is supposed to be fully contained within the pores of the filter paper without any movement across the top surface of the filter paper (Meeten and Smeulders, 1995). The funnel used with the standard CST apparatus, however, is made from metal, and leakage of filtrate can be seen coming from between the edge of the metal funnel and the filter paper. Leakage of the filtrate across the surface of the filter paper is expected to interfere with test results because it is uncontrolled and will vary during the test.

It has been suggested that an elastic sealant should be fixed to the edge of the funnel to eliminate or at least significantly reduce unwanted leakage and subsequently to improve testing precision and repeatability. Preventing unwanted leakage through sealing the gap between the funnel and the filter paper is expected to reduce the variability of the CST and to improve the precision, reliability and confidence in the CST test results and, therefore, the product and its methodology as a whole. However, as the amount of filtrate will be constrained to that within the pores of the filter paper, the mean CST test time is expected to increase with the sealant compared to the standard test. A slightly longer CST test, which is inconvenient for the user, is the drawback in return for better test results.

6.2.4 Conclusions and Recommendations

The precision of the CST test can be greatly improved by the introduction of a rubber sealant attached to the funnel. The variability of the CST test data as measured by the coefficients of variation and the standard deviation was much less with the use of a funnel sealant for both the standard circular and the rectangular funnels.

The SRF can be predicted from the linear regression of the multi-radii CST test values. The prediction is more accurate if the funnel sealant is used. More research work is planned to relate the CST to the SRF at different environmental boundary conditions such as temperature. Furthermore, additional work to standardize the CST test for different conditions (solid content, temperature and sludge age) will be carried out using the funnel sealant.

6.3 Sludge Floc Size and Water Composition Impact on Dewaterability

6.3.1 Introduction

The water content of sludge (also called biosolids) is approximately 95%. The excess water needs to be reduced prior to disposal to reduce material volumes. Consequently, almost half of the treatment costs can be committed to dewatering and disposal (Chen et al. 2010). Sludge dewatering has been considered as one of the most expensive, but one of the least understood elements of the water and wastewater treatment processes (Bruus et al. 1992). Since the early 1970s, the key test to measure sludge dewaterability is the capillary suction time (CST) test. The CST testing apparatus provides a simple, flexible, rapid and inexpensive method to measure sludge dewaterability, which particularly supports the design of filter presses in industry (Bruus et al. 1992; Chen et al. 2010; Lee and Wang 2000).

In the United Kingdom (UK), large volumes (>11 m^3 s^{-1}) of raw water (usually reservoir water) and domestic wastewater are processed every day in water and wastewater treatment plants (United Utilities 2024). The treatment of raw water will produce tap water, while the treatment of domestic wastewater will separate contaminants from water, and subsequently enhance the quality of the resultant discharge. In water treatment, coagulation processes, which are followed by further treatment steps for liquid and solid separation, are the most used processes to remove particles from water (Slavik et al. 2012).

Coagulation increases the tendency of particles to attach to one another forming larger-size agglomerates. Particle size is therefore an important factor in the coagulation process (Zhan et al. 2011), where it has an important role in influencing the settling process following coagulation. The larger the agglomerate (or floc) particle size, the easier it can be removed from water (Besra et al. 2000).

In relation to sludge dewaterability, sludge characteristics including particle size have an important role in sludge dewaterability assessment tests (Razi and Molla 2007). Sludge dewaterability is very dependent on sludge floc characterisation; in particular, the particle-size distribution and the presence of small particles (Jin et al. 2004), which are determined by coagulation mechanisms.

Furthermore, the composition of sludge is very dependent on the treatment process and wastewater composition (Gale and Baskerville 1970; Pallier et al. 2010; Wang et al. 2009; Zhang et al. 2004). The quality of treated water and sludge is dependent on the quality of the source of water (American Water Works Association 1999; Jin et al. 2004; Zhang et al. 2012). Moreover, the efficiency of the dewatering process is highly dependent on the nature of the sludge (Jin et al. 2004).

The quality of raw water (source of drinking water) will determine the selection of the treatment process in a water treatment plant. The quality of the drinking water sources is dependent on a range of factors such as natural geology, land use and pollution (Gray 2005). The raw water content is often dominated by its inorganic matter such as clay particles (Lee and Wang 2000). In contrast, domestic wastewater is relatively high in organic content such as human faeces (Metcalf and Eddy 2003; Sanin et al. 2011). Furthermore, organic sludge contains fats, fibres, protein and sugars (Baudez et al. 2007).

To obtain a sample with consistent water quality characteristics for laboratory tests, synthetic raw and domestic water should be used (at least for benchmarking purposes) in all experiments. The use of synthetic waters minimises differences in experimental conditions, which are particularly important for most laboratory-scale tests (Baudez et al. 2007). The properties of 'natural' or 'real' water samples can often be highly variable and very dynamic. These properties depend on the treatment plant operational conditions and may change over time during transport, handling and storage (Baudez et al. 2007). Although water and sludge compositions impact on sludge characteristics, there has been no detailed investigation of the effect of different water compositions on sludge floc size and on the dewaterability of the associated sludge. Chapter 6.3 seeks to assess these parameters.

Using ferric chloride as a coagulant, a range of experiments have been conducted at the bench scale to assess the impact of different water compositions on floc size and sludge dewaterability using a range of mixer shapes and applying different rapid mixing velocities to make the findings representative for various industrial scenarios. The main objectives of Chapter 6.3 are to assess the impact of different (i) water compositions (synthetic raw water and synthetic domestic wastewater) on the relationship between floc sizes and sludge dewaterability; (ii) mixer shapes (radial, axial and magnetic stirrers) on the link between sludge floc sizes and sludge dewaterability and (iii) water compositions (synthetic raw water, synthetic domestic wastewater and natural domestic wastewater) on floc size distributions.

The findings are of relevance for practitioners involved in sludge management and treatment associated particularly with water, wastewater, biotechnological and chemical industrial applications. Fitria et al. (2014) provided the scientific basis for Chapter 6.3.

6.3.2 Materials and Methods

6.3.2.1 Dewaterability Tests

Coagulation experiments were undertaken to explore the influence of each variable on the CST. This section provides an overview of the dewaterability tests, the coagulants and mixers as well as assesses how samples were prepared and outlines both the coagulation and floc size measurements.

The key measure of efficiency with respect to the sludge dewatering process is traditionally the CST test value (Sawalha and Scholz 2009; Scholz 2005). The capillary suction pressure generated by a chromatographic paper is used to suck liquid from the sludge sample. The rate at which water permeates through the paper varies depending on the sludge characteristics and the filterability of the sludge cake formed on top of the paper. The CST is obtained from two electrodes placed at a specific interval from the funnel containing the sample. The time taken for the advancing liquid front to pass between these two electrodes is the CST value. A relatively low CST value indicates good sludge dewaterability (Sanin et al. 2011).

The specific resistance to filtration (SRF) test, which is another common but more time-consuming and inflexible dewaterability test, was performed for some experiments to confirm CST findings (Metcalf and Eddy 2003). The test was determined by pouring a flocculated sample into a Büchner funnel operated at 80 kPa negative vacuum pressure with a Whatman number 1 filter paper (Whatman International, Maidstore, Rudolf Place, London, England, UK).

6.3.2.2 Coagulants and Mixers

The coagulant ferric chloride (Sigma Aldrich Company Ltd., The Old Brickyard, New Road, Gillingham, Dorset SP8 4XT, UK) has been investigated. Ferric stock solutions were prepared by mixing ferric concentrates with distilled water to produce a concentration of 1000 mg L^{-1}. Utilising distilled water minimised quality variation and ensured consistent and repeatable process performance. These solutions were stored in the fridge and renewed every three weeks to obtain fresh solutions.

Three shapes of mixers and mixing methods have been used (radial, axial and magnetic) to disperse the coagulant into the water to be treated and subsequently tested. The influence of all mixers on the sludge dewaterability process was also assessed. The mixer types and shapes have been chosen based on the information provided by companies producing and/or selling standard mixers used by the water and wastewater industry such as Chemineer Ltd. (Cranmer Road, Derby, UK) and Promix Mixing Equipment and Engineering Ltd. (Columbus Road, Mississauga, Canada).

The radial and axial mixers were obtained from Monmouth Scientific Ltd. (Units 5 and 6, Kilnside, East Quay, Bridgwater, Somerset, UK). The axial mixer represents the shape of a jar test paddle, whilst the radial mixer is a small version of a radial mixer used in the water industry. The radial mixer has two blades, which are 1.2 cm long, 0.8 cm wide and located at a 45° angle from the mixer shaft. In comparison, the axial mixer has the same blade dimensions, but the blades are located at a 90° angle from the shaft. The magnetic stirrer IKA REO was purchased from Sartorius Instrumental Ltd. (Avenue Road, Belmont, Surrey, UK). The magnetic stirrer comprises an elongated rod, which operates at the base of the chamber, whereas the other mixers operate at a higher elevation (1.5 cm from the bottom) within the test beaker. The magnetic stirrer is 3 cm long and 0.5 cm wide. The edges of the stirrer are rounded. All mixers have diameters of 3 cm. Mixing took place in a 250-mL round glass beaker with internal dimensions of 6.5-cm length and 9.0-cm height.

6.3.2.3 Sample Preparation

Real raw water and raw wastewater and their corresponding biosolids have unstable properties; i.e., their water quality parameters are constantly changing. Therefore, synthetic raw water (representing reservoir water contaminated with inorganic solids) and synthetic domestic wastewater (simulating domestic wastewater containing both organic and inorganic matter) sludges have been used for the CST tests to simulate consistent water properties, which is essential for research purposes. Table 6.14 shows the components of synthetic domestic wastewater. For comparison only, natural domestic wastewater was also used. The wastewater was obtained from the United Utilities Wastewater Treatment Plant at Davyhulme in Manchester. Samples were taken from the approach channel before the sedimentation unit.

A kaolin solution was used in this study to simulate the sludge content of synthetic raw water (Zouboulis et al. 2005). Sample preparation was undertaken by adding 1 g of kaolin (Sigma Aldrich Company Ltd., The Old Brickyard, New Road, Gillingham, Dorset SP8 4XT, UK) to 100 mL of distilled water. This solution was mixed at 1200 rpm using a magnetic stirrer for five minutes to produce a well-mixed solution. Samples were freshly prepared for each experimental run.

The synthetic domestic wastewater recipe followed recently published guidelines (Hu et al. 2011) using kaolin to simulate total suspended solids (TSS; Table 6.14). All chemicals were supplied by Sigma Aldrich Company Limited (The Old Brickyard, New Road, Gillingham, Dorset, UK). The wastewater contained about 1% of TSS. This solution was prepared freshly each work day and always stored in a fridge to avoid the uncontrolled growth of microorganisms that might influence the wastewater quality.

6.3.2.4 Coagulation Experiments

The experimental coagulation methods applied in Chapter 6.3 are based on previous methods (Zhan et al. 2011). The influence of rapid mixing velocity on coagulant performance using ferric was assessed. A 100-mL water sample was poured into a glass beaker. The coagulant ferric chloride was subsequently added. After adjusting the pH value with sulphuric acid (H_2SO_4) or sodium hydroxide (NaOH) to reach a value of approximately 6.5, the sample was mixed rapidly by using a mixer at variable high rates (from 60 to 100 rpm at 5 rpm intervals) for 1 minute at each step, and then at a moderate rate of 50 rpm for 15 minutes.

Table 6.14 Composition of the synthetic domestic wastewater sample.

No.	Constituent	Concentration (mg L^{-1})
1	Dextrin	150
2	Ammonium chloride	130
3	Yeast extract	120
4	Glucose	100
5	Soluble starch	100
6	Sodium Carbonate	150
7	Detergent (commercial)	10
8	Sodium di-hydrogen orthophosphate	100
9	Potassium sulphate	8.3
10	Kaolin (soft white clay)	10 000

Sedimentation was permitted for 15 minutes, and the sludge (dense flocs) was carefully separated from the supernatant by discarding the supernatant so that only sludge remained in the glass beaker. After switching-on of the CST apparatus (Triton Electronics, Bigods Lane, Great Dunmow, Essex, UK) and using Whatman 17 chromatographic paper, the sludge was poured into the funnel of the CST apparatus.

The particle characteristics of the sludge were determined using a particle-size analyser (Horiba Laser Scattering Particle Size Analyzer LA-950, Horiba Instruments, Bunsen Drive, Insine, California, USA). The instrument calculates the correlation between the intensity and the angle of light scattered from a particle, and subsequently determines the particle size based on Mie-scattering theory (scattering of electromagnetic radiation by a sphere). Floc size shearing was minimised during the experiment, because of careful mixing during the measurement process.

Three replicate measurements were undertaken for each experiment described above. All values mentioned or displayed in Chapter 6.3 are mean values unless specified otherwise.

6.3.3 Results and Discussions

6.3.3.1 Impact of Synthetic Raw Water Flocs on Dewaterability

The impact of different synthetic flocs on sludge dewaterability was evaluated (Figures 6.7 and 6.8). Synthetic raw water together with ferric as the coagulant was assessed in Chapter 6.3. The median

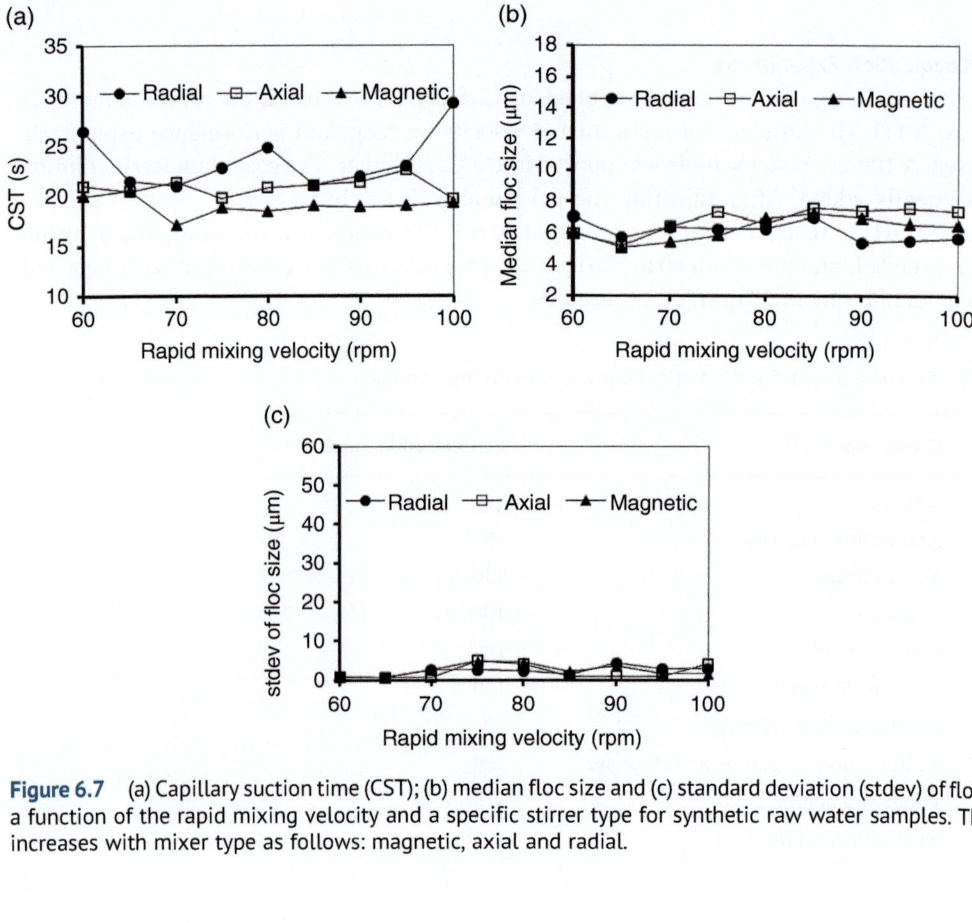

Figure 6.7 (a) Capillary suction time (CST); (b) median floc size and (c) standard deviation (stdev) of floc size as a function of the rapid mixing velocity and a specific stirrer type for synthetic raw water samples. The CST increases with mixer type as follows: magnetic, axial and radial.

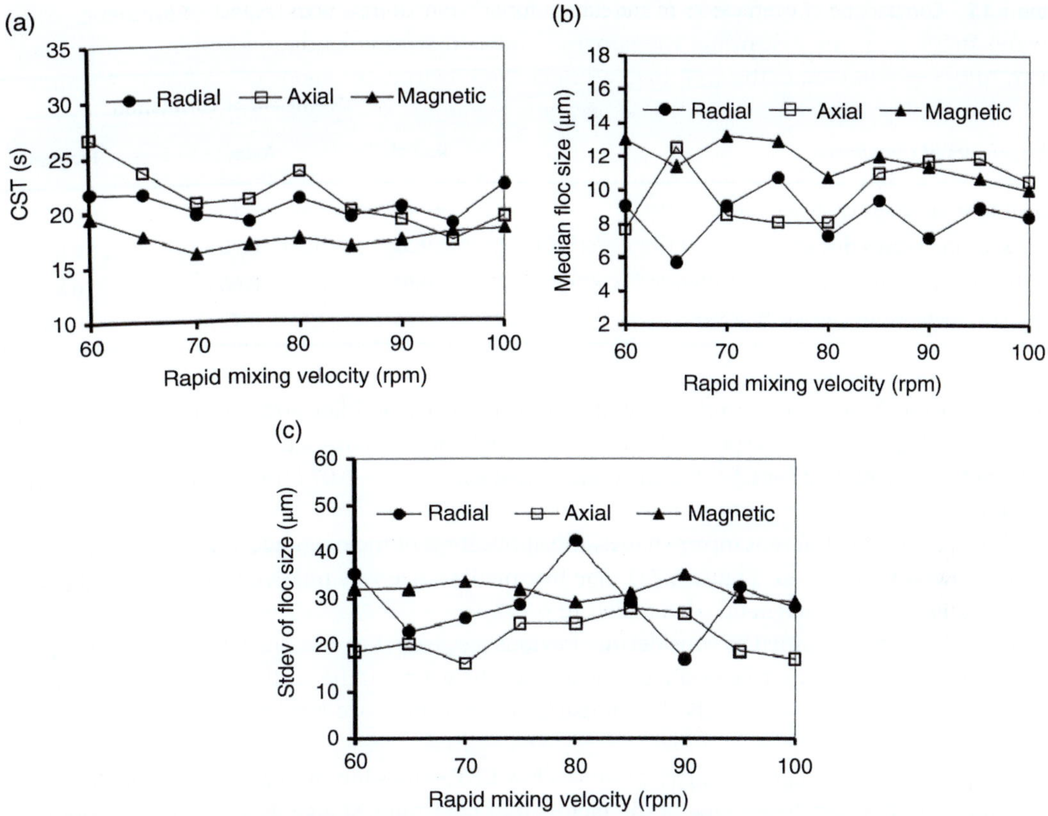

Figure 6.8 (a) Capillary suction time (CST); (b) median floc size and (c) standard deviation (stdev) of floc size as a function of the rapid mixing velocity and a specific stirrer type for domestic wastewater samples. The CST is lower for the magnetic mixer type compared to both axial and radial.

floc size was determined after coagulation, flocculation and sedimentation. Figure 6.7a–c indicates the CST values, floc median sizes and floc size standard deviations, respectively. Scales have been chosen to allow for an easy comparison between components of Figures 6.7 and 6.8. Findings indicate that different rapid mixing velocities correlate well with sludge dewaterability, but poorly with floc median size and floc size standard deviation.

Pearson's product–moment correlation coefficients, which indicate the strength of the relationship between two variables, are shown in Table 6.15. The coefficient is a measure of the linear correlation (dependence) between two variables, giving a value between 1 and −1. It is widely used as a measure of the strength of linear dependence between two variables; the higher the correlation value, the stronger is the relationship between the variables. The correlation values between different rapid mixing intensities and sludge dewaterability values for three shapes of mixers were −0.73 for the radial, −0.87 for the axial and −0.89 for the magnetic stirrers (see also Figure 6.7a). The coefficients of correlation for different rapid mixing intensities and median floc sizes were −0.18, 0.75 and 0.44 for the radial, axial and magnetic stirrers, respectively (see also Figure 6.7b). The coefficients vary considerably, if Figure 6.7a, b is compared with each other. This can be explained by the poor correlation between the CST value and floc size for synthetic raw water. In comparison, the coefficients for different rapid mixing velocities and floc size standard deviations were 0.64 for the radial, −0.63 for the axial and 0.40 for the magnetic stirrers (see also Figure 6.7c).

Table 6.15 Comparison of coefficients of correlation for different stirrers with respect to synthetic domestic water.

Components of comparison	Coefficient of correlation		
	Radial	Axial	Magnetic
Rapid mixing versus capillary suction time	−0.85	−0.83	−0.87
Rapid mixing versus floc size	0.52	0.72	0.62
Rapid mixing versus standard deviation of floc size	0.57	0.66	0.83
Capillary suction time versus floc size	−0.69	−0.72	−0.88

The CST value had a poor correlation with floc median size and floc size standard deviations as illustrated in Figure 6.7b, c, respectively, and by the coefficient of correlation data (Table 6.15). The coefficients of correlation for CST and the median floc sizes were −0.07 for the radial, −0.46 for the axial and −0.16 for the magnetic stirrers.

With respect to the different mixer shapes, the application of the magnetic stirrer was associated with the lowest CST values (Figure 6.7a). The floc median size was relatively small, regardless of using a radial, axial or magnetic stirrer (Figure 6.7b).

This result can be explained by considering previous research (Lee and Liu 2001; Zhao 2003). Floc strength plays an important role in sludge dewaterability testing. If the particle is large but has little strength, it disintegrates very easily. In comparison, small flocs are harder to break up (Lee and Liu 2001).

The application of a magnetic stirrer produces low CST values but does not yield large floc sizes (Turchiulli and Fargues 2004). One of the factors that determine sludge dewaterability is the floc structure and associated physical characteristics such as size and density. Smaller flocs are created when there is a high concentration of relatively small solids with low bound water content. Sludges produced by using a magnetic stirrer tend to have denser flocs compared to those associated with other mixer shapes (Turchiulli and Fargues 2004). Therefore, the magnetic stirrer produces lower CST values even though it is associated with larger floc sizes.

Figure 6.7b shows that increasing the rapid mixing velocity has an impact on floc size. Floc median and mean size figures indicated that lower rapid mixing velocities produce smaller floc sizes. However, as the rapid mixing velocity increased, the floc sizes became larger, and once the optimum rapid mixing was reached, the floc size decreased in response to the increase in rapid mixing velocity.

It is likely that rapid mixing velocity has an important role in the formation of flocs and their corresponding floc sizes (Coufort et al. 2005; Yu et al. 2011; Yukselen and Gregory 2004). Floc size depends on hydrodynamic processes due to the size changes, if the mixing intensity is modified (Coufort et al. 2005). Lower rapid mixing velocities have less impact on the formation of flocs, which is due to the conditions produced by turbulence. Slow turbulence does not encourage an appropriate coagulant distribution within the water so that floc formation cannot occur perfectly and homogenously.

The requirement of rapid mixing to foster floc size creation is due to the formation of coagulant hydrolysis products in less than seven seconds from the addition of the coagulant to the water (Amirtharajah and Mills 1982). Therefore, mixing should be very rapid to achieve the optimum mix between the coagulant and particles. Moreover, in the destabilisation process, high-intensity mixing will have additional benefits supporting floc formations (American Water Works

Association 1999). After reaching optimum conditions indicated by the size of the flocs, the enhancement of rapid mixing intensity brings about a decrease in the floc size. This result is like findings of previous research (Coufort et al. 2005; Rossini et al. 1990). Rapid mixing velocities (using alum and ferric as a coagulant) were assessed by Rossini et al. (1990). When using ferric, primary iron particles are retained and the hydroxide flocs are eroded and split due to high-intensity mixing and the corresponding impact on floc size.

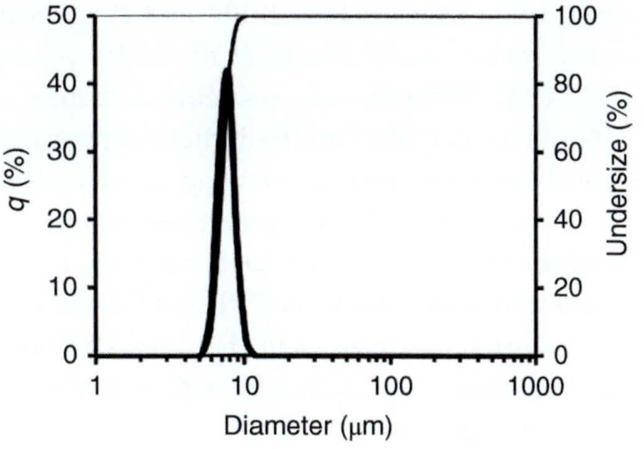

Figure 6.9 The percentage of similar particle size (q) and the accumulative of the q value (undersize) as a function of floc size diameter distribution for synthetic raw water. The particle-size distribution peaks at a diameter of about 7 μm.

Furthermore, using synthetic raw water as the water sample in the coagulation process produces particle sizes with a narrow range standard deviation. Using kaolin as the only ingredient creates a more uniform floc size during the coagulation process. Moreover, the dewatering characteristics of kaolin are very much dependent on particle size and distribution, sphericity (measure of how round an object is) of the particles, bed porosity, water-retaining capacities and variation in the dispersion properties (Besra et al. 2000). The use of kaolin in the water sample results in rather uniform floc sizes (Figure 6.9).

The SRF test was performed to evaluate CST findings. The correlation coefficients between CST and SRF for the radial, axial and magnetic stirrer applications were 0.88, 0.99 and 0.99, respectively. These findings confirm that both tests provide similar findings.

6.3.3.2 Impact of Synthetic Domestic Wastewater Flocs on Dewaterability

Figure 6.8 shows the CST, median floc size and standard deviation of the floc size as a function of the rapid mixing velocity and a specific stirrer type for domestic wastewater samples. The floc sizes in Figure 6.8b were like those in Figure 6.7b. In general, the application of a magnetic stirrer resulted in the lowest CST. This finding is supported by the particle-size analyser results indicating that the magnetic stirrer produced the largest floc size. The standard deviation of the floc size using synthetic domestic wastewater is higher than using synthetic raw water.

Table 6.15 presents the coefficient of correlation values of different parameters while using synthetic domestic wastewater. The coefficients between different rapid mixing intensities and sludge dewaterability are −0.85 for the radial, −0.83 for the axial and −0.87 for the magnetic stirrers. In comparison, the coefficients of correlation for different rapid mixing intensities and median floc sizes are 0.52, 0.72 and 0.62 for the radial, axial and magnetic stirrer, correspondingly. The coefficients for different rapid mixing velocities and floc size standard deviations are 0.52 for the radial, 0.66 for the axial and 0.83 for the magnetic stirrer. Finally, the corresponding values for CST and median floc sizes are −0.69 for the radial, −0.72 for the axial and −0.88 for the magnetic stirrers.

Furthermore, the highest coefficient of correlation between the CST and the median floc sizes was noted when the magnetic stirrer was used. This can be explained by the floc median size influencing the sludge dewaterability. Data show that large median floc sizes result in lower sludge dewaterability and vice versa.

The floc size assessment undertaken in Chapter 6.3 supports findings by Guo et al. (2009) indicating that a decrease in CST value correlates to an increase in floc size. A lower CST value means that it is easier for the sludge to release water. Normally, the larger the floc size, the easier it is for water to be released (Turchiulli and Fargues 2004). Smaller flocs with narrow capillaries do not easily release water (Basra et al. 2000). A large floc size makes it easier for particles to settle (Guo et al. 2009), thereby reducing turbidity.

Synthetic clay floc studies indicated poor correlations between CST and floc size. This effect is probably due to the relatively high density of the small particles, which influences the dewaterability of sludge (Turchiulli and Fargues 2004). Concerning synthetic domestic wastewater, the larger the floc size, the easier it is to dewater the floc (Larve and Vorobiev 2003). This usually results in a linear correlation between CST and floc size.

The correlation coefficients between CST and SRF for the radial, axial and magnetic stirrer applications were −0.93, 0.66 and 0.97, respectively. These findings indicate that both tests provide rather similar results.

6.3.3.3 Synthetic Raw Water Floc Size Distribution

Synthetic raw water represents reservoir water where the main contaminants are of inorganic nature. Particle-size data of this study allow for the assessment of the distribution of floc sizes. Each data point is based on three particle-size analyser readings. Particle-size statistical distribution graphs illustrate the floc size distribution influenced by the coagulation process.

Kaolin was the only ingredient in the synthetic raw water recipe. Figure 6.9 presents the corresponding general particle-size analysis indicating a short range of uniform floc distribution (concentrated around 7 μm) with kurtosis and skewness values of 3.51 and 0.47, respectively. The kurtosis value indicates that the particle-size distribution of this sample is extremely leptokurtic (i.e., a narrow and sharp peak). Moreover, there is very little skewness in the distribution.

6.3.3.4 Synthetic Domestic Wastewater Floc Size Distribution

Synthetic domestic wastewater represents real domestic wastewater where the main contaminants are of organic nature. In this investigation, the recipe for synthetic domestic wastewater encompassed three different ingredients (Hu et al. 2011). Figure 6.10 indicates the corresponding general particle-size distribution for synthetic domestic wastewater, which has no clear peak, if compared to the distribution of raw water (Figure 6.9). Synthetic domestic wastewater has a wider range of particle sizes, and a larger mean floc size compared to synthetic raw water. The distribution can be described as platykurtic (i.e., a wide and flat profile) with an asymmetric particle-size distribution and a tendency towards coarse characteristics. The synthetic domestic wastewater may also have a different impact on the floc formation process. A synthetic domestic wastewater floc is characterised more by its size (Table 6.15), while a synthetic raw water floc is influenced considerably by its density as indicated by the reduced correlation between the CST value and floc size (Table 6.16).

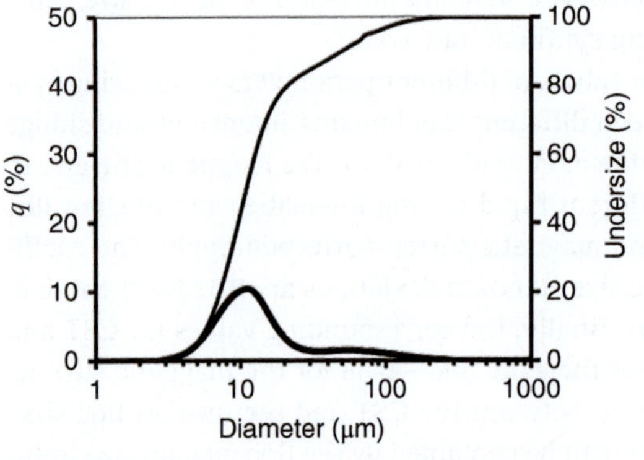

Figure 6.10 The percentage of similar particle size (q) and the accumulative of the q value (undersize) as a function of floc size diameter distribution for synthetic domestic wastewater. The particle-size distribution peaks at a diameter of about 10.05 μm, but the corresponding distribution curve is rather flat.

Synthetic domestic wastewater produced generally a wider range of particle sizes and larger flocs. In comparison, synthetic raw water, which has only one instead of three ingredients, produces a narrow particle-size range

Table 6.16 Comparison of coefficients of correlation for different stirrers with respect to synthetic raw water.

Components of comparison	Coefficient of correlation		
	Radial	Axial	Magnetic
Rapid mixing versus capillary suction time	−0.73	−0.87	−0.89
Rapid mixing versus floc size	−0.18	0.75	0.44
Rapid mixing versus standard deviation of floc size	0.64	−0.63	0.40
Capillary suction time versus floc size	−0.07	−0.46	−0.16

and relatively small flocs. This might be explained by the more likely presence of naturally developing microorganisms within synthetic domestic wastewater compared to the synthetic raw water. The presence of any microorganism is associated with a relatively large area and a higher viscosity (Sawalha and Scholz 2009), ultimately impacting on the floc size distribution and CST value.

6.3.3.5 Natural Water Floc Size Distribution

To obtain further information like variability and stability about the influence of floc size, a sample of natural domestic wastewater was also assessed in terms of floc size using the particle-size analyser. This comparison was made to better interpret the impact of the floc size distribution on sludge dewaterability.

The typical floc size distribution for natural sludge is shown in Figure 6.11. The distribution is platykurtic and asymmetric. Natural sludge has usually a larger floc size, and a more uniform particle-size distribution compared to synthetic sludge. This is due to the ingredients in natural sludge being more diverse in terms of their sizes than synthetic sludge. The agglomeration process seems to be more efficient when using natural wastewater, rather than synthetic water.

TSS within natural sludge are usually relatively large and dense (Verrelli et al. 2009). Natural sludge has typically also more hydrophobic (repelling water) content than synthetic sludge, so that it is more easily coagulated, producing bigger floc sizes than synthetic sludge (Pallier et al. 2010). Hydrophobic components can also support the formation of ferric hydroxide flocs (Matilanen et al. 2010; Zhan et al. 2010). Moreover, the hydrophobic fraction has a higher molecular weight and lower repulsion against the flocculant, so that the natural sludge has a bigger floc size and a more uniform floc size distribution than the synthetic wastewater sludge (Kim et al. 2006).

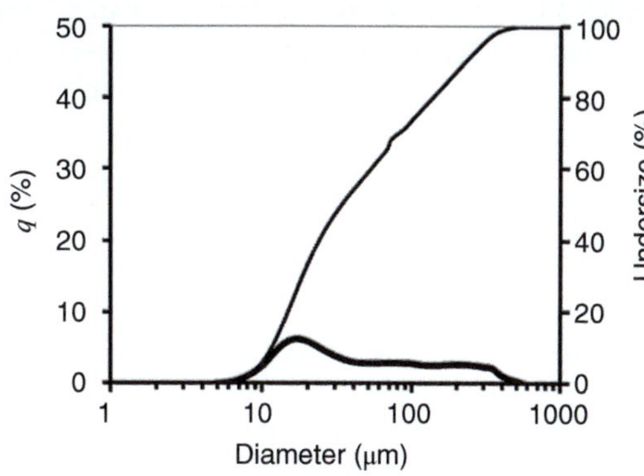

Figure 6.11 The percentage of similar particle size (q) and the accumulative of the q value (undersize) as a function of floc size diameter distribution for natural sludge. The particle-size distribution peaks at a diameter of about 10.07 μm, but the corresponding distribution curve is very flat.

6.3.4 Conclusions and Recommendations

The CST findings were similar for both synthetic raw and synthetic domestic wastewater, and were confirmed by the SRF test. The application of a magnetic stirrer resulted in the lowest CST.

Synthetic raw water had relatively small floc size and a narrow range standard deviation compared to synthetic domestic wastewater. Nevertheless, real wastewater has the widest and more uniform floc size distribution, indicating that synthetic waters, which are good for benchmarking purposes, might, however, not always be very representative of real waters. Findings for synthetic raw water also indicate that floc size does not correlate with sludge dewaterability in contrast to synthetic domestic wastewater.

It seems that synthetic raw water is affected more by its floc density than by its floc size. In contrast, the synthetic domestic wastewater is affected by floc size. This may reflect the presence of microorganisms, which have a large diameter and thus influencing the floc size and subsequently the sludge dewaterability process.

Further research using more complex (and therefore realistic) wastewater recipes representing discharges from different industrial sectors would be advantageous. The main alternative dewaterability test, which is the SRF, should also be assessed for waters with different particle-size distributions and compared with CST findings for a wide range of mixer types.

References

Abu-Orf, M.M. and Dentel, S.K. (1999). Rheology as tool for polymer dose assessment and control. *Journal of Environmental Engineering* 125 (12): 1133–1141.

Abu-Orf, M.M. and Örmeci, B. (2005). Measuring sludge network strength using rheology and relation to dewaterability, filtration, and thickening – laboratory and full-scale experiments. *Journal of Environmental Engineering* 131 (8): 1139–1146.

Abu-Orf, M.M., Walker, C.A., and Dentel, S.K. (2003). Centrate viscosity for continuous monitoring of polymer feed in dewatering applications. *Advances in Environmental Research* 7 (3): 687–694.

American Water Works Association (1999). *Coagulation and Flocculation; Water Quality and Treatment*, 5e. New York: Mc Graw Hill.

Amirtharajah, A. and Mills, P. (1982). Rapid-mix design for mechanisms of alum coagulation. *Journal of American Water Works Association* 74 (4): 210–216.

Barnes, H.A., Hutton, J.F., and Walters, K. (1993). *An Introduction to Rheology*. Amsterdam, Netherlands: Elsevier.

Baudez, J.C., Ginisty, P., and Spinosa, L. (2007). The preparation of synthetic sludge for lab testing. *Water Science and Technology* 56 (9): 67–74. https://doi.org/10.2166/wst.2007.714.

Besra, L., Sengupta, D.K., and Roy, S.K. (2000). Particle characteristics and their influence on dewatering of kaolin, calcite and quartz suspensions. *International Journal of Mineral Processing* 59 (2): 89–112: https://doi.org/10.1016/S0301-7516(99)00065-4.

Bruus, J.H., Nielsen, P.H., and Keiding, K. (1992). On the stability of activated sludge flocs with implications to dewatering. *Water Research* 26 (12): 1597–1604: https://doi.org/10.1016/0043-1354(92)90159-2.

Buyukkamaci, N. (2004). Biological sludge conditioning by Fenton's reagent. *Process Biochemistry* 39 (11): 1503–1506.

Chen, G.W., Lin, W.W., and Lee, D.J. (1996). Capillary suction time (CST) as a measure of sludge dewaterability. *Water Science and Technology* 34: 443–448.

Chen, B.H., Lee, S.J., and Lee, D.J. (2005). Rheological characteristics of the cationic polyelectrolyte flocculated wastewater sludge. *Water Research* 39 (18): 4429–4435.

Chen, C., Zhang, P., Zeng, G. et al. (2010). Sewage sludge conditioning with coal fly ash modified by sulfuric acid. *Chemical Engineering Journal* 158 (3): 616–626. https://doi.org/10.1016/j.cej.2010.02.021.

Chu, C.P. and Lee, D.J. (2005). Comparison of dewaterability and floc structure of synthetic sludge with activated sludge. *Environmental Technology* 26 (1): 1–10.

Coufort, C., Bouyer, D., and Line, A. (2005). Flocculation related to local hydrodynamics in a Taylor–Couette reactor and in a jar. *Chemical Engineering Science* 60 (8–9): 2179–2192. https://doi.org/10.1016/j.ces.2004.10.038.

Dentel, S.K. and Abu-Orf, M.M. (1995). Laboratory and full-scale studies of liquid stream viscosity and streaming current for characterization and monitoring of dewaterability. *Water Research* 29 (12): 2663–2672.

Dursun, D., Ayol, A., and Dentel, S.K. (2004). Physical characteristic of a waste activated sludge: conditioning responses and correlations with a synthetic surrogate. *Water Science and Technology* 50 (9): 129–136.

Fitria, D., Scholz, M., Swift, G.M., and Hutchinson, S.M. (2014). Impact of sludge floc size and water composition on dewaterability. *Chemical Engineering and Technology* 37 (3): 471–477.

Forster, C.F. (2002). The rheological and physico-chemical characteristics of sewage sludges. *Enzyme and Microbial Technology* 30 (3): 340–345.

Gale, R.S. and Baskerville, R.C. (1968). A simple automatic instrument for determining the filterability of sewage sludge. *Water Pollution Control* 67: 233–241.

Gray, N.F. (2005). *Water Technology: An Introduction for Environmental Scientists and Engineers*. Oxford: Elsevier Butterworth-Heinemann.

Guan, J., Amal, R., and Waite, T.D. (2004). Effect of floc size and structure on biosolids capillary suction time. *Water Science and Technology* 47: 255–260.

Guo, L., Zhang, D., Xu, D. et al. (2009). An experimental study of low concentration sludge settling velocity under turbulent condition. *Water Research* 43 (9): 2383–2390. https://doi.org/10.1016/j.watres.2009.02.032.

Hasar, H., Kinaci, C., Ünlü, A. et al. (2004). Rheological properties of activated sludge in a sMBR. *Biochemical Engineering Journal* 20 (1): 1–6.

Herwijn, A.J.M., Laheij, E.J., Ijzermans, J.J. et al. (1995). Determination of specific cake resistance with a new capillary suction time apparatus. *Industrial and Engineering Chemistry Research* 34: 1310–1319.

Hu, B., Wheatley, A., Ishtchenko, V., and Huddersman, K. (2011). The effect of shock loads on SAF bioreactors for sewage treatment works. *Chemical Engineering Journal* 166 (1): 73–80. https://doi.org/10.1016/j.cej.2010.10.005.

Jin, B., Marie, W.B., and Lant, P. (2004). Impacts of morphological, physical and chemical properties of sludge flocs on dewaterability of activated sludge. *Chemical Engineering Journal* 98 (1–2): 115–126. https://doi.org/10.1016/j.cej.2003.05.002.

Kim, H.C., Hong, J.H., and Lee, S. (2006). *Journal of Membrane Science* 283 (1–2): 266–272. https://doi.org/10.1016/j.memsci.2006.06.041.

Lai, C.K., Chen, G., and Lo, M.C. (2004). Salinity effect on freeze/thaw conditioning of activated sludge with and without chemical addition. *Separation and Purification Technology* 34 (1–3): 155–164.

Larve, O. and Vorobiev, E. (2003). *International Journal of Mineral Processing* 71 (1–4): 1–15. https://doi.org/10.1016/S0301-7516(03)00026-7.

Lee, D.J. (1994). Modified rectangular capillary suction apparatus test. *Journal-Chinese Institute of Chemical Engineers* 25 (6): 341–399.

Lee, D.J. and Hsu, Y.H. (1992). Fluid-flow in capillary suction apparatus. *Industrial and Engineering Chemistry Research* 31: 2379–2385.

Lee, D.J. and Hsu, Y.H. (1993). Cake formation in capillary suction apparatus. *Industrial and Engineering Chemistry Research* 32 (6): 1180–1185.

Lee, D.J. and Hsu, Y.H. (1994a). A rectangular capillary suction apparatus. *Industrial and Engineering Chemistry Research* 33 (6): 1593–1599.

Lee, D.J. and Hsu, Y.H. (1994b). Use of capillary suction apparatus for estimating the averaged specific resistance of filtration cake. *Journal of Chemical Technology and Biotechnology* 59 (1): 45–51.

Lee, C.H. and Liu, J.C. (2001). *Advances in Environmental Research* 5 (2): 129–136. https://doi.org/10.1016/S1093-0191(00)00049-6.

Lee, D.J. and Wang, C.H. (2000). *Water Research* 34 (1): 1–20. https://doi.org/10.1016/S0043-1354(99)00096-2.

Lee, D.J., Chen, G.W., and Hsu, Y.H. (1994). On some aspects of capillary suction apparatus tests. *Journal of the Chinese Institute of Chemical Engineers* 25 (1): 35–44.

Leu, W.F. (1981). Cake filtration, PhD dissertation, University of Houston.

Matilanen, A., Vepsäläinen, M., and Silappa, M. (2010). *Advances in Colloid and Interface Science* 159 (2): 189–197. https://doi.org/10.1016/j.cis.2010.06.007.

Meeten, G.H. and Lebreton, C. (1993). A filtration model of the capillary suction time method. *Journal of Petroleum Science and Engineering* 9 (2): 155–162.

Meeten, G.H. and Smeulders, J.B.A.F. (1995). Interpretation of filterability measured by the capillary suction time method. *Chemical Engineering Science* 50 (8): 1273–1279.

Metcalf and Eddy (2003). *Wastewater Engineering: Treatment and Reuse*. New York: Mc Graw Hill.

Örmeci, B. and Vesilind, P.A. (2000). Development of an improved synthetic sludge: a possible surrogate for studying activated sludge dewatering characteristics. *Water Research* 34 (4): 1069–1078.

Pallier, V., Cathalifaud, G.F., Serpaud, B., and Bollinger, J.C. (2010). *Journal of Colloid and Interface Science* 342 (1): 26–32. https://doi.org/10.1016/j.jcis.2009.09.068.

Pollice, A., Giordano, C., Laera, G. et al. (2007). Physical characteristics of the sludge in a complete retention membrane bioreactor. *Water Research* 41 (8): 1832–1840.

Razi, A.F. and Molla, A.H. (2007). *Journal of Hazardous Materials* 147 (1–2): 250–356. https://doi.org/10.1016/j.jhazmat.2007.01.060.

Rossini, M., Garrido, G., and Galluzo, M. (1990). *Water Research* 33 (8): 1817–1826. https://doi.org/10.1016/S0043-1354(98)00367-4.

Sanin, D.F. and Vesilind, A.P. (1996). Synthetic sludge: a physical/chemical model in understanding biofloculation. *Water Environment Research* 68 (5): 927–933.

Sanin, F.D. and Vesilind, P.A. (1999). A comparison of physical properties of synthetic sludge with activated sludge. *Water Environment Research* 71 (2): 191–196.

Sanin, F.D., Clarkson, W.W., and Vesilind, P.A. (2011). *The Treatment and Disposal of Wastewater Sludge*. Lancaster, Pensylvania: DEStech Publications, Inc.

Sawalha, O. and Scholz, M. (2007). Assessment of capillary suction time (CST) test methodologies. *Environmental Technology* 28 (12): 1377–1386.

Sawalha, O. and Scholz, M. (2009). Innovative enhancement of the design and precision of the capillary suction time testing device. *Water Environment Research* 81 (11): 2344–2352. https://doi.org/10.2175/106143009X407276.

Scholz, M. (2005). Review of recent trends in capillary suction time (CST) dewaterability testing research. *Industrial & Engineering Chemistry Research* 44 (22): 8157–8163. https://doi.org/10.1021/ie058011u.

Scholz, M. (2006). Revised capillary suction time (CST) test to reduce consumable costs and improve dewaterability interpretation. *Journal of Chemical Technology and Biotechnology* 81 (3): 336–344.

Scholz, M. and Martin, R.J. (1998). Control of bio-regenerated granular activated carbon by spreadsheet modelling. *Journal of Chemical Technology and Biotechnology* 71: 253–261.

Seyssiecq, I., Ferasse, J.H., and Roche, N. (2003). State-of-the-art: rheological characterization of waste water treatment sludge. *Biochemical Engineering Journal* 16 (1): 41–56.

Singh, B.P., Bhattacharjee, S., Besra, L., and Sengupta, D.K. (2004). Comparison between techniques based on charge characterisation and capillary suction time for assessing the dispersion characteristics of concentrated slurry. *Journal of Materials Science* 39: 2437–2442.

Slavik, I., Muller, S., Mokosh, R. et al. (2012). *Water Research* 46 (19): 6543–6553. https://doi.org/10.1016/j.watres.2012.09.033.

Smiles, D.E. (1998). Water flow in filter paper and capillary suction time. *Chemical Engineering Science* 53: 2211–2218.

Tiller, F.M. and Wenping, P.L. (2001). Modified capillary suction theory with effects of sedimentation for rectangular cells. *Journal of the Chinese Institute of Chemical Engineers* 32 (5): 391–399.

Tiller, F.M., Shen, Y.L., and Adin, A. (1990). Capillary suction theory for rectangular cells. *Research Journal of the Water Pollution Control Federation* 60 (2): 215–220.

Tixier, N., Guibaud, G., and Baudu, M. (2003). Determination of some rheological parameters for the characterization of activated sludge. *Bioresource Technology* 90 (2): 215–220.

Turchiulli, C. and Fargues, C. (2004). *Chemical Engineering Journal* 103 (1–3): 123–131. https://doi.org/10.1016/j.cej.2004.05.013.

United Utilities (2024). www.unitedutilities.com (accessed 30 May 2013).

Verrelli, D.I., Dixon, D.R., and Scales, P.J. (2009). *Colloids and Surfaces A: Physicochemical and Engineering Aspects* 348 (1–3): 14–23. https://doi.org/10.1016/j.colsurfa.2009.06.013.

Wang, Y., Gao, B.Y., Xu, X.M. et al. (2009). *Journal of Colloid and Interface Science* 332 (2): 354–359. https://doi.org/10.1016/j.jcis.2009.01.002.

Yu, W., Gregory, J., Campos, L., and Li, G. (2011). *Chemical Engineering Journal* 171 (2): 425–430. https://doi.org/10.1016/j.cej.2011.03.098.

Yukselen, M.A. and Gregory, J. (2004). *International Journal of Mineral Processing* 73 (2–4): 251–259. https://doi.org/10.1016/S0301-7516(03)00077-2.

Zhan, X., Gao, B., Yue, Q. et al. (2010). *Separation and Purification Technology* 75 (1): 61–68. https://doi.org/10.1016/j.seppur.2010.07.012.

Zhan, X., Gao, B., Yue, Q. et al. (2011). *Chemical Engineering Journal* 166 (2–4): 116–121. https://doi.org/10.1016/j.cej.2010.10.037.

Zhang, S., Wang, S., Shan, X., and Mu, H. (2004). *Biology and Fertility of Soils* 40: 237–242. https://doi.org/10.1007/s00374-004-0771-1.

Zhang, K., Achari, G., Sadiq, R. et al. (2012). *Water Research* 46 (6): 1673–1683. https://doi.org/10.1016/j.watres.2011.12.006.

Zhao, Y.Q. (2003). *Chemical Engineering Journal* 92 (1–3): 227–235. https://doi.org/10.1016/S1385-8947(02)00253-X.

Zouboulis, A.I., Moussas, P.A., and Vasilakou, F. (2005). *Journal of Hazardous Materials* 155 (3): 459–468. https://doi.org/10.1016/j.jhazmat.2007.11.108.

Sutler, D.L. (1998). Water flow in filter paper and capillary-driven ... Chemical Engineering Science 53: 2141–2153.
Tilley, J.M. and Wertheim, M. (200?). Modified capillary-suction theory with effects of ... length ... rectangular cells. Journal of the Chinese Institute of Chemical Engineers 36 (?): 151–158.
Tiller, F.M., Shen, Y.L. and Adin, A. (1990). Capillary suction theory for rectangular cells. Research Journal of the Water Pollution Control Federation 62 (?): 130–136.
Tiller, F.M. ... and Rushton, A. (1983). Determination of some theoretical parameters for the characterization of activated sludge. Water Research ... 60 (?): 215–22.
Tuchbrock, C. and Fahrbach, C. (2004). ... Research Engineering News ... 10 (?): 123–131. https://doi.org/10.1002/cr.2309.05102.
United Utilities (2021). www.unitedutilities.com (accessed 30 May 2023).
Vesilind, P.A. (1979). Sludge and Sediment ... Ann Arbor, MI ... Chemical Engineering ... 1(?): 14–58. https://doi.org/10.1002/...
Wang, X., Shen, Y., Xu, X. et al. (2007). Journal of Colloid and Interface Science 314: 45–50. https://doi.org/10.1016/j.jcis.2007.05.002.
Xu, W., Fletcher, J., ... and ... (2021). ... Research ... 1 (?): 435–440. https://doi.org/10.1016/... 2021.05.026.
Yueshan, ..., and ..., L. (2006). ... consolidation ... Water Research 52 (?): ... https://doi.org/10.1016/...
Zhan, X., ..., Y. et al. (2020). ... Environmental Technology ... https://doi.org/10.1016/j....
Zhang, X., Fan, B., ... et al. (2021). Chemical Engineering Journal ... https://doi.org/10.1016/j.cej.2021.1352.
Zhang, X., Wang, S., Shao, Y. et al. (2021). ... https://doi.org/10.1016/j....
Zhang, K., Adams, ... et al. (2021). Water Research ... https://doi.org/10.1016/j.watres.2021.117...
Zhao, Y.Q. (2003). Chemical Engineering Journal ... https://doi.org/10.1016/S1385-8947(...)
Zoulounis, A., Mathews, C. and Waite ... (2023). ... https://doi.org/10.1016/... (accessed ... 2023).

7

Water Availability and Public Health

7.1 Introduction

Since the independence of Nigeria, successive governments have allocated great proportions of their budgets for water resources development projects. Some international organisations have also made contributions towards the improvement of water resources and adequate sanitation programmes. The various forms in which the money has been used include: (i) educating and informing local communities about the importance of safe water and adequate sanitation; (ii) training of relevant personnel; viz. engineers, mechanics, chemists, geologists, hydrogeologists and sociologists; and (iii) creating and funding institutional frameworks such as water boards, corporations, agencies as well as river basins and rural development authorities (Sandodoyin 1991).

Between 1979 and 1983, the total expenditure of the River Basin and Rural Development Authorities, which were charged with the responsibility of developing both surface and ground water resources for multi-purpose application, was 2.1 billion Naira (₦) or approximately 21 million US dollars (Salau 1986). Despite these efforts and expenditures, there are still many public health problems associated with the development and utilisation of Nigeria's water resources in rural areas. More investment is usually concentrated in urban areas than in rural areas.

Rural areas suffer from two other disadvantages, which have hindered the introduction of improved water supply schemes: The settlement pattern is dispersed so that house-to-house water connections are prohibitively expensive. Coupled with this is the fact that rural people have low disposable incomes. They find it difficult to fund, maintain or contribute meaningfully to modern water supply projects which are relatively costly (Haile 1980). Rural communities often lack the necessary skills to maintain such projects when they are established. Thus, any improved water scheme that would be suitable for rural areas must be simple to operate, relatively cheap to construct and maintain and take account of the dispersed pattern of settlement and population distribution (Ayode 1988).

For a community water supply to be effective, firstly, the community must recognise the potential benefits of the improved water supply. Secondly, the community must be able to pay for at least the operation and maintenance costs. Thirdly, members of the community must have the skills to sustain the water resources programme in the long term and access to spare parts, materials and tools.

Modern water supply facilities use sophisticated techniques, which make them expensive for communities in rural areas. Moreover, reservoir dams used for water supply and irrigation purposes are expensive for developing countries. For example, in 1991 about ₦20 million was the estimate for a dam proposed for the Lamorde River that passes across the Jalingo Local Government boundaries (Figure 7.1) according to the Taraba State Ministry of Water Resources (1995). In comparison, a borehole supplying 500 m^3 water per day costs about ₦1 million.

Sustainable Water Systems, First Edition. Miklas Scholz.

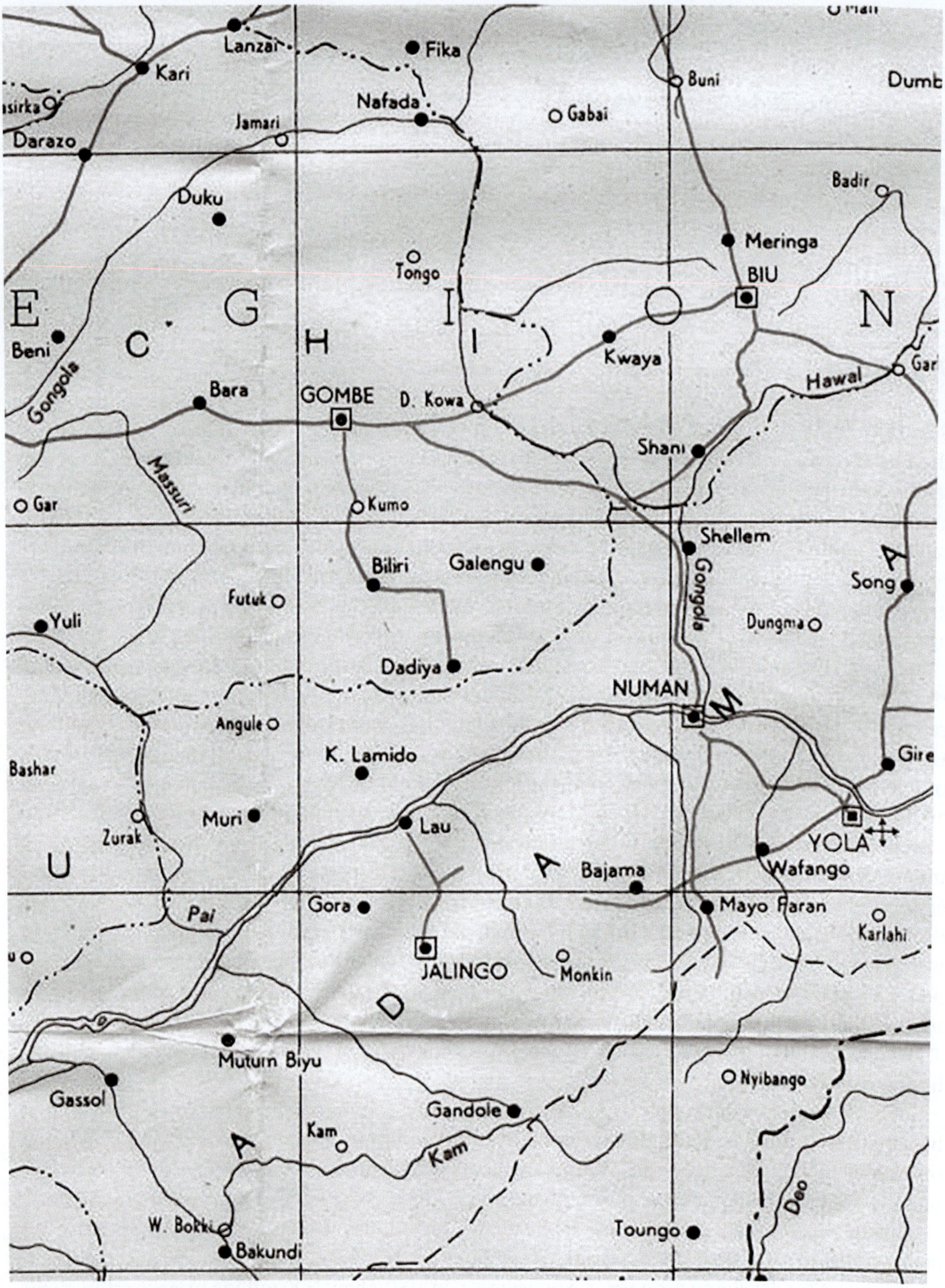

Figure 7.1 Map showing the study area near Jalingo (grid lines are 1° apart from each other). The study area is located to the South-west of Numan.

Table 7.1 Discussion of water-related disease classes.

Disease class	Discussion
Water-washed	Water-washed infections of the intestinal tract and of the skin can be reduced (particularly in the tropics) following improvements in domestic and personal hygiene. These improvements rather depend on the water availability than the water quality. Diarrhoea diseases (particularly bacillary dysentery) decrease with the availability of water and with the volume of water used, but do not decrease significantly with improvements in the microbiological quality of the water. Water-washed diseases can also relate to skin and eye infections. Examples are skin sepsis, scabies, fungal infections of the skin and trachoma of the eye.
Water-borne	A water-borne transmission occurs when the pathogen is part of the water which is drunk by a person who may then become infected. Examples are cholera, typhoid, infectious hepatitis, diarrhoea and dysentery.
Insect vector mechanism	This is a disease that is spread by insects which either breed or bite near water. Examples include malaria, yellow fever, dengue and onchocerciasis.
Water-based	A disease where the corresponding pathogen spends part of its life cycle in a water snail or other aquatic animals. Schistosomiasis (bilharziasis) and the Guinea worm are examples. Provision of safe water supplies is the primary measure to control these diseases.

Some agencies like the Directorate of Food and Rural Infrastructures and frameworks including the Family Economic Advancement Programme were established to enhance the extension of basic amenities and to create better living conditions for rural communities. One of the areas that requires attention was water supply and sanitation. Concerning water supply schemes, there is the need to establish rural community participation in conjunction with support by the government (Elabi 1984).

In many countries of the tropics, water supply schemes providing portable water serve only the urban centres. In rural areas, water shortage is common, especially during the dry season. The rural dwellers resort to traditional sources of water supply such as rainwater storage tanks, springs, ponds and hand-dug wells. This practice had long been abandoned in urban areas following the commissioning of modern water supply schemes. Traditional water supply has its associated water-related diseases, which eventually become a problem in the concerned area.

A water-related disease is one that is related to water or impurities in water. There is a distinction between infectious water-related diseases and those related to some chemical properties of the water. Water-related diseases are classified according to the transmission mechanism or the route through which the disease gains entrance into the human body. A discussion of water-related disease classes is summarised in Table 7.1.

This study investigates the water availability and quality in rural areas controlled by the Jalingo Local Government (Nigeria). Consequently, it takes public health implications of the available water resources on the village communities into account. Chapter 7.1 is based on the original article by Uzomah and Scholz (2002).

7.2 Methodology

Jalingo is the capital of Taraba State (Nigeria). The Jalingo study area is located between longitudes 10°35′N and 11°15′N and latitudes 8°50′E and 11°25′E (Figure 7.1). This area has borders with Yorro in the east, with Ardo Koal in the south and with Karim Lamido in the South-west as well as with areas controlled by the Lau Local Government (Barbour et al. 1982). The region has an estimated population of 98 592 and a land size of 57.4 km^2. Rainfall generally occurs between March and October. In 1979, the average annual rainfall was 1350 mm (Elabi 1984).

The area of this case study is a vast plain that developed on sedimentary rock of the Yolde formation. The geological setting of the area is dominated by underlying rocks, which are largely undifferentiated, and referred to as basement complex. These rocks consist of granites, quatzites, schist, diorites, amphibolities and meta-sediments (Taraba State Ministry of Water Resources 1995). The regional water table varies from about 6 m in Ardo-Kola (east of Jalingo) to about 30 m near Jalingo. The vegetation consists mainly of short grasses and small trees. This is a characteristic of the sub-Sudan (Barbour et al. 1982).

During the wet season, access via van to some villages was not easy, because the roads were narrow. This made the administration of the questionnaires difficult for three villages. The study area comprises 19 villages, out of which 8 villages were conveniently chosen for this case study. The villages were selected by local officials based on their estimated size; viz. small (40–60 houses), intermediate (61–120 houses) and large (121–300 houses) villages. Three sites were selected from each small- and intermediate-size villages, while two villages were selected from those considered to be large. Many of the inhabitants of these villages understand only their tribal language, while the remaining population can also speak the Hausa language. There were four different tribes living within the study area (Fulani, Mumuye, Kona and Nyandan). Four persons were used as interpreters of the questionnaire, one from each of the four tribes.

A total of 60 households were sampled in this area. A typical household (usually one extended family) comprises between 7 and 15 members (Appendix: Section A). Considering the relatively small sample size of ≤60 households and potentially low sample accuracy, all numerical results including percentages are expressed in integers. However, the sample size in terms of participants (≤653 people) is large considering very difficult study conditions like tribal conflicts.

Households were selected using systematic convenience sampling depending on the pattern of the village. The house of the village head is usually located in the centre of the village. The village head's house and the fifth or sixth house from his building (in all major footpath directions) were included for sampling. Figure 7.2 shows a traditional residence typical for Eastern Nigeria

Figure 7.2 A traditional residence typical for Eastern Nigeria. A photograph showing a simple traditional house made of locally available natural material.

(Barbour et al. 1982). The area near to the village head's house usually contains the location of the community well. Alternatively, where the location of the village head's house did not permit this method, the selection of houses was undertaken on consideration of their distance from the market square.

The questionnaire was designed according to international standards considering local conditions (Converse and Presser 1988; Oppenheim 1996). Questionnaires as well as oral interviews were administered on household bases. Principal participants (usually the owners of the properties) were questioned in their homes at a convenient time when maximum co-operation could be expected. However, other family members gave also feedback to the interviewers. A copy of the full questionnaire (translated into English) can be found in the Appendix.

7.3 Results and Discussion

7.3.1 Water Resources and Usage

The source of domestic water supply in the sampled villages was largely the potentially polluted community well. It was discovered that 35% of the villagers obtain water from wells during the wet season. However, other potentially contaminated sources were available: Approximately 11% of the participants collect water for domestic use (other than drinking water) from nearby ponds, either in form of dug pits or natural depressions. These depressions are subsequently widened to increase their capacity. Direct collection of rainwater was another major source (16%) of water. Three of the sampled villages have seasonal streams; approximately 28% of their inhabitants depend on them for domestic use (Appendix: Section B).

However, during the dry season, most of the villagers depend on shallow community well water (69%). During this season the streams and ponds usually dry up. However, 17% get their water from streambed pits (usually dug very close to the streams).

Deep boreholes, which usually represent a clean water source, are serving only 10% of the inhabitants of the study area during the wet and 14% during the dry season. These low figures could be attributed to the scanty number of boreholes in the area (Appendix: Section B).

There is an indication that the level of water supply development is low (Barbour et al. 1982). In the eight sampled villages, there are only 35 artesian wells (fed by unconfined aquifers), with three villages having just a single well each for the whole community.

Water use in the study area is minimal. The average amount of water used daily per person was estimated based on the response made during the survey. Table 7.2 indicates that about 80% of the households have access to less than 30 L of water per person per day (Appendix: Section B). This is a very low water consumption rate compared with an average of 86 L per person per day as estimated for Ibadan, one of the largest urban cities in Nigeria. However, these figures are low if compared with an average water consumption rate of 290 litres per person per day for England (Ayoye 1988).

Table 7.2 Average daily water-use per person in litre.

Water use (litre)	Households (%)	Number of people
<11	13	85
11–20	44	282
21–30	25	160
31–40	12	78
>40	6	39

7.3.2 Water Availability and Public Health

Approximately 66% of the sample population were dissatisfied with the available water sources. The low consumption rate of water has serious implications on community welfare and health. There are constraints faced by individuals regarding bathing, washing and watering animals. Water-related (particularly water-borne and water-washed) diseases are high. Table 7.3 summarises responses regarding diseases suffered by the villagers. The proportions for malaria and the Guinea worm could be greater than identified because these diseases were not listed on the questionnaire but have been specified by some participants (Appendix: Section C; Question 4) according to Oppenheim (1996).

Most of these diseases were reported to be endemic (occurring throughout the year) except for cholera and typhoid which are predominant at the beginning of the wet and the dry seasons, respectively. The lack of surface water sources causes reliance on ground water. Five of the sampled villages rely only on community wells during the dry season. Two out of these five villages have only one well each. None of these wells yields sufficient water for more than half of the village population during the dry season. Approximately 50% of the community wells usually develop problems. These wells either dry up (or yield little) during the dry or collapse during the wet season. The failure of wells coupled with an increase in drinking water demand during the dry season usually creates acute water availability problems (Appendix: Section D).

Distances travelled to get water are particularly high during the dry season. Table 7.4 presents a summary of findings on individual distances covered. Members of only 27% of the households need

Table 7.3 Water-related diseases suffered by the villagers.

Disease (common name)	Class of water-related diseases	Seasonal occurrence		Victims (%)	Number of People
		Wet	Dry		
Malaria[a]	Insect vector mechanism	Yes	Yes	22	73
Diarrhoea	Water-borne/water-washed	Yes	Yes	20	66
Typhoid	Water-borne	No	Yes	17	56
Dysentery	Water-washed/water-borne	No	Yes	15	50
Guinea worm[a]	Water-based	Yes	Yes	13	43
Cholera	Water-borne	Yes	No	7	23
Schistosomiasis (Bilharziasis)	Water-based	Yes	Yes	6	20

[a] Specified by the participants themselves (Appendix: Section C; Question 4).

Table 7.4 Distance covered in fetching water.

Distance (km)	Villagers			
	Wet season (%)	Number of people	Dry season (%)	Number of people
<0.5	72	470	25	163
0.5–1.0	19	124	36	236
1.1–2.0	9	59	14	91
>2.0	0		25	163

less than 30 minutes to fetch their water during the wet and 17% during the dry seasons (Appendix: Section B). Natural footpaths and local roads as well as simple footwear and equipment are not ideal for undertaking this daily routine efficiently and safely according to the standards of developed countries.

7.3.3 Potential for Community Involvement

Approximately 88% of the community wells were constructed by the villagers themselves. Moreover, 11% of the inhabitants have self-developed wells within their compounds. The community wells are also maintained by the community (collective responsibility). Wells and boreholes constructed by the Government are also taken care of by the communities (Appendix: Section D).

Development associations exist only in few villages. These associations are involved in the construction and maintenance of wells and ponds (Appendix: Section E). Approximately 96% of the people agreed that self-help could improve their community's water supply problem. The contributions they were likely to make include labour (45%), materials (18%), money (27%) and services (10%). The proportion for services could be greater than identified because it was not listed as a default option on the questionnaire. Furthermore, these responses also indicate the willingness of the community to co-operate in solving their current water supply problems. However, the proportion for money would probably be smaller if the request for contribution was not hypothetical (Appendix: Section D; Question 14) (Oppenheim 1996).

Springs discovered in the northern part of the study area (village in a valley near Gongon Maliki Hill; Badawaira; villages in Wuro Mallam) could be properly developed to obtain an additional water supply. Developing springs does not require high capital investment. Expenditure usually arises from periodic bacteriological examinations, purchasing of treatment chemicals and fencing.

A know-how and technology transfer from developed to developing countries is required to improve the water availability problem. More boreholes need to be drilled to provide clean water. A geostatistical analysis of the spatial water distribution might help to reduce both sampling costs and effort (Scholz et al. 1999). Furthermore, the application of low-cost vertical-flow constructed wetlands technology for the treatment of polluted river and wastewater might help to improve public health conditions by reducing water-borne and water-based diseases (Scholz and Xu 2001).

Federal and Government Agencies should consider to co-ordinate their activities effectively and extend their services to rural areas. The Family Economic Advancement Programme could be extended to all rural communities.

7.4 Conclusions

The development of safe water resources like deep boreholes in the case study area is low. The results indicate that a great proportion of the rural dwellers in the Jalingo Local Government area depend on community wells during both the wet and dry seasons as their main source of water supply. This has led to water-related diseases caused by water shortages despite a high community participation potential regarding water resources projects. There is a potential for international agencies to improve the situation by low-cost know-how transfer to a population willing to be involved in water resources projects.

Appendix

The Questionnaire

Section A: Personal Data

1 Name of respondent:

2 Village:

3 Gender: Male ( ) Female ( )

4 Age (Years):

5 Type of family: Nuclear ( ) Extended ( )

6 Size of family (Number of persons):

7 Number of children helping in household activities:

Section B: Household Water Collection and Use

1 How do you meet your water need in the wet season?
Stream ( )
Pond ( )
Well ( )
Rainwater ( )
Borehole ( )
Pipe borne water ( )
Others, please specify

2 How do you meet your water need in the dry season?
Stream ( )
Pond ( )
Well ( )
Rain water ( )
Borehole ( )
Pipe borne water ( )
Others, please specify

3 Is the water collected used for drinking before ( ) or after ( ) treatment?

4 If the water is treated (Question 3), what process is used?
Sedimentation ( )
Sand filter ( )
Others, please specify

5 How many jerry cans (approximately four litres each) full of water do you use on average per day and household for drinking, cooking and washing?

6 Do you have a well located in your compound?
Yes ( )
No ( )

7 If not (Question 6), from where do you get your water?
Community well ( )
Pond ( )
Dam ( )
Others, please specify

8 If yes (Question 6), is the well lined?

9 How far is the source of water from your house?
During the wet season:
0.5 km ( )
1 km ( )
2 km ( )
More than 2 km ( )
During the dry season:
0.5 km ( )
1 km ( )
2 km ( )
More than 2 km ( )

10 How much time does it take to get enough water?
During the wet season:
0.5 hour ( )
1 hour ( )
2 hours ( )
3 hours ( )
More than 3 hours ( )
During the dry season:
0.5 hour ( )
1 hour ( )
2 hours ( )
3 hours ( )
More than 3 hours ( )

Section C: Water and Water-Related Diseases

1 Do you treat raw water taken from a source?
Yes ( )
No ( )

2 If not (Question 1), why do you not treat the raw water? Because ...

3 Have there been cases of water-related diseases in the household? Yes ( ) No ( ) How many household members were ill? Who was ill, in particular?

4 If yes (Question 3), which of the following diseases were recorded?
Diarrhoea ( )
Dysentery ( )
Typhoid ( )
Cholera ( )
Bitharziasis ( )
Others ( ), please specify

5 During what period of the year have you recorded severe cases?
Wet season ( )
Harmattan ( )
Hot dry season ( )

Section D: Community Water

1 Are public wells located in the village?
Yes ( )
No ( )

2 If yes (Question 1), how many public wells are located in the village?

3 Do the wells usually dry-up during the dry season?
Yes ( )
No ( )

4 If yes (Question 3), what does the community do?
Deepen the wells ( )
Dig new wells ( )
Look for other sources ( )

5 Is/are borehole(s) located in the village?
Yes ( )
No ( )

6 If yes (Question 5), do you get enough water from it/them?
Yes ( )
No ( )

7 Has a borehole collapsed in the past?
Yes ( )
No ( )

8 Who maintains the borehole(s)?
Government only ( )
Government and community, jointly ( )
Community only ( )

9 If the borehole was not functioning, what was the cause?
Lack of trained personnel ( )
Lack of spare parts ( )

10 What alternative sources of water supply do you think should be developed for the village?
Pipe borne water ( )
Borehole ( )
Dam ( )
Well ( )
Spring ( )
Others, please specify

11 Why is this your opinion?
They are cheaper ( )
They are easier to construct and maintain ( )
They can yield more water ( )
Others, please specify

12 Why have they not been developed?
Lack of founds ( )
Lack of technical experts ( )
Lack of community interest ( )

13 Do you think self-help effort can improve your water supply?
Yes ( )
No ( )

14 What contribution would you like to make if self-help is encouraged?
Money ( )
Labour ( )
Materials ( )
Others, please specify

Section E: Community Participation in Water Supply Development

1 Is there any development association, club or union active in the village?
Yes ( )
No ( )

2 Has the organisation (Question 1) ever been engaged in water supply projects?
Yes ( )
No ( )

3 If yes (Question 2), in what type of work was the organisation engaged?
Well ( )
Bore hole ( )
Impoundment ( )

4 If not (Question 2), does the organisation still exist or has ever existed before?
Yes ( )
No ( )

5 If yes (Question 4), why did the organisation stop to function?
Lack of support from community leaders ( )
Lack of support from government ( )
Cheating by club leaders ( )
Mistrust among members ( )

References

Ayode, J.O. (1988). *Tropical Hydrology and Water Resources.* London, UK: McMillan Publishers Co Ltd.

Barbour, K.M., Oguntoyinbo, J.S., Onyemelukwe, J.O.C., and Nwafor, J.C. (1982). *Nigeria in Maps.* Hong Kong, China: Hodder and Stoughton.

Converse, J.M. and Presser, S. (1988). *Survey Questions – Handcrafting the Standardized Questionnaire.* Beverley Hills, USA: Stage Publications Inc.

Elabi, D. D. (1984). A socially appropriate water resources development plan proposal for lafia area of plateau state. A seminar paper presented at the Institute of Water Resources, Kaduna, Nigeria.

Haile, I. Community Participation in Water Supply Development. *Journal of Rural Water Supply in Developing Countries.* IDRC, 1980, 167C, 85–91.

Oppenheim, A.N. (1996). *Questionnaire Design, Interviewing and Attitude Measurement*, 2e. London, UK: Pinter Publishers Ltd.

Salau, A.T. (1986). River basin planning as a strategy for rural development in Nigeria. *Journal of Rural Studies* 2 (4): 320–335.

Sandodoyin, A.Y. (1991). Water quality, influence and maintenance of rural boreholes in Nigeria. *International Journal of Environmental Studies* 37: 97–107.

Scholz, M. and Xu, J. (2001). Comparison of constructed wetlands treating wastewater contaminated with heavy metals. *Proceedings of the 3rd International Conference on Microorganisms in Activated Sludge and Biofilm Processes held between 13th and 15th June 2001 in Rome*, Italy (ed. V. Tandoi, R. Passino, and C.M. Blundo). International Water Association, poster paper on CD Rom (Number 174).

Scholz, M., Oliver, M.A., Webster, R. et al. (1999). Sampling to monitor soil in England and wales. *Proceedings of the Second European Conference on Geostatistics for Environmental Applications held between 18th and 20th November 1998 in Valencia*, Spain, Kluwer Academic Publishers, Dordrecht, ISBN 0-7923-5783-3, 465–476.

Taraba State Ministry of Water Resources (1995). Jalingo Local Government Diary, Nigeria.

Uzomah, V.C. and Scholz, M. (2002). Water availability assessment and corresponding public health implications for a rural area in Nigeria. *Journal of the Chartered Institution of Water and Environmental Management* 16 (4): 296–299.

Schulte, N.O., Oliver, M.A., Webster, R. et al. (1993). Sampling of weather soil in farmland and water. Proceedings of the Second European Conference on Geostatistics for Environmental Applications held between 20th and 22nd November 1998 in Valencia, Spain. Kluwer Academic Publishers, Dordrecht. ISBN 0-7923-5783-3, 465–476.

Telangana State Ministry of Water Resources (1999). Village-level Government Diary. Nigeria.

Ottomah, V.C. and Schulz, M. (2007). Water availability, movement and corresponding public health implications for communities in Nigeria from the perspective of the Millennium. Water and Environment Management 18 (4): 296–305.

Index

a

adsorption
  diesel oil 288, 290
  oil 284, 288
aggregate 72, 86, 92, 94, 106–108, 119, 120, 124, 134, 178, 180, 259, 269, 271–274, 280
agricultural waste product 281–290

b

benzene removal 269–280
best management practice 108, 117, 163
bioassay 291, 292
biochemical oxygen demand
  biochemical oxygen demand *(general)* 65, 72, 76–78, 98, 99, 101, 102, 104–105, 109, 123–125, 131, 136, 137, 200, 211, 212, 223, 231, 238, 244, 257, 271, 278
  prediction 290–304
biodegradation 15, 100, 107, 118, 132, 178, 240, 258, 264, 265, 270, 271, 273–274, 280, 281
biomass, monitoring and control of 224–226

c

capacity *(system volume analysis)* 114–115
capillary suction time *(also CST)* 311–344
  biosolids 334, 337
  coagulant (active mixing) 335–338, 340–341
  dewaterability *(of sludge)* 322, 326, 328, 331, 334–344
  dewaterability test *(of sludge)* 336
  filtration (of sludge) 333, 336
  multi-radii *(CST device)* 312, 319, 323–324, 326, 328–331, 333, 334
  natural water floc size distribution 343
  paper *(for the CST test)* 316–319, 321
  particle size distribution 335, 341–343
  product design *(CST test or similar)* 324, 331
  product development *(CST test)* 311, 321
  sealant *(its effect on test accuracy)* 327–332
  sedimentation *(of sludge)* 319
  synthetic domestic wastewater flocs 341–342
  synthetic domestic wastewater floc size distribution 342–343
  synthetic raw water flocs 338–341
  synthetic raw water floc size distribution 342
  synthetic sludge *(for CST benchmarking purposes)* 320–321
carbon
  activated 70–72, 74, 80–82
  biological activated 70, 221–231
  dioxide *(general)* 32, 36–43, 45–48, 50, 51, 127, 129, 131
  dioxide *(sink function)* 32, 33, 37, 45–47, 49, 50
case study 1, 2, 15, 17, 63, 106, 108–112, 115, 117, 145, 148, 149, 153, 164, 177, 257, 265, 269, 294–296, 304, 352, 355
chemical oxygen demand 85, 88, 136–138, 141, 151, 181–185, 190–192, 194, 195, 200, 209–212, 221, 222, 224–231, 242–253, 257, 260–269, 271, 273, 275–279, 291, 292, 294–296, 298, 299
climate
  chamber *(mesocosm experiment)* 33
  change *(general)* 34–51, 153, 154
  change *(scenario simulation)* 33, 34
  cold 94, 107
  variable*(s)* 33, 35, 37–39

Sustainable Water Systems, First Edition. Miklas Scholz.

Coliforms *(total)* 73, 76, 78, 81, 136–141, 201, 205, 209
Common Reed *(Phragmites australis)* 94, 96, 109, 178, 234, 266
community *(involvement)* 355
concentration pathway *(representative)* 32, 33, 35, 38–50
condition
  boundary 18, 130, 143, 150, 271
  environmental 70–71, 97–98, 107, 119, 120, 130, 178, 189, 195, 210, 213, 226, 270, 271, 273, 280
copper 66–70, 72, 74–76, 78–79, 82, 93–95, 97–101, 103, 105–108, 132
corn stalk 282, 284–290
cost
  comparison 115
  saving 1, 115

d

data variability 122, 130, 166, 272, 288, 312, 320, 332, 333
decision-making 23
denitrification 12, 30, 186, 189, 236, 240, 257, 258, 260
diatom 81, 248
diesel oil spillage removal 281–290
diffuse pollution. *see* pollution
disease*(s) (prevention of water-related)* 128–130
dissolved air flotation 257, 258. 261–262, 268
dissolved oxygen *(control of)* 222
ditch *(open and groundwater-fed)* 1–2, 4–7, 9, 11–13
drainage 3, 11, 13–17, 21, 26, 63–154
drought 32, 36, 45–48

e

economics *(sustainable drainage rebate)* 115
ecosystem exchange *(net)* 33, 36, 37, 39–43, 45–49
Eider *(river)* 2–7, 9–13
empty bed contact time 222, 223, 226–227, 229, 230, 244
Escherichia coli 130, 131, 137, 139, 140, 201, 205
experiment
  adsorption isotherm 86–88
  column 88–93

f

fertiliser 15, 25, 26, 28, 29, 31, 71, 78, 79, 271–274, 280
filter
  design 70
  media *(composition)* 72, 94–97
  media *(cost)* 74
  media *(material)* 70, 74, 81, 108
floating performance 281, 283, 285–286
flood *(also flooding)*
  control 142–154
  risk management 175–177
  risk of 116, 117, 144
  summer 2, 12
Freundlich isotherm 85, 87, 88, 93

g

Gauckler-Manning-Strickler's equation *(for open channels)* 5
geostatistics
  disjunctive kriging 164, 167–168, 171–175, 177
  kriging 166–168
  ordinary kriging 168–172
  variogram 166
ground source heating pump 117–131
groundwater
  analysis 200–201
  contamination 194
  groundwater *(general)* 196–213
  monitoring 200–201
  monitoring well 199–200
  quality 196–213
  remediation 196–213

h

heavy metal*(s)* 14, 63, 69, 72, 74, 79, 93, 94, 96–98, 103, 210, 292
hydraulic
  hydraulic*(s)* 1–13
  loading rate 2, 180, 197, 231, 233, 236, 239, 241
  retention time 83, 85, 88–93, 233, 236, 241
hydrocarbon 118, 128, 132, 270–274, 280
hydrograph
  piezometer 202–204
  seasonal fluctuation *(also variation)* 202–204
hydroxyl radical 133–134

i

indicator *(biological)* 226–231
infiltration
  infiltration *(general)* 13–18, 31, 110–112, 116–118, 132, 194–197
  system *(made of planted soil)* 13

k

Kohonen self-organizing map 293, 301
kriging. *see* geostatistics

l

land use *(change)* 18–31
land use *(planning)* 22
lead 67, 68, 70, 72, 74–76, 78, 79, 82, 94, 105, 244
liner
  artificial *(like plastic or concrete)* 196
  soil *(natural)* 179

m

macrophyte 1, 2, 11–13, 69, 70, 82, 93–94, 106, 108, 109, 113, 114, 178, 259, 269, 271
management *(environmental)* 31–51
mass balance *(equation)* 23, 27
membrane bioreactor 257, 258, 260, 262–264, 266–268
mesocosm *(experiment)* 33
  managed 33, 39, 41–44, 48–50
  unmanaged 33, 36, 39, 41–50
metal analysis 78–79
microbiology *(environmental)* 69–82
microorganism *(also micro-organism)* 72, 78, 81, 122, 126, 130, 131, 136–137, 178, 190, 222, 225–227, 230–231, 245, 269, 279, 280, 291, 322–323, 343, 344

n

neural network
  model 245–247
  neural network *(general)* 242–253
nickel 66–69, 93–95, 97–107, 223, 264
Nigeria 349, 351–353
nitrification 65, 96, 99, 122, 186–189, 236, 239–241, 257, 258, 260, 265, 268, 269
nitrogen. *see* nutrient
nutrient
  ammonia 5, 9, 27, 28, 85, 98–100, 122–125, 127, 130, 132, 181–195, 200, 204, 205, 211, 212, 231, 236, 238, 241–244, 257, 259, 261–268, 273, 278, 279, 295, 296, 298, 299
  nitrate 1, 5, 7, 9, 10, 13, 15, 17, 19, 25, 26, 28–30, 65, 66, 85, 94, 95, 98–101, 103, 106, 122–127, 130, 132, 151, 181–184, 186, 188–190, 194, 196, 200, 204, 205, 207, 208, 211, 212, 221, 223, 231, 236, 238, 241, 243, 259, 261, 262, 264–268, 270, 273, 274, 278–280
  nitrogen 1, 5, 7, 9, 10, 13–15, 17–19, 22, 23, 25–31, 65, 66, 71, 85, 88, 98–101, 103, 123–127, 130, 151, 181–196, 200, 204, 207, 208, 211, 212, 223, 231, 235, 236, 238–243, 257, 259, 261–262, 264–268, 270–274, 278, 279, 295, 296, 298, 299
  nutrient *(general)* 1, 9, 12, 13, 18–32, 36, 46, 48, 73, 80, 96, 108, 114, 118, 122, 130, 136, 151, 154, 177–196, 222, 224, 226, 230, 231, 240–243, 248, 260, 265, 266, 270, 271, 273
  ortho-phosphate 5, 9, 65, 66, 122–126, 128, 130, 151, 235, 238, 239, 243, 273, 278, 279
  phosphorus *(general)* 14–19, 21–25, 27–31, 67, 68, 71, 79, 82–86, 88–93, 105, 106, 122–126, 128, 130, 151, 182–184, 189–190, 192–196, 200, 207, 208, 223, 231, 235, 236, 238–243, 257, 261, 265, 266, 270, 272, 273, 278, 279
  phosphorus retention 82, 90, 93

o

oil *(diesel)* 281–290
orthogonal design *(experiment)* 283–284

p

particle size distribution. *see* capillary suction time
peatland 31–51
performance comparison 105–106
permeable pavement 117–142
pH *(control of)* 222, 224
phosphorus. *see* nutrient
photocatalytic
  disinfection model 136–137
  titanium dioxide oxidation 131–141
photochemical
  bacterial inactivation 79
  experiment 135–136
  microbial deactivation 138–141
Phragmites australis. *see* Common Reed
plant *(biological)*
  analysis 37

plant *(biological)* *(cont'd)*
community 70
distribution 43–45
examination 73
vascular 32, 45, 47, 48, 50
pollution
diffuse 1, 12, 19, 21, 30, 69, 70, 154, 163, 176, 178
diffuse *(control of)* 30
groundwater 196, 203
microbial 70, 137, 142
pond
dry 108–117
wet 109, 115, 116, 118
process control 269, 291, 292
protozoa 70, 80–82, 135, 221, 224
public health 131, 196, 349–355

q
questionnaire *(evaluation)* 353, 354

r
release 19, 27, 37, 71, 73, 82, 106, 118, 177–195, 240, 341
rendering plant effluent *(or wastewater)* 221–253, 257–304, 341–343
respiration *(biological)* 32, 36, 39–43, 45–50, 79, 240, 291, 304
road salting 102, 107
roughness *(total)* 5–7, 12–13
runoff
farm yard 178–180, 184
log yard *(treatment of runoff)* 13–18
roof water 18
stormwater (*see* stormwater runoff)
surface water 63, 64, 69, 104, 106, 108
urban 21, 117, 131, 132, 134–137, 139, 141

s
salt 18, 28, 103, 106–108, 122, 189, 190, 323
sampling scheme optimisation 266, 269
sawdust 281, 282, 284–290
sawmill *(wood industry)* 14, 15
schmutzdecke (means layer of dirt) 71, 74, 78
screen system
automated *(screen cleaning)* 144, 148, 150
design 143–144
flood control 142–145
maintenance 143, 150–153
refurbishment *(or improvement)* 142–154
renovation prioritisation 145–147, 153
water protection 42–154
sediment
contaminated 66, 69, 195, 270
management 194–195
size distribution 66
trap 143, 144, 146, 149, 151, 153
sensor
biosensor 291, 304
software 303, 304
silt trap 63, 65–69, 114, 151
solids
suspended 18, 28, 63, 66, 72, 76–78, 94, 105, 107, 122, 125, 130, 224, 257, 296, 298, 326
total 68, 72, 76–79, 100, 104, 106, 109, 111, 311, 314
sorption capacity test 281–285, 287
specific resistance to filtration 20, 311, 322, 333–334, 336
spreadsheet modelling 229–230
stormwater
management 93, 107
pipeline system 64, 65, 106, 107
quality 68
runoff 66, 104, 292
Streptococci *(faecal)* 136–141
substrate 82–86, 91, 94, 186, 196, 234, 240, 269
sustainability 106, 108, 116
Sustainable Water Industry Asset Resource Decisions Analysis 116–117
synthetic waters. *see* capillary suction time
systems engineering *(sustainable)* 132

t
temperature 5, 9, 33, 46, 113, 118, 121, 130, 200, 267, 271, 292, 295, 322
total
organic carbon 14, 15, 17, 221, 228, 231
total phosphorus 14–18, 27, 71, 85, 92, 200, 236, 261
tracer *(studies)* 2, 5, 7–8, 12, 13
treatment
efficiency 15, 17, 18, 63, 135, 231
process 94, 197, 259, 335